Multichain Immune Recognition Receptor Signaling
From Spatiotemporal Organization to Human Disease

ADVANCES IN EXPERIMENTAL MEDICINE AND BIOLOGY

Multichain Immune Recognition Receptor Signaling

From Spatiotemporal Organization to Human Disease

Edited by

Alexander B. Sigalov, PhD

Department of Pathology, University of Massachusetts Medical School,
Worcester, Massachusetts, USA

Springer Science+Business Media, LLC
Landes Bioscience

Springer Science+Business Media, LLC
Landes Bioscience

Copyright ©2008 Landes Bioscience and Springer Science+Business Media, LLC

Printed in the USA.

Springer Science+Business Media, LLC, 233 Spring Street, New York, New York 10013, USA
http://www.springer.com

Please address all inquiries to the publishers:
Landes Bioscience, 1002 West Avenue, Austin, Texas 78701, USA
Phone: 512/ 637 5060; FAX: 512/ 637 6079
http://www.landesbioscience.com

*Multichain Immune Recognition Receptor Signaling: From Spatiotemporal Organization to Human
Disease*, edited by Alexander B. Sigalov, Landes Bioscience / Springer Science+Business Media, LLC
dual imprint / Springer series: Advances in Experimental Medicine and Biology

ISBN: 978-0-387-09788-6

While the authors, editors and publisher believe that drug selection and dosage and the specifications and
usage of equipment and devices, as set forth in this book, are in accord with current recommendations
and practice at the time of publication, they make no warranty, expressed or implied, with respect to
material described in this book. In view of the ongoing research, equipment development, changes in
governmental regulations and the rapid accumulation of information relating to the biomedical sciences,
the reader is urged to carefully review and evaluate the information provided herein.

Library of Congress Cataloging-in-Publication Data

Multichain immune recognition receptor signaling : from spatiotemporal organization to human disease
/ edited by Alexander B. Sigalov.
　　p. ; cm. -- (Advances in experimental medicine and biology ; v. 640)
　Includes bibliographical references and index.
　ISBN 978-0-387-09788-6
　1. Immune recognition. 2. T cells--Receptors. 3. Killer cells. 4. Cellular signal transduction. 5. Cell
receptors. I. Sigalov, Alexander B. II. Series.
　[DNLM: 1. Receptors, Immunologic. 2. Models, Immunological. 3. Signal Transduction. W1 AD559
v.640 2008 / QW 570 M961 2008]
　QR185.95.M85 2008
　571.9'646--dc22
　　　　　　　　　　　2008025647

DEDICATION

This book is dedicated in loving memory to my parents, Galina Ya. Sigalova and Boris L. Sigalov, who are the source of all great things in my life. Without their love, wisdom, understanding, faith, support, and guidance, I would not be the person I am today.

FOREWORD

Immunological recognition is a central feature of the adaptive immunity of vertebrates. With the exception of agnathans, which developed an entirely distinct set of immunologically-specific molecules, all vertebrates use a recognition system based on what Achsah Keegan and I suggested in 1992 be termed *multichain immune recognition receptors* (MIRRs). MIRRs consist of ligand-binding molecules that are immunoglobulin supergene family members associated with signal transducers and enhancers in such a way as both insure precise ligand recognition, discrimination and amplification of the signal.

Two of the prototypic sets of MIRRs, the T-cell and B-cell receptors, are among the most remarkable recognition molecules known. These are extraordinarily diverse molecules in which the range of ligands that can be potentially recognized probably exceeds the actual numbers of lymphocytes in the body. The discovery of the genetic basis of assembling these receptors and understanding how they bind to their cognate antigens are among the most stunning of scientific achievements. Yet these immensely specific binding chains (the heavy/light chain pair for immunoglobulin and the α/β chain pair for most T cells), when expressed as membrane molecules, have no obvious mechanism of signaling. For example, the μH chain cytosolic domain consists of three amino acids (lysine-valine-lysine) and the L chain is not even embedded in the membrane. Furthermore, there is no known direct mechanism to propagate information from the binding domain of the B-cell or T-cell receptors to the membrane-proximal domains of the same chains.

The solution is based on the assemblage of the multichain receptor complex, in which pairing of key chains can occur in the membrane, often based on the presence of oppositely charged residues at critical locations in the transmembrane portion of partner chains. The signaling process then depends on aggregation and/or structural rearrangement induced by binding of multivalent ligands and on the properties of the partner signaling chains, such as Igα and Igβ for the B cell receptor; the CD3 γ, δ and ϵ and the ζ chains for the T-cell receptor and FcϵRI β and γ for the high affinity IgE receptor. These partners contain one or more immunoreceptor tyrosine-based activation motifs (ITAMs). The phosphorylation of tyrosines within the ITAM motif (YxxL/Ix6-8YxxL/I) is a key early event in signal transduction as a result of engagement and aggregation of the ligand-binding domain.

The multichain system links extraordinarily powerful ligand recognition mechanisms with highly effective signaling pathways capable of both immense amplification and precise discrimination between stimulatory ligands (agonists), inhibitory ligands (antagonists) and even partial agonists.

In *Multichain Immune Recognition Receptor Signaling: From Spatiotemporal Organization to Human Disease*, Alexander Sigalov and his colleagues present a comprehensive examination of the full range of MIRRs, of how they propagate signals, how they discriminate between classes of ligands, and the roles they play in physiologic responses and in various diseases. An outstanding set of scientist scholars have provided their expertise in dealing with virtually every aspect of this fascinating set of receptors and in thus providing a comprehensive treatment of this important and exciting area. The nature of this remarkable set of receptors, how they mediate their functions and the potential for abnormal function due to defects in signal transduction, amplification and discrimination, places this family of molecules at the center of the immune response. Dr. Sigalov is to be congratulated for undertaken this important task. Readers interested in this subject will benefit enormously from this important volume.

William Paul, PhD
Laboratory of Immunology, National Institute of Allergy and Infectious Diseases
National Institutes of Health, Bethesda, Maryland, USA

PREFACE

Multichain immune recognition receptors (MIRRs) represent a family of surface receptors expressed on different cells of the hematopoietic system and function to transduce signals leading to a variety of biologic responses. These receptors share common structural features including extracellular ligand-binding domains and intracellular signaling domains intriguingly carried on separate subunits. Another important feature that links members of the MIRR family is the presence of one or more copies of a cytoplasmic structural module termed the immunoreceptor tyrosine-based activation motif (ITAM). ITAMs consist of conserved sequences of amino acids that contain two appropriately spaced tyrosines (YxxL/Ix6-8YxxL/I, where x denotes non-conserved residues). Following receptor engagement, phosphorylation of ITAM tyrosine residues represents one of the earliest events in the signaling cascade. Although the MIRR-mediated ligand recognition and the MIRR-triggered downstream signaling cascades are believed to be among the best studied in biology in recent years, at present the spatial organization of the MIRRs, its reorganization in response to ligand binding as well as the molecular mechanisms underlying the initiation of MIRR signaling remain to be elucidated.

MIRR-mediated signal transduction plays an important role in both health and disease, making these receptors attractive targets for rational intervention in a variety of immune disorders. Thus, future therapeutic strategies depend on our detailed understanding of the molecular mechanisms underlying MIRR triggering and subsequent transmembrane signal transduction. In addition, knowing these mechanisms would provide a new handle in dissecting the basic structural and functional aspects of the immune response.

The central idea of this book is to show that the structural similarity of the MIRRs determines the general principles underlying MIRR-mediated transmembrane signaling mechanisms and also provides the basis for existing and future therapeutic strategies targeting MIRRs. The reviews assembled in this book detail the progress in defining and controlling the spatiotemporal organization of key events in immune cell activation. An improved understanding of MIRR-mediated signaling has numerous potential practical applications, from the rational design of drugs and vaccines to the engineering of cells for biotechnological purposes. Section I reviews the spatial organization and physiological function of MIRR family members such as T-cell receptor, B-cell receptor, Fc receptors, natural killer cell receptors and the platelet collagen receptor glycoprotein VI. Section II focuses on current models of MIRR triggering and highlights modern technologies available to visualize cell-cell interaction contacts such as immunological synapse and also to measure protein-protein interactions in space in real time. Potential therapeutic strategies targeting MIRR-mediated signaling are briefly reviewed in Section III.

This book summarizes current knowledge in this field and illustrates how control of MIRR-triggered signaling could become a potential target for medical intervention, thus bridging basic and clinical immunology. Describing the molecular basis of MIRR-mediated transmembrane signaling, this volume addresses a broad audience ranging from biochemists and molecular and structural biologists to basic and clinical immunologists and pharmacologists.

Alexander B. Sigalov, PhD

ABOUT THE EDITOR...

ALEXANDER SIGALOV, PhD, is a Research Assistant Professor in the Department of Pathology at the University of Massachusetts Medical School in Worcester, Massachusetts, USA. His main research interests include protein intrinsic disorder and oligomericity in the context of transmembrane signal transduction, the molecular mechanisms underlying immune receptor-mediated signaling and ways to control these processes and thus to modulate the immune response, as well as the development and applications of novel targets and strategies for innovative immune therapy. He discovered and investigated a very unusual and unique biophysical phenomenon, the homooligomerization of intrinsically disordered proteins, thus providing the first evidence for the existence of specific interactions between unfolded protein molecules. In the field of immunology, he unraveled a long-standing mystery of transmembrane signaling and immune cell activation triggered by multichain immune recognition receptors. Later, he developed a novel concept of platelet inhibition and invented a novel class of platelet inhibitors. In the field of immune therapy, he proposed new therapeutic strategies for a variety of malignancies and immune disorders, including immunodeficiencies, inflammatory and autoimmune diseases, allergy and HIV. He is a member of the American Association for the Advancement of Science and the Biophysical Society USA. Alexander Sigalov received his academic degrees (MSc in Chemistry and a PhD in Organic Chemistry) from Moscow State University, Russia.

PARTICIPANTS

Balbino Alarcón
Centro de Biología Molecular
    Severo Ochoa
CSIC-Universidad Autónoma
    de Madrid
Cantoblanco, Madrid
Spain

Marina Ali
Department of Rheumatology
Westmead Hospital
Westmead, New South Wales
Australia

Michael Amon
Department of Rheumatology
Westmead Hospital
Westmead, New South Wales
Australia

Vasso Apostolopoulos
Burnet Institute at Austin
Heidelberg
Australia

Raquel Bello
Centro de Investigaciones Biológicas
CSIC
Madrid
Spain

Veronika Bender
Department of Rheumatology
Westmead Hospital
Westmead, New South Wales
Australia

Roberto Biassoni
Molecular Medicine
Istituto Giannina Gaslini
Genova
Italy

Randall J. Brezski
Department of Pathology
    and Laboratory Medicine
University of Pennsylvania
    School of Medicine
Philadelphia, Pennsylvania
USA

Daniel Coombs
Department of Mathematics
University of British Columbia
Vancouver, British Columbia
Canada

Elaine P. Dopfer
Department of Molecular Immunology
Max Planck-Institute
   for Immunobiology
University of Freiburg
Freiburg
Germany

Michael L. Dustin
Program in Molecular Pathogenesis
Skirball Institute of Biomolecular
   Medicine
and
Department of Pathology
New York University School
   of Medicine
New York, New York
USA

James R. Faeder
Theoretical Biology and Biophysics
   Group
Theoretical Division
Los Alamos National Laboratory
Los Alamos, New Mexico
USA

Tamas Fülöp
Research Center on Aging
   and Immunology Program
University of Sherbrooke
Sherbrooke, Quebec
Canada

Byron Goldstein
Theoretical Biology and Biophysics
   Group
Theoretical Division
Los Alamos National Laboratory
Los Alamos, New Mexico
USA

William S. Hlavacek
Center for Nonlinear Studies
   and Theoretical Biology
   and Biophysics Group
Los Alamos National Laboratory
Los Alamos, New Mexico
USA

P. Mark Hogarth
The MacFarlane Burnet Institute for
   Medical Research and Public Health
Austin Health
Heidelberg
and
Department of Pathology
The University of Melbourne
Parkville
Australia

Stephanie M. Jung
Department of Protein Biochemistry
Institute of Life Science
Kurume University
Kurume, Fukuoka
Japan

Darja Kanduc
Department of Biochemistry
   and Molecular Biology
University of Bari
Bari
Italy

Walter M. Kim
Department of Pathology
University of Massachusetts
   Medical School
Worcester, Massachusetts
USA

Tomohiro Kubo
Department of Experimental
   Immunology
Institute of Development,
   Aging and Cancer
Tohoku University
Sendai
Japan

Anis Larbi
Center for Medical Research (ZMF)
Tübingen Aging and Tumour
   Immunology Group
Section for Transplantation Immunology
   and Immunohematology
University of Tübingen
Tübingen
Germany

Eliada Lazoura
Burnet Institute at Austin
Heidelberg
Australia

Nicholas Manolios
Department of Rheumatology
Westmead Hospital
Westmead, New South Wales
Australia

Stephen D. Miller
Department of Microbiology-
   Immunology
Northwestern University Feinberg
   School of Medicine
Chicago, Illinois
USA

Susana Minguet
Department of Molecular Immunology
Max Planck-Institute
   for Immunobiology
University of Freiburg
Freiburg
Germany

Eszter Molnar
Department of Molecular Immunology
Max Planck-Institute
   for Immunobiology
University of Freiburg
Freiburg
Germany

John G. Monroe
Department of Pathology
   and Laboratory Medicine
University of Pennsylvania
   School of Medicine
Philadelphia, Pennsylvania
USA

Masaaki Moroi
Department of Protein Biochemistry
Institute of Life Science
Kurume University
Kurume, Fukuoka
Japan

Akira Nakamura
Department of Experimental
   Immunology
and
CREST Program of Japan Science
   and Technology Agency
Institute of Development, Aging
   and Cancer
Tohoku University
Sendai
Japan

Angel R. Ortiz
Centro de Biología Molecular
   Severo Ochoa
CSIC-Universidad Autónoma
   de Madrid
Cantoblanco, Madrid
Spain

William Paul
Laboratory of Immunology
National Institute of Allergy
  and Infectious Diseases
National Institutes of Health
Bethesda, Maryland
USA

Graham Pawelec
Center for Medical Research (ZMF)
Tübingen Aging and Tumour
  Immunology Group
Section for Transplantation Immunology
  and Immunohematology
University of Tübingen
Tübingen
Germany

Joseph R. Podojil
Department of Microbiology-
  Immunology
Northwestern University Feinberg
  School of Medicine
Chicago, Illinois
USA

Pilar Portolés
Centro Nacional de Microbiología
Instituto de Salud Carlos III
Majadahonda, Madrid
Spain

Maree S. Powell
The MacFarlane Burnet Institute for
  Medical Research and Public Health
Austin Health
Heidelberg
and
Department of Pathology
The University of Melbourne
Parkville
Australia

Jacob Rachmilewitz
Goldyne Savad Institute of Gene
  Therapy
Hadassah University Hospital
Jerusalem
Israel

Michael Reth
Department of Molecular Immunology
Max Planck-Institute
  for Immunobiology
University of Freiburg
Freiburg
Germany

Ruth M. Risueño
Centro de Biología Molecular
  Severo Ochoa
CSIC-Universidad Autónoma
  de Madrid
Cantoblanco, Madrid
Spain

Jose M. Rojo
Centro de Investigaciones Biológicas
CSIC
Madrid
Spain

Wolfgang W. A. Schamel
Department of Molecular Immunology
Max Planck-Institute
  for Immunobiology
University of Freiburg
Freiburg
Germany

Gabrielle M. Siegers
Department of Molecular Immunology
Max Planck-Institute
  for Immunobiology
University of Freiburg
Freiburg
Germany

Alexander B. Sigalov
Department of Pathology
University of Massachusetts
    Medical School
Worcester, Massachusetts
USA

Mahima Swamy
Department of Molecular Immunology
Max Planck-Institute
    for Immunobiology
University of Freiburg
Freiburg
Germany

Toshiyuki Takai
Department of Experimental
    Immunology
Institute of Development,
    Aging and Cancer
Tohoku University
Sendai
Japan

Danielle M. Turley
Department of Microbiology-
    Immunology
Northwestern University Feinberg
    School of Medicine
Chicago, Illinois
USA

Jianying Yang
Department of Molecular Immunology
Max Planck-Institute
    for Immunobiology
University of Freiburg
Freiburg
Germany

Minmin Yu
Physical Biosciences Division
Lawrence Berkeley National
    Laboratory
Berkeley, California
USA

Tomasz Zal
Department of Immunology
University of Texas MD Anderson
    Cancer Center
Houston, Texas
USA

CONTENTS

Section I. MIRRs: Structure and Physiological Function

Roberto Biassoni

Stephanie M. Jung and Masaaki Moroi

Section II. MIRR Signaling: Possible Mechanisms and the Techniques to Study and Visualize

Wolfgang W.A. Schamel and Michael Reth

Elaine P. Dopfer, Mahima Swamy, Gabrielle M. Siegers, Eszter Molnar,
    Jianying Yang and Wolfgang W.A. Schamel

8. KINETIC PROOFREADING MODEL 82

Byron Goldstein, Daniel Coombs, James R. Faeder and William S. Hlavacek

9. SERIAL TRIGGERING MODEL.. 95

Jacob Rachmilewitz

10. CONFORMATIONAL MODEL 103

Ruth M. Risueño, Angel R. Ortiz and Balbino Alarcón

11. PERMISSIVE GEOMETRY MODEL 113

Susana Minguet and Wolfgang W.A. Schamel

Section III. MIRR Signaling and Therapy of Immune Disorders

19. MHC AND MHC-LIKE MOLECULES: STRUCTURAL PERSPECTIVES ON THE DESIGN OF MOLECULAR VACCINES

Vasso Apostolopoulos, Eliada Lazoura and Minmin Yu

20. SCHOOL MODEL AND NEW TARGETING STRATEGIES

Alexander B. Sigalov

21. IMMUNE RECEPTOR SIGNALING, AGING AND AUTOIMMUNITY

Anis Larbi, Tamas Fülöp and Graham Pawelec

22. VIRAL PATHOGENESIS, MODULATION OF IMMUNE RECEPTOR SIGNALING AND TREATMENT 325

Walter M. Kim and Alexander B. Sigalov

ACKNOWLEDGEMENTS

I would like to acknowledge all those individuals behind the scenes who have contributed to this publication: to numerous scientists, many unrecognized, who have diligently added to the body of knowledge from which this book draws. I wish to express my gratitude to my respective spouse and son who have supported me in this time-consuming endeavor throughout the years. Thanks to all those who have contributed in any way to the publication of this book.

CHAPTER 1

T-Cell Receptor

Jose M. Rojo,* Raquel Bello and Pilar Portolés

Abstract

The T-cell antigen receptor complex (TCR/CD3) is a cell surface structure that defines the T lymphocyte lineage, where it fulfills two basic functions, namely antigen recognition and triggering of signals needed to mount adequate responses to foreign aggression and/or to undergo differentiation. Knowing the precise structure of the complex in terms of its components and their relative arrangement and interactions before and after antigen recognition is essential to understand how ligand binding transforms into functionally relevant T-cell responses. These include not only full responses to foreign peptide antigens by mature T-cells, but also other phenomena like modulation of T-cell activation with altered peptide ligands, positive and negative selection of thymocytes, alloreactivity and autoimmune reactions.

A wealth of new data has accumulated in recent years on the structure of TCR/antigen complexes and CD3 polypeptides and on the stoichiometry of the TCR/CD3 complex and intersubunit interactions. In this review, we discuss how these data fit into a meaningful model of the TCR/CD3 function.

Introduction

In the TCR/CD3 complex, antigen recognition and signal triggering functions are carried out by two distinct molecular modules: the TCR chains are responsible for antigen recognition and the invariant CD3 (CD3ε, CD3γ, CD3δ) and CD247 (ζ) chains are in charge of signal transduction (Fig. 1) (reviewed by refs. 1-6).

The TCR antigen recognition unit exists in three distinct molecular species. In humans and mice, most mature T-lymphocytes express TCRs composed of two class I membrane glycosylated polypeptides termed α and β (αβTCRs). The overall organization of the extracellular region of these TCRs is similar to that of antibody Fab fragments. Each chain contains one variable (V) and one constant (C) Ig domain linked by a disulfide bridge. Some peculiarities of the chain include the flexibility of the external sheet (CFG face) of the small Cα which does not adopt a standard Ig structure, the high interaction surfaces between C domains and intrachain C-V domains and the small C-V angle of the TCR β chain.[7] The Ig-like domains of the TCR α and β chains are followed by a stalk of 19 (α chain) or 15 (β chain) residues, a 22-residue long transme mbrane (TM) domain containing two (α) or one (β) basic residues and a short 4-10 residue long intracellular region (Fig. 1).

On the cell surface, the TCR antigen recognition module is noncovalently associated with the invariant CD3 ε, δ and γ polypeptides and the ζ (CD247) homodimer (Figs. 1,2). These chains are needed to transform ligand binding by the TCR module into signals inside the cell. The CD3 and ζ chains are also involved in regulating the expression of the TCR/CD3 complex on the cell surface (reviewed by refs. 1,2,8).

*Corresponding Author: Jose M. Rojo—Centro de Investigaciones Biológicas, CSIC, Ramiro de Maeztu, 9, E-28040 Madrid, Spain. Email: jmrojo@cib.csic.es

Multichain Immune Recognition Receptor Signaling: From Spatiotemporal Organization to Human Disease, edited by Alexander B. Sigalov. ©2008 Landes Bioscience and Springer Science+Business Media.

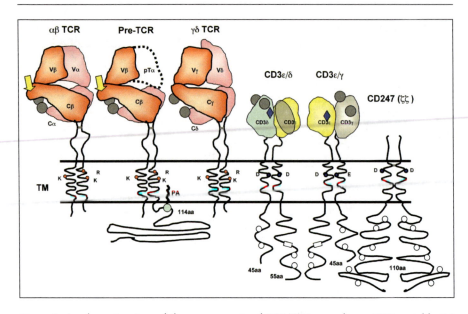

Figure 1. A schematic view of the components of TCR/CD3 complexes. TCR variable (V) and constant (C) domains are as indicated. Grey circles represent N-linked glycans (human Cα $Asn_{/185}$ Cβ $Asn_{/186}$ pTα $Asn_{/51}$ Cγ $Asn_{/184}$ or Cδ $Asn_{/135}$ CD3γ Asn_{30} and $Asn_{/70}$ CD3δ Asn_{16} and Asn_{52}). Arrows indicate the position of the extended F-G loop of TCR β chains. Disulfide links are indicated by grey bars. Acidic and basic residues in the transmembrane (TM) region are indicated by circles. PA: palmitic acid. The binding site for OKT3 and UCHT1 in CD3ε is shadowed. Diamonds indicate the approximate location of N-termini in CD3 chains. The proline-rich region in the cytoplasmic tail of CD3ε is indicated by a white box; open circles indicate ITAM Tyr residues. Where indicated, the number of amino acid residues (aa) refers to the cytoplasmic domain.

CD3 polypeptides possess an Ig-like ectodomain, a ten-residue long connecting peptide and the TM region that contains one acidic residue. The disulfide-linked homodimeric ζ chains also contain one acid residue in their TM region and have a short nine-residue long extracellular domain. Unlike TCR α and β chains, CD3 and ζ chains possess relatively large 45-110-residue long intracytoplasmic domains that contain immunoreceptor tyrosine-based activation motifs (ITAMs), polyproline motifs, endoplasmic reticulum (ER) retention and endocytosis motifs involved in transmembrane signaling[9,10] and cell surface receptor expression[1,2,8] (Fig. 1). However, it is unlikely that these intracytoplasmic domains have a significant role in the noncovalent interactions between the short-tailed TCR α and β chains and the CD3 and ζ chains, at least in those needed for surface expression of the TCR/CD3 complex.[11-15]

Although most T-lymphocytes express αβ TCRs, a minor subpopulation of functionally distinct mature T-cells expresses TCRs that contain γ and δ polypeptides homologous to the TCR β and α chains, respectively (Fig. 1). Like in immunoglobulins, the observed diversity in the N-terminal V domains of the TCR α, β, γ and δ chains is due to clonotypic rearrangement of V, D and J segments of the relevant genes. T-cells are developmentally selected to express αβ TCRs that specifically recognize short antigenic peptides bound to Major Histocompatibility Complex molecules (peptide-MHCs; pMHCs) on the surface of antigen-presenting cells (APCs). The γδ TCR-expressing cells (γδ T-cells) also bind antigens on the surface of APCs. However, unlike αβ T-cells, γδ T-cells do not show restriction by polymorphic MHCs but recognize nonclassical MHC alone or in complexes with small phosphate-containing bacterial antigens.[5,16]

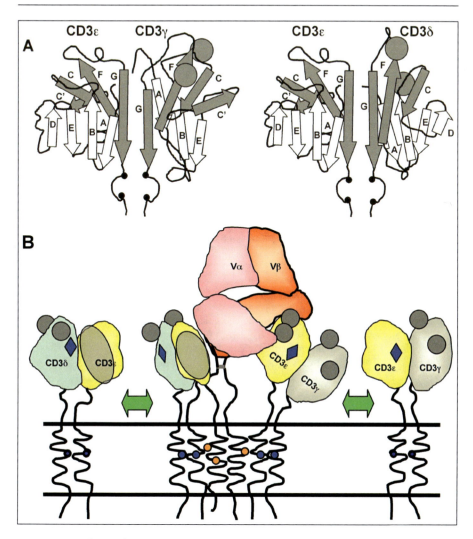

Figure 2. CD3 heterodimer structure and interactions. A) Scheme of the Ig-domain β-sheet structure in human CD3εγ and CD3εδ dimers. Black dots represent Cys residues in RxCxxCxD motifs of CD3 stalks. Circles represent N-linked glycans. B) A model of TCR-CD3 subunit interactions. ζ dimers and the cytoplasmic domains of CD3 are omitted for clarity. Other symbols and indications are as in Figure 1. Modified from references 18, 20, 21.

During differentiation, double negative (DN) thymocytes express TCR β chains complexed to a polypeptide with a long intracellular domain and a single extracellular C domain homologous to Cα (pTα) (Fig. 1). All αβ and γδ TCR/CD3 complexes share the CD3ε and ζ polypeptides and expression of these signaling modules best characterizes the T-cell lineage. This is possibly due to the unique, nonredundant function of CD3ε and ζ in blocking ER retention signals.

Minimal Components and Stoichiometry of the TCR/CD3 Complex

The minimal components of TCR/CD3 complexes, their number and their organization within the complex are essential to understanding the mechanisms of ligand-induced activation.

In this regard, the structure of the $\alpha\beta$ TCR/CD3 complex is among the best studied examples. In addition to the α and β antigen recognition units, mature $\alpha\beta$ TCR complexes contain CD3ϵ, CD3δ and CD3γ polypeptides as well as ζ homodimers. CD3ϵ binds noncovalently with CD3δ and CD3γ in a mutually exclusive manner, yielding CD3$\epsilon\delta$ and CD3$\epsilon\gamma$ heterodimers.[17-21] Results from the coprecipitation experiments performed using human CD3ϵ-transfected mouse T-cells or human CD3ϵ-transgenic mice suggest that the TCR/CD3 complexes contain two CD3ϵ chains.[22,23] Gel-shift analysis of the TCR-CD3 complexes bound by antiCD3 Fab fragments supports this notion.[24]

The number of TCR antigen recognition units in each minimal TCR/CD3 complex has been the matter of long debate. Charged TM residues are known to be important for the stability of the complex and, as expected theoretically, two $\alpha\beta$ TCR units should be presented in the complex $(\alpha\beta)_2$:$\epsilon\gamma$:$\epsilon\delta$:$\zeta\zeta$ in order to mantain electrostatic equilibrium in the TM region. Estimation of the binding sites for antiTCR or anti CD3 antibodies yielded conflicting TCR:CD3 ratios between 1:1 and 1:2.[25-28] Association of the TCR α and β chains with either CD3$\epsilon\delta$ or CD3$\epsilon\gamma$ during the TCR/CD3 complex assembly indirectly supported a 1:1 ratio for TCR:CD3 chains.[29] This ratio was also supported by the finding of a unique lodging site for CD3 dimers in a "cave" beneath the F-G loop of the C domain of the TCR β chain (see below).[30] Further experimental evidence was obtained from co-immunoprecipitation and fluorescent resonance energy transfer (FRET) studies performed on the T-cells expressing two different TCRs.[31]

However, gel-shift studies of complete TCR complexes solubilized in digitonin provide convincing data that under these experimental conditions the ratio of TCR $\alpha\beta$ chains to CD3 heterodimers is 1:2.[24] Analysis of the TCR/CD3 complex assembly in vitro shows an interaction of TCR β chain with CD3$\epsilon\gamma$ and TCR α chain with CD3$\epsilon\delta$ and $\zeta\zeta$. These data suggest that in the TM milieu each basic residue in the TCR chains interacts with two acidic charges of the CD3 or ζ dimers.[6,32,33] Thus, current data strongly favor a minimal monovalent $\alpha\beta$:$\epsilon\gamma$:$\epsilon\delta$:$\zeta\zeta$ TCR/CD3 complex.

TCR Clusters on the Cell Surface

By blue native electrophoresis of isolated TCR/CD3 complexes, immunostaining and electron microscopy (EM) of fixed cells, TCRs have been shown to exist as monovalent complexes and multivalent clusters.[24,34] On the cell surface, these cholesterol extraction-sensitive clusters form linear structures of closely packed TCR and CD3 units.[24] The mean valency of these complexes varies among different T-cells and this variability is not determined by the nature of the TCRs or their antigen specificity.[24,34] The degree of multivalency is likely to impact the initiation and maintenance of TCR-mediated signals.[35] Consequently, it will be of great interest to determine the factor(s) regulating the multivalency of TCR complexes and the relative orientation of the monovalent units within the multimers. Functionally, it is striking that the presence of these high-order structures of closely packed TCRs on the cell surface does not lead to a permanent state of T-cell activation. Cross-linking of TCRs is essential to signal triggering and it has been proposed as the main, if not the only, factor in TCR activation.[36] The fact that TCR multimers exist in the absence of detectable activation argues in favor of a nonrandom, ordered structure within these linear multimers. This organization may preclude spontaneous TCR activation, suggesting the importance of intermolecular orientation in modulation of cell activation.

Topology of Chain Interactions within TCR/CD3 Complexes

The structural analysis of a large, multichain structure like the TCR/CD3 complex is a formidable task. Recently, invaluable information on the molecular and structural mechanisms of antigen recognition was obtained from structural studies of the ectodomains of CD3 components and about 40 $\alpha\beta$ TCR units and more than 20 $\alpha\beta$ TCR/pMHC complexes.[5] These structural data, together with biochemical and functional data, shed light on the position and relative orientation of each chain within the complex.

The crystal structure of the TCR, alone or in a complex with the Fab fragment of the H57 antiTCR antibody, localized one possible docking site for CD3 dimers in a "cave" beneath the F-G loop of TCR β chain, sided by the Cα A-B loop that contains an exposed lysine residue and the glycan at the Asn_{185}.[7,30,37] The size of this cave seems to be sufficient to harbor one small, nonglycosylated Ig domain like that of CD3ε. Furthermore, it contains basic residues that could interact with the negatively charged surfaces of CD3ε. H57 antibody has been shown to bind to the F-G loop of TCR Cβ and inhibit the binding of antiCD3 antibodies to at least one of the two CD3 dimers in the TCR/CD3 complex.[37] Partial inhibition of antiCD3 binding by clonotypic antiTCR antibodies has been also observed in other systems.[26] These data confirm the intimate relationship between the TCR and CD3 ectodomains.

The interactions between the CD3 and TCR units can help reconcile other experimental data. For instance, the ectodomains of CD3ε, CD3γ and CD3δ all have an elongated shape; sized about 40x25x25 Å for CD3ε and CD3γ, while the CD3δ molecule is slightly wider.[18-21] Mouse CD3ε and CD3γ ectodomains have a C2-set Ig-fold, whereas human CD3ε and CD3δ have a C1-set Ig-fold (Fig. 2).[18-21] The ectodomains interact mostly through their G strands and the contacts in a continuous β sheet along the dimerization interface result in a rigid "paddle-like," 50-55Å wide structure.[18,21] The short 10-residue stalk region connecting CD3 ecto- and TM domains, contains the RxCxxCxE motif conserved in all CD3 chains. This motif may contribute to the interactions between the CD3 chains and add rigidity to the extracellular CD3 structure, bringing the CD3 TM regions in close proximity to the TCR TM regions to allow interactions between relevant acidic and basic residues.[18,19,38] Together, these data suggest that the CD3 ectodomains are very close to and underneath the TCR α and β C domains. However, despite the close proximity of these ectodomains, no direct interactions between soluble ectodomains of the TCR α and β chains and CD3 heterodimers have been detected.[18,19]

Three- and four-chain assembly studies using an in vitro translation system show that the TCR α chain interacts with one CD3εδ heterodimer, whereas the TCR β chain binds to one CD3εγ heterodimer.[32] A unique role of the extracellular domain of CD3γ chain in the TCR/CD3 complex assembly has been also suggested.[39] In addition, association of the CD3εδ heterodimer with the TCRαβ unit is lost in cells expressing the TCR α chain in which an original stalk region with the FETDxNLN motif is substituted by the shorter TEKVN sequence presenting in a TCR δ chain.[40,41] The favored association of CD3εδ with TCR α and the proximity of CD3εγ to TCR β, as shown by chemical cross-linking,[42] and mutational analysis of TCR Cβ F-G loop,[43] indicate that the CD3εγ heterodimer usually occupies the site close to the TCR β chain. According to docking models, the probable location of CD3εγ in one specific site suggests that another CD3 heterodimer, CDεδ, might be located on the opposite, free of interfering glycan side of the TCR/CD3 complex. This potential site of CD3εγ location could include part of the exposed faces of the TCR α and β chains[18,19] and conserved regions of the TCR α chain facing the membrane or close to it.[21]

Mouse γδ TCR/CD3 complexes have been reported to contain only CD3εγ, but not CD3εδ heterodimer.[44,45] This is in agreement with the normal development of γδ T-cells in CD3δ[-/-] mice.[46] Interestingly, γδ TCR/CD3 complexes still maintain the stoichiometry of two CD3ε-containing CD3 heterodimers per complex.[46] Perhaps because of the common evolutionary origin of CD3γ and CD3δ,[4] there is a certain degree of functional and structural redundancy of these chains varying among species. For instance, CD3δ[-/-] mice develop γδ T-cells, whereas humans with CD3δ deficiency do not.[47] Quite the contrary, CD3γ-deficient humans, but not mice, produce some γδ T-cells (Regueiro, J.R. personal communication). The eventual incorporation of ζ dimers into the partial TCR/CD3 complex allows export of the mature complexes to the cell surface (reviewed by ref. 8). Charged residues in the ζ TM domain interact with the TM region of the TCR α chain.[6] Additionally, the short extracellular region of ζ and a conserved TM Tyr residue of the TCR β chain also have been suggested to contribute to the TCR/CD3 complex assembly and function.[48-50]

The structures of human CD3εγ and CD3εδ complexed to the anti-CD3 antibodies OKT3 and UCHT1 have been determined.[20,21] Both antibodies bind exposed overlapping regions mainly located in the CD3ε C'-CFG sheet. In human CD3γ and CD3δ, N-glycosylation sites are located in the CD3γ B-C loop, on top of CD3γ G strands and in the CD3δ F-G loop (Fig. 2).[18-21] This suggests that the conserved CD3ε ABE strands are most likely to interact with the TCR unit under Cβ FG loop. This arrangement brings the acidic residues conserved in the N-terminal sequence and D-E loop of human CD3ε (or in the C'-D loop in mouse CD3ε) close to basic residues in the TCR unit (Fig. 2). Additionally, in this arrangement much of the CD3γ or CD3δ extracellular domains face the cell membrane and thus remain inaccessible to antibody binding.

Binding of monovalent Fab fragments of anti-CD3 antibodies OKT3 or UCHT1, but not anti-TCR β chain JOVI.1 Fab fragment, to the TCR/CD3 complex induces association of the adaptor protein Nck to the cytoplasmic domain of the CD3ε chain.[9] This suggests that the binding of the monovalent anti-CD3 antibodies may change TCR-CD3 interactions in a way resembling physiological TCR ligands.[9] Thus, although it is assumed that OKT3 or UCHT1 antibodies bind to an exposed face of CD3ε, it is possible that full exposition of the relevant epitopes is achieved upon conformational change during the binding process as suggested by Kjer-Nielsen et al.[20]

Interactions between the TCR and Antigen—Role of CD4 and CD8 Coreceptors

Antigen peptides recognized by αβ TCR are located in a "groove" formed by two alpha helices and a beta-sheet floor in domains α1 and α2 of class I MHCs or in the homologous domains α1 and β1 of class II MHCs. The position of αβ TCRs is approximately perpendicular to the plane defined by the peptide and the alpha helices on the top of α1 and α2 domains or α1 and β1 domains of MHCs class I and II, respectively.[5] T-cells that recognize peptides bound to class I MHCs express the CD8αβ coreceptor, whereas those recognizing peptides bound to class II MHCs express the CD4 coreceptor (Fig. 3). Coreceptors are strongly associated with lck tyrosine kinases and provide these enzymes to the TCR-mediated signaling pathways, thus setting a biochemical basis of the linkage between CD8 and CD4 expression and MHC class I or class II restriction.[51]

The interaction between the relatively flat and oblong—with a size of about 40x20 Å—CDR surface of αβ TCR V domains and the pMHC complex takes place in a precisely oriented fashion. With some angle variation among different TCR/pMHC pairs, the long axis of this surface is centered diagonally to the groove formed by the two alpha helices of the N-terminal domains of MHC molecules[7] (reviewed by ref. 5). The highly variable CDR3 loops are the main, but not exclusive, zone contact with solvent-exposed side chains of the antigenic peptide, whereas CDR1 and CDR2 tend to interact with the less variable α helices of MHC molecules. The reasons for the diagonal orientation of the TCR to the pMHC in the complex are not known. The mode of interaction of a given CDR loop with MHC or peptide residue varies among known TCR/pMHC complexes, making unlikely an intrinsic affinity of CDR1 and CDR3 of each V domain for MHC molecules. CD8 and CD4 interactions with MHC molecules can restrict the orientation of the TCR recognition unit toward the antigen peptide complexed to the same MHC molecule, thus setting permissive limits for optimal activation. In class I MHCs, the major binding site for CD8 V-like domains of CD8αα dimer is located in the C-D loop of the α3 domain, close to the APC membrane.[52] The about 40-residue long disulfide-linked connecting peptide of CD8 is thought to have relative flexibility and might be located close to the TCR/CD3 complex. In fact, there is biochemical evidence for an interaction between CD8αβ dimers and the CD3δ chain.[53,54]

CD4 has four IgSF ectodomains and interacts with MHC molecules through its N-terminal D1 V-like domain. The CD4 binding site on class II MHC molecules is located at the junction of the α2 and β2 domains and is oriented similarly to that for binding with CD8.[55] The intact soluble CD4 crystal structure reveals that D1-D2 and D3-D4 form rigid rods and two CD4 molecules dimerize through the D4 membrane-proximal domains.[56] Based on the crystal structure of a complex containing the human CD4 N-terminal two-domain fragment and the MHC class II molecule, ternary complexes of TCR and CD4 bound to one single peptide-MHC molecule were modeled, suggesting that both TCR and CD4 are tilted to the T-cell surface rather than oriented vertically.[55] The relative orientation of TCR and CD4 and the position of the membrane-proximal

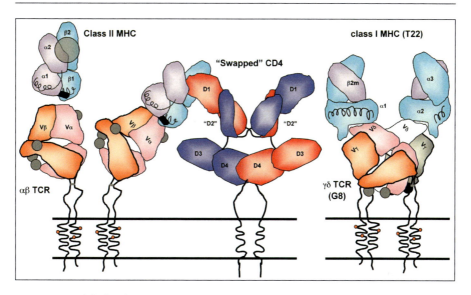

Figure 3. Models for the intercomponent interactions in the αβ TCR-peptide-MHC class II-dimeric CD4 complex (left panel) and in the γδ TCR(G8) dimer-peptide-MHC class I (T22) complex (right panel). The region of MHC-CD4 interaction interface is shown as a shadowed circle. The αβ TCR and γδ TCR units are as in Figures 1 and 2. CD4 dimers rearrange and exchange part of their D2 domains, forming "swapped" disulfide-linked domains. The γδ-γδ domain interactions are mediated by Vδ domains facing each other. The TCR γ and δ chains in the back are shown in grey and white, respectively, with glycans shown as black circles. Adapted from refs. 16,58.

D3-D4 CD4 domains have been thought to prevent TCR-CD4 interactions in the same complex. However, D4- and D2-mediated CD4 dimerization has been recently reported to be very relevant to the CD4 coligand and coreceptor functions.[57,58] In the D2-mediated dimerization, a large conformational change takes place whereby, in a CD4 dimer, D2 domains swap their parts (the so-called D2 swapping) and form two interchain disulfide bonds involving Cys_{130} and Cys_{159} of each CD4 monomer (Fig. 3).[58] This rearrangement does not interfere with the D1 binding to MHC or with the CD4 dimerization through D4 domains. Under these conditions, in ternary TCR-pMHC-CD4 complex, the CD4 D3-D4 domains are oriented towards the pMHC-TCR so that Lys_{279} in CD4 D3 is in close range with Glu_{59} located in the MHC β1 alpha helix (Fig. 3).[58] Thus, CD4 might orientate close to the δ chain of the CD3δε heterodimer[54] and distal to the cave beneath the TCRβ chain FG loop.[58] This might explain the functional data suggesting the proximity of CD3 and CD4 during TCR-mediated cell activation.[26,59] Formation of this trimolecular TCR/CD3/CD4 complex with the TCR unit tilted to the T-cell membrane upon pMHC recognition suggests that CD4 dimers may function as active cross-linkers between nearby TCR/pMHC complexes (Fig. 3). As mentioned above, the assumption that TCR clusters exist on the cell surface as ordered structures could impose a restriction on the orientation of αβ TCR cross-linking upon antigen recognition and this restriction would be independent, but not mutually exclusive, on the restrictions posed by coreceptors.

Other TCRs

The recently reported structure of the G8 γδ TCR in complex with its ligand, the nonclassical MHC molecule T22[16] (reviewed by ref. 5), suggests that antigen- mediated ordered cross-linking of TCRs and their following reorientation (tilting) toward the T-cell surface may both play a role in productive T-cell activation. Compared to αβ TCRs, the γδ TCRs have some distinctive features, including a canonical Ig C-fold of δ chain C domains and a normal F-G loop in γ chains which have a very acute interdomain angle of 41°. The interaction of G8 with T22 is dominated by the prominent CDR3 δ-chain and is neither centered nor vertical on top of T22. Furthermore, there is no diagonal orientation of the CDRs toward the MHC molecule. Still, T22-interacting G8 TCRs are likely to dimerize so that the productive antigen recognition may be a result of a precisely ordered structure formation with bridging noncontiguous MHC molecules and re-orientation of the γδ TCRs relative to the T-cell surface (Fig. 3), as for αβ TCRs in the presence of coreceptors. Interestingly, many γδ T-cells do not express coreceptors which reinforces the idea that coreceptors contribute to constraints on the orientation of TCR/MHC interactions permissive for productive activation.

No ligand has been defined for preTCR complexes and the structure of the pTα ectodomain complexed with the TCR β chain has not been determined, although it is assumed that this structure should be similar to that of αβ TCRs. It is known that pTα is susceptible to palmitoylation (Fig. 1) and preTCRs are resident in rafts with constitutive signaling.[60]

Are All TCRs Equal, or Are Some TCRs More Equal Than Others?

Aside from the differences in the TCR V domains, some qualitative differences have been noted in the TCR/CD3 complexes expressed by T-cells of different lineages or in distinct differentiation steps. For instance, the relative abundance of CD3δ and CD3γ within the cells might have an impact on the proportion of these chains in TCR/CD3 complexes.[17,61] As mentioned before, in mouse γδ T-cells the TCR complexes contain CD3εγ but not CD3εδ heterodimers.[44,45] Also, the CD3γ chain in activated mouse γδ T-cells is so heavily glycosylated that it was mistaken for the CD3δ chain. In addition, these γδ T-cells have been shown to incorporate FcεRIγ chains instead of ζ chains.[44,45] Ectodomain glycosylation can have an important impact on the shape and size of the TCR unit and CD3γ and CD3δ chains, setting physical limits to their interactions with other molecules.[62] Thus, while anti-CD3 WT31 antibody binds to αβ T-cells, its binding to γδ T-cells is only possible upon deglycosylation. Epitope scanning of anti-CD3 binding sites in normal CD4+ or CD8+ human cells have also detected strong dependence of binding on glycosylation.[63] The UCHT1 antibodies bind to CD4+ cells better than to CD8+ cells whereas the RW2-8C8 antibodies bind better to CD8+ than to CD4+ cells. These differences are linked to differential glycosylation of the TCR/CD3 chains in each subset.[63]

Another source of variability among different TCR/CD3 complexes comes from the results on stepwise proteolytic degradation of the acidic residue-rich, N-terminal sequence of mouse or human CD3ε (ref. 64 and Bello R et al, unpublished data). As analyzed by isoelectric focusing, different T-cells present a distinct profile of CD3ε isoforms (Bello R et al, unpublished results). Loss of the negative N-terminal charges weakens interactions between the TCR and CD3 units,[64] facilitates the recognition by certain anti-CD3 antibodies,[64] and might lower the threshold for TCR activation (Rojo J.M et al, unpublished results). Because of the weak association between the ectodomains of the CD3 and TCR units, all these variations might affect the quaternary changes upon TCR ligation and thus contribute to fine-tuning of the TCR-mediated responses.

Future Directions

Although future developments cannot be anticipated, it is safe to say that current interests include establishing the precise topology of the TCR/CD3 complex ectodomains and particularly the exact sites and modes of interaction between the TCR antigen recognition unit and the CD3 signaling heterodimers. Other important topics include the nature and mechanisms of the structural changes/re-arrangements in the TCR/CD3 complex organization upon ligand stimulation with qualitatively different TCR ligands. The past history of structure-function studies of antigen recognition by T-cells has been a surprise box to immunologists and surely there are still many surprises to come in the future.

Acknowledgements

Many important publications have been cited indirectly due to space limitations and we apologize to the authors. Supported by grants ISCIII-05/054 (Ministerio de Sanidad y Consumo, Spain) and SAF2004-06852 (Ministerio de Ciencia y Tecnología, Spain). R.B. is recipient of a Predoctoral Fellowship of Ministerio de Ciencia y Tecnología, Spain.

References

1. Clevers H, Alarcón B, Wileman T et al. The T-cell receptor/CD3 complex: A dynamic protein ensemble. Ann Rev Immunol 1988; 6:629-662.
2. Klausner RD, Lippincott-Schwartz J, Bonifacino JS. The T-cell antigen receptor: Insight into organelle biology. Annu Rev Cell Biol 1990; 6:403-431.
3. Kuhns MS, Davis MM, Garcia KC et al. Deconstructing the form and function of the TCR/CD3 Complex. Immunity 2006; 24:133-9.
4. Göbel TWF, Bolliger L. Evolution of the T-cell receptor signal transduction units. Curr Top Microbiol Immunol 2000; 248:303-320.
5. Rudolph MG, Stanfield RL, Wilson IA. How TCRs bind MHCs, peptides and coreceptors. Annu Rev Immunol 2006; 24:419-466.
6. Call ME, Wucherpfennig KW. The T-cell receptor: Critical role of the membrane environment in receptor assembly and function. Annu Rev Immunol 2005; 23:101-125.
7. García KC, Degano M, Stanfield RL et al. An αβ T-cell receptor structure at 2.5 Å and its orientation in the TCR-MHC complex. Science 1996; 274:209-219.
8. Alarcón B, Gil D, Delgado P et al. Initiation of TCR signaling: Regulation within CD3 dimers. Immunol Rev 2003; 191:38-46.
9. Gil D, Schamel WWA, Montoya M et al. Recruitment of Nck by CD3ε reveals a ligand-induced conformational change essential for T-cell receptor signaling and synapse formation. Cell 2002; 109:901-912.
10. Cambier JC. New nomenclature for the Reth motif (or ARH1/TAM/ARAM/YxxL). Immunol Today 1995; 16:110.
11. Buferne M, Luton F, Letorneur F et al. Role of CD3δ in surface expression of the TCR/CD3 complex and in activation for killing analyzed with a CD3δ-negative cytotoxic T-lymphocyte variant. J Immunol 1992; 148:657-664.
12. Dietrich J, Neising A, Hou X et al. Role of CD3γ in T-cell receptor assembly. J Cell Biol 1996; 132:299-310.
13. Luton F, Buferne M, Legendre V et al. Role of CD3γ and CD3δ cytoplasmic domains in cytolytic T-lymphocyte functions and TCR/CD3 down-modulation. J Immunol 1997; 158:4162-4170.
14. Transy C, Moingeon P, Stebbins C et al. Deletion of the cytoplasmic region of the CD3 epsilon subunit does not prevent assembly of a functional T-cell receptor. Proc Natl Acad Sci USA 1989; 86:7108-7112.
15. Wegener AMK, Letourneur F, Hoeveler A et al. The T-cell receptor/CD3 complex is composed of at least two autonomous transduction modules. Cell 1992; 68:83-95.
16. Adams EJ, Chien Y-H, Garcia KC. Structure of a γδ T-cell receptor in complex with the nonclassical MHC T22. Science 2005; 308:227-231.
17. Alarcón B, Ley SC, Sánchez-Madrid F et al. The CD3-γ and CD3-δ subunits of the T-cell antigen receptor can be expressed within distinct functional TCR/CD3 complexes. EMBO J 1991; 10:903-912.
18. Sun Z-YJ, Kim ST, Kim IC et al. Solution structure of the CD3εδ ectodomain and comparison with CD3εγ as a basis for modeling T-cell receptor topology and signaling. Proc Natl Acad Sci USA 2004; 101:16867-16872.
19. Sun ZJ, Kim KS, Wagner G et al. Mechanisms contributing to T-cell receptor signaling and assembly revealed by the solution structure of an ectodomain fragment of the CD3εγ heterodimer. Cell 2001; 105:913-923.
20. Kjer-Nielsen L, Dunstone MA, Kostenko L et al. Crystal structure of the human T-cell receptor CD3εγ heterodimer complexed to the therapeutic mAb OKT3. Proc Natl Acad Sci USA 2004; 101:7675-7680.
21. Arnett KL, Harrison SC, Wiley DC. Crystal structure of a human CD3-ε/δ dimer in complex with a UCHT1 single-chain antibody fragment. Proc Natl Acad Sci USA 2004; 101:16268-16273.
22. Blumberg RS, Ley S, Sancho J et al. Structure of the T-cell antigen receptor: Evidence for two CD3 ε subunits in the T-cell receptor complex. Proc Natl Acad Sci USA 1990; 87:7220-7224.
23. de la Hera A, Müller U, Olsson C et al. Structure of the T-cell antigen receptor (TCR): Two CD3ε subunits in a functional TCR/CD3 complex. J Exp Med 1991; 173:7-17.

24. Schamel WWA, Arechaga I, Risueno RM et al. Coexistence of multivalent and monovalent TCRs explains high sensitivity and wide range of response. J Exp Med 2005; 202:493-503.
25. Meuer S, Acuto O, Hussey RE et al. Evidence for the T3-associated 90K heterodimer as the T-cell antigen receptor. Nature 1983; 303:808-810.
26. Portolés P, Rojo J, Golby A et al. Monoclonal antibodies to murine CD3ε define distinct epitopes, one of which may interact with CD4 during T-cell activation. J Immunol 1989; 142:4169-4175.
27. Punt JA, Roberts JL, Kearse KP et al. Stoichiometry of the T-cell antigen receptor (TCR) complex: Each TCR/CD3 complex contains one TCR α, one TCR β and two CD3ε chains. J Exp Med 1994; 180:587-593.
28. Thibault G, Bardos P. Compared TCR and CD3ε expression on αβ and γδ T-cells. Evidence for the association of two TCR heterodimers with three CD3ε chains in the TCR/CD3 complex. J Immunol 1995; 154:3814-3820.
29. San José E, Sahuquillo AG, Bragado R et al. Assembly of the TCR/CD3 complex: CD3ε/δ and CD3ε/γ dimers associate indistinctly with both TCRα and TCRβ chains. Evidence for a double TCR heterodimer model. Eur J Immunol 1998; 28:12-21.
30. Wang J, Lim K, Smoylar A et al. Atomic structure of an αβ T-cell receptor (TCR) heterodimer in complex with an anti-TCR Fab fragment derived from a mitogenic antibody. EMBO J 1998; 17:10-26.
31. Fernández-Miguel G, Alarcón B, Iglesias A et al. Multivalent structure of an αβ T-cell receptor. Proc Natl Acad Sci USA 1999; 96:1547-1552.
32. Call ME, Pyrdol J, Wiedmann M et al. The organizing principle in the formation of the T-cell receptor-CD3 complex. Cell 2002; 111:967-979.
33. Call M, Pyrdol J, Wucherpfennig K. Stoichiometry of the T-cell receptor-CD3 complex and key intermediates assembled in the endoplasmic reticulum. EMBO J 2004; 23:2348-2357.
34. Hellwig S, Schamel WWA, Pflugfelder U et al. Differences in pairing and cluster formation of T-cell receptor α- and β-chains in T-cell clones and fusion hybridomas. Immunobiology 2005; 210:685-694.
35. Schamel WWA, Risueno RM, Minguet S et al. A conformation- and avidity-based proofreading mechanism for the TCR-CD3 complex. Trends Immunol 2006; 27:176-182.
36. Cochran JR. Cochran JR, Cameron TO et al. Receptor proximity, not intermolecular orientation, is critical for triggering T-cell activation. J Biol Chem 2001; 276:28068-28074.
37. Ghendler Y, Smolyar A, Chang H-C et al. One of the CD3ε subunits within a T-cell receptor complex lies in close proximity to the Cβ FG loop. J Exp Med 1998; 187:1529-1536.
38. Borroto A, Mallabiabarrena A, Albar JP et al. Characterization of the region involved in CD3 pairwise interactions within the T-cell receptor complex. J Biol Chem 1998; 273:12807-12816.
39. Wegener AMK, Hou X, Dietrich J et al. Distinct domains of the CD3-γ chain are involved in the surface expression and function of the T-cell antigen receptor. J Biol Chem 1995; 270:4675-4680.
40. Bäckström BT, Milia E, Peter A et al. A motif within the T-cell receptor α chain constant region connecting peptide domain controls antigen responsiveness. Immunity 1996; 5:437-447.
41. Werlen G, Hausmann B, Palmer E. A motif in the αβ T-cell receptor controls positive selection by modulating ERK activity. Nature 2000; 406:422-426.
42. Brenner MB, Trowbridge IS, Strominger JL. Cross-linking of human T-cell receptor proteins: Association between the T-cell idiotype β subunit and the T3 glycoprotein heavy subunit. Cell 1985; 40:183-190.
43. Touma M, Chang H-C, Sasada T et al. The TCR Cβ FG loop regulates αβ T-cell development. J Immunol 2006; 176:6812-6823.
44. Hayes SM, Laky K, El-Khoury D et al. Activation-induced modification in the CD3 complex of the γδ T-cell receptor. J Exp Med 2002; 196:1355-1361. Erratum in: J. Exp. Med. 196: 1653.
45. Hayes SM, Love PE. Distinct structure and signaling potential of the γδ TCR complex. Immunity 2002; 16:827-838.
46. Hayes SM, Love PE. Stoichiometry of the murine γδ T-cell receptor. J Exp Med 2006; 203:47-52.
47. Dadi HK, Simon AJ, Roifman CM. Effect of CD3δ deficiency on maturation of α/β and γ/δ T-cell lineages in severe combined immunodeficiency. New Eng J Med 2003; 349:1821-1828.
48. Johansson B, Palmer E, Bolliger L. The extracellular domain of the ζ-chain is essential for TCR function. J Immunol 1999; 162:878-885.
49. Rodríguez-Tarduchy G, Sahuquillo AG, Alarcón B et al. Apoptosis but not other activation events is inhibited by a mutation in the transmembrane domain of T-cell receptor β that impairs CD3ζ association. J Biol Chem 1996; 271:30417-30425.
50. Teixeiro E, Daniels MA, Hausmann B et al. T-cell division and death are segregated by mutation of TCRβ chain constant domains. Immunity 2004; 21:515-526.
51. Janeway Jr. CA. The T-cell receptor as a multicomponent signalling machine: CD4/CD8 Coreceptors and CD45 in T-cell activation. Ann Rev Immunol 1992; 10:645-674.

52. Gao FG, Tormo J, Gerth UC et al. Crystal structure of the complex between human CD8αα and HLA-A2. Nature 1997; 387:630-634.
53. Doucey M-A, Goffin L, Naeher D et al. CD3δ establishes a functional link between the T-cell receptor and CD8. J Biol Chem 2003; 278:3257-3264.
54. Suzuki S, Kupsch J, Eichmann K et al. Biochemical evidence of the physical association of majority of the CD3δ chains with the accessory/co receptor molecules CD4 and CD8 on nonactivated T-lymphocytes. Eur J Immunol 1992; 22:2475-2479.
55. Wang J, Meijers R, Xiong Y et al. Crystal structure of the human CD4 N-terminal two-domain fragment complexed to a class II MHC molecule. Proc Natl Acad Sci USA 2001; 98:10799-10804.
56. Wu H, Kwong PD, Hendrickson WA. Dimeric association and segmental variability in the structure of human CD4. Nature 1997; 387:527-530.
57. Moldovan M-C, Yachou A, Levesque K et al. CD4 dimers constitute the functional component required for T-cell activation. J Immunol 2002; 169:6261-6268.
58. Maekawa A, Schmidt B, Fazekas de St. Groth B et al. Evidence for a domain-swapped CD4 dimer as the coreceptor for binding to Class II MHC. J Immunol 2006; 176:6873-6878.
59. Dianzani U, Shaw A, Al-Ramadi BK et al. Physical association of CD4 with the T-cell receptor. J Immunol 1992; 148:678-688.
60. Saint-Ruf C, Panigada M, Azogi O et al. Different initiation of preTCR and γδTCR signalling. Nature 2000; 406:524-527.
61. Wilson A, MacDonald HR. Expression of genes encoding the preTCR and CD3 complex during thymus development. Int Immunol 1995; 7:1659-1664.
62. Rudd PM, Wormald MR, Stanfield RL et al. Roles for glycosylation of cell surface receptors involved in cellular immune recognition. J Mol Biol 1999; 293:351-366.
63. Zapata DA, Schamel WWA, Torres PS et al. Biochemical differences in the alpha beta TCR-CD3 surface complex between CD8+ and CD4+ human mature T-lymphocytes. J Biol Chem 2004; 279:24485-24492.
64. Criado G, Feito MJ, Ojeda G et al. Variability of invariant mouse CD3ε chains detected by anti-CD3 antibodies. Eur J Immunol 2000; 30:1469-1479.

CHAPTER 2

B-Cell Receptor

Randall J. Brezski and John G. Monroe*

Abstract

The subunit structure of the B-cell antigen receptor (BCR) and its associated compartmentalization of function confer enormous flexibility for generating signals and directing these toward specific and divergent cell fate decisions. Like all the multichain immune recognition receptors discussed in this volume, assembly of these multi-unit complexes sets these receptors apart from almost all other cell surface signal transduction proteins and affords them the ability to participate in almost all of the diverse aspects of, in this case, B-cell biology. We discuss here the structural aspects of the BCR and its associated coreceptors and relate these mechanistically to how BCR signaling can be directed towards specific fate decisions. By doing so, the BCR plays a pivotal role in ensuring the effective and appropriate B-cell response to antigen.

Introduction

Antigen receptors on B- and T-cells exhibit enormous flexibility with regards to not only the diversity of ligands (antigens) that they can engage but also their ability to trigger signals that result in very different cellular fates. This flexibility is possible in part by separating ligand-binding components from signal transduction units. This separation of functions allows the former to be genetically diversified in order to confer on each developing B-cell a unique specificity without affecting the signals generated following ligand engagement. The latter are generated by invariable protein units that associate noncovalently with the ligand binding unit. This multi-unit design (termed through out this book as multichain immune recognition receptors, MIRRs) is common to receptors involved with triggering responses in cells of the immune system.

Once ligand facilitates oligomerization of BCR units, the signals generated lead to remarkably different responses dependent upon the maturation or developmental stage at which the B-cell resides when antigen is introduced. These responses can range from rapid induction of apoptotic programs to initiation of strong survival and activation programs. From an immune system perspective, the choice between these cell fates determines whether the ligand engaged B-cell (and its unique specificity) will be eliminated from the immune system repertoire of reactive B-cells. On the other hand, positive signaling initiates processes that will ultimately result in generation of antibody-secreting cells, cell division and clonal expansion of B-cells with identical specificity and the development of memory B-cells. In part, this decision is guided by the precise plasma membrane lipid environment to which the oligomerized BCR complexes are localized.

Finally, it has been very recently appreciated that BCR complexes can generate biologically meaningful signals even in the absence of ligand. These signals, termed tonic signals, may play a role in driving early B-cell development and later in peripheral survival of B-cells by selecting for

*Corresponding Author: John G. Monroe—Department of Pathology and Laboratory Medicine, University of Pennsylvania School of Medicine, Philadelphia, PA, USA.
Email: monroej@mail.med.upenn.edu

Multichain Immune Recognition Receptor Signaling: From Spatiotemporal Organization to Human Disease, edited by Alexander B. Sigalov. ©2008 Landes Bioscience and Springer Science+Business Media.

cells with functional BCR and eliminating those that fail to assemble BCR properly or later lose its expression at the cell surface.

Each of these characteristics of the BCR will be discussed in this chapter. In addition, where appropriate we will point out challenges that remain in order for us to fully understand in mechanistic terms how BCR signals are initiated, regulated and directed to trigger specific B-cell fate decisions.

Structure of the BCR

The B-cell antigen receptor (BCR) provides B-cells with the capacity for antigen recognition, thereby allowing them to engage potentially dangerous pathogens and is directly responsible for generating signals that result in an appropriate antigen-induced B-cell response. The BCR is a transmembrane protein complex comprised of multiple subunits, including an Immunoglobulin Heavy Chain (IgHC) covalently linked by disulfide bonds to an Ig Light Chain (IgLC). The amino-terminal region of the IgHC and IgLC is characterized by extensive sequence variability brought about by gene recombination, nucleotide addition and somatic hypermutation. It is this region, termed the complementarity-determining region (CDR) that is responsible for the clonal specificity of antigen recognition by the receptor as well as for the affinity of ligand binding. In addition to the ligand binding components, the BCR contains Ig alpha (Igα) and Ig beta (Igβ), the subunits responsible for signal transduction. A single surface BCR unit contains two IgHC, two IgLC, one Igα and one Igβ.[1] The carboxy-terminal invariable portion of the IgHC is termed the constant region and it is this region that defines the 5 different isotypes of BCR, IgM, IgD, IgG, IgA and IgE (see Fig. 1). IgM and IgD are expressed by newly generated and naive B-cells whereas the other isotypes are typically expressed only on activated or memory B-cells.

Efficient surface expression of the BCR complex requires the concomitant expression of each of the individual components. In part, this appears due to the unusual nature of the transmembrane domains of IgHC, Igα and Igβ. Transmembrane domains are generally comprised of neutral and nonpolar amino acids, allowing them to fold into alpha helices that are thermodynamically stable within the hydrophobic environment of the plasma membrane. However, the heavy chain of IgM (Igμ) contains 9 polar amino acids in its transmembrane domain, while the heavy chain of IgD (Igδ) contains 7 and Igα and Igβ each contain 3.[2,3] Noncovalent interactions between the transmembrane domains of IgHC and Igα/β are postulated to shield the polar amino acids from the hydrophobic lipid environment, thereby allowing the BCR complex to exist stably in the plasma membrane.[4,5]

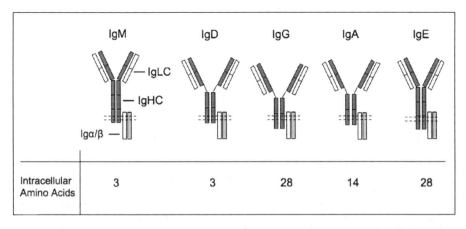

Figure 1. Structure of the monomers of the 5 isotypes of BCR. Each isotype is paired with an Igα/β heterodimer. The isotypes differ in the number of immunoglobulin domains as well as the number of intracellular amino acids.

Once at the cell surface, the BCR complex is capable of transducing signals in response to ligand (antigen)-induced aggregation. The relative paucity of amino acids in the cytoplasmic portion of IgHC (e.g., Igμ and Igδ each contain only 3 amino acids)[6] led to the early hypothesis that the IgHC alone would be incapable of signal transduction. This hypothesis was confirmed in a series of studies in which IgHC was expressed on the surface of B-cells in the absence of the Igα/β heterodimer. Expression of the IgHC alone in this case occurred following mutation of a YS to VV sequence in the transmembrane domain of Igμ[7] In the absence of Igα/β heterodimer association, BCR complexes containing the mutated Igμ were unable to transduce signals, indicating that IgHC-associated Igα/β heterodimers directly participated in BCR-induced signal transduction. The subsequent identification by Reth and colleagues of a signaling motif, termed an Immunoreceptor Tyrosine-Based Activation Motif (ITAM), in both Igα and Igβ supported the idea that the Igα/β heterodimer mediates BCR signaling.[8] The canonical ITAM is Yxx(L/I) x_{6-8} Yxx(L/I) and ITAM motifs are ubiquitously expressed in the MIRR signal transduction components.[9] As described further below, phosphorylation of tyrosines embedded within the Igα/β ITAMs plays a critical role in initiating BCR-induced signal transduction.[10]

While it is quite clear that Igα/β heterodimers play a critical role in initiating and propagating BCR-induced signals, it has also been postulated that higher order complexes may regulate BCR-induced signal transduction. In particular, while it is known that monomers of surface BCR have a stoichiometry of 2 IgHC/LC to 1 Igα/Igβ heterodimer, one study has shown that several of these units can be grouped together to form an oligomeric complex on the surface of resting B-cells.[1] Based on these data, it was postulated that disruption of preformed oligomeric complexes by ligand binding would promote BCR-induced signal transduction. However, the existence of preformed oligomeric complexes was not observed in fluorescence resonance energy transfer (FRET) analysis in which it was demonstrated that the BCR existed as a monomer on the surface of resting B-cells.[11] Thus, in the absence of aggregation by multimeric ligand, the role that higher order BCR structures play in BCR-induced signal transduction remains an issue warranting further investigation.

B-Cell Development

B-cell development occurs in a stepwise manner and the developmental stages are defined by the ordered and sequential assembly of the preBCR followed by the BCR (see Fig. 2). At each stage, the functionality of the receptor complex is assessed and only those B-cells that express signaling competent receptors are allowed to continue development.[12] Indeed, genetic alterations that abrogate the surface expression or signal transduction capacity of any of the PreBCR/BCR components block B-cell development.[13]

To a large extent, expression of the IgHC and IgLC is regulated by a process known as V(D)J recombination. During V(D)J recombination, gene segments known as the variable (V), diversity (D) and joining (J) segments within the IgHC and IgLC genes are brought together to create the mature ligand-binding form of the protein. During the initial stages of B-cell development, recombination of the IgHC and IgLC genes has not yet occurred. However, proB-cells, the first cells committed to the B-cell lineage, do express an Igα and Igβ-containing complex known as the proBCR on their surface.[14] In the proBCR, Igα and Igβ are associated with the endoplasmic reticulum (ER) chaperone calnexin, presumably to mask the polar amino acids found within their transmembrane domains.

As development progresses, the IgHC is the first to recombine. Expression of the mature form of the IgHC marks the transition to the preB-cell stage. However, as IgLC gene recombination has not yet occurred, preB-cells do not express IgLC. Instead, preB-cells express a surrogate light chain (SLC) composed of VpreB and Lamba5 (λ5) which together with IgHC, Igα and Igβ comprise the preBCR.[15] Expression of SLC is essential to get the IgHC to the cell surface as it masks ER retention signals located at the N-terminal portion of the IgHC that retain unassociated IgHC within the ER.[16-18]

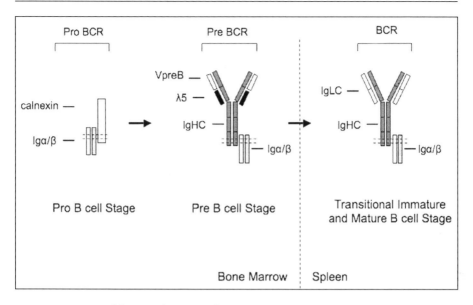

Figure 2. Structure of the BCR during B-cell development. As B-cells develop, the BCR undergoes series of changes in its structure and the subunits expressed. At the proB-cell stage, the proBCR contains an Igα/Igβ heterodimer associated with calnexin. The preB-cell expresses the preBCR composed of IgHC associated with the SLC (VpreB and λ5) and a Igα/Igβ heterodimer. The BCR expressed on transitional immature and mature B-cells contains two IgHC, two IgLC and the Igα/Igβ heterodimer. ProB and preB-cells are localized in the bone marrow, while transitional immature and mature B-cells are localized in the spleen.

Following recombination of the IgLC, the mature form of the BCR can now be assembled and it is the expression of the ligand-binding form of the BCR that marks the transition to the immature B-cell stage. This newly acquired capacity for antigenic recognition provides the first opportunity for selection against cells reactive to endogenous or "self" antigens, a process known as negative selection.[19] Non-autoreactive immature B-cells that have evaded negative selection emerge in the periphery and are known as transitional B-cells. Transitional B-cells are considered direct precursors to mature follicular B-cells.[20]

On mature B-cells the BCR serves two functions. The first is to maintain the survival of BCR-expressing B-cells by generating signals in the apparent absence of a requirement for antigen binding. Evidence for this function comes from studies where conditional ablation of components of the BCR signaling complex at the mature B-cell stage results in apoptosis and loss of the peripheral B-cell population,[21,22] suggesting that ligand-independent signaling is required for the survival of mature B-cells (discussed further below). The second function is to facilitate antigen-specific responsiveness by the mature B-cell. In this case, ligand-induced aggregation of the BCR on mature B-cells elicits effector functions required for an effective humoral response such as proliferation, increased antigen presentation and upregulation of costimulatory molecules required to elicit cognate T cell help. Thus, in the absence of ligand-induced aggregation, low-level BCR-induced signal transduction is sufficient to maintain B-cell survival but is insufficient to fully activate the B-cell, whereas aggregation by multimeric ligand elicits B-cell effector function.

Molecular Aspects of Ligand-Induced BCR Signal Transduction

The tyrosines embedded in the ITAMs of Igα and Igβ are critical for preBCR/BCR-induced signaling function. As some evidence indicates that Src family kinases are constitutively associated with the BCR in resting B-cells[23,24] they are generally thought to initiate BCR-induced signaling.

Following ligand-induced aggregation, Src kinases phosphorylate tyrosines within the ITAMs of Igα and Igβ, with preference for the proximal tyrosine due to the presence of an isoleucine or leucine at the -1 position.[25] This initial phosphorylation creates sites for proteins with Src homology-2 (SH2) domains (highly conserved phosphotyrosine-binding domains) to dock. Following this initial signal amplification, dually phosphorylated ITAMs recruit and activate Syk family kinases by virtue of their tandem SH2 domains, leading to their allosteric activation and engaging a positive feedback loop that further amplifies BCR-induced signal transduction.[26] Indeed, Syk activation is considered essential for optimal BCR-induced signal transduction as SH2 domain mutations that block recruitment to the BCR and mutations in its autophosphorylation site that block function or genetic ablation compromise BCR-induced signaling.[27,28]

Syk activation is responsible for the recruitment and phosphorylation of adaptor molecules that serve as scaffolds to recruit downstream effector molecules through SH2 and phospho-tyrosine-binding (PTB) domain binding. These events allow for the generation of a stabilized signalosome to engage second messenger pathways that direct changes in gene expression that ultimately dictate the cellular response. A key component of this complex is the adapter protein SH2 domain containing leukocyte protein of 65 kDa (SLP-65), also known as BLNK. The SH2 domains on SLP-65 bind to phosphorylated tyrosines and recruit SLP-65 to the BCR. While the ITAM-embedded tyrosines appear necessary for Syk recruitment, early studies in cell lines suggest that SLP-65 is recruited to the complex via non-ITAM-associated phosphotyrosines in Igα.[29,30] In more recent "knock-in" studies, generation of genetically altered mice in which the non-ITAM tyrosines (tyr-204 and tyr-176) of Igα were mutated to phenylalanine revealed normal recruitment and activation of Syk to the BCR signalosome, but reduced SLP-65 recruitment (most notable in the tyr-204 mutant) following BCR engagement.[31] While the inability to recruit SLP-65 was associated with reduced IkBα degradation, impaired JNK and ERK phosphorylation and a relative inability to enter the cell cycle, other aspects of BCR signaling (Syk, PLCγ2 and Btk phosphorylation) were unimpaired. Furthermore, the deficiencies in BCR-induced signal transduction observed in the Igα[tyr-204-phe/tyr-204-phe] mice were not as severe as those displayed by SLP-65-deficient B-cells, suggesting that SLP-65 can function in a tyr-204-independent manner, perhaps by binding to the adaptor linker for activation of B-cells (LAB).[32] However, based on these studies, it is likely that non-ITAM-dependent phosphotyrosine-mediated recruitment of SLP-65 to the BCR signalosome plays a critical role in the induction of multiple elements required for optimal functional responses of B-cells.

Finally, for B-cell effector function to be elicited, BCR-induced signals must be transduced for 20-24 hours.[33] However, the BCR internalizes within minutes of antigen encounter so that bound antigen can be delivered to endosomal compartments for processing and presentation to major histocompatibility complex (MHC) class II-restricted CD4 T cells. It is not clear how signal transduction is maintained in the face of concomitant internalization. However, recent studies propose mechanisms that may help to resolve this paradox. The first model proposes that when IgHC:antigen is internalized, Igα/β heterodimers remain on the cell surface, maintaining the signalosome and thereby allowing for continued signal propagation. This hypothesis is supported by data obtained using a variety of approaches indicating that the IgHC is physically separated from Igα and Igβ following ligand-induced aggregation.[34-36] However, it is not immediately apparent how a receptor complex that requires its components to associate during transport to the cell surface can dissociate following ligand binding, particularly as it is likely that the polar amino acids in the transmembrane domains will now be exposed. While it was observed that in some circumstances Igα and Igβ may associate with MHC class II molecules following their dissociation from IgHC,[37] Igβ-containing complexes can still be observed on the cell surface after IgHC internalization in cells devoid of MHC class II.[34] Alternatively, it has been proposed that while the vast majority of the receptor complexes are indeed internalized following ligand binding, the small subset of receptors that is inducibly phosphorylated is selectively retained at the cell surface to propagate BCR-induced signals.[38] In this model, each receptor undergoes one of two mutually exclusive fates that is determined by its phosphorylation status. Clearly further analysis of the mechanisms involved in the endocytosis of the BCR complex following ligand binding is required to provide greater insight into the interrelationships between receptor internalization and BCR-induced signal transduction.

Membrane Compartmentalization of the BCR

Recent evidence suggests that the spatial organization of the BCR, its signaling effector molecules and positive and negative coregulatory molecules plays a critical role in the initiation and regulation of BCR-induced signal transduction.[39] In part, this spatial organization is postulated to occur by the association of ligand-aggregated BCR into lipid-ordered, cholesterol- and sphingolipid-rich plasma membrane microdomains commonly referred to as lipid rafts.[40] The plasma membrane of the cell is composed of lipid-disordered and lipid-ordered regions. In lipid-ordered domains, the acyl chains of the sphingolipids are saturated, promoting a rigid organization mediated by van der Waals interactions between adjacent acyl chains. Cholesterol further stiffens these domains by packing between the saturated acyl chains of the sphingolipids. The association of cholesterol and the sphingolipids of the lipid-ordered domain makes them resistant to solubilization by non-ionic detergents such as Triton X-100 and confers upon them increased buoyancy in discontinuous sucrose gradients. Based on the latter characteristic, the lipid-ordered domain is referred to as the lipid raft fraction.[41]

Lipid rafts are hypothesized to concentrate proteins necessary for the propagation of signals and exclude proteins and receptors that attenuate signals. In this regard, they are often referred to as platforms for signal transduction. Recent studies on the structure of lipid rafts suggest that in resting cells lipid rafts are small (diameters ranging from 4 to 200 nm) but coalesce into larger structures when cell surface receptors are engaged.[42] While the BCR is predominantly detected in the lipid-disordered, detergent soluble domains of the plasma membrane in mature resting B-cells, it is unclear at the present time if the BCR is located in the nonraft fraction or in submicroscopic lipid raft structures (described above) that are too small to be isolated by sucrose density centrifugation. Regardless, multiple studies suggest that BCR cross-linking in mature B-cells promotes stable association of the BCR with the lipid raft domain. This association is thought to enhance BCR-induced signal transduction as the Src kinase Lyn preferentially localizes to lipid rafts due to the addition of myristoyl and palmitoyl posttranslational modifications.[43] As described further in the section below, there is also evidence that negative coregulatory transmembrane proteins such as CD45 and CD22 that down-modulate signal transduction through the Igα/β-organized signalosome are selectively excluded from lipid rafts. Thus, association of the BCR with lipid rafts promotes sustained signal transduction by providing a microenvironment rich in signal effector molecules and devoid of negative coregulatory molecules.

As lipid rafts directly influence the ability of the BCR to interact with signal transduction effector molecules, their presence can affect BCR-induced signal transduction in both a quantitative and qualitative manner. The influence of lipid rafts on BCR-induced signal transduction seems most apparent when evaluating the differential response of transitional and mature B-cells to BCR crosslinking. While mature B-cells proliferate and upregulate costimulatory molecules involved in cognate T cell interactions, transitional immature B-cells fail to sustain BCR triggered pathways that have been linked to survival and proliferation and instead undergo BCR-induced apoptosis.[44,45] Specifically, transitional B-cells manifest an impaired ability to signal through the PLCγ/ PKCβ/NF-κB/*c-myc* pathway.[46,47] The inability of transitional B-cells to sustain these pathways is associated with a relative inability to detectably colocalize their BCR with cholesterol-enriched lipid rafts following BCR cross-linking.[47,48] As transitional B-cells maintain distinctly lower levels of membrane-associated cholesterol than do follicular mature B-cells, it has been proposed that the differential ability of transitional B-cells and mature B-cells to colocalize their BCRs with cholesterol-enriched lipid rafts following BCR engagement is due to developmentally regulated differences in membrane-associated unesterified cholesterol levels. In support of this idea, increasing the membrane-associated cholesterol levels of transitional B-cells to levels equivalent to that of follicular mature B-cells results in colocalization of the BCR into cholesterol enriched domains and enhances signal transduction through the PLCγ2/PKCβ/NFκB/*c-myc* pathway to an extent that resembles that of the mature B-cell.[47] These results argue that the relative ability of the BCR to associate with lipid rafts has profound effects on BCR-induced signaling and a significant impact on the physiologic response of the cell.

Balance between Positive and Negative Regulators of BCR Signaling

BCR-induced signal transduction is subject to both positive and negative regulation and the spatial repositioning of BCR relative to either negative coregulators or effector molecules plays a critical role in the determining the final B-cell response. For example, the transmembrane protein CD22 is a negative regulator of BCR signaling and contains an Immunoreceptor Tyrosine-based Inhibitory Motif (ITIM). Stochastic activation of Src kinases constitutively associated with the BCR results in phosphorylation of the tyrosines embedded in the ITIM of CD22. These phosphotyrosines subsequently recruit the tyrosine phosphatase SHP-1 via its SH2 domain. SHP-1 attenuates BCR-induced signals by dephosphorylating the Igα/β heterodimer, thereby dissociating the BCR signalosome.[49] As noted above, ligand-induced aggregation of the BCR promotes its association with lipid rafts and this translocation spatially segregates it from CD22.[40] Such spatial segregation potentially allows for more prolonged BCR-induced signaling events. The importance of both CD22 and SHP-1 as negative regulators in B-cell signaling is underscored by the observation that B-cells deficient in either of these proteins are constitutively hyperactive.[50]

Alternatively, coregulatory molecules can affect BCR-induced signal transduction by promoting colocalization with downstream effector molecules. For example, the transmembrane protein CD19 is a positive regulator of BCR signaling and cocrosslinking of the BCR with CD19 lowers the threshold for BCR activation.[51] One mechanism by which CD19 has been postulated to mediate this effect is by reducing BCR internalization, thereby prolonging the residency of the BCR in lipid rafts.[52] The persistence of BCR in lipid rafts at the cell surface correlates with prolonged signaling as revealed by the extended duration of activated signaling molecules in the lipid raft fractions.[52,53]

Finally, recent studies have shown that cross-linking of the BCR results in a Ca^{2+}-dependent generation of reactive oxygen intermediates (ROIs).[54] ROIs such as hydrogen peroxide have the potential to oxidize a conserved cysteine residue within the catalytic active site of tyrosine phosphatases, such as SHP-1 and CD45, rendering them inactive. Only BCR-proximal phosphatases were inactivated and it was proposed that a short pulse of localized phosphatase inactivation favored the initiation of BCR-induced signal transduction. While the mechanism by which H_2O_2 is produced following BCR ligation has not yet been established, it likely involves a plasma membrane-associated NADPH oxidase.[55] As NADPH oxidase is localized to lipid rafts in neutrophils,[56] it is reasonable to predict that BCR translocation into rafts enhances BCR-induced signal transduction by providing an ROI-enriched environment in which inhibitory phosphatases will be inactivated.

Ligand-Independent BCR-Induced Tonic Signaling

While signal transduction through the BCR is generally thought to be initiated by ligand-induced aggregation, the BCR can also transduce tonic signals in a ligand-independent manner. The existence of ligand-independent BCR-mediated tonic signal transduction was originally suggested in studies in which BCR-expressing cells were treated with the tyrosine phosphatase inhibitor pervanadate.[57] In the absence of any known ligand, induction of a pervanadate-induced phosphotyrosine "footprint" that resembled that induced by BCR-crosslinking suggested that, by inhibiting phosphatases, low-level ligand-independent BCR-induced tyrosine kinase activity can be revealed. More recent studies suggest that tonic signaling through both the BCR and preBCR provide functionally relevant signals for mature B-cell survival and developmental progression, respectively. For example, despite any obvious requirement for antigen, conditional deletion of the BCR signaling complex in mature B-cells results in a rapid loss of the peripheral B-cell compartment, suggesting a requirement for continued BCR signaling for mature B-cell survival.[21,22] In less mature

B-cell populations, tonic BCR-induced signaling is thought to maintain developmental progression by preventing back-differentiation.[58] Finally, the inability of preBCR to bind conventional antigen (due to the lack of IgLC expression) suggested that preBCR-induced signal transduction driving B-cell development occurred in a ligand-independent fashion.[39] A direct evaluation of the ligand-independence for preBCR function was assessed in studies in which a chimeric protein consisting of the cytoplasmic regions Igα and Igβ was stargeted to the plasma membrane using the membrane-targeting sequence of Lck that contains both myristoylation and palmitoylation sites. Expression of this protein, which lacked ectodomains, was sufficient to generate signals for IgH allelic exclusion, progression through the pro-B-preB-cell checkpoint, IL-7-dependent expansion and IgLC gene recombination both in vitro and in vivo.[59-61] Together with earlier studies using amino-terminal truncated IgHC proteins,[62] these studies establish the existence of biologically relevant ligand-independent tonic signals by Igα/Igβ-containing complexes.

Conclusion

While much is known about the biochemical events that are initiated following activation of B-cells with multimeric ligand, the molecular events that trigger these signal transduction pathways remain more enigmatic. In particular, it is quite clear that spatial proximity with signal transduction effector molecules following association with lipid rafts, the balance of positive and negative coregulatory molecules and the oligomeric status of the BCR all play an essential role in determining the final outcome of BCR-induced signal transduction. The further identification and functional evaluation of molecular interactions responsible for these effects will be greatly enhanced by new, more quantitative techniques such as FRET, two photon microscopy and Total Internal Reflection Fluorescence (TIRF). In addition, such techniques may provide insight into the mechanisms that permit sustained BCR-induced signal transduction under conditions where the vast majority of the receptor is internalized.

Finally, while signal transduction through MIRRs has traditionally been considered ligand-induced, it is becoming apparent that several of these receptors are able to signal in a ligand-independent tonic manner.[39] Identifying the molecular events that distinguish tonic ligand-independent from multimeric ligand-induced signal transduction may well aid in designing therapeutic treatments to alleviate autoimmune disease.

Acknowledgements
The authors wish to thank other members of the laboratory for thoughtful discussions especially Dr. Leslie King for help in the organization of the manuscript. We also thank Sarah Hughes for editorial assistance in the preparation of the manuscript.

References
1. Schamel WW, Reth M. Monomeric and oligomeric complexes of the B-cell antigen receptor. Immunity 2000; 13(1):5-14.
2. Campbell KS, Hager EJ, Cambier JC. Alpha-chains of Igm and Igd antigen receptor complexes are differentially n-glycosylated mb-1-related molecules. J Immunol 1991; 147(5):1575-1580.
3. Reth M. Antigen receptors on B lymphocytes. Annu Rev Immunol 1992; 10:97-121.
4. Hombach J, Lottspeich F, Reth M. Identification of the genes encoding the Igm-alpha and Ig-beta components of the Igm antigen receptor complex by amino-terminal sequencing. Eur J Immunol 1990; 20(12):2795-2799.
5. Reth M, Wienands J, Tsubata T et al. Identification of components of the B-cell antigen receptor complex. Adv Exp Med Biol 1991; 292:207-214.
6. Mitchell RN, Shaw AC, Weaver YK et al. Cytoplasmic tail deletion converts membrane immunoglobulin to a phosphatidylinositol-linked form lacking signaling and efficient antigen internalization functions. J Biol Chem 1991; 266(14):8856-8860.
7. Grupp SA, Campbell K, Mitchell RN et al. Signaling-defective mutants of the B lymphocyte antigen receptor fail to associate with Ig-alpha and Ig-beta/gamma. J Biol Chem 1993; 268(34):25776-25779.
8. Reth M. Antigen receptor tail clue. Nature 1989; 338(6214):383-384.
9. Sigalov AB. Multichain immune recognition receptor signaling: Different players, same game? Trends Immunol 2004; 25(11):583-589.

10. Kurosaki T. Genetic analysis of B-cell antigen receptor signaling. Annu Rev Immunol 1999; 17:555-592.
11. Tolar P, Sohn HW, Pierce SK. The initiation of antigen-induced B-cell antigen receptor signaling viewed in living cells by fluorescence resonance energy transfer. Nat Immunol 2005; 6(11):1168-1176.
12. Fuentes-Panana EM, Monroe JG. Ligand-dependent and -independent processes in B-cell-receptor-mediated signaling. Springer Semin Immunopathol 2001; 23(4):333-350.
13. Chung JB, Silverman M, Monroe JG. Transitional B-cells: Step by step towards immune competence. Trends Immunol 2003; 24(6):343-349.
14. Koyama M, Ishihara K, Karasuyama H et al. CD79 alpha/CD79 beta heterodimers are expressed on pro-B-cell surfaces without associated mu heavy chain. Int Immunol 1997; 9(11):1767-1772.
15. Rolink A, Haasner D, Melchers F et al. The surrogate light chain in mouse B-cell development. Int Rev Immunol 1996; 13(4):341-356.
16. Grupp SA, Mitchell RN, Schreiber KL et al. Molecular mechanisms that control expression of the B lymphocyte antigen receptor complex. J Exp Med 1995; 181(1):161-168.
17. Shaw AC, Mitchell RN, Weaver YK et al. Mutations of immunoglobulin transmembrane and cytoplasmic domains: Effects on intracellular signaling and antigen presentation. Cell 1990; 63(2):381-392.
18. Wu Y, Pun C, Hozumi N. Roles of calnexin and Ig-alpha beta interactions with membrane igs in the surface expression of the B-cell antigen receptor of the igm and Igd classes. J Immunol 1997; 158(6):2762-2770.
19. King LB, Monroe JG. Immunobiology of the immature B-cell: Plasticity in the B-cell antigen receptor-induced response fine tunes negative selection. Immunol Rev 2000; 176:86-104.
20. Srivastava B, Lindsley RC, Nikbakht N et al. Models for peripheral B-cell development and homeostasis. Semin Immunol 2005; 17(3):175-182.
21. Kraus M, Alimzhanov MB, Rajewsky N et al. Survival of resting mature B lymphocytes depends on BCR signaling via the Igalpha/beta heterodimer. Cell 2004; 117(6):787-800.
22. Lam KP, Kuhn R, Rajewsky K. In vivo ablation of surface immunoglobulin on mature B-cells by inducible gene targeting results in rapid cell death. Cell 1997; 90(6):1073-1083.
23. Campbell MA, Sefton BM. Association between B-lymphocyte membrane immunoglobulin and multiple members of the src family of protein tyrosine kinases. Mol Cell Biol 1992; 12(5):2315-2321.
24. Yamanashi Y, Kakiuchi T, Mizuguchi J et al. Association of B-cell antigen receptor with protein tyrosine kinase lyn. Science 1991; 251(4990):192-194.
25. Schmitz R, Baumann G, Gram H. Catalytic specificity of phosphotyrosine kinases blk, lyn, c-src and syk as assessed by phage display. J Mol Biol 1996; 260(5):664-677.
26. Rolli V, Gallwitz M, Wossning T et al. Amplification of B-cell antigen receptor signaling by a syk/itam positive feedback loop. Mol Cell 2002; 10(5):1057-1069.
27. Kurosaki T, Johnson SA, Pao L et al. Role of the syk autophosphorylation site and SH2 domains in B-cell antigen receptor signaling. J Exp Med 1995; 182(6):1815-1823.
28. Takata M, Sabe H, Hata A et al. Tyrosine kinases lyn and syk regulate B-cell receptor-coupled Ca2+ mobilization through distinct pathways. EMBO J 1994; 13(6):1341-1349.
29. Engels N, Wollscheid B, Wienands J. Association of slp-65/BLNK with the B-cell antigen receptor through a non-itam tyrosine of Ig-alpha. Eur J Immunol 2001; 31(7):2126-2134.
30. Kabak S, Skaggs BJ, Gold MR et al. The direct recruitment of BLNK to immunoglobulin alpha couples the B-cell antigen receptor to distal signaling pathways. Mol Cell Biol 2002; 22(8):2524-2535.
31. Patterson HC, Kraus M, Kim YM et al. The B-cell receptor promotes B-cell activation and proliferation through a non-itam tyrosine in the Igalpha cytoplasmic domain. Immunity 2006; 25(1):55-65.
32. Janssen E, Zhu M, Zhang W et al. Lab: A new membrane-associated adaptor molecule in B-cell activation. Nat Immunol 2003; 4(2):117-123.
33. DeFranco AL, Raveche ES, Paul WE. Separate control of B lymphocyte early activation and proliferation in response to anti-Igm antibodies. J Immunol 1985; 135(1):87-94.
34. Kremyanskaya M, Monroe JG. Ig-independent ig beta expression on the surface of B lymphocytes after B-cell receptor aggregation. J Immunol 2005; 174(3):1501-1506.
35. Kim JH, Cramer L, Mueller H et al. Independent trafficking of Ig-alpha/Ig-beta and mu-heavy chain is facilitated by dissociation of the B-cell antigen receptor complex. J Immunol 2005; 175(1):147-154.
36. Vilen BJ, Nakamura T, Cambier JC. Antigen-stimulated dissociation of BCR mIg from Ig-alpha/Ig-beta: Implications for receptor desensitization. Immunity 1999; 10(2):239-248.
37. Lang P, Stolpa JC, Freiberg BA et al. TCR-induced transmembrane signaling by peptide/MHC class II via associated Ig-alpha/beta dimers. Science 2001; 291(5508):1537-1540.
38. Hou P, Araujo E, Zhao T et al. B-cell antigen receptor signaling and internalization are mutually exclusive events. PLoS Biol 2006;4(7):e200
39. Monroe JG. ITAM-mediated tonic signalling through pre-BCR and BCR complexes. Nat Rev Immunol 2006; 6(4):283-294.

40. Pierce SK. Lipid rafts and B-cell activation. Nat Rev Immunol 2002; 2(2):96-105.
41. Laude AJ, Prior IA. Plasma membrane microdomains: Organization, function and trafficking. Mol Membr Biol 2004; 21(3):193-205.
42. Edidin M. Shrinking patches and slippery rafts: Scales of domains in the plasma membrane. Trends Cell Biol 2001; 11(12):492-496.
43. Cheng PC, Dykstra ML, Mitchell RN et al. A role for lipid rafts in B-cell antigen receptor signaling and antigen targeting. J Exp Med 1999; 190(11):1549-1560.
44. Chung JB, Sater RA, Fields ML et al. CD23 defines two distinct subsets of immature B-cells which differ in their responses to T cell help signals. Int Immunol 2002; 14(2):157-166.
45. Norvell A, Mandik L, Monroe JG. Engagement of the antigen-receptor on immature murine B lymphocytes results in death by apoptosis. J Immunol 1995; 154(9):4404-4413.
46. King LB, Norvell A, Monroe JG. Antigen receptor-induced signal transduction imbalances associated with the negative selection of immature B-cells. J Immunol 1999; 162(5):2655-2662.
47. Karnell FG, Brezski RJ, King LB et al. Membrane cholesterol content accounts for developmental differences in surface B-cell receptor compartmentalization and signaling. J Biol Chem 2005; 280(27):25621-25628.
48. Chung JB, Baumeister MA, Monroe JG. Cutting edge: Differential sequestration of plasma membrane-associated B-cell antigen receptor in mature and immature B-cells into glycosphingolipid-enriched domains. J Immunol 2001; 166(2):736-740.
49. Nitschke L. The role of CD22 and other inhibitory coreceptors in B-cell activation. Curr Opin Immunol. 2005; 17(3):290-297.
50. Cornall RJ, Cyster JG, Hibbs ML et al. Polygenic autoimmune traits: Lyn, CD22 and SHP-1 are limiting elements of a biochemical pathway regulating bcr signaling and selection. Immunity 1998; 8(4):497-508.
51. Dempsey PW, Allison ME, Akkaraju S et al. C3d of complement as a molecular adjuvant: Bridging innate and acquired immunity. Science 1996; 271(5247):348-350.
52. Cherukuri A, Cheng PC, Sohn HW et al. The CD19/CD21 complex functions to prolong B-cell antigen receptor signaling from lipid rafts. Immunity 2001; 14(2):169-179.
53. Cherukuri A, Shoham T, Sohn HW et al. The tetraspanin CD81 is necessary for partitioning of coligated CD19/CD21-B-cell antigen receptor complexes into signaling-active lipid rafts. J Immunol 2004; 172(1):370-380.
54. Singh DK, Kumar D, Siddiqui Z et al. The strength of receptor signaling is centrally controlled through a cooperative loop between Ca2+ and an oxidant signal. Cell 2005; 121(2):281-293.
55. Reth M. Hydrogen peroxide as second messenger in lymphocyte activation. Nat Immunol 2002; 3(12):1129-1134.
56. Shao D, Segal AW, Dekker LV. Lipid rafts determine efficiency of NADPH oxidase activation in neutrophils. FEBS Lett 2003; 550(1-3):101-106.
57. Wienands J, Larbolette O, Reth M. Evidence for a preformed transducer complex organized by the B-cell antigen receptor. Proc Natl Acad Sci USA 1996; 93(15):7865-7870.
58. Tze LE, Schram BR, Lam KP et al. Basal immunoglobulin signaling actively maintains developmental stage in immature B-cells. PLoS Biol 2005; 3(3):e.82.
59. Bannish G, Fuentes-Panana EM, Cambier JC et al. Ligand-independent signaling functions for the B lymphocyte antigen receptor and their role in positive selection during B lymphopoiesis. J Exp Med 2001; 194(11):1583-1596.
60. Fuentes-Panana EM, Bannish G, Shah N et al. Basal Igalpha/Igbeta signals trigger the coordinated initiation of pre-B-cell antigen receptor-dependent processes. J Immunol 2004; 173(2):1000-1011.
61. Fuentes-Panana EM, Bannish G, van der Voort D et al. Ig alpha/Ig beta complexes generate signals for B-cell development independent of selective plasma membrane compartmentalization. J Immunol 2005; 174(3):1245-1252.
62. Shaffer AL, Schlissel, MS. A truncated heavy chain protein relieves the requirement for surrogate light chains in early B-cell development. J Immunol 1997; 159(3):1265-75.

CHAPTER 3

Fc Receptors

Maree S. Powell* and P. Mark Hogarth

Abstract

The aggregation of cell surface Fc receptors by immune complexes induces a number of important antibody-dependent effector functions. It is becoming increasingly evident that the organization of key immune proteins has a significant impact on the function of these proteins. Comparatively little is known, however, about the nature of Fc receptor spatiotemporal organization. This review outlines the current literature concerning human Fc receptor spatial organization and physiological function.

Introduction

Like the T- and B-cell antigen receptors (TCR and BCR, respectively), Fc receptors for immunoglobulins (FcRs) belong to a group of cell surface glycoproteins known as the multichain immune recognition receptors (MIRRs). FcRs are key immune regulatory receptors, connecting humoral immune responses to cellular effector mechanisms. Our current understanding of Fc receptor function has been brought about by more than 25-years of work defining the molecular expression, functional outcomes following receptor cross-linking and the structures of these receptors complexed with ligand. This review focuses on human Fc receptor spatial organization and physiological function as it becomes increasingly evident that the organization of Fc receptors has a significant impact on their function. Further aspects of Fc receptor biology, such as the cellular expression, Ig subclass specificity and functional characterization have been extensively reviewed elsewhere[1-6] and are briefly overviewed in Tables 1-3.

Human Receptors for Immunoglobulin

The cross-linking and aggregation of FcRs are critically important for leukocyte activation. Key immunological functions such as macrophage phagocytosis, inhibition of B-cell activation, respiratory burst, pro-inflammatory cytokine secretion and antibody-dependent cellular cytotoxicity are all initiated as a result of Fc receptor aggregation. Receptors for all classes of immunoglobulins, including FcγR (IgG), FcεRI (IgE), FcαRI (IgA), FcμR (IgM) and FcδR (IgD), have been identified. This review examines the spatial organization and physiological functions of receptors for IgG, IgE and IgA.

Human IgG Receptors

There are three classes of receptors for human IgG (FcγRs) found on leukocytes: CD64 (FcγRI), CD32 (FcγRIIa, FcγRIIb, FcγRIIc) and CD16 (FcγRIIIa, FcγRIIIb) (Fig. 1, Table 1). FcγRI sets itself apart from FcγRII and FcγRIII as it binds ligand with high affinity (10^8-10^9 M^{-1}) and is

*Corresponding Author: Maree S. Powell—The MacFarlane Burnet Institute for Medical Research and Public Health, Austin Health, Studley Road, Heidelberg and Department of Pathology, The University of Melbourne, Parkville 3052, Australia.
Email: mpowell@burnet.edu.au

Multichain Immune Recognition Receptor Signaling: From Spatiotemporal Organization to Human Disease, edited by Alexander B. Sigalov. ©2008 Landes Bioscience and Springer Science+Business Media.

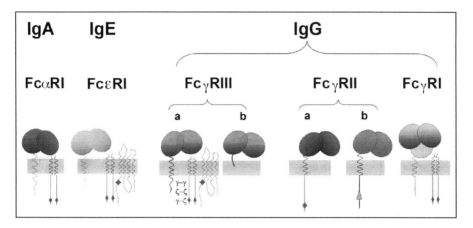

Figure 1. Schematic representation of the human Fc receptor family. All receptors exhibit two (or three), Ig-like domains and are arranged on the cell surface so that the immunoglobulin-binding surface is positioned away from the membrane (for FcγRI, there is no defined structure to date). With the exception of FcγRIIIb, which is a glycosylphosphatidyl inositol-anchored protein, all other Fc receptors are type I integral proteins containing cytoplasmic domains. The FcRs initiate activatory or inhibitory signaling processes via the signaling motifs intrinsic to the receptor (FcγRII) or found on the associated FcR γ–chain (FcγRI, FcγRIIIa, FcεRI), ζ–ζ homodimers or γ–ζ heterodimers (FcγRIIIa on NK cells). Signaling via the FcεRI is also mediated by the FcR β-chain which acts as an amplifier of the FcεRI activation signals. On the surface of mast cells, FcγRIIIa also associates with the FcR β-chain. Within the receptor cytoplasmic domain, immunoreceptor tyrosine-based activation motif (ITAM) is represented by diamonds and immunoreceptor tyrosine-based inhibition motif (ITIM) is represented by triangles.

composed of three Ig-like domains. FcγRII and FcγRIII are composed of two Ig-like domains and bind IgG with low to intermediate affinity (10^6-10^7 M^{-1}). For some time, the division of Fc receptor into classes has been largely based around ligand specificity and affinity. However, more recently this classification has been expanded by key in vivo analyses that describe Fc receptors as having either an activating or inhibitory function, characterized by the presence of an immunoreceptor tyrosine-based activation[7] or inhibition[8] motifs (ITAM or ITIM, respectively) within the cytoplasmic domain of the receptor or by association with ITAM-containing signaling subunits, the FcR γ-chain or the ζ–chain homodimers. Upon receptor cross-linking, FcγRI, FcγRIIa and FcγRIII all initiate activation processes via ITAMs in the noncovalently associated γ-chain (FcγRI, FcγRIII) or via intrinsic ITAMs (FcγRIIa,[9] FcγRIIIc[10]). Notably, for FcγRIIIa, γ–ζ heterodimers or ζ–ζ homodimers are also capable of inducing activatory signals upon FcγRIIIa, cross-linking on the surface of natural killer (NK) cells.[11-13] FcγRIII on mast cell is also known to associate with the FcR β-chain of the FcεRI oligomeric complex.[14] The ITAM found in FcγRIIa differs from the canonical ITAM with 12 amino acids (rather than 7 residues) separating the essential tyrosine residues within the signaling motif.[9] The intracellular signaling pathways following activatory receptor cross-linking are induced by the phosphorylation of tyrosine residues within the ITAM via the cooperative recruitment of nonreceptor tyrosine kinase, src, which in turn induces the recruitment of Src homology 2 (SH2) domain-containing signaling molecules such as the syk kinase to the phosphorylated ITAM. These early events in the signaling pathway induce the phosphorylation of numerous intracellular substrates leading to the generation of inositol triphosphate (IP3), diacylglycerol (DAG) and intracellular calcium mobilization or, depending on the cell type activated, the induction of gene expression (reviewed by refs. 2,3,15).

Unlike all other FcRs, FcγRIIb is a negative regulator of activation. This low-affinity receptor shares a high degree of homology with the activatory FcγRIIa molecule, but contains the ITIM sequence (I/V/L/SxYxxL/V) within the cytoplasmic domain.[16] Two isoforms of FcγRIIb are

Table 1. Molecular and functional characteristics of human FcγRs

	FcγRI	FcγRII	FcγRIII
CD	CD64	CD32	CD16
Isoforms	FcγRIa	FcγRIIa, FcγRIIb1, FcγRIIb2, FcγRIIc	FcγRIIIa, FcγRIIIb
Alleles		FcγRIIa HR; FγRIIa LR	FcγRIIIb NA1; FcγRIIIb NA2
Affinity for monomeric IgG	High, $(10^8-10^9\ M^{-1})$	Low, $(10^6\ M^{-1})$	Low, $(10^6-10^7\ M^{-1})$
Human IgG isotype specificity	$3 > 1 > 4 >> 2$	FcγRIIa HR; $3 = 1 >> 2 > 4$ FcγRIIa LR; $3 = 1 = 2 >> 4$ FcγRIIb; $3 > 1 > 4 > 2$	$1 = 3 >> 2, 4$
Cellular distribution	Monocytes, neutrophils, macrophages	FcγRIIa; monocytes, neutrophils, macrophages, eosinophils, basophils, platelets, dendritic cells FcγRIIb; B-cells, monocytes, macrophages, mast cells FcγRIIc; NK-cells	FcγRIIIa; subpopulations of monocytes, NK cells, macrophages, mast cells FcγRIIIb; neutrophils,
Associated signaling subunit	FcR γ-chain, but not absolutely required for expression	none known to date; signaling motif located within cytoplasmic domain	FcγRIIIa; FcR γ-chain or γζ heterodimer on NK-cells FcγRIIIb; none, is a GPI-anchored protein and uses FcγRIIa to signal
Functional characteristics	Phagocytosis, endocytosis ADCC, cytokine release	FcγRIIa; phagocytosis, ADCC, cytokine release, endocytosis (FcγRIIb1 incapable of this function). FcγRIIb; blockade of BCR-induced B-cell activation	Phagocytosis, endocytosis ADCC, cytokine release
Polymorphisms	FcγRI 'null' family	FcγRIIa, HR, LR; influences IgG2 and mouse IgG1 binding FcγRIIb, I232T; raft exclusion	FcγRIIIa; F176, V176; influences IgG1, IgG3 binding FcγRIIIb; NA1, NA2; influences phagocytosis
Modulation of expression	IFN-γ, IL-10, G-CSF↑ TGF-β, IL-4 ↓	FcγRIIa; IL-4 ↓ FcγRIIb2; IFN-γ ↓, IL-4 ↑	IFN-β, GM-CSF, G-CSF↑ TGF-β, IL-4, TNF-α (FcγRIIIb only) ↓

known, FcγRIIb1 and FcγRIIb,2 which are distinguished from each other by the insertion of a 19-amino-acid sequence within FcγRIIb1,[17] which renders the receptor incapable of endocytosis (unlike FcγRIIb2).[18] FcγRIIb1 is found on B lymphocytes while FcγRIIb2 expression is restricted to myeloid cells. The cytoplasmic ITIM is sufficient to inhibit a number of key immunological processes. When coligated with receptors containing the cytoplasmic ITAM sequence, both FcγRIIb1 and FcγRIIb2 are negative regulators of cellular activation. FcγRIIb is also known to modulate cellular responses triggered by coaggregation with either the BCR or the high-affinity IgE receptor (FcεRI).[8]

The expression of FcγRs on the surface of leukocytes has been well documented (Table 1 and reviewed by ref. 1), with the receptor expression levels differentially modulated by cytokine secretion during immune responses. Cytokines, such as interferon gamma (IFN-γ) and granulocyte colony stimulating factor, both increase the expression of FcγRI and FcγRIII (reviewed by ref. 1). Interleukin 10 (IL-10) is known to induce the up-regulation of FcγRI, whereas IL-4 inhibits expression of all activatory receptors.[19-21] Interestingly, FcγRIIb2 expression is decreased by IFN-γ, while IL-4 up regulates the receptor level.[22] Transforming growth factor-β (TGF-β) has been known to modulate the expression of FcγRIII on the surface of monocytes.[21] Recently, it was demonstrated that myeloid cells cultured in the presence of TGF-β have reduced expression of FcγRI and FcγRIII with a concomitant reduction in the levels of detectable FcR γ-chain.[23] Interestingly, TGF-β has also been reported to inhibit the expression of FcεRI on mast cells, with TGF-β affecting the rate of mRNA translation of the FcεRI β-chain.[24] This cytokine has long been recognized as having an immunosuppressive phenotype with these later studies helping to elucidate the molecular mechanisms by which this cytokine can exert such effects on varying cell types. It has been suggested that through the ability to differentially regulate FcγR expression, cytokines may potentially act to modulate effector cell functions in an autocrine and paracrine manner.[25] As the activatory and inhibitory FcγRs are often co-expressed within the cell and bind ligand with comparable affinities, the regulation of FcγR expression is biologically significant. The numerous in vivo studies of mice null for a particular FcR gene encoding the FcR γ-chain (reviewed by refs. 3,26-28), or carrying the FcγRIIa transgene,[29,30] have identified that disruptions in the balance of activatory and inhibitory receptors results in potent pathological responses as well as possible increased susceptibility to infection (reviewed by ref. 25).

Human IgE Receptor

The FcεRI is a complex of three distinct polypeptides and comprises the ligand-binding α subunit and two signaling subunits, FcR β- and γ chains. Two isoforms of the FcεRI complex are

Table 2. Molecular and functional characteristics of human FcεR

	FcεR
CD	No CD antigen number assigned as yet
Isoform	FcεRI
IgE specificity	Human IgE, rat IgE, mouse IgE
Affinity for monomeric IgE	High (10^{10} M^{-1})
Cellular distribution	Monocytes (activated), mast cells, basophils, Langerhan cells, eosinophils
Receptor forms	$\alpha\beta\gamma_2$, detected on the cell surface of mast cells and basophils
	$\alpha\gamma_2$ detected on the surface of monocytes, macrophages, Langerhan and dendritic cells
Functional characteristics	Degranulation, phagocytosis, endocytosis ADCC, release of histamine and leukotriene, cytokine production

known. Monocytes, macrophages, Langerhans and dendritic cells express the αγ2 complex, while mast cells and basophils express the classical αβγ$_2$ complex (Fig. 1, Table 2). The α, β and γ chains are maintained within the plasma membrane by a combination of hydrophobic and electrostatic noncovalent interactions (reviewed extensively by ref. 4). Signaling and cellular activation are initiated by the high-affinity binding of specific multivalent antigens to receptor-bound IgE, thus inducing receptor clustering. Upon clustering, the nonreceptor tyrosine kinase lyn, constitutively associated with the FcεRI β-chain, trans-phosphorylates the ITAMs of both γ- and β-chains and stimulates further recruitment of lyn and trans-phosphorylation of receptors within the cluster. Phosphorylation of the ITAM allows lyn to phosphorylate syk which in turn initiates the phosphorylation of a number of key intracellular substrates, thus leading to a number of different functional outcomes. These outcomes include the release of intracellular calcium stores; antigen presentation; degranulation and release of histamine, leukotrienes, cytokines and other inflammatory mediators from mast cells; or anti-parasitic responses following activation of eosinophils. In cells expressing the αγ$_2$ complex, it is the membrane-associated lyn that orchestrates the phosphorylation of the FcR γ-chain and the activation of syk. However, the level of activation and functional outcomes (calcium signaling and degranulation) occur at reduced levels compared to αβγ2 complexes. The FcεRI β-chain has been shown to act as an amplifier of the FcεRI activation signals mediated through the FcR γ-chain.[31]

Human IgA Receptor

There are several well-characterized receptors for human IgA: the polymeric Ig receptor (pIgR), Fcα/μR and FcαRI.[5,6] This review will only focus on the leukocyte Fc receptor for IgA, FcαRI (CD89) (Fig. 1, Table 3). The FcαR gene is found in the leukocyte receptor cluster on chromosome 19 which is known to also encode the leukocyte immunoglobulin-like receptor (LIR) and killer cell immunoglobulin-like receptor (KIR). These immunoreceptors are only distantly related to the leukocyte FcR gene family which is located on chromosome 1. However, despite this notable difference, the FcαR does share structural similarities with the other members of the FcR family. FcαRI is a type I integral membrane protein consisting of two Ig-like domains, a transmembrane region and a short cytoplasmic tail. On the cell surface, FcαRI noncovalently associates (via a transmembrane domain interface containing the crucial arginine residue) with the FcR γ-chain[32-34] which transduces activatory signals upon receptor clustering by IgA-immune complexes. The intracellular signaling pathway following FcαRI cross-linking is induced by the phosphorylation of tyrosine residues within the FcR γ-chain ITAM, results in the phosphorylation of intracellular intermediates and culminates in a number of different functional outcomes including phagocytosis, antibody-dependent cellular cytotoxicity (ADCC), cytokine release and respiratory burst.

The FcαRI receptor is set apart from other FcRs by three features. First, there is no known mouse homolog of the FcαRI protein. FcαRI is the only Ig-like receptor identified as binding IgA immune complexes. In mice, a gene translocation between chromosomes 1 and X is thought to

---

Table 3. Molecular and functional characteristics of human FcαR

	FcαR
CD	CD89
Isoform	FcαRI
IgA specificity	Serum and secretory forms of IgA1 and IgA2
Affinity for monomeric IgA	Low (10^7 M^{-1})
Cellular distribution	Monocytes (activated), mast cells, basophils, Langerhan cells, eosinophils
Receptor forms	α, αγ$_2$
Functional characteristics	Phagocytosis, endocytosis, respiratory burst and cytokine release, ADCC

lead to a loss of the equivalent FcαR gene.[35,36] Second, the nature of ligand-binding by the FcαRI and other FcRs differs significantly.[37,38] Site-directed mutagenesis of FcαRI identified domain 1 (D1) as the focal point of interaction with ligand (see below).[37] Homology modeling and biosensor analysis suggested that the stoichiometry of FcαR:IgA interaction is 2:1,[38] which was confirmed by crystallographic analysis of the FcαRI:IgA complex.[39] The crystal structure of the receptor:ligand complex demonstrated that FcαRI has a folding topology similar to other FcR proteins.[39] However, the relative orientation of domains 1 and 2 is opposite to that observed for the FcR family. Finally, the third unique feature of FcαRI is its capacity to induce negative cellular regulation following activation via the associated FcRγ-chain ITAM[40] (reviewed by ref. 41). The binding of complexed IgA to FcαRI is well documented as inducing numerous inflammatory cell functions that are crucial to immune regulation. However, binding of monomeric IgA (serum IgA) can inhibit IgG-mediated activation of myeloid cells, thereby exerting an anti-inflammatory function through FcαRI.[40] Both in vitro and in vivo analyses of binding of serum IgA or anti-FcαRI Fab fragments to FcαRI demonstrated a dual signaling role for this receptor. Moreover, it has been postulated that this dual function is regulated by the avidity of ligand: receptor interactions, whereby low avidity binding causes inhibition of receptor signaling and high avidity interaction induces cellular activation.[41]

Interaction between Fc Receptor and Immunoglobulin

The determination of receptor structures has allowed us to closely examine the interaction with ligand and better understand how these interactions may impact on cellular activation.

The orientation of FcγRII, FcγRIII, FcεRI and FcαRI extracellular D1 and D2 is highly conserved across the leukocyte receptor family: an acute angle (approximately 50-90 degrees) separates the Ig domains and the ligand binding surface is positioned away from the cell membrane.[39,42-46] Notably, for FcαRI, the relative D1-D2 orientation is also bent (approximately 90 degrees), but the orientation of the domains is opposite to that of other FcRs.[38,39] To date there are no structural data for FcγRI. Mutagenesis studies have identified that D2 is crucial for the interaction with ligand, while D3 is responsible for the high affinity interaction between FcγRI and ligand.[45,47]

For FcγRII, FcγRIII and FcεRI, the B-C, C'-E and F-G loops of D2 as well as the D1 and D2 linker region and the D2 C' β strand have been shown to directly interact with the lower hinge region of immunoglobulin.[42-44,48-50] Based on the crystal structures of FcγRIIIa and FcεRI complexed with their ligands, the receptor-ligand interaction is known to involve a combination of salt bridges, hydrogen bonds and hydrophobic interactions. The interface between FcγRIII and IgG Fc is dominated by hydrogen-bonding interactions within one Ig-heavy chain as well as hydrophobic interactions occurring at the binding interface of the other Ig-heavy chain.[43,44] Hydrogen-bonding networks are thought to contribute to the stability of receptor:ligand complexes as evident from alanine scanning mutagenesis of FcγRIIa which found that mutation of His134 culminates in the loss of Ig binding, probably due to the loss of key hydrogen bonds between receptor and ligand.[42,49] For the FcεRI:IgE interaction, two binding sites within the FcεRI D2 domain involve overlapping but non-identical sets of IgE residues.[46] The binding surface dominated by potential salt bridges, hydrogen bonds and extensive hydrophobic interactions is much larger than that of the FcγRIII. The extensive hydrophobic and electrostatic interactions have been suggested to contribute to the observed differences in high-ligand affinity for FcεRI compared to the lower affinity for FcγRIII.[43] A comparative analysis of FcR crystal structures[43,44,46] has formed the basis of a model that describes the mode of interaction between receptor and ligand. Receptor tryptophan residues and a proline residue within the Ig-CH_2 domain form a proline sandwich that acts as the primary site of receptor: ligand interaction followed by the interaction between the hinge proximal peptide sequences (Leu234-239)[51] within Ig and the D2 loop regions (B-C, C'-E and F-G) of the receptor. Mutagenesis studies of the hinge peptide sequence have highlighted the crucial importance of this region for the interaction with FcRs.[45,52-54]

The structure of FcαRI:IgA1-Fc[39] demonstrated that the B-C loop, D strand and the D-E and F-G loops of D1 form the ligand-binding site. Also, the FcαRI:IgA1-Fc structure confirmed that the stoichiometry of the complex is 2:1 whereby two FcαRI molecules bind the IgA at each Cα2-Cα3 junction.[38,39] This is strikingly different when compared with the determined 1:1 stoichiometry of the receptor-ligand complexes of FcγRIII and FcεRI.[55,56] Dimerization of the FcαRI molecules within the receptor:ligand complex is unlikely to induce spontaneous activation as the C-termini are separated by 124Å which is too far to initiate trans-phosphorylation of the receptors.[39] It is also speculated that the rapid kinetics of receptor:ligand interaction may also prevent receptor-mediated cellular activation.[39]

For FcγRIIa, a receptor:ligand complex structure has not been defined. However, the crystal structure of the unligated FcγRIIa revealed the receptor can form a crystallographic dimer.[42] Mutagenesis of the dimer interface highlighted the functional importance of Ser129 which is predicted to make a main chain contact between adjacent receptor molecules. Dimerization of FcγRIIa is speculated to bring the cytoplasmic domains sufficiently close (Fig. 2A) to initiate trans-phosphorylation of the cytoplasmic ITAMs, thereby mediating downstream signaling processes common to the MIRR family (see above). However, dimerization of FcγRIIa alone was unable to induce signal transduction,[57] supporting the well-established dogma that aggregation of FcγRIIa is required to induce cellular activation.[58] This work demonstrates that receptor spatial organization is a crucial aspect of receptor activation. Prior to ligand-induced aggregation, the organized association of FcγRIIa dimers is an essential component of the receptor signaling cascade.

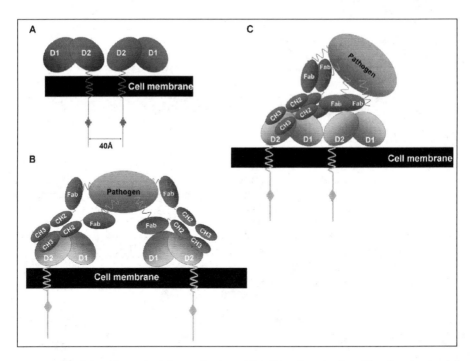

Figure 2. Models of FcγR spatial organization. A) FcγRIIa dimerization. The dimerization of FcγRIIa[42] brings the cytoplasmic domains of this receptor within approximately 40 Å of each other which is sufficiently close to initiate transphosphorylation of the cytoplasmic ITAM and is speculated to contribute to receptor activation. FcγRIII-ligand induced models of activation.[43] B) The model of simple avidity assumes binding of multiple antibodies clusters receptors in close proximity and induces activation. C) An alternative model of activation proposes that binding of oligomeric antigens leads to the formation of an ordered receptor-ligand aggregation which contributes to the receptor activation complex. Panels B and C are adapted from Radaev et al.[43]

Two models of receptor activation have been outlined based on the FcγRIII:ligand structural data (Figs. 2B and 2C).[43] The avidity model (Fig. 2B) suggests that the multiple binding of oligomeric antigen by antibodies binds and clusters the FcRs to induce activation. The second model (Fig. 2C) suggests that the binding of oligomeric antigens leads to the formation of an ordered receptor-ligand aggregation as a distinct event in the activation process. Considering structural data and examples of ordered aggregation of immunological proteins, it has been suggested that the ordered receptor cluster model may also be a plausible mechanism for activation.[43] Despite the recent studies[43,57] showing that receptor organization is a crucial aspect of receptor function and ultimately cellular activation, it remains to be elucidated in detail how FcRs become organized within the cell membrane following ligand binding.

Spatial Organization of FcRs

Studies of FcR spatial arrangements have centered on the translocation of the FcRs to detergent-resistant microdomains and the analysis of ligand-binding subunit association with signaling subunits whereas the physical protein:protein interactions between the FcRs have been measured in a limited number of studies. Ligand-dependent aggregation of FcRs initiates the receptor translocation to detergent-resistant microdomains (DRMs, or lipid rafts). The DRMs have been shown to bring key enzymes and signaling proteins in close proximity to the FcRs including FcγRIIa, FcεRI and FcαRI to initiate downstream processes.[59-61] For FcγRIIa, palmitoylation of the Cys208 residue within the juxtamembrane region is critical to the receptor localization within the DRM.[59] Notably, the equivalent mutation in the FcR γ-chain had little effect on FcαRI function, suggesting that different factors control DRM localization of the FcR γ-chain.[62] Only recently, the importance of FcγRII association with DRMs was reinforced by the data showing that the systemic lupus erythematosus (SLE)-associated FcγRIIb allele carrying the Ile232Thr polymorphism is not localized in lipid rafts.[63,64] The exclusion of FcγRIIb from DRMs results in failure to inhibit the activatory signals generated in rafts by FcγRI cross-linking[63] as well as activatory signals from the BCR,[64,65] highlighting that the localization of FcγRs to DRMs is crucial for receptor organization and regulatory functions. Association of FcRs (in particular, FcεRI) with DRMs is extensively reviewed elsewhere.[66]

Within the cell membrane, FcR organization is thought to be mediated by transmembrane domain interactions (reviewed by refs. 34, 67-71). For example, in FcεRI, charged residues within the transmembrane domain act as a focal point of association between the ligand-binding α–chain and the FcεRI β or FcR γ signal transduction subunits.[72] The FcγRIII and FcγRI transmembrane domains also contain the charged residues[67,73] necessary for interaction with the FcR γ-chain. Notably, despite reported association of FcγRII with the FcR γ–chain, this association is not required for FcγRII expression or signaling function.[9,74] Considering the absence of charged residues within the FcγRII transmembrane domain and the data on FcγRII dimerization, it is difficult to rationalize this association. It also remains to be determined if FcγRII associates with any other signaling subunits, such as the FcR β-chain or an FcR β-chain-like molecule. As this particular signaling subunit is known to amplify signals,[31] one may speculate on the possible recruitment of such a molecule to form a complex with FcγRII.

FcR spatial organization within the cell membrane is not well-studied and only limited studies have examined the physical interaction between FcRs themselves or with associated proteins. Using protein-fragment complementation assays to monitor receptor-dependent reassembly of complementary, nonfunctional fragments of the enzyme, dihydrofolate reductase (DHFR),[75] it was determined that FcγRIIa is organized in such a manner that the receptors are in close association with each other.[57] Using fluorescence recovery after photobleaching (FRAP) technique, FcγRIIa has also been observed to have lateral mobility within the cell membrane whereby the diffusion of receptors become diminished with the progressive truncation of the cytoplasmic domain,[76] suggesting that the cytoplasmic tail can be critically important for receptor spatial organization. Studies of FcR:ligand complexes have shown a 1:1 stoichiometry for FcγRIII[55] and FcεRI[56]. However, the stoichiometry between FcR ligand-binding and signaling subunits has not been analyzed in these studies. Fluorescence correlation spectroscopy data describing the real-time interaction between FcεRI and the src family tyrosine kinase lyn within intact cells, demonstrated that there are possibly multiple associations between FcεRI and lyn.[77] Recent stoichiometric measurements of the activating NKG2D receptor associated with the DAP10 signaling homodimer, found that the receptor is dimerized at the cell membrane, forming a hexameric protein complex.[78] Physiologically, the association of the NKG2D receptor and DAP10 signaling molecules may facilitate the phosphorylation of up to four ITAM motifs, potentially improving the sensitivity of NKG2D signaling. The stoichiometry of other key immunological receptors such as the BCR[79] and TCR[80] complexes has also been undertaken, thus proving that spatial arrangements are not always as predicted.

Physiological Function of Fc Receptors

The tissue-specific expression of FcRs and the pairing of activation and inhibitory receptors[3,25,27,28] are a crucial part of processes such as phagocytosis, endocytosis, release of inflammatory mediators, as well as the inhibition of cellular activation (Table 1).

Phagocytosis and endocytosis are both initiated by the cross-linking and aggregation of receptors on the cell surface of macrophages and neutrophils. Both are crucial processes responsible for the clearance of antibody-coated particles[81] or small soluble complexes[82] from the circulation, subsequent antigen presentation and for the killing of pathogens. While phagocytosis and endocytosis are fundamentally similar, they are functionally distinct. This difference is based not only on the size of particle ingested but also on the cytoskeleton requirements. Phagocytosis requires the assembly of F-actin structures whereas endocytosis is dependent on clathrin and the ubiquitylation of cytoplasmic lysine residues.[83] Phagocytosis has been extensively studied using the aggregation of FcRs to map the processes that underpin this function. All activatory FcγRs[84], the glycosylphosphatidyl inositol (GPI)-anchored FcγRIIIb,[85] FcαRI,[5,6] and FcεRI[4] have phagocytic function. Differences in the phagocytic function of the FcγRs appear to be affected by the recruitment of tyrosine kinases[67,86-88] via direct association with the receptor ITAM or through FcR γ-chain association. Interestingly, both phagocytosis and endocytosis are also modulated by the sequences within the cytoplasmic tails of FcγRs.[89] The FcγRI cytoplasmic tail possesses no intrinsic signaling function and the modulation of phagocytosis and endocytosis is known to occur via intracellular sequences[89] or via the association with the intracellular protein periplakin.[90] It remains to be elucidated if other FcγRs associate with intracellular nonsignaling proteins and what (if any) effect this has on the spatial organization of the receptor within the cell.

While processes such as FcR-mediated phagocytosis and endocytosis are recognized as essential effector functions that culminate in antigen presentation, FcR activation is also a potent inducer of pro-inflammatory cytokines, in particular IL-1[91] and TNF-α.[92] The deleterious effects of these cytokines in vivo are noted in the extensive literature outlining the efficacy of anti-cytokine therapies for the treatment of diseases such as rheumatoid arthritis (RA) (reviewed by ref. 93). For RA patients, the treatment can involve the combination of steroids and antagonists of TNF-α function[94] and to date over one million patients receive anti-TNF-α treatments.[93] Interestingly, FcγRIIa transgenic mice[29] develop spontaneous autoimmune disease and are hypersensitive to antibody-induced inflammatory reactions, demonstrating that this receptor plays a critical role in the modulation of inflammatory reactions in vivo.[30] As the structure of FcγRIIa has a spatial arrangement that potentially sets it apart from other FcRs, the development of small chemical entities or monoclonal antibodies that block receptor function could make this receptor a valid target for immunotherapy. Therapies designed to block immune complex binding to FcγRIIa are strengthened by findings that immune complex-mediated inflammatory reactions can be blocked using recombinant soluble FcγRIIa.[95] For FcεRI, it has been well established that this receptor plays a crucial role in IgE-mediated allergic responses.[4] The activation of FcεRI on the cell surface of mast cells and basophils triggers cellular degranulation and release of pro-inflammatory histamine from the granules. The potency of FcεRI in immediate allergic responses is demonstrated in mice null for the FcεRI α-or FcR γ-chains showing no immediate hypersensitivity reactions (reviewed by ref. 1). IgA nephropathy (IgAN), is the most common form of glomerulonephritis which affects a vast number of people globally and can lead to renal failure. It has been suggested that impaired FcαRI function or changes in the IgA antibody contribute to the pathology of this disease.[5,6]

Concluding Comments

Receptors for the Fc portion of immunoglobulins are a diverse family of cell surface glycoproteins capable of connecting humoral immune responses to cellular effector mechanisms. Elucidation of receptor structures together with physiological studies using FcR gene knock-out mice or FcR transgenics have provided useful insights to the nature of receptor spatial organization on the cell surface as well as the physiological involvement of these receptors in disease. The balance between activatory and inhibitory receptor functions is crucial to immune homeostasis as disruptions to this balance are often evident by the ensuing potent pathological responses. The recent analyses of receptor organization within the cell add significantly to the exciting field of Fc receptor function, highlighting that the spatial organization of receptors is a critically important aspect of receptor function and cellular activation. However, much work still needs to be done to elucidate the precise organization of the Fc receptors within the cell and what changes of the receptor arrangement occur upon activation.

Reference

1. Hulett MD, Hogarth PM. Molecular basis of Fc receptor function. Adv Immunol 1994; 57:1-127.
2. Daeron M. Fc receptor biology. Annu Rev Immunol 1997; 15:203-234.
3. Ravetch JV, Bolland S. IgG Fc receptors. Annu Rev Immunol 2001; 19:275-290.
4. Kinet JP. The high-affinity IgE receptor (Fc epsilon RI): From physiology to pathology. Annu Rev Immunol 1999; 17:931-972.
5. Wines BD, Hogarth PM. IgA receptors in health and disease. Tissue Antigens 2006; 68:103-114.
6. Gomes MM, Herr AB. IgA and IgA-specific receptors in human disease: Structural and functional insights into pathogenesis and therapeutic potential. Springer Semin Immunopathol 2006.
7. Reth M. Antigen receptor tail clue. Nature 1989; 338:383-384.
8. Daeron M, Latour S, Malbec O et al. The same tyrosine-based inhibition motif, in the intracytoplasmic domain of Fc gamma RIIB, regulates negatively BCR-, TCR- and FcR-dependent cell activation. Immunity 1995; 3:635-646.
9. Van den Herik-Oudijk IE, Ter Bekke MW, Tempelman MJ et al. Functional differences between two Fc receptor ITAM signaling motifs. Blood 1995; 86:3302-3307.
10. Metes D, Manciulea M, Pretrusca D et al. Ligand binding specificities and signal transduction pathways of Fc gamma receptor IIc isoforms: the CD32 isoforms expressed by human NK cells. Eur J Immunol 1999; 29:2842-2852.
11. Lanier LL, Yu G, Phillips JH. Co-association of CD3 zeta with a receptor (CD16) for IgG Fc on human natural killer cells. Nature 1989; 342:803-805.
12. Anderson P, Caligiuri M, O'Brien C et al. Fc gamma receptor type III (CD16) is included in the zeta NK receptor complex expressed by human natural killer cells. Proc Natl Acad Sci USA 1990; 87:2274-2278.
13. Letourneur F, Klausner RD. T-cell and basophil activation through the cytoplasmic tail of T-cell-receptor zeta family proteins. Proc Natl Acad Sci USA 1991; 88:8905-8909.
14. Kurosaki T, Gander I, Wirthmueller U et al. The beta subunit of the Fc epsilon RI is associated with the Fc gamma RIII on mast cells. J Exp Med 1992; 175:447-451.
15. Isakov N. ITIMs and ITAMs. The Yin and Yang of antigen and Fc receptor-linked signaling machinery. Immunol Res 1997; 16:85-100.
16. Muta T, Kurosaki T, Misulovin Z et al. A 13-amino-acid motif in the cytoplasmic domain of Fc gamma RIIB modulates B-cell receptor signalling. Nature 1994; 369:340.
17. Brooks DG, Qiu WQ, Luster AD et al. Structure and expression of human IgG FcRII(CD32). Functional heterogeneity is encoded by the alternatively spliced products of multiple genes. J Exp Med 1989; 170:1369-1385.
18. Van den Herik-Oudijk IE, Capel PJ, van der Bruggen T et al. Identification of signaling motifs within human Fc gamma RIIa and Fc gamma RIIb isoforms. Blood 1995; 85:2202-2211.
19. Pan LY, Mendel DB, Zurlo J et al. Regulation of the steady state level of Fc gamma RI mRNA by IFN-gamma and dexamethasone in human monocytes, neutrophils and U-937 cells. J Immunol 1990; 145:267-275.
20. te Velde AA, Huijbens RJ, de Vries JE et al. IL-4 decreases Fc gamma R membrane expression and Fc gamma R-mediated cytotoxic activity of human monocytes. J Immunol 1990; 144:3046-3051.

21. Welch GR, Wong HL, Wahl SM. Selective induction of Fc gamma RIII on human monocytes by transforming growth factor-beta. J Immunol 1990; 144:3444-3448.
22. Pricop L, Redecha P, Teillaud JL et al. Differential modulation of stimulatory and inhibitory Fc gamma receptors on human monocytes by Th1 and Th2 cytokines. J Immunol 2001; 166:531-537.
23. Tridandapani S, Wardrop R, Baran CP et al. TGF-beta 1 suppresses [correction of supresses] myeloid Fc gamma receptor function by regulating the expression and function of the common gamma-subunit. J Immunol 2003; 170:4572-4577.
24. Gomez G, Ramirez CD, Rivera J et al. TGF-beta 1 inhibits mast cell Fc epsilon RI expression. J Immunol 2005; 174:5987-5993.
25. Salmon JE, Pricop L. Human receptors for immunoglobulin G: Key elements in the pathogenesis of rheumatic disease. Arthritis Rheum 2001; 44:739-750.
26. Schmidt RE, Gessner JE. Fc receptors and their interaction with complement in autoimmunity. Immunol Lett 2005; 100:56-67.
27. Takai T. Roles of Fc receptors in autoimmunity. Nat Rev Immunol 2002; 2:580-592.
28. Takai T. Fc receptors and their role in immune regulation and autoimmunity. J Clin Immunol 2005; 25:1-18.
29. McKenzie SE, Taylor SM, Malladi P et al. The role of the human Fc receptor Fc gamma RIIA in the immune clearance of platelets: A transgenic mouse model. J Immunol 1999; 162:4311-4318.
30. Tan Sardjono C, Mottram PL, van de Velde NC et al. Development of spontaneous multisystem autoimmune disease and hypersensitivity to antibody-induced inflammation in Fcgamma receptor IIa-transgenic mice. Arthritis Rheum 2005; 52:3220-3229.
31. On M, Billingsley JM, Jouvin MH et al. Molecular dissection of the FcRbeta signaling amplifier. J Biol Chem 2004; 279:45782-45790.
32. Pfefferkorn LC, Yeaman GR. Association of IgA-Fc receptors (Fc alpha R) with Fc epsilon RI gamma 2 subunits in U937 cells. Aggregation induces the tyrosine phosphorylation of gamma 2. J Immunol 1994; 153:3228-3236.
33. Morton HC, van den Herik-Oudijk IE, Vossebeld P et al. Functional association between the human myeloid immunoglobulin A Fc receptor (CD89) and FcR gamma chain. Molecular basis for CD89/FcR gamma chain association. J Biol Chem 1995; 270:29781-29787.
34. Wines BD, Trist HM, Ramsland PA et al. A common site of the Fc receptor gamma subunit interacts with the unrelated immunoreceptors FcalphaRI and FcepsilonRI. J Biol Chem 2006; 281:17108-17113.
35. Maruoka T, Nagata T, Kasahara M. Identification of the rat IgA Fc receptor encoded in the leukocyte receptor complex. Immunogenetics 2004; 55:712-716.
36. Reljic R. In search of the elusive mouse macrophage Fc-alpha receptor. Immunol Lett 2006; 107: 80-81.
37. Wines BD, Hulett MD, Jamieson GP et al. Identification of residues in the first domain of human Fc alpha receptor essential for interaction with IgA. J Immunol 1999; 162:2146-2153.
38. Wines BD, Sardjono CT, Trist HH et al. The interaction of Fc alpha RI with IgA and its implications for ligand binding by immunoreceptors of the leukocyte receptor cluster. J Immunol 2001; 166:1781-1789.
39. Herr AB, Ballister ER, Bjorkman PJ. Insights into IgA-mediated immune responses from the crystal structures of human FcalphaRI and its complex with IgA1-Fc. Nature 2003; 423:614-620.
40. Pasquier B, Launay P, Kanamaru Y et al. Identification of FcalphaRI as an inhibitory receptor that controls inflammation: Dual role of FcRgamma ITAM. Immunity 2005; 22:31-42.
41. Hamerman JA, Lanier LL. Inhibition of immune responses by ITAM-bearing receptors. Sci STKE 2006; 2006:re1.
42. Maxwell KF, Powell MS, Hulett MD et al. Crystal structure of the human leukocyte Fc receptor, Fc gammaRIIa. Nat Struct Biol 1999; 6:437-442.
43. Radaev S, Motyka S, Fridman WH et al. The structure of a human type III Fcgamma receptor in complex with Fc. J Biol Chem 2001; 276:16469-16477.
44. Sondermann P, Huber R, Oosthuizen V et al. The 3.2-A crystal structure of the human IgG1 Fc fragment-Fc gammaRIII complex. Nature 2000; 406:267-273.
45. Woof JM, Burton DR. Human antibody-Fc receptor interactions illuminated by crystal structures. Nat Rev Immunol 2004; 4:89-99.
46. Garman SC, Wurzburg BA, Tarchevskaya SS et al. Structure of the Fc fragment of human IgE bound to its high-affinity receptor Fc epsilonRI alpha. Nature 2000; 406:259-266.
47. Hulett MD, Hogarth PM. The second and third extracellular domains of FcgammaRI (CD64) confer the unique high affinity binding of IgG2a. Mol Immunol 1998; 35:989-996.

48. Hulett MD, McKenzie IF, Hogarth PM. Chimeric Fc receptors identify immunoglobulin-binding regions in human Fc gamma RII and Fc epsilon RI. Eur J Immunol 1993; 23:640-645.
49. Hulett MD, Witort E, Brinkworth RI et al. Identification of the IgG binding site of the human low affinity receptor for IgG Fc gamma RII. Enhancement and ablation of binding by site-directed mutagenesis. J Biol Chem 1994; 269:15287-15293.
50. Hulett MD, Witort E, Brinkworth RI et al. Multiple regions of human Fc gamma RII (CD32) contribute to the binding of IgG. J Biol Chem 1995; 270:21188-21194.
51. Kato K, Sautes-Fridman C, Yamada W et al. Structural basis of the interaction between IgG and Fcgamma receptors. J Mol Biol 2000; 295:213-224.
52. Duncan AR, Woof JM, Partridge LJ et al. Localization of the binding site for the human high-affinity Fc receptor on IgG. Nature 1988; 332:563-564.
53. Lund J, Winter G, Jones PT et al. Human Fc gamma RI and Fc gamma RII interact with distinct but overlapping sites on human IgG. J Immunol 1991; 147:2657-2662.
54. Chappel MS, Isenman DE, Everett M et al. Identification of the Fc gamma receptor class I binding site in human IgG through the use of recombinant IgG1/IgG2 hybrid and point-mutated antibodies. Proc Natl Acad Sci USA 1991; 88:9036-9040.
55. Ghirlando R, Keown MB, Mackay GA et al. Stoichiometry and thermodynamics of the interaction between the Fc fragment of human IgG1 and its low-affinity receptor Fc gamma RIII. Biochemistry 1995; 34:13320-13327.
56. Keown MB, Ghirlando R, Mackay GA et al. Basis of the 1:1 stoichiometry of the high affinity receptor Fc epsilon RI-IgE complex. Eur Biophys J 1997; 25:471-476.
57. Powell MS, Barnes NC, Bradford TM et al. Alteration of the Fc gamma RIIa dimer interface affects receptor signaling but not ligand binding. J Immunol 2006; 176:7489-7494.
58. Pribluda VS, Pribluda C, Metzger H. Transphosphorylation as the mechanism by which the high-affinity receptor for IgE is phosphorylated upon aggregation. Proc Natl Acad Sci USA 1994; 91:11246-11250.
59. Barnes NC, Powell MS, Trist HM et al. Raft localisation of FcgammaRIIa and efficient signaling are dependent on palmitoylation of cysteine 208. Immunol Lett 2006; 104:118-123.
60. Katsumata O, Hara-Yokoyama M, Sautes-Fridman C et al. Association of FcgammaRII with low-density detergent-resistant membranes is important for cross-linking-dependent initiation of the tyrosine phosphorylation pathway and superoxide generation. J Immunol 2001; 167:5814-5823.
61. Kwiatkowska K, Frey J, Sobota A. Phosphorylation of FcgammaRIIA is required for the receptor-induced actin rearrangement and capping: the role of membrane rafts. J Cell Sci 2003; 116:537-550.
62. Wines BD, Trist HM, Monteiro RC et al. Fc receptor gamma chain residues at the interface of the cytoplasmic and transmembrane domains affect association with FcalphaRI, surface expression and function. J Biol Chem 2004; 279:26339-26345.
63. Floto RA, Clatworthy MR, Heilbronn KR et al. Loss of function of a lupus-associated FcgammaRIIb polymorphism through exclusion from lipid rafts. Nat Med 2005; 11:1056-1058.
64. Kono H, Kyogoku C, Suzuki T et al. FcgammaRIIB Ile232Thr transmembrane polymorphism associated with human systemic lupus erythematosus decreases affinity to lipid rafts and attenuates inhibitory effects on B-cell receptor signaling. Hum Mol Genet 2005; 14:2881-2892.
65. Li X, Wu J, Carter RH et al. A novel polymorphism in the Fcgamma receptor IIB (CD32B) transmembrane region alters receptor signaling. Arthritis Rheum 2003; 48:3242-3252.
66. Holowka D, Gosse JA, Hammond AT et al. Lipid segregation and IgE receptor signaling: A decade of progress. Biochim Biophys Acta 2005; 1746:252-259.
67. Kim MK, Huang ZY, Hwang PH et al. Fcgamma receptor transmembrane domains: role in cell surface expression, gamma chain interaction and phagocytosis. Blood 2003; 101:4479-4484.
68. Lanier LL, Yu G, Phillips JH. Analysis of Fc gamma RIII (CD16) membrane expression and association with CD3 zeta and Fc epsilon RI-gamma by site-directed mutation. J Immunol 1991; 146:1571-1576.
69. Miller L, Alber G, Varin-Blank N et al. Transmembrane signaling in P815 mastocytoma cells by transfected IgE receptors. J Biol Chem 1990; 265:12444-12453.
70. Dombrowicz D, Flamand V, Miyajima I et al. Absence of Fc epsilonRI alpha chain results in upregulation of Fc gammaRIII-dependent mast cell degranulation and anaphylaxis. Evidence of competition between Fc epsilonRI and Fc gammaRIII for limiting amounts of FcR beta and gamma chains. J Clin Invest 1997; 99:915-925.
71. Dombrowicz D, Lin S, Flamand V et al. Allergy-associated FcRbeta is a molecular amplifier of IgE- and IgG-mediated in vivo responses. Immunity 1998; 8:517-529.
72. Cosson P, Lankford SP, Bonifacino JS et al. Membrane protein association by potential intramembrane charge pairs. Nature 1991; 351:414-416.

73. Varin-Blank N, Metzger H. Surface expression of mutated subunits of the high affinity mast cell receptor for IgE. J Biol Chem 1990; 265:15685-15694.

74. Masuda M, Roos D. Association of all three types of Fc gamma R (CD64, CD32 and CD16) with a gamma-chain homodimer in cultured human monocytes. J Immunol 1993; 151:7188-7195.

75. Remy I, Michnick SW. Clonal selection and in vivo quantitation of protein interactions with protein-fragment complementation assays. Proc Natl Acad Sci USA 1999; 96:5394-5399.

76. Zhang F, Yang B, Odin JA et al. Lateral mobility of Fc gamma RIIa is reduced by protein kinase C activation. FEBS Lett 1995; 376:77-80.

77. Larson DR, Gosse JA, Holowka DA et al. Temporally resolved interactions between antigen-stimulated IgE receptors and Lyn kinase on living cells. J Cell Biol 2005; 171:527-536.

78. Garrity D, Call ME, Feng J et al. The activating NKG2D receptor assembles in the membrane with two signaling dimers into a hexameric structure. Proc Natl Acad Sci USA 2005; 102:7641-7646.

79. Schamel WW, Reth M. Monomeric and oligomeric complexes of the B-cell antigen receptor. Immunity 2000; 13:5-14.

80. Call ME, Pyrdol J, Wiedmann M et al. The organizing principle in the formation of the T-cell receptor CD3 complex. Cell 2002; 111:967-979.

81. Odin JA, Edberg JC, Painter CJ et al. Regulation of phagocytosis and [Ca2+]i flux by distinct regions of an Fc receptor. Science 1991; 254:1785-1788.

82. Indik Z, Kelly C, Chien P et al. Human Fc gamma RII, in the absence of other Fc gamma receptors, mediates a phagocytic signal. J Clin Invest 1991; 88:1766-1771.

83. Booth JW, Kim MK, Jankowski A et al. Contrasting requirements for ubiquitylation during Fc receptor-mediated endocytosis and phagocytosis. EMBO J 2002; 21:251-258.

84. Anderson CL, Shen L, Eicher DM et al. Phagocytosis mediated by three distinct Fc gamma receptor classes on human leukocytes. J Exp Med 1990; 171:1333-1345.

85. Bredius RG, Fijen CA, De Haas M et al. Role of neutrophil Fc gamma RIIa (CD32) and Fc gamma RIIIb (CD16) polymorphic forms in phagocytosis of human IgG1- and IgG3-opsonized bacteria and erythrocytes. Immunology 1994; 83:624-630.

86. Cox D, Greenberg S. Phagocytic signaling strategies: Fc(gamma)receptor-mediated phagocytosis as a model system. Semin Immunol 2001; 13:339-345.

87. Kim MK, Pan XQ, Huang ZY et al. Fc gamma receptors differ in their structural requirements for interaction with the tyrosine kinase Syk in the initial steps of signaling for phagocytosis. Clin Immunol 2001; 98:125-132.

88. Indik ZK, Park JG, Hunter S et al. The molecular dissection of Fc gamma receptor mediated phagocytosis. Blood 1995; 86:4389-4399.

89. Edberg JC, Yee AM, Rakshit DS et al. The cytoplasmic domain of human FcgammaRIa alters the functional properties of the FcgammaRI.gamma-chain receptor complex. J Biol Chem 1999; 274:30328-30333.

90. Beekman JM, Bakema JE, van de Winkel JG et al. Direct interaction between FcgammaRI (CD64) and periplakin controls receptor endocytosis and ligand binding capacity. Proc Natl Acad Sci USA 2004; 101:10392-10397.

91. Remvig L, Thomsen BS, Baek L et al. Interleukin 1, but not interleukin 1 inhibitor, is released from human monocytes by immune complexes. Scand J Immunol 1990; 32:255-261.

92. Debets JM, Van de Winkel JG, Ceuppens JL et al. Cross-linking of both Fc gamma RI and Fc gamma RII induces secretion of tumor necrosis factor by human monocytes, requiring high affinity Fc-Fc gamma R interactions. Functional activation of Fc gamma RII by treatment with proteases or neuraminidase. J Immunol 1990; 144:1304-1310.

93. Feldmann M, Steinman L. Design of effective immunotherapy for human autoimmunity. Nature 2005; 435:612-619.

94. Feldmann M, Brennan FM, Foxwell BM et al. Anti-TNF therapy: Where have we got to in 2005? J Autoimmun 2005; 25 Suppl:26-28.

95. Ierino FL, Powell MS, McKenzie IF et al. Recombinant soluble human Fc gamma RII: Production, characterization and inhibition of the Arthus reaction. J Exp Med 1993; 178:1617-1628.

CHAPTER 4

Natural Killer Cell Receptors

Roberto Biassoni*

Abstract

Natural killer (NK) cells are an important arm of the innate immune response that are directly involved in the recognition and lysis of virus-infected and tumor cells. Such function is under the control of a complex array of germline-encoded receptors able to deliver either inhibitory or activating signals. The majority of inhibitory receptors expressed by NK cells are major histocompatibility complex (MHC) class I-specific and display clonal and stochastic distribution on the cell surface. Thus, a given NK cell expresses at least one self class I inhibitory receptor. Under normal conditions, the strength of inhibitory signals delivered by multiple interactions always overrides the activating signals, resulting in NK cell self-tolerance. Under certain pathological conditions, such as viral infections or tumor transformation, the delicate balance of inhibition versus activation is broken, resulting in downregulation or loss of MHC class I expression. In general, the degree of inhibition induced by class I-specific receptors is proportional to the amount of these molecules on the cell surface. Thus, in transformed cells, this inhibition can be overridden by the triggering signal cascades, leading to cell activation. The majority of triggering receptors expressed by NK cells belong to the multichain immune recognition receptor (MIRR) family and use separate signal-transducing polypeptides similar to those used by other immune receptors such as the T-cell antigen receptor, the B-cell antigen receptor and other receptors expressed by myeloid cells. Inhibitory receptors are not members of the MIRR family but they are relevant for a better understanding the exquisite equilibrium and regulatory crosstalk between positive and negative signals.

Introduction

Natural killer (NK) cells are effector lymphocytes of the innate immune system. They are involved in the continuous surveillances and early defenses against "nonself" allogeneic or autologous cells underwent cell transformation induced by pathogen infection or tumor proliferation.[1] To accomplish this task, NK cells use multiple opposingly acting, either activating or inhibitory, receptors. An impaired balance between positive or negative signaling may result in the direct cytotoxic attack towards transformed target cells. NK cells are known to display multiple, clonally distributed major histocompatibility complex (MHC) class I-specific receptors. These receptors sense any modification or transformation occurring in autologous cells and usually resulting in downregulation or loss of MHC class I expression (reviewed in refs. 2 and 3). If this happens, the strength of inhibitory signaling induced by the receptor engagements with MHC molecules is not sufficient to control the signaling cascades induced by activating receptors that are specific for stress-inducible and/or constitutive ligands expressed at the target cell surface (Fig. 1).

*Roberto Biassoni—Molecular Medicine, Istituto Giannina Gaslini, Largo G. Gaslini 5, 16147 Genova, Italy. Email: robertobiassoni@ospedale-gaslini.ge.it

Multichain Immune Recognition Receptor Signaling: From Spatiotemporal Organization to Human Disease, edited by Alexander B. Sigalov. ©2008 Landes Bioscience and Springer Science+Business Media.

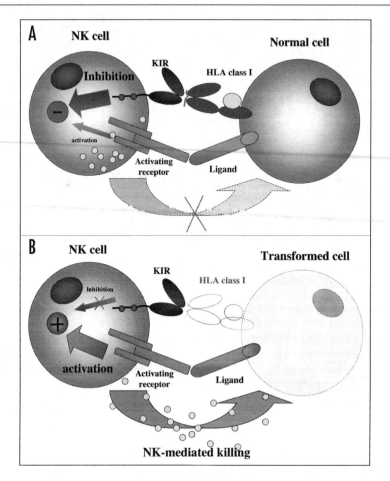

Figure 1. A) Recognition of self MHC class I molecules by specific inhibitory NK receptors protects normal cell from NK cell-mediated lysis. B) Cells underwent transformation induced by viral infection or cancer loose MHC class I molecules or decrease their surface expression. NK cells see these transformed cells as nonself and kill them.

Thus, NK cells spare normal autologous cells since they sense appropriate levels of surface-expressed MHC class I molecules that induce an inhibitory receptor-mediated negative signaling while NK cell triggering is mediated by nonMHC-restricted activating receptors (reviewed in ref. 4). In inhibitory receptors, extracellular MHC-recognition domains and intracellular signaling immunoreceptor tyrosine-based inhibitory motifs (ITIMs; V/I/L/SxYxxL/V) are located on the same protein chain. Thus, inhibitory receptors do not belong to the multichain immune recognition receptor (MIRR) family,[5] however, they are important for understanding NK-mediated functions. They are also important for understanding genetic and structural aspects of the proteins encoded by multigene families and underwent to heavy structural modification during evolution.

Human and primates express inhibitory receptors belonging to the killer Ig-like receptor (KIR) superfamily. In contrast, rodents express receptors from a completely different family known as Ly49 and belonging to the C-type lectin superfamily (reviewed in ref. 6). In addition, all the above-mentioned species express an additional receptor known as CD94/NKG2A that belongs to the type II transmembrane proteins and functions to broadly patrol for MHC class I surface expression (reviewed in ref. 7). The inhibitory signal is initiated by the receptor/MHC interaction that induces the phosphorylation of the ITIM tyrosine and results in the recruitment of specific tyrosine-phosphatases that may dampen triggering receptor-induced phosphorylation (reviewed in ref. 8). Activating NK cell receptors belong to the MIRR family of surface receptors. NK-mediated lysis is controlled by the natural cytotoxicity receptors (NCRs) like NKp46, NKp30, NKp44 that belong to the Ig-like superfamily and by the lectin-like NKG2D receptor.

In addition to the induction of cytolytic machinery, NK cells are involved in the tuning of the adaptive immune system through different mechanisms. They release several cytokines and chemokines that are important in the induction of antigen-specific immune response and in the maturation of professional antigen-presenting cells such as dendritic cells (DCs).[9] In addition, NK cells are responsible for an additional pathway of regulation of the adaptive immune system through the NK-mediated lysis of immature DCs (iDCs).[10] The above-mentioned functions indicate that NK cells are important not only in the regulation of innate immune responses but also in tuning the adaptive immune response by direct interactions with iDCs and by the secretion of different soluble factors induced upon activation (reviewed in ref. 11).

Inhibitory Receptors

In humans, inhibitory receptors are encoded by the multigene families located in the leukocyte receptor complex (LRC) on human chromosome 19q13.4 and in the NK gene complex (NKC) on human chromosome 12 (Table 1).[12] The main inhibitory receptors encoded in the LRC are KIRs (CD158) specific for groups of HLA class I alleles, the leukocyte Ig-like receptor 1 (LIR1) or Ig-like transcript 2 (ILT2) receptor member of the LIR (or LILR) receptor family,[13] that display a broad HLA class I specificity,[14] while CD94/NKG2A, the heterodimeric C-type lectin receptor, is encoded in the NKC.[7] The latter receptor is specific for nonclassical MHC class I (class Ib) molecules (HLA-E and Qa-1 in humans and mice, respectively)[7,15] that are efficiently expressed on the cell surface upon presentation of leader sequence-derived peptides cleaved of classical MHC class I molecules. Thus, the CD94/NKG2A receptor may not only distinguish transformed cells with overall decreased expression of MHC class I alleles but also exhibit a certain degree of peptide selectively.[16-18]

KIRs are the major HLA class I inhibitory receptors (termed KIR2DL and KIR3DL followed by an additional number), which are structurally characterized by two or three extracellular Ig-like C2-type domains, a transmembrane region containing nonpolar amino acid residues and a long cytoplasmic tail containing two ITIM sequences. KIR2DL1 is specific for HLA-C alleles (-Cw2, -Cw4 -Cw5 and -Cw6) characterized by Lys80 while KIR2DL2,3 recognize alleles represented by -Cw1 -Cw3 -Cw7 and -Cw8 that have Asn80.[19-21] KIR3DL1 and KIR3DL2 are specific for HLA-Bw4 and HLA-A3/A11 alleles, respectively (reviewed in ref. 22). In addition, KIRs display peptide specificity.[23] Upon ligand recognition, the ITIM sequences are tyrosine-phosphorylated by src-kinases and associate with intracellular protein tyrosine phosphatases 1 or 2 (SHP1 or SHP2, respectively) containing src homology 2 (SH2) domain.[8] Thus, ligand-induced co-aggregation of inhibitory receptors with activating receptors results in inhibition of the NK function and the degree of this inhibition depends upon the relative strength of inhibitory signals (Fig. 1). Vav family proteins, known to regulate the organization of the actin cytoskeleton and promote the activation of GTPases, are suggested to play a role in the downstream triggering mechanisms dampened by inhibitory receptors.[24,25] In humans, KIR/HLA class I interaction has been shown to promote binding of SHP-1 phosphatase to Vav-1.[26,27] SHP-1-mediated dephosphorylation of Vav-1 blocks actin polymerization induced by Vav1-Rac1 signal which is known to regulate adhesion, granule exocytosis and cellular cytotoxicity.[28-30] This Chapter mainly focuses on the activating NK receptors controlling cell function in humans as these receptors are potential targets for new therapeutic strategies of intervention. It should be noted that function of MHC class I-specific inhibitory NK receptors is also fundamental in the NK cell biology. In rodents, the surveillance for class I allele expression is under control of structures like the C-type lectin Ly49 receptors or other lectin-like

Table 1. Inhibitory HLA class I-specific human NK cell receptors encoded by two distinct multigene families mapping to chromosomes 19 and 12, respectively. The unique exception is KIR2DL4 characterized by the presence of the Arg residue in the transmembrane region, an apparently nonfunctional immunoreceptor tyrosine-based inhibitory motif (ITIM) in the cytoplasmic tail and by the association with the signaling ζ polypeptide.[125] On the contrary to the other KIR2DLs that have a D1-D2 configuration of the immunoglobulin domains of the C2-type KIR2DL4 and KIR2DL5 have a D0-D2 organization, where D0 shares a higher number of identities with the amino terminal Ig-C domain of KIR3DLs receptors.

Receptor	Alternative Name(s)	Associated Signaling Chain(s)	ITIMs	Cellular Ligand	Viral Ligand	Chromosome
KIR2DL1	p58.1, CD158A, 47.11, CL42, NKAT1, KIR-K3, KIR-K9, KIR221, KIR-K64, KIR-K65	No	2	HLA-Cw2, 4, 5, 6		19q13.42-LRC
KIR2DL2	p58.2, CD158B1, CL43, NKAT6, CD158k,	No	2	HLA-Cw1, 3, 7, 8, 15		19q13.42-LRC
KIR2DL3	p58.2, CL6, CD158b, GL183, NKAT2, NKAT2A, NKAT2B, CD158B2, KIR-K15, KIR-023GB	No	2	HLA-Cw1, 3, 7, 8, 15		19q13.42-LRC
KIR2DL4[125]	15.212, CD158D, 103AS, KIR103, KIR103AS	ζ	1	HLA-G		19q13.42-LRC
KIR2DL5[126]	KIR2DL5A, KIR2DL5B	No	1+one TxYxxL motif	?		19q13.42-LRC
KIR3DL1	p70, CD158E1/2, CL-2, NKB1, AMB11, CL-11, NKAT3, NKB1B, KIR-G1, NKAT10	No	2	HLA-Bw4		19q13.42-LRC
KIR3DL2	p140, CD158K, CL5, NKAT4, NKAT4A, NKAT4B,	No	2	HLA-A3, A11		19q13.42-LRC
KIR3DL3	KIR44, KIRC1, CD158Z, KIR3DL7	No	1	?		19q13.42-LRC
CD94/NKG2A	KLRC1v3 CD159A	No	2	HLA-E		12p13-NKC
LILRB1	LIR-1, CD85, ILT2, LIR1, MIR7, CD85J, MIR-7	No	4	various HLA class I	HCMV-UL18[13]	19q13.42-LRC

Table 2. *Mouse NK receptors involved in the MHC class I recognition. In contrast to human NK receptors, they all belong to the C-type lectin superfamily and map to the NK gene complex on the chromosome 6. Only partial sequences are available for Ly49K and Ly49N, however, translated nucleotide sequences of the transcript fragments suggest possible activating function*

Receptor	Function	Associated Signaling Chain(s)	Cellular Ligand	Viral Ligand
Ly49A	Inhibitory	No	D^b, D^d, D^k	
Ly49B	Inhibitory	No	?	
Ly49C	Inhibitory	No	K^b, K^d, K^k, D^d, D^b	
Ly49D	Activating	DAP-12	D^d	
Ly49E	Inhibitory	No	?	
Ly49F	Inhibitory	No	D^d	
Ly49G	Inhibitory	No	D^d, L^d	
Ly49H	Activating	DAP-12	D^b	MCMV-m157
Ly49I	Inhibitory	No	K^d	MCMV-m157
Ly49J	Inhibitory	No	K^b	
Ly49K	Activating	DAP-12	?	
Ly49L	Activating	DAP-12	K^k	
Ly49M	Activating	DAP-12	?	
Ly49N	Activating	DAP-12	?	
Ly49O	Inhibitory	No	D^b, D^d, D^k, L^d	
Ly49P	Activating	DAP-12	D^d	
Ly49Q	Inhibitory	No	?	
Ly49R	Activating	DAP-12	D^d, D^k, L^d	
Ly49S	Inhibitory	No	?	
Ly49T	Inhibitory	No	?	
Ly49U	Activating	DAP-12	?	
Ly49V	Inhibitory	No	D^b, D^d, K^k	
Ly49W	Activating	DAP-12	D^d, K^k	
CD94/NKG2A	Inhibitory	No	Qa-1(b)	
CD94/NKG2C	Activating	DAP-12	Qa-1(b)[128]	
CD94/NKG2E	Activating	DAP-12	Qa-1(b)[128]	

receptors, such as CD94/NKG2A in humans.[6,15,30] These receptors are structurally different from KIRs and encoded by a different multigene family. Table 2 summarizes these receptors, their functions and ligands in mice. Interestingly, all inhibitory receptors share the same inhibitory signaling cascades induced by ITIM phosphorylation and SHP phosphatases recruitment.[8]

Activating Receptors

Natural Cytotoxicity Receptors

Activating receptors are organized as multichain complexes of ligand-binding and signal-transducing subunits. Short cytoplasmic tail of ligand-binding subunits is irrelevant for the signal-transducing pathway(s) while a transmembrane region contains a charged amino acid that facilitates the association with signal-transducing polypeptides (Table 3). Thus, in contrast to single-chain inhibitory receptors, the triggering function in NK cells is provided by receptor complexes where the signal transducing elements are present on accessory polypeptides, such as

Table 3. Human activating NK cell receptors

Receptor	Alternative Name(s)	Associated Signaling Chain(s)	Cellular Ligand	Viral Ligand	Chromosome
NKp46	NCR1, CD335	ζ, FcεRIγ	HSPGs	IV-HA, SV-HN	19q13.42-LRC
NKp30	NCR3, CD337, 1C7, LY117	ζ, FcεRIγ	HSPGs[56] (not confirmed by others)	HCMV-pp65	6p21.3-HLA class III
NKp44	NCR2, CD336, LY95	DAP-12		IV-HA, SV-HN	6p21.1-HLA class III
NKG2D	KLRK1, CD314, KLR,	DAP-10	MICA/B, ULBP1-3, ULBP4/RAET1E, RAET1G		12p13-NKC
KIR2DS1	p50.1, CD158H, EB6ActI, EB6ActII	DAP-12	HLA-Cw4[95]		19q13.42-LRC
KIR2DS2	p50.2, CD158J, cl-49, NKAT5, 183ACTI	DAP-12	HLA-Cw3[129] (cocrystallization data do not support direct binding)		19q13.42-LRC
KIR2DS3	NKAT7, 59C/K3	DAP-12			
KIR2DS4	p50.3, CD158I,NKAT8, PAX, KKA3, KIR1D, cl-39, KIR412	DAP-12	HLA-Cw4 (but not Cw6)[96]		19q13.42-LRC
KIR2DS5	NKAT9, CD158G	DAP-12			19q13.42-LRC
KIR3DS1	p70, NKAT10	DAP-12	HLA-Bw4		19q13.42-LRC
CD94/NKG2C	CD94/CD159c, KLRC2	DAP-12	HLA-E		12p13-NKC
CD94/NKG2E	CD94/KLRC3	DAP-12	HLA-E		12p13-NKC

ζ, FcεRγ and DAP-12[31] molecules bearing the immunoreceptor tyrosine-based activation motifs (ITAMs) and the DAP-10 molecule with the tyrosine-based motif YINM (Fig. 2).[32-34]

NCRs have been defined as molecules involved in the NK-mediated cytolysis and able to recognize ligand(s) different from MHC class I and thus kill MHC class I-deficient or negative targets (reviewed in refs. 4,35). Three major NCRs, NKp46 (NCR1), NKp30 (NCR3) and NKp44 (NCR2), together with NKG2D, an additional triggering receptor characterized by complex assembly, are known to control the induction of the NK cell cytotoxicity. All these receptors are also known to participate in the integration of activating signals and show a functional crosstalk aimed to amplify NCR-mediated activating signals.[36]

The *NCR1* gene is located in the telomeric region of LRC and encodes NKp46 (CD335).[4,12,37,38] This gene is maintained during speciation, at least in rodents, bovine and primates.[39-44] The 46 kDa NKp46 receptor is characterized by a short cytoplasmic tail, the transmembrane segment containing one Arg residue and two extracellular Ig-like C2-type domains connected to the transmembrane domain by a short, 25-residue-long peptide. The Arg residue located in the amino terminal portion of the transmembrane region forms a salt bridge with the Asp residue located in a similar topological context of the transmembrane domains of ζ and FcεRγ (Fig. 2A).[37] NKp46 shares three-dimensional structural similarities with LIR-1 (also known as ILT2, LILRB1, CD85j) and with KIR2D receptors, but none of the residues involved in the interaction of these receptors with ligands is conserved in NKp46.[45,46]

Among NCRs, the NKp46 receptor expressed by both resting and activated NK cells is the major receptor controlling NK-mediated cytolysis of cancer cell lines.[47-52] NKp46 cellular ligand is still elusive while hemagglutin and hemagglutinin-neuroaminidase molecules of influenza or parainfluenza viruses have been defined as its ligands in virally infected cells.[53] Recently, NKp46 has been shown to control the eradication of lethal influenza infection in a mouse model,[54] suggesting an active role of this receptor in fighting influenza infection. Other authors have suggested that vimentin expressed on the surface of a *M. tuberculosis*-infected monocyte cell lines, participates in the binding of NKp46 to these cells and contributes to their lysis.[55]

In some studies, membrane-associated heparan sulfate proteoglycans (HSPGs) have been found to associate with NKp46- and NKp30-mediated tumor cell lysis,[56] but others confuted this recognition (at least for NKp30).[57] Thus, cellular ligand(s) for these receptors are still to be elucidated. In addition, the target cell recognition mediated by NKp46 is apparently dependent on the presence of the Thr225-bound alpha 2,6-linked sialic acid residue in the membrane proximal portion of the receptor.[58]

NKp30 (CD337) is characterized by a single Ig-V-like domain followed by a short stalk region, a transmembrane domain containing one Arg residue and a short cytoplasmic tail without any signaling motifs. Like in NKp46, the transmembrane Arg residue is involved in the stabilization of multichain assembly of NKp30 with the associated ζ and FcεRIγ chains by salt bridge formation in the hydrophobic environment of the cellular membrane (Fig. 2A).[4,35] In addition, both NKp30 and NKp46 are associated with glycosylphosphatidylinositol (GPI)-linked CD59 molecules[59] that specifically induce phosphorylation of the ITAM tyrosines of the ζ chains associated with NKp30 and NKp46 but not CD16 upon mAb-mediated crosslinking.[59] NKp30-mediated NK cell activation appears to be altered by human cytomegalovirus (HCMV) infection since the HCMV tegument protein (pp65) has been reported to bind NKp30 and induce the dissociation of ζ from NKp30, thus provoking the signal transduction pathway/receptor uncoupling.[60] Apart from the HCMV pp65 viral proteins and HSPGs mentioned above, no other cellular ligand has been characterized so far. On the other hand, it is clear that at least DCs express a putative NKp30 ligand on their surface.[10] The NKp30 receptor is encoded by the *NCR3* gene located on human chromosome 6p21.3 in the HLA class III region close to the *lymphotoxin β* and *allograft inflammatory factor 1* (*AIF1*) genes.[61] Its expression is conserved during speciation in macaques, chimpanzee and rodents.[43,44,62] Interestingly, in humans, NKp30 is expressed by both resting and activated NK cells, whereas chimpanzee NK cells express this receptor only upon activation.[44]

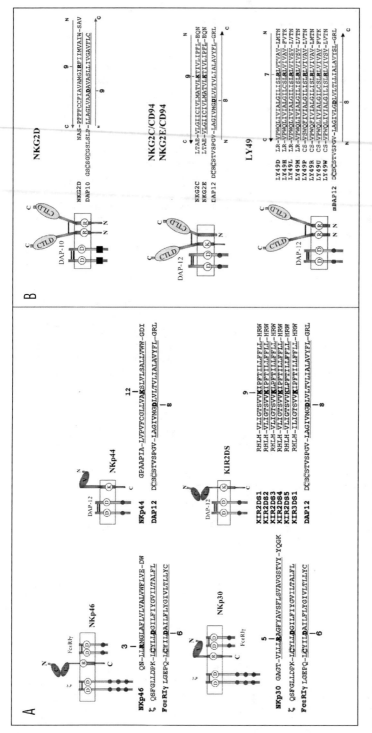

Figure 2. Graphic representation of activating NK cell receptors assembled as MIRRs. The transmembrane (TM) sequences of receptors and associated signal-transducing polypeptides are underlined. The position of the transmembrane charged residue involved in the receptor complex assembly is indicated. Cys residues involved in the formation of interchain disulfide bonds are shown in bold. In panel A all the TM sequences shown are represented in "parallel orientation" N amino -) to C (Carboxy- termini), while in panel B the "antiparallel" orientation (N-C/C-N) is indicated. Abbreviation: mouse DAP-12, mDAP-12.

NKp30 is the receptor that links innate and adaptive immune systems, due to crosstalk between NK cells and DCs at sites of primary inflammation and in secondary lymphoid tissues.[9,11,63] NKp30 has been shown to be responsible for the recognition and killing of iDCs, thus participating in modulation and editing the intensity and quality of adaptive immune responses.[10]

NKp44 (CD336) is expressed by activated NK and T γ/δ cells but not by resting NK cells.[64,65] This receptor is characterized by a single Ig V-like domain containing an additional disulphide bridge. The bridge is a characteristic structural feature of other receptors having the Ig V-like domain, such as CD300-like molecules, polymeric Ig-like receptors 1-4 and Fcα/μR, as well.[66] The Ig V-like domain is linked by a 55-amino acid stalk region containing 10 prolines, to a transmembrane portion containing a Lys residue (reviewed in ref. 37). In contrast to other NCRs, NKp44 displays a nonfunctional ITIM sequence in the cytoplasmic tail.[64,67] However, similarly to activating KIRs and other triggering receptors, NKp44 transduces signals using a disulfide-linked homodimer of the DNAX-activating protein (DAP)-12, also known as killer cell activating receptor-associated protein, KARAP (Fig. 2A).[4,35,37] NKp44 is encoded by a gene on chromosome 6p21.1.[68] In addition, it has been recently found in chimpanzee and both macaca mulatta and fascicularis (Biassoni R et al unpublished data) but not in the mouse genome.[68] The extracellular domain of NKp44 shares structural similarities with sialoadhesins and has a characteristic groove rich in basic residues. This groove contains an Arg residue possibly involved in sialic acid recognition in similar manner to that described for Siglec-1 and Siglec-7.[69-71] At present, only viral hemagglutinin has been proposed as a ligand for NKp44.[72]

NKG2D

Human NKG2D (CD314), or killer cell lectin-like receptor, subfamily k, member 1 (KLRK1) according to the Human Genome Organization classification, is a receptor that is structurally unrelated to the NCRs. This receptor features a diversified signal transduction pathway induced by the non-ITAM-bearing homodimeric polypeptide DAP-10, also known as killer activating protein 10, KAP-10 (Fig. 2B).[33,34] In mice but not in humans, NKG2D also exists as an alternative form (NKG2D-S) that lacks 13 amino-terminal residues and associates with both DAP-10 and DAP-12.[73] Upon NKG2D/ligand interaction, the tyrosine-based motif (YINM) of DAP-10 becomes tyrosine-phosphorylated and binds to a phosphatidylinositol 3-kinase (PI3K). Thus, DAP-10 recruits the p85 subunit of the PI3K and the adapter protein Grb-2 and induces activation of the signaling pathway mediated by activated extracellular signal-regulated kinase 1/2 (ERK1/2), a mitogen-activated protein (MAP) family kinase, JAK2 kinase and signal transducers and activators of transcription (STATs).[74] The DAP-10-mediated NK cell activation may be amplified by additional effectors. Thus, the signaling pathway induced by phosphorylation of the DAP-10 YINM motif is apparently independent upon a src-like kinase Syk that is normally involved in the ITAM-dependent activation.[29] In contrast, the JAK2/STAT5 signaling pathway is not typically involved in the ITAM-dependent cell activation.[75] The mentioned differences between YINM- and ITAM-mediated signaling may explain, at least in part, observed differences in the cytokine release upon YINM- and ITAM-dependent cell activation.[76,77] Also, in contrast to ITAM-mediated activation, YINM-mediated signaling has apparently less-pronounced sensitivity to inhibitory signals induced by KIRs.[78]

NKG2D is a C-type II lectin-like receptor that does not associate with CD94 like other members of the NKG2 family. This receptor is constitutively expressed on the surface of all human NK cells and γ δ and CD8+ T lymphocytes (reviewed in ref. 74). The *NKG2D* gene maps to the NKC on human chromosome 12p13.2, where other genes encoding members of the C-type II lectin family are located and it is expressed in different species including human, rodents and primates.[44,74,79] In contrast to other NCRs, at least 7 different ligands of cellular origin have been characterized for human NKG2D. These molecules have been also demonstrated to express differently in response to different stresses. The ligands are the MHC class I-related molecules, MICA and MICB (MICA/B) and the human cytomegalovirus UL16-binding protein/RAE-like transcript 1 (ULBP/RAET1) molecules.[75,76,80-82] Also, murine NKG2D is known to recognize at least 7 different surface ligands represented by retinoic acid-inducible gene molecules (RAE1α-ε), H60 and the murine ULBP-like transcript 1 (MULT1). In contrast to RAE, H60 and MULT1 do not have human homologues.[74,83]

In humans, the genes coding for the NKG2D ligands, map on chromosome 6 or, more precisely, on 6p21.3 and 6q24 for MICA/B and ULBP/RAET, respectively (reviewed in ref. 83) and all code for MHC class I-like α1 and α2 domains important for the interaction with NKG2D. Similar to ULBP4/RAET1E and RAET1G, the MICA/B molecules also display an additional α3 domain. However, the MICA/B ligands do not associate with β2-microglobulin and do not bind any peptide. In addition, they have both transmembrane and cytoplasmic domains, while ULBP-1 (RAET1I), ULBP-2 (RAET1H), ULBP-3 (RAET1N) and RAET1L are GPI-linked surface molecules.[81] MICA/B expression is upregulated in response to cellular stresses such as infections or tumorigenesis,[84] whereas ULBP are differentially expressed in a variety of cell lines and in several tumors (not only of epithelial origin). Also, expression of RAET1G and ULBP1-3 but not ULBP4/RAET1E is induced in primary fibroblasts upon HCMV but not by herpex simplex virus (HSV) infection.[85] Considering that different viral infections do not result in identical expression of NKG2D ligands and that Epstein-Barr virus (EBV)-infected B-cells rarely express NKG2D ligands, it can be suggested that ULBP/RAET expression requires pathogen-specific factor(s). However, we should consider that these data are limited to gene expression and do not indicate surface expression of NKG2D ligands that can be controlled by additional factors. Thus, it is possible that the mechanism of adaptation leading NKG2D to see multiple ligands is under selective pressure of infectious diseases that may downregulate the expression of NK receptor ligands using different protein factors. For example, murine cytomegalovirus (MCMV) is known to interfere with NKG2D recognition using different factors. One of them, the gp40 protein, encoded by the m152 gene, has been shown to downregulate mouse NKG2D ligands and other molecules[86] while the m145 and m155 gene products are known to prevent the NKG2D/MULT1-induced activation[87] and downregulate H60 ligand expression,[88] respectively. In humans, the human cytomegalovirus glycoprotein UL142 has been reported to downregulate MICA expression.[89]

In addition to a variety of NKG2D ligands, we should take into account that MICA and MICB are known to be polymorphic with about 60 and 25 alleles, respectively and that MICA and RAET1G genes may encode soluble protein variants acting as decoy ligands and thus representing a possible mechanism of tumour immune escape.[82,83] Within the proposed "induce-to-fit" model,[90] the existence of a profusion of ligands for a single receptor does not have simple structural explanation and recently reported thermodynamic and kinetic data can be much better describe by a mechanism of rigid protein-protein recognition using different receptor/ligand interface interactions.[91]

Other Activating NK Receptors

Within the LRC, the KIR and ILT/LIR gene families also code for activating receptors. Sharing highly conserved extracellular domains with inhibitory receptors, activating KIRs are, however, characterized by a transmembrane region containing the charged amino acid residue (typically, Lys) and a short cytoplasmic tail lacking tyrosine-based motifs (Table 3).[92] Activating KIRs (KIR2DS and KIR3DS) probably originate from a common ancestral gene[93] and induce positive intracellular signalling through the associated homodimeric signal-transducing DAP-12 polypeptide (Fig. 2A).[4,31] The absence of the transmembrane Lys residue prevents the association with the DAP-12 polypeptide and generates a nonfunctional triggering receptor.[94] MHC receptors displaying either inhibitory or activating function (Fig. 2B) also exist in rodents.[6] The sequence homology of extracellular regions between inhibitory and activating KIRs suggest that activating KIRs may share, at least in part, HLA class I allele specificity with the inhibitory counterpart. For KIR2DS1, this is confirmed by NK-mediated cytotoxicity and soluble KIR2DS1 Ig-Fc receptor binding studies using the same HLA-Cw4 transfected cells.[95] Soluble KIR2DS4 Ig-Fc receptor has also been found to stain HLA-Cw4 but not HLA-Cw6 cell transfectants.[96] In addition, a possible nonMHC ligand governing cytotoxic killing melanoma cell lines has been suggested for the KIR2DS4 receptor.[97] For other activating KIRs, such as KIR2DS2 and KIR3DS1, the functional data are intriguing but still elusive. Essentially, despite functional correlation with gene typing suggests MHC class I specificity, failure to detect direct interaction by cytofluorometric analysis using soluble molecules may reside in much weaker HLA binding affinity of activating KIRs compared to that of inhibitory receptors.[98,99]

Similar to activating KIRs, CD94/NKG2 heterodimers are receptors that activate but not inhibit NK receptor-mediated cell function. These lectin-like receptors use the common CD94 chain to form a functional receptor with the ITAM-bearing DAP-12 polypeptide transducing the activation signal upon tyrosine phosphorylation (Fig. 2B).[100]

Similar to their inhibitory counterpart, both activating receptors, NKG2C/CD94 and NKG2E/CD94, bind to HLA-E,[101,102] recognizing partially overlapping amino acid sequences.[102] Activating NKG2E/CD94 and inhibitory NKG2A/CD94 receptors display similar affinities in binding to HLA-E, whereas NKG2C/CD94 receptor is characterized by weaker affinity.[16]

Conclusions

Signaling in NK cells is regulated by a balance between activating and inhibitory receptors. The majority of activating NK receptors are surface molecules that belong to the MIRR family and transduce the activation signal through the associated signaling subunits. Other triggering receptors, such as 2B4,[103-106] NTB-A,[107-108] CRACC (CS1),[109] NKp80,[110,111] NKRP1,[112-114] and adhesion molecules like DNAM1[115-117] and CEACAM-1[118-120] may transduce signals using different pathways. Table 4 briefly describes additional NK receptors involved in modulating NK cell-mediated functions.

The structural principles used by the activating NK receptors to assembly are similar to those described for the TCR assembly (Fig. 2).[121] Three of the NK receptor-associated signaling polypeptides transduce signals using classical ITAM sequences, whereas the fourth polypeptide, DAP-10, uses a different tyrosine-based motif. The recognition of ligands by more than one of the triggering MIRR-like NK receptors results in a crosstalk of signaling pathways that amplifies the triggering cascade events.[36]

The association of ligand-binding receptor subunit with the signaling polypeptides is based on the formation of salt bridges between positively (Lys and Arg) and negatively (Asp) charged residues present in the transmembrane regions of the receptor and signaling partner chains, respectively. The assembly of MIRRs is characterized by the formation of three-helix transmembrane interaction interface involving two acidic residues of the hetero- or homodimeric signal-transducing polypeptides and one basic residue of the ligand-interacting receptor subunit.[121,122] Interestingly, the stoichiometry of the NKG2D/DAP10 receptor indicates a hexameric complex with a homodimeric NKG2D bound to two homodimeric DAP-10 signaling polypeptides.[122] Thus, the mechanism of association found in the TCR/CD3 complex[123] is also valid for different, Ig-like or C-type lectin, NK receptors. These NK receptors differ not only in their overall tertiary structure but, more importantly, in the antiparallel orientation of their transmembrane domains with the amino- to carboxy-terminal end orientation in the outside-in direction for Ig-like molecules and reverse orientation for lectin receptors.[121] Preferences for receptor-specific assembly are apparently governed by the charged residue position within the transmembrane region and the particular basic residue present. The contribution of other residues surrounding the charged residue within the transmembrane region is apparently irrelevant. In contrast, steric hindrance in the extracellular domains of both the ligand-recognizing and signaling receptor chains may prevent the association of these subunits.[123,124] Interestingly, in human NK receptors, the transmembrane Lys residue interacting with the transmembrane Asp residue of DAP-12 (position 8, Fig. 2) is located in the middle of the transmembrane regions (positions 9-12, Fig. 2), whereas in mouse LY49 receptors, the transmembrane Asp residue at the position 8 of the associated DAP-12 chain interacts with the Arg residue at the position 7 in the antiparallel-oriented receptor transmembrane domain (Fig. 2B).

Table 4. Human activating and inhibitory NK receptors different from the MIRR-like assembled NK receptors

Receptor	Alternative Name(s)	Associated Signaling Chain(s)	Function	Cellular Ligand	Chromosome
p75/AIRM1	Siglec,7 CDw,328 p75, QA79, AIRM1	No	Inhibitory	α(2, 8)- sialilated saccharides,[130] Sialyl-Lewis,[131,132]	19q13.3-LRC
IRp60	IRC1, IRC2, CD300a IGSF12, CMRF35H, CMRF35H9	No	Inhibitory	?	17q25
CEACAM1	CD66a, C-CAM, BGP	No	Inhibitory	CEA-related cell adhesion molecule CEACAM-1[120] Carcinoembryonic antigen (CEA)[118]	19q13.2
DNAM1	PTA1, CD226 TLiSA1	No	Activating	PVR (CD155)[116,117] and Nectin-2 (CD112)	18q22.3
2B4	NAIL, CD244, Nmrk, NKR2B4, SLAMF4		Activating Inhibitory*	CD48[103]	1q23.3
NTB-A	SLAMF6, KALI, KALIb, Ly108, SF2000	No	Activating Inhibitory*	NTB-A[108]	1q23.3
CRACC	SLAMF7, CD319, 19A, CS1	No	Activating Inhibitory*	CRACC[109]	1q23.1-q24.1
NKp80	KLRF1, CLEC5C	No	Activating	AICL[111]	12p12.3-13.2
NKRP-1A	KLRB1, CD161, CLEC5B, NKRP1A, NKR-P1A, hNKR-P1A,	No	?	lectin-like transcript-1 (LLT1)[113,114]	12p13

*Transduce signals through the immunoreceptor tyrosine-based switch motif (ITSM, TxYxxV/I) that, upon phosphorylation, recruits the so-called signaling lymphocyte activation molecule (SLAM)-associated protein (SAP/SH2D1A), triggering NK cells. In patients with X-linked lymphoproliferative syndrome (XLP), a functional SAP molecule is absent and, upon stimulation, the ITSMs may recruit SHP-phosphatases inducing inhibitory but not activating signals.[104,105,107,133]

It is interesting to speculate that activating KIRs (KIR2DS and KIR3DS) proposed to have arisen from a common ancient inhibitory ancestor within the last 35 million years (my), have been evolved in the association with more ancient (>400 my) signal-transducing polypeptides.[93] The evolution of activating NK receptors, such as KIRs in humans and Ly49 receptors in rodents, from inhibitory ancestor genes is probably a recurrent competitive process for new receptor variants. Since activating NK receptors are under balanced selection conditions, they are apparently short-lived species. These receptors are frequently lost in populations because of their detrimental role in autoimmunity. However, as suggested, the activating NK receptors are often reinvented during evolution because of their favorable effects in reproduction and host-resistance to pathogens.

In terms of immune defense, innate immunity is believed to be more ancient than the adaptive one. Nevertheless, both systems use the same triggering and signal transduction pathway for more than 400 my. In contrast, the gene families coding KIRs appeared relatively recently in evolution (at least, they appeared after speciation of rodents, <40-50 my) and the activating KIRs are continuously lost and re-acquired indicating that KIR multigene family is still under evolution. This is in agreement with the rapid evolution of KIR single haplotype loci in primates.[127] Thus, in the immune system, the different, old and new, effectors are evolved continuously to fight and defeat pathogens and tumor degenerative transformations in order to preserve the different species to compete for life.

Acknowledgements

I would like to thank "Associazione Italiana Ricerca sul Cancro" grant for financial support and Professor Lorenzo Moretta (Istituto Giannina Gaslini) for his scientific support.

References

1. Ljunggren H-G, Karre K. In search of the missing self. MHC molecules and NK cell recognition. Immunol Today 1990; 11:237-44.
2. Biassoni R, Dimasi N. Human Natural Killer cell receptor functions and their implication in diseases. Expert Rev Clin Immunol 2005; 1:405-17.
3. Moretta A, Bottino C, Vitale M et al. Receptors for HLA class I molecules in human natural killer cells. Annu Rev Immunol 1996; 14:619-48.
4. Moretta A, Bottino C, Vitale M et al. Activating receptors and coreceptors involved in human natural killer cell-mediated cytolysis. Annu Rev Immunol 2001; 19:197-23.
5. Sigalov AB. Multichain immune recognition receptor signaling: Different players, same game? Trends Immunol 2004; 25:583-9.
6. Dimasi N, Biassoni R. Structural and functional aspects of the Ly49 natural killer cell receptors. Immunol Cell Biol 2005; 83:1-8.
7. Lopez-Botet M, Llano M, Navarro F et al. NK cell recognition of nonclassical HLA class I molecules. Semin Immunol 2000; 12:109-19.
8. Long EO. Regulation of immune responses through inhibitory receptors. Annu Rev Immunol 1999; 17:875-04.
9. Vitale M, Della Chiesa M, Carlomagno S et al. NK-dependent DC maturation is mediated by TNFalpha and IFNgamma released upon engagement of the NKp30 triggering receptor. Blood 2005; 106:566-71.
10. Ferlazzo G, Tsang ML, Moretta L et al. Human dendritic cells activate resting natural killer (NK) cells and are recognized via the NKp30 receptor by activated NK cells. J Exp Med 2002; 195:343-51.
11. Moretta A. The dialogue between human natural killer cells and dendritic cells. Curr Opin Immunol 2005; 17:306-11.
12. Kelley J, Walter L, Trowsdale J. Comparative genomics of natural killer cell receptor gene clusters. PLoS Genet 2005; 1:129-39.
13. Cosman D, Fanger N, Borges L et al. A novel immunoglobulin superfamily receptor for cellular and viral MHC class I molecules. Immunity 1997; 7:273-82.
14. Vitale M, Castriconi R, Parolini S et al. The leukocyte Ig-like receptor (LIR)-1 for the cytomegalovirus UL18 protein displays a broad specificity for different HLA class I alleles: analysis of LIR-1 + NK cell clones. Int Immunol 1999; 11:29-35.
15. Vance RE, Kraft JR, Altman JD et al. Mouse CD94/NKG2A is a natural killer cell receptor for the nonclassical major histocompatibility complex (MHC) class I molecule Qa-1(b). J Exp Med 1998; 188:1841-8.
16. Kaiser BK, Barahmand-Pour F, Paulsene W et al. Interactions between NKG2x immunoreceptors and HLA-E ligands display overlapping affinities and thermodynamics. J Immunol 2005; 174:2878-84.
17. Llano M, Lee N, Navarro F et al. HLA-E-bound peptides influence recognition by inhibitory and triggering CD94/NKG2 receptors: preferential response to an HLA-G-derived nonamer. Eur J Immunol 1998; 28:2854-63.

18. Brooks CR, Elliott T, Parham P et al. The inhibitory receptor NKG2A determines lysis of vaccinia virus-infected autologous targets by NK cells. J Immunol 2006; 176:1141-7.
19. Ciccone E, Pende D, Viale O et al. Involvement of HLA class I alleles in NK cell specific functions: Expression of HLA-Cw3 confers selective protection from lysis by alloreactive NK clones displaying a defined specificity (specificity 2). J Exp Med 1992; 176:963-71.
20. Colonna M, Borsellino G, Falco M et al. HLA-C is the inhibitory ligand that determines dominant resistance to lysis by NK1- and NK2-specific natural killer cells. Proc Natl Acad Sci USA 1993; 90:12000-4.
21. Biassoni R, Falco M, Cambiaggi A et al. Amino acid substitutions can influence the NK-mediated recognition of HLA-C molecules. Role of Serine-77 and lysine-80 in the target cell protection from lysis mediated by group 2 or group 1 NK clones. J Exp Med 1995; 182:605-9.
22. Moretta A, Biassoni R, Bottino C et al. Major histocompatibility complex class I-specific receptors on human natural killer and T lymphocytes. Immunol Rev 1997; 155:105-17.
23. Malnati MS, Peruzzi M, Parker KC et al. Peptide specificity in the recognition of MHC class I by natural killer cell clones. Science 1995; 267:1016-8.
24. Costello PS, Walters AE, Mee PJ et al. The Rho-family GTP exchange factor Vav is a critical transducer of T-cell receptor signals to the calcium, ERK and NF-B pathways. Proc Natl Acad Sci USA 1999; 96:3035-40.
25. Penninger JM, Crabtree GR. The actin cytoskeleton and lymphocyte activation. Cell 1999; 96:9-12.
26. Kon-Kozlowski M, Pani G, Pawson T et al. The tyrosine phosphatase PTP1C associates withVav, Grb,2 andmSos1 in hematopoietic cells. J Biol Chem 1996; 271:3856-62.
27. Stebbins CC, Watzl C, Billadeau DD et al. Vav1 dephosphorylation by the tyrosine phosphatase SHP-1 as a mechanism for inhibition of cellular cytotoxicity. Mol Cell Biol 2003; 23:6291-9.
28. Hall A. Rho GTPases and the actin cytoskeleton. Science 1998; 279:509-14.
29. Billadeau DD, Brumbaugh KM, Dick CJ et al. The Vav-Rac1 pathway in cytotoxic lymphocytes regulates the generation of cell-mediated killing. J Exp Med 1998; 188:549-59.
30. Yokoyama WM, Kim S, French AR. The dynamic life of natural killer cells. Annu Rev Immunol 2004; 22:405-29
31. Lanier LL, Corliss BC, Wu J et al. Immunoreceptor DAP12 bearing a tyrosine-based activation motif is involved in activating NK cells. Nature 1998; 391:703-7.
32. Jiang K, Zhong B, Gilvary DL et al. Pivotal role of phosphoinositide-3 kinase in regulation of cytotoxicity in natural killer cells. Nat Immunol 2000; 1:419-25.
33. Chang C, Dietrich J, Harpur AG et al. Cutting edge: KAP10, a novel transmembrane adapter protein genetically linked to DAP12 but with unique signaling properties. J Immunol 1999; 163:4651-4.
34. Wu J, Song Y, Bakker ABH et al. An activating receptor complex on natural killer and T cells formed by NKG2D and DAP10. Science 1999; 285:730-2.
35. Moretta A, Biassoni R, Bottino C et al. Natural cytotoxicity receptors that trigger human NK-cell-mediated cytolysis. Immunol. Today 2000; 21:228-34.
36. Augugliaro R, Parolini S, Castriconi R et al. Selective crosstalk among natural cytotoxicity receptors in human natural killer cells. Eur J Immunol 2003; 33:1235-41.
37. Biassoni R, Cantoni C, Marras D et al. Human natural killer cell receptors: Insights into their molecular function and structure. J Cell Mol Med 2003; 7:376-87.
38. Pessino A, Sivori S, Bottino C et al. Molecular cloning of NKp46: A novel member of the immunoglobulin superfamily involved in triggering of natural cytotoxicity. J Exp Med 1998; 188:953-60.
39. Biassoni R, Pessino A, Bottino C et al. The murine homologue of the human NKp46, a triggering receptor involved in the induction of natural cytotoxicity. Eur J Immunol 1999; 29:1014-20.
40. Falco M, Cantoni C, Bottino C et al. Identification of the rat homologue of the human NKp46 triggering receptor. Immunol Lett 1999; 68:411-4.
41. Westgaard IH, Berg SF, Vaage JT et al. Rat NKp46 activates natural killer cell cytotoxicity and is associated with FcepsilonRIgamma and CD3zeta. J Leukoc Biol 2004; 76:1200-6.
42. Storset AK, Kulberg S, Berg I et al. NKp46 defines a subset of bovine leukocytes with natural killer cell characteristics. Eur J Immunol 2004; 34:669-76.
43. De Maria A, Biassoni R, Fogli M et al. Identification, molecular cloning and functional characterization of NKp46 and NKp30 natural cytotoxicity receptors in Macaca fascicularis (Macaca rhesus) NK cells. Eur J Immunol 2001; 31:3546-56.
44. Rutjens E, Mazza S, Biassoni R et al. Differential NKp30 inducibility is associated with conserved NK cell phenotype and function and maintenance of function in AIDS resistant chimpanzees. J Immunol 2007; 178:1702-12.
45. Ponassi M, Cantoni C, Biassoni R et al. Structure of the human NK cell triggering receptor NKp46 ectodomain. Biochem Biophys Res Comm 2003; 309:317-23.

46. Foster CE, Colonna M, Sun PD. Crystal structure of the human natural killer (NK) cell activating receptor NKp46 reveals structural relationship to other leukocyte receptor complex immunoreceptors. J Biol Chem 2003; 278:46081-6.

47. Sivori S, Pende D, Bottino C et al. NKp46 is the major triggering receptor involved in the natural cytotoxicity of fresh or cultured human NK cells. Correlation between surface density of NKp46 and natural cytotoxicity against autologous, allogeneic or xenogeneic target cells. Eur J Immunol 1999; 29:1656-66.

48. Sivori S, Parolini S, Marcenaro E et al. Triggering receptors involved in natural killer cell-mediated cytotoxicity against choriocarcinoma cell lines. Hum Immunol 2000; 61:1055-58.

49. Sivori S, Parolini S, Marcenaro E et al. Involvement of natural cytotoxicity receptors in human natural killer cell-mediated lysis of neuroblastoma and glioblastoma cell lines. J Neuroimmunol 2000; 107:220-25.

50. Weiss L, Reich S, Mandelboim O at el. Murine B-cell leukemia lymphoma (BCL1) cells as a target for NK cell-mediated immunotherapy. Bone Marrow Transplant 2004; 33:1137-41.

51. Spaggiari GM, Carosio R, Pende D et al. NK cell-mediated lysis of autologous antigen-presenting cells is triggered by the engagement of the phosphatidylinositol 3-kinase upon ligation of the natural cytotoxicity receptors NKp30 and NKp46. Eur J Immunol 2001; 31:1656-65.

52. Costello RT, Sivori S, Marcenaro M et al. Defective expression and function of natural killer cell-triggering receptors in patients with acute myeloid leukemia. Blood 2002; 99:3661-67.

53. Mandelboim O, Lieberman N, Lev M et al. Recognition of haemagglutinins on virus-infected cells by NKp46 activates lysis by human NK cells. Nature 2001; 409:1055-60.

54. Gazit R, Gruda R, Elboim M et al. Lethal influenza infection in the absence of the natural killer cell receptor gene Ncr1. Nat Immunol 2006; 7:517-23.

55. Garg A, Barnes PF, Porgador A et al. Vimentin expressed on Mycobacterium tubercolosis-infected human monocytes is involved in binding to the NKp46 receptor. J Immunol 2006; 177:6192-8

56. Bloushtain N, Qimron U, Bar-Ilan A et al. Membrane-associated heparan sulfate proteoglycans are involved in the recognition of cellular targets by NKp30 and NKp46. J Immunol 2004; 173:2392-401.

57. Warren HS, Jones AL, Freeman C et al. Evidence that the cellular ligand for the human NK cell activation receptor NKp30 is not a heparan sulfate glycosaminoglycan. J Immunol 2005; 175:207-12.

58. Arnon TI, Achdout H, Lieberman N et al. The mechanisms controlling the recognition of tumor- and virus-infected cells by NKp46. Blood 2004; 103:664-72.

59. Marcenaro E, Augugliaro R, Falco M et al. CD59 is physically and functionally associated with natural cytotoxicity receptors and activates human NK cell-mediated cytotoxicity. Eur J Immunol 2003; 33:3367-76.

60. Arnon TI, Achdout H, Levi O et al. Inhibition of the NKp30 activating receptor by pp65 of human cytomegalovirus. Nat Immunol 2005; 6:515-23.

61. Trowsdale J. Genetic and functional relationships between MHC and NK receptor genes. Immunity 2001; 15(3):363-74.

62. Hollyoake M, Campbell RD, Aguado B. NKp30 (NCR3) is a pseudogene in 12 inbred and wild mouse strains, but an expressed gene in Mus caroli. Mol Biol Evol 2005; 22:1661-72.

63. Ferlazzo G, Pack M, Thomas D et al. Distinct roles of IL-12 and IL-15 in human natural killer cell activation by dendritic cells from secondary lymphoid organs. Proc Natl Acad Sci USA 2004; 101:16606-11.

64. Cantoni C, Bottino C, Vitale M et al. NKp44, a triggering receptor involved in tumor cell lysis by activated human natural killer cells, is a novel member of the immunoglobulin superfamily. J Exp Med 1999; 189:787-96.

65. von Lilienfeld-Toal M, Nattermann J, Feldmann G et al. Activated gammadelta T-cells express the natural cytotoxicity receptor natural killer p44 and show cytotoxic activity against myeloma cells. Clin Exp Immunol 2006; 144:528-33.

66. Cantoni C, Ponassi M, Biassoni R et al. The three-dimensional structure of NK cell receptor NKp44, a triggering partner in natural cytotoxicity. Structure 2003; 11:725 34.

67. Campbell KS, Yusa S, Kikuchi-Maki A et al. NKp44 triggers NK cell activation through DAP12 association that is not influenced by a putative cytoplasmic inhibitory sequence. J Immunol 2004; 172:899-06.

68. Allcock RJ, Barrow AD, Forbes S et al. The human TREM gene cluster at 6p21.1 encodes both activating and inhibitory single IgV domain receptors and includes NKp44. Eur J Immunol 2003; 33:567-77.

69. May AP, Robinson RC, Vinson M et al. Crystal structure of the N-terminal domain of sialoadhesin in complex with 3' sialyllactose at 1.85 A resolution Mol Cell 1998; 1:719-28.

70. Swaminathan CP, Wais N, Vyas VV et al. Entropically assisted carbohydrate recognition by a natural killer cell surface receptor. Chembiochem 2004; 5:1571-75.

71. Attrill H, Takazawa H, Witt S et al. The structure of siglec-7 in complex with sialosides: leads for rational structure-based inhibitor design. Biochem J 2006; 397:271-8.
72. Arnon TI, Lev M, Katz G et al. Recognition of viral hemagglutinins by NKp44 but not by NKp30. Eur. J Immunol 2001; 31:2680-89.
73. Rosen DB, Araki M, Hamerman JA et al. A Structural basis for the association of DAP12 with mouse, but not human, NKG2D. J Immunol 2004; 173:2470-8.
74. Watzl C. The NKG2D receptor and its ligands-recognition beyond the missing self? Microbes Infect 2003; 5:31-37.
75. Sutherland CL, Chalupny NJ, Schooley K et al. UL16-binding proteins, novel MHC class I-related proteins, bind to NKG2D and activate multiple signaling pathways in primary NK cells. J Immunol 2002; 168:671-9.
76. Diefenbach A, Tomasello E, Lucas M et al. Selective associations with signaling proteins determine stimulatory versus costimulatory activity of NKG2D. Nat Immunol 2002; 3:1142-9. Erratum in: Nat Immunol 2004; 5:658.
77. Zompi S, Hamerman JA, Ogasawara K et al. NKG2D triggers cytotoxicity in mouse NK cells lacking DAP12 or Syk family kinases. Nat Immunol 2003; 4:565-72.
78. Bauer S, Groh V, Wu J et al. Activation of NK cells and T-cells by NKG2D, a receptor for stress-inducible MICA. Science 1999; 285:727-9.
79. Biassoni R, Fogli M, Cantoni C et al. Molecular and Functional Characterization of NKG2D, NKp80, and NKG2C Triggering NK Cell Receptors in Rhesus and Cynomolgus Macaques: Monitoring of NK Cell Function during Simian HIV Infection. J Immunol 2005; 174:5695-705.
80. Cosman D, Mullberg J, Sutherland CL et al. ULBPs, novel MHC class I-related molecules, bind to CMV glycoprotein UL16 and stimulate NK cytotoxicity through the NKG2D receptor. Immunity 2001; 14:123-33.
81. Jan Chalupny N, Sutherland CL, Lawrence WA et al. ULBP4 is a novel ligand for human NKG2D. Biochem Biophys Res Commun 2003; 305:129-35.
82. Bacon L, Eagle RA, Meyer M et al. Two human ULBP/RAET1 molecules with transmembrane regions are ligands for NKG2D. J Immunol 2004; 173:1078-84.
83. Bahram S, Inoko H, Shiina T et al. MIC and other NKG2D ligands: From none to too many. Curr Opin Immunol 2005; 17:505-9.
84. Gleimer M, Parham P. Stress management: MHC class I and class I-like molecules as reporters of cellular stress. Immunity 2003; 19(4):469-77.
85. Eagle RA, Traherne JA, Ashiru O et al. Regulation of NKG2D ligand gene expression. Hum Immunol 2006; 67:159-69.
86. Krmpotic A, Busch DH, Bubic I et al. MCMV glycoprotein gp40 confers virus resistance to CD8+ T-cells and NK cells in vivo. Nat Immunol 2002; 3:529-35.
87. Krmpotic A, Hasan M, Loewendorf A et al. NK cell activation through the NKG2D ligand MULT-1 is selectively prevented by the glycoprotein encoded by mouse cytomegalovirus gene m145. J Exp Med 2005; 201:211-20.
88. Hasan M, Krmpotic A, Ruzsics Z et al. Selective down-regulation of the NKG2D ligand H60 by mouse cytomegalovirus m155 glycoprotein. J Virol 2005; 79:2920-30.
89. Chalupny NJ, Rein-Weston A, Dosch S et al. Down-regulation of the NKG2D ligand MICA by the human cytomegalovirus glycoprotein UL142. Biochem Biophys Res Commun 2006; 346:175-81.
90. Radaev S, Kattah M, Zou Z et al. Making sense of the diverse ligand recognition by NKG2D. J Immunol 2002; 169:6279-85.
91. McFarland BJ, Strong RK. Thermodynamic analysis of degenerate recognition by the NKG2D immunoreceptor: not induced fit but rigid adaptation. Immunity 2003; 19:803-12.
92. Biassoni R, Cantoni C, Falco M et al. The Human Leukocyte Antigen (HLA)-C-specific Activatory or Inhibitory Natural Killer cell receptors display highly homologous extracellular domains but differ in their transmembrane and intracytoplasmic portions. J Exp Med 1996; 183:645-650.
93. Abi-Rached L, Parham P. Natural selection drives recurrent formation of activating killer cell immunoglobulin-like receptor and Ly49 from inhibitory homologues. J Exp Med 2005; 201:1319-32.
94. Bottino C, Falco M, Sivori S et al. Identification and molecular characterization of a natural mutant of the p50.2/KIR2DS2 activating NK receptor that fails to mediate NK cell triggering. Eur J Immunol 2000; 30:3569-3574.
95. Biassoni R, Pessino A, Malaspina A et al. Role of amino acid position 70 in the binding affinity of p50.1 and p58.1 receptors for HLA-Cw4 molecules. Eur J Immunol 1997; 27:3095-9.
96. Katz G, Markel G, Mizrahi S et al. Recognition of HLA Cw4 but not HLA-Cw6 by the NK cell receptor killer cell Ig-like receptor two-domain short tail number 4. J Immunol 2001; 166:7260-7.
97. Katz G, Gazit R, Arnon TI et al. MHC class I-independent recognition of NK-activating receptor KIR2DS4. J Immunol 2004; 173:1819-25.

98. Vales-Gomez M, Reyburn HT, Erskine RA et al. Differential binding to HLA-C of p50-activating and p58-inhibitory natural killer cell receptors. Proc Nat Acad Sci USA 1998; 95:14326-31.
99. Maenaka K, Juji T, Nakayama T et al. Killer cell immunoglobulin receptors and T-cell receptors bind peptide-major histocompatibility complex class I with distinct thermodynamic and kinetic properties. J Biol Chem 1999; 274:28329-34.
100. Lanier LL, Corliss B, Wu J et al. Association of DAP12 with activating CD94/NKG2C NK cell receptors. Immunity 1998; 8:693-01.
101. Braud VM, Allan DS, O'Callaghan CA et al. HLA-E binds to natural killer cell receptors CD94/NKG2A, B and C. Nature 1998; 391:795-9.
102. Wada H, Matsumoto N, Maenaka K et al. The inhibitory NK cell receptor CD94/NKG2A and the activating receptor CD94/NKG2C bind the top of HLA-E through mostly shared but partly distinct sets of HLA-E residues. Eur J Immunol 2004; 34:81-90.
103. Kubin MZ, Parshley DL, Din W et al. Molecular cloning and biological characterization of NK cell activation-inducing ligand, a counterstructure for CD48. Eur J Immunol 1999; 29:3466-77.
104. Parolini S, Bottino C, Falco M et al. X-linked lymphoproliferative disease: 2B4 molecules displaying inhibitory rather than activating function are responsible for the inability of NK cells to kill EBV-infected cells. J Exp Med 2000; 192:347-58.
105. Bottino C, Augugliaro R, Castriconi R et al. Analysis of the molecular mechanism involved in 2B4-mediated NK cell activation: evidence that human 2B4 is physically and functionally associated with the linker for activation of T-cells (LAT). Eur J Immunol 2000; 30:3718-22.
106. Eissmann P, Beauchamp L, Wooters J et al. Molecular basis for positive and negative signaling by the natural killer cell receptor 2B4 (CD244). Blood 2005; 105:4722-9
107. Bottino C, Falco M, Parolini S et al. NTB-A, a novel SH2D1A-associated surface molecule contributing to the inability of NK cells to kill EBV-infected B-cells in X-linked lymphoproliferative disease. J Exp Med 2001; 194:235-46.
108. Flaig RM, Stark S, Watzl C. Cutting edge: NTB-A activates NK cells via homophilic interaction. J Immunol 2004; 172:6524-7.
109. Stark S, Watzl C. 2B4 (CD244), NTB-A and CRACC (CS1) stimulate cytotoxicity but no proliferation in human NK cells. Int Immunol 2006; 18:241-7.
110. Vitale M, Falco M, Castriconi R et al. Identification of NKp,80 a novel triggering molecule expressed by human natural killer cells. Eur J Immunol 2001; 31:233-42.
111. Welte S, Kuttruff S, Waldhauer I et al. Mutual activation of natural killer cells and monocytes mediated by NKp80-AICL interaction. Nat Immunol 2006; 7:1334-42.
112. Lanier LL, Chang C, Phillips JH. Human NKR-P1A. A disulfide-linked homodimer of the C-type lectin superfamily expressed by a subset of NK and T lymphocytes. J Immunol 1994; 153:2417-28.
113. Rosen DB, Bettadapura J, Alsharifi M et al. Cutting edge: Lectin-like transcript-1 is a ligand for the inhibitory human NKR-P1A receptor. J Immunol 2005; 175:7796-9.
114. Aldemir H, Prod'homme V, Dumaurier MJ et al. Cutting edge: Lectin-like transcript 1 is a ligand for the CD161 receptor. J Immunol 2005; 175:7791-5.
115. Shibuya A, Campbell D, Hannum C et al. DNAM-1, a novel adhesion molecule involved in the cytolytic function of T lymphocytes. Immunity 1996; 4:573-81.
116. Bottino C, Castriconi R, Pende D et al. Identification of PVR (CD155) and Nectin-2 (CD112) as cell surface ligands for the human DNAM-1 (CD226) activating molecule. J Exp Med 2003; 198:557-67.
117. Tahara-Hanaoka S, Shibuya K, Onoda Y et al. Functional characterization of DNAM-1 (CD226) interaction with its ligands PVR (CD155) and nectin-2 (PRR-2/CD112). Int Immunol 2004; 16:533-8.
118. Markel G, Lieberman N, Katz G et al. CD66a interactions between human melanoma and NK cells: a novel class I MHC-independent inhibitory mechanism of cytotoxicity. J Immunol 2002; 168:2803-10.
119. Stern N, Markel G, Arnon TI et al. Carcinoembryonic antigen (CEA) inhibits NK killing via interaction with CEA-related cell adhesion molecule 1. J Immunol 2005; 174:6692-701.
120. Gray-Owen SD, Blumberg RS. CEACAM1: Contact-dependent control of immunity. Nat Rev Immunol 2006; 6:433-46.
121. Feng J, Garrity D, Call ME et al. Convergence on a distinctive assembly mechanism by unrelated families of activating immune receptors. Immunity 2005; 22:427-38.
122. Garrity D, Call ME, Feng J et al. The activating NKG2D receptor assembles in the membrane with two signaling dimers into a hexameric structure. Proc Natl Acad Sci USA 2005; 102:7641-6.
123. Call ME, Wucherpfennig KW. The T-cell receptor: Critical role of the membrane environment in receptor assembly and function. Annu Rev Immunol 2005; 23:101-25.
124. Feng J, Call ME, Wucherpfennig KW. The assembly of diverse immune receptors is focused on a polar membrane-embedded interaction site. PLoS Biol 2006; 4:e142.

125. Faure M, Long EO. KIR2DL4 (CD158d), an NK cell-activating receptor with inhibitory potential. J Immunol 2002; 168:6208-14.
126. Estefania E, Flores R, Gomez-Lozano N et al. Human KIR2DL5 is an inhibitory receptor expressed on the surface of NK and T lymphocyte subsets. J Immunol. 178; 2007:4402-10.
127. Sambrook JG, Bashirova A, Palmer S et al. Single haplotype analysis demonstrates rapid evolution of the killer immunoglobulin-like receptor (KIR) loci in primates. Genome Res 2005; 15:25-35.
128. Vance RE, Jamieson AM, Raulet DH. Recognition of the class Ib molecule Qa-1(b) by putative activating receptors CD94/NKG2C and CD94/NKG2E on mouse natural killer cells. J Exp Med 1999; 190:1801-12.
129. Saulquin X, Gastinel LN, Vivier E. Crystal structure of the human natural killer cell activating receptor Kir2Ds2 (Cd158J). J Exp Med 2003; 197:933-8.
130. Attrill H, Takazawa H, Witt S et al. The structure of siglec-7 in complex with sialosides: L eads for rational structure-based inhibitor design. Biochem J 2006; 397:271-8.
131. Swaminathan CP, Wais N, Vyas VV et al. Entropically assisted carbohydrate recognition by a natural killer cell surface receptor. Chem Biochem 2004; 5:1571-75.
132. Miyazaki K, Ohmori K, Izawa M et al. Loss of disialyl Lewis(a), the ligand for lymphocyte inhibitory receptor sialic acid-binding immunoglobulin-like lectin-7 (Siglec-7) associated with increased sialyl Lewis(a) expression on human colon cancers. Cancer Res 2004; 64:4498-505.
133. Eissmann P, Beauchamp L, Wooters J et al. Molecular basis for positive and negative signaling by the natural killer cell receptor 2B4 (CD244). Blood 2005; 105:4722-9.

CHAPTER 5

Platelet Glycoprotein VI

Stephanie M. Jung and Masaaki Moroi*

Abstract

Glycoprotein VI (GPVI) is a membrane glycoprotein unique to platelets and has been identified as a physiological receptor for collagen. Damage to a vessel wall exposes the subendothelial component collagen to platelets in the blood flow. Interaction of platelets with collagen via the receptor GPVI results in platelet activation and adhesion—the processes that are essential for thrombus formation. On the platelet surface, GPVI is present as a complex with the homodimeric Fc receptor γ-chain (FcRγ with a possible stoichiometry of two GPVI molecules and one FcRγ dimer). When collagen binds to GPVI, a platelet activation cascade is initiated by tyrosine phosphorylation of the immunoreceptor tyrosine-based activation motif of FcRγ and this phosphorylation induces the formation of a large complex composed from many signal-transducing proteins. In flow adhesion experiments that closely approximate physiological conditions, GPVI is essential for the formation of large platelet aggregates on collagen. However, GPVI-deficient patients or mice do not show any severe bleeding tendency. This suggests that a GPVI inhibitor would be able to inhibit thrombus formation but still not cause a significant bleeding tendency. Such an inhibitor would show promise as an anti-thrombotic agent for clinical use.

Introduction

Platelet glycoprotein VI (GPVI) is a platelet-specific collagen receptor with a molecular weight of about 62 kDa. Structural model for the interaction of GPVI with collagen based on the crystal structure of GPVI has been recently suggested.[1] This receptor was first identified as a protein deficient in patients' platelets that did not aggregate in response to the physiologically important agonist collagen.[2-6] The GPVI-deficient platelets, however, aggregated normally when induced by other agonists, including thrombin, ADP and ristocetin. These observations suggested that GPVI acts in a critical step specifically related to the reaction with collagen. Although the initial findings on GPVI were tantalizing because of its possible role in thrombotic processes, for the first five years or so research progressed slowly because the only means to identify this glycoprotein was to use the autoantibody from one patient. In addition, platelets also contain another, well-established collagen receptor, integrin $\alpha_2\beta_1$ ($\alpha_2\beta_1$). By using functional studies, such as aggregometry, to assess collagen responsiveness, it is difficult to differentiate the effects of GPVI from those of $\alpha_2\beta_1$. However, the contribution of GPVI could be differentiated from that of $\alpha_2\beta_1$ by using of an autoantibody against GPVI or GPVI-deficient patients' platelets, indicating that GPVI is a platelet collagen-specific receptor involved in platelet activation.[7-9] In 1995-97, research on GPVI was greatly accelerated by the discovery of GPVI-specific agonists. Searching for the specific structure of collagen that induces platelet activation, Dr. Barnes' group at Cambridge found that a triple helical collagen-mimetic peptide containing 10 repeats of the Gly-Pro-Hyp sequence induces platelet

*Corresponding Author: Masaaki Moroi—Department of Protein Biochemistry, Institute of Life Science, Kurume University, 1-1 Hyakunenkoen, Kurume, Fukuoka, 839-0864, Japan. Email: mmoroi@lsi.kurume-u.ac.jp

Multichain Immune Recognition Receptor Signaling: From Spatiotemporal Organization to Human Disease, edited by Alexander B. Sigalov. ©2008 Landes Bioscience and Springer Science+Business Media.

aggregation independently of $\alpha_2\beta_1$, which the authors called collagen-related peptide (CRP).[10,11] CRP strongly activates platelets, particularly when it is cross-linked. The CRP-induced activation is inhibited by the Fab fragment of an antibody against GPVI, verifying that CRP activates platelets by specifically interacting with GPVI. The tropical rattlesnake *Crotalus durissus terrificus* was found to have a platelet activating venom protein, Convulxin (Cvx), that is a specific agonist of GPVI.[12,13] Cvx is a C-type lectin composed of two subunits (α and β) cross-linked by disulfide bonds to form a heterotetramer $(\alpha\beta)_4$.[14] Similar to collagen, these two agonists (CRP and Cvx) induce many of the platelet activation reactions, including secretion, phosphorylation and intracellular Ca^{2+} release. However, they react only with GPVI, strongly suggesting that collagen-induced platelet activation is mediated mainly by GPVI rather than other collagen receptor, integrin $\alpha_2\beta_1$. One of the most interesting findings is that these agonists induce a strong tyrosine phosphorylation of platelet proteins such as the tyrosine kinase Syk, phospholipase Cγ2 (PLCγ2) and the scaffolding protein, linker for activation of T-cells (LAT). This raises questions about possible signal-transducing mechanisms of platelet activation initiated by the collagen binding to GPVI. The relevant studies will be discussed later in this chapter.

Discovery of the GPVI-specific agonists facilitated the cloning and elucidation of the structure of this receptor. Low content of GPVI in platelet membranes had made it difficult to purify and study the protein without these molecular and structural biology tools. In 1999, Clemetson's group reported the first successful purification and cloning of GPVI.[15] The deduced structure of GPVI indicates that GPVI is a member of the immunoglobulin (Ig) superfamily with two Ig-like domains. The Arg residue in its transmembrane domain has been shown to interact with an Asp residue of the Fc receptor γ-chain (FcRγ), forming a receptor complex on the platelet surface.[16,17] After gene cloning, it became possible to perform detailed analyses of GPVI function at the molecular level by using a recombinant protein, GPVI-expressing cells and monoclonal antibodies against GPVI. Finally, the crystal structure of the GPVI extracellular Ig-like domain has been recently reported,[1] thus providing a reasonable explanation for the structural and functional features of this receptor.

In contrast to ubiquitous integrin $\alpha_2\beta_1$, GPVI is a platelet-specific receptor that has a unique physiological function in thrombus formation. GPVI reacts specifically with collagen fibrils and activates platelets through a tyrosine phosphorylation-dependent pathway. Clinical and animal studies indicate that the deficiency in GPVI does not induce any severe bleeding tendency, although the thrombus formation in GPVI-deficient mice is impaired. Thus, specific inhibitors against GPVI are potentially ideal anti-thrombotic drugs that would not induce significant bleeding as a side effect and nowadays, much effort is being put into finding a specific GPVI inhibitor suitable for clinical use.

Structure of GPVI

From its cDNA sequence, GPVI is identified as a glycoprotein composed of 319 amino acid residues and a signal sequence of 20 amino acids.[15,18,19] The GPVI gene consists of eight exons and is located on chromosome 19q13.4 of the human genome.[20] The extracellular region of GPVI contains two Ig-like domains and a mucin-like Ser/Thr-rich domain is present between the Ig-like and transmembrane domains. The Ig-like domain is responsible for the binding of GPVI to collagen. Also, many O-linked carbohydrate chains are expected to conjugate to the Ser/Thr residues of the Ser/Thr-rich domain. One putative glycosylation site for the N-linked carbohydrate chain is identified at Asn72. Thus, it might be hypothesized that the Ig-like domain extends out from the polysaccharide layer over the platelet surface, forming a structure similar to platelet GPIb. Another characteristic structural feature of GPVI is the presence of the charged residue in its transmembrane domain (Arg252). The presence of the charged amino acid residue in the transmembrane domain is known to be characteristic for the proteins that associate with FcRγ containing a negatively charged Asp residue in its transmembrane domain. Thus, a salt bridge formed between GPVI and FcRγ, stabilizes the receptor complex. The importance of this salt bridge was confirmed by expression experiments.[16,17] The model of the GPVI-FcRγ complex presented in Figure 1 is consistent with the experimental data.

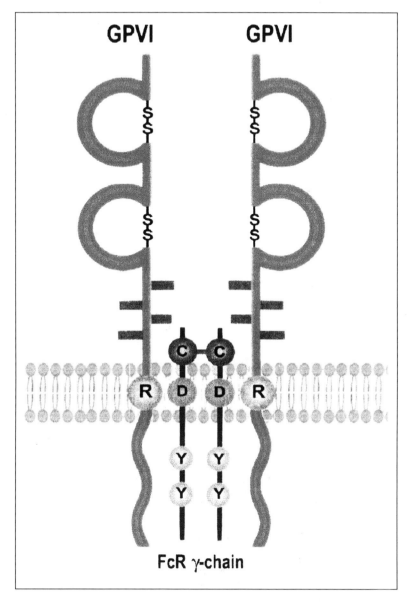

Figure 1. Structure of GPVI. GPVI is a dimeric complex with the homodimeric FcR γ-chain. (Reprinted from Thromb Res, 114, 221-233; Moroi M, Jung SM. Platelet glycoprotein VI: its structure and function, Copyright 2004 with permission from Elsevier.)

The two Ig-like domains of GPVI show homology to other proteins belonging to the paired Ig-like receptor family such as the Fcα receptor, mouse mast cell receptor and both inhibitory and activatory members of the natural killer cell receptors.[15] Since crystal structures of Ig-like domains from several of these proteins had already been determined, simulated models of the three-dimensional structure of GPVI were calculated from the structures of the homologous proteins. Recently, the crystal structure of the GPVI Ig-like domain has been reported by Horii et al[1]

Although the general structure of GPVI reported by these authors is similar to the simulated models, the crystal structure clearly indicates that GPVI forms a back-to-back dimer and the docking algorithm identifies two parallel collagen-binding sites on the GPVI dimer surface (Fig. 2). This structure presents the basic model explaining the unique binding characteristics of GPVI and collagen, thus marking progress towards the eventual identification of the exact collagen-binding site of GPVI.

The cytoplasmic region of GPVI contains 51 amino acids and shows no apparent homology with other receptor molecules. This GPVI tail can be divided into 4 regions: juxtamembrane, basic, proline-rich and C-terminal regions. The cytoplasmic region of mouse GPVI contains only 27 residues and the domain corresponding to 24 residues of the human C-terminal region is absent. Thus, the C-terminal region does not seem to be important for GPVI function. In contrast, the juxtamembrane region is necessary for the interaction with the associated FcRγ signaling subunit. The basic region next to the juxtamembrane region is rich in basic amino acids and analogous to the calmodulin-binding motifs found in GPIbβ, L-selectin and calmodulin-binding control peptide (CBCP).[21] This region binds to calmodulin and also interacts with FcRγ.[22] The adjacent Pro-rich region contains a consensus Pro-rich motif that binds to the Src homology 3 (SH3) domain. Src family tyrosine kinases Fyn and Lyn are also shown to bind to this region.[23]

Interaction of GPVI with Collagen

In ex vivo flow adhesion and in vitro stirring conditions (e.g., platelet aggregometry), collagen binds to platelets and activates them to form aggregates. Platelets have two collagen receptors, GPVI and integrin $\alpha_2\beta_1$, that interact with collagen through different mechanisms. Under physiological conditions, collagen is present as polymerized, insoluble collagen fibrils. Platelets are activated by these polymeric fibrils, whereas soluble monomeric collagen has little or no activity. Soluble collagen can bind to the active form of integrin $\alpha_2\beta_1$ expressed in agonist-activated platelets but can not bind to the inactive $\alpha_2\beta_1$ in resting platelets.[24,25] Since we observed that GPVI shows no

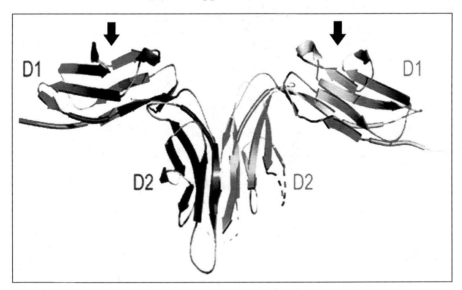

Figure 2. Crystal structure of GPVI immunoglobulin-like extracellular domain. Two GPVI immunoglobulin-like domains form a dimer with a dimerization interface located on the D2 domain. The upper surface of the D1 domain forms a shallow groove (indicated by arrows) and collagen triple helix is simulated to bind to this groove. (Reproduced from Horii K et al. Blood 2006; 108:936-942; with permission from Blood.)

discernible binding to soluble collagen,[25,26] GPVI binds almost exclusively to insoluble collagen fibrils. Thus, the binding of GPVI to collagen fibrils is the main molecular mechanism underlying the GPVI-mediated induction of platelet activation.

In platelets, the functional form of GPVI is hypothesized to be a dimer composed of two GPVI molecules (Fig. 1). This hypothesis is supported by the following experimental data: (1) the dimeric form of the recombinant GPVI extracellular Ig-like domain shows much higher binding affinity to collagen than its monomeric form,[26,19] (2) a high level expression of GPVI in RBL-2H3 cells is necessary for the cells to adhere to a collagen surface,[17,27] and (3) the monomeric GPVI Ig-like domain is shown to form a dimer in the crystal (Fig. 2).[1] In addition, the dimeric form of the GPVI extracellular Ig-like domain shows higher binding affinity towards CRP than its monomeric form.[26] Generally speaking, a dimeric monovalent ligand has higher affinity to its receptor than its monomeric form. However, for the dimeric GPVI Ig-like domain, the binding affinity to collagen ($K_D = 5.78 \times 10^{-7}$ M) is much higher than that of the monomeric form ($K_D > 10^{-4}$ M), whereas much smaller difference (~6-fold) has been observed in their affinities to CRP (5.27×10^{-6} M and 8.5×10^{-5} M, for dimeric and monomeric forms, respectively).[26] Also, we recently developed an antibody specific for the dimeric form that exhibits good inhibitory activity against collagen-induced platelet activation (Stephanie M. Jung and Masaaki Moroi, unpublished data). Taken together, all these data suggest that the dimeric form of GPVI is likely to have a specific conformation that allows more efficient binding with collagen fibrils and differs from that of the monomeric form.

The molecular mechanism of the interaction between GPVI and collagen or CRP has been studied by mutational analyses and confirmed by the recently reported crystal structure of the GPVI extracellular Ig-like domain. In contrast to homologous proteins that have a paired Ig-like domain, GPVI lacks 11-13 amino acids between Glu-49 and Gln-50 and this deletion induces a unique shallow hydrophobic groove on the surface of the Ig-like domain 1 (D1) of GPVI. Simulation of binding around this groove was made by using collagen model peptides. From the mutational analysis, the charged and polar residues of GPVI, such as Lys59, Arg60 and Arg166, have been shown to be involved in the binding to collagen.[28,29] These residues are oriented around the contact sites of the collagen model peptides, thus confirming the collagen-binding site. However, the amino acids Val34 and Leu36 have also been reported to bind to collagen or CRP.[30] However, these residues are present at the distal end of D1, suggesting that these residues might form another binding site for collagen or that the mutation of these residues might induce a conformational change that affects the binding to collagen or CRP. The ligand-binding sites in some of the homologous immunoreceptors, such as Ig-like transcript 2 (ILT-2) and FcαRI, are also located in the distal end of D1,[31] suggesting that this region of GPVI might partly contribute to collagen binding. However, single mutations of these functional amino acids of GPVI, Val34 and Leu36, do not induce a strong decrease of the binding activity, suggesting that GPVI can specifically interact with collagen by using multiple binding sites. In contrast, the specific interaction of integrin $\alpha_2\beta_1$ with collagen is mostly provided by the interaction of the Glu residue of collagen with the metal ion of $\alpha_2\beta_1$.[32]

Based on the crystal structure, the model of the collagen-binding site depicts a back-to-back dimer of the GPVI Ig-like domain, so that two nearly parallel putative collagen-binding sites are separated by a distance of 55 Å. This space is sufficient to fit 3 triple helical collagen molecules between these two collagen-binding sites.[1] Thus, the model can explain why GPVI binds preferably to polymeric collagen fibrils rather than monomeric collagen. Since collagen helices polymerize to each other with a gap of about a quarter of the collagen length, two collagen-binding sites of the GPVI dimer could bind the same collagen sites containing the Gly-Pro-Hyp sequences separated by 3 triple helices, suggesting that the binding of GPVI to collagen triple helices oriented parallel to the collagen fibrils might be more efficient.

GPVI-Mediated Signal Transduction

GPVI is the major signaling receptor for collagen on the platelet surface.[33-35] It is present as a complex with the FcRγ homodimer (Fig. 1). FcRγ is essential for both expression and function of GPVI. Each FcRγ contains one copy of an immunoreceptor tyrosine-based activation motif (ITAM) with two Tyr residues that undergo phosphorylation upon activation of GPVI. Then, the tyrosine kinase Syk binds to the phosphorylated ITAM, gets activated and induces the downstream signaling events.

Cross-linking of GPVI induced by its interaction with multivalent ligands, such as antibodies, CRP and Cvx, results in receptor triggering and platelet activation. Similarly, having numerous GPVI-binding sites, collagen fibrils cross-link GPVI that results in its triggering. Upon receptor cross-linking, the Src family protein kinases, Fyn and Lyn, phosphorylate the ITAM Tyr residues and thus initiate the GPVI-mediated signaling pathway.[36-38] The SH3 domains of these kinases have been shown to associate with the Pro-rich domain of the cytoplasmic tail of GPVI upon stimulation by a GPVI agonist.[73] This association is necessary for the phosphorylation of the ITAM Tyr residues. Thus, the association of Fyn and Lyn to the cytoplasmic domain of GPVI can be the first step of the activation mechanism induced by a GPVI agonist. Mouse platelets deficient in Fyn have a partial reduction in CRP-induced activation while those deficient in Lyn show a slight stimulation of CRP-induced activation. Mouse platelets deficient in both kinases exhibit strong impairment of platelet activation but are still able to induce a low level of activation, suggesting the possible contribution of another still not identified kinase.[38] Thus, the contribution of Fyn and Lyn kinases to the GPVI-mediated platelet activation remains to be elucidated.

Binding of the tyrosine kinase Syk to phosphorylated ITAM induces phosphorylation of Syk and its activation. Then, the activated Syk initiates the downstream signaling cascade by phosphorylating other proteins, such as the transmembrane adapter protein LAT, cytosolic adapter protein SLP-76 and phospholypase Cγ2 (PLCγ2). LAT has many phosphorylation sites and the phosphorylated LAT binds to the SH2- and/or other domains of many proteins, such as PLCγ2, PI 3-kinase, adaptor protein Gads, protein kinase Btk and small G-protein exchange factors Vav1 and Vav3.[39,40] Using LAT as a scaffold, these proteins form a large complex, the so-called LAT signalosome.[33] Formation of the LAT signalosome has many similarities to the signal-transducing mechanisms used by immune receptors, such as T- and B-cell receptors on lymphocytes. Among the components of the complex, PLCγ2 is thought to be the most important one because of its enzymatic activity. Upon the complex formation, PLCγ2 localizes near the plasma membrane and becomes activated through the phosphorylation by kinases of the complex. In the membrane, the activated PLCγ2 cleaves phosphatidylinositol 4,5-diphosphate (PIP2) into 1,2-diacylglycerol (DG) and inositol 1,4,5-triphosphate (IP3). DG is recognized as an activator of protein kinase C (PKC), whereas IP3 binds to IP3 receptors in plasma and intracellular membranes, leading to the release of Ca^{2+}. These two events, the activation of PKC and intracellular Ca^{2+} release, cause the release reaction and integrin-mediated activation of platelets. Based on the activation scheme proposed by Watson et al,[35,33] the signaling pathways induced by GPVI/FcRγ triggering are outlined in Figure 3. For some of the indicated proteins, their contribution in this complex mechanism was confirmed by studies of mouse platelets deficient in one of these proteins. Thus, platelets deficient in FcRγ, Syk, PLCγ2, or SLP-76 show lack of collagen-induced platelet activation without major effects on thrombin-induced platelet aggregation.[41-43] Although LAT-null platelets show a specific decrease in the activation induced by collagen or CRP, they aggregate normally at high concentration of the agonist, suggesting the existence of a LAT-independent signaling pathway.[44,45] Thus, these proteins play an important role in GPVI/FcRγ-mediated platelet activation. These proteins are all tyrosine-phosphorylated upon platelet activation, indicating that tyrosine-phosphorylation is a key step in signal-transducing mechanisms of the GPVI-induced platelet activation.

The phosphoinositide 3-kinase (PI 3-kinase)-specific inhibitors, such as wortmannin or LY294002, have been shown to inhibit the GPVI-induced platelet activation, suggesting that this enzyme is involved in this activation pathway. PI 3-kinase phosphorylates PIP2, converting it into phosphatidylinositol 3,4,5-triphosphate (PIP3). The pleckstrin homology (PH) domain-containing

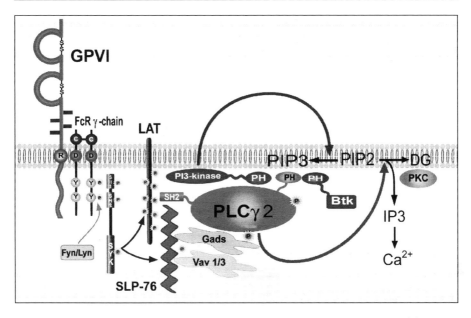

Figure 3. Scheme of the GPVI/FcRγ-induced activation pathways. The activation of the GPVI/ FcRγ receptor complex induces tyrosine phosphorylation and results in the formation of a large complex containing many signaling molecules, the so-called LAT signalosome. Tyrosine phosphorylation is indicated by P inside a star and each arrow indicates an enzyme reaction. Abbreviations: PH, pleckstrin-homology domain; SH2, Src-homology 2 domain; LAT, linker for activation of T-cells; SLP-76, SH2 domain-containing leukocyte protein of 76 kDa.

proteins, like PLCγ2, bind to PIP3 in the membrane and, thereby facilitate processes occurring near the membrane. The regulatory subunit of PI 3-kinase, p85, has Src homology (SH) 2, SH3 and Pro-rich domains and acts as an adaptor protein in the LAT signalosome. P85α-deficient mouse platelets show specifically decreased aggregation in response to collagen while having intact ADP-induced aggregation.[46] Since another type of PI 3-kinase, PI 3-kinaseγ, has been reported to be involved specifically in ADP-induced platelet activation,[47] these results suggest that different subtypes of PI 3-kinase are involved in different activation pathways.

Physiological Function of GPVI

GPVI-deficient patients are known to have mildly prolonged bleeding times.[2-5] Also, GPVI-depleted mice have been reported to show moderately increased or normal bleeding time.[48,49] These data suggest that in vivo thrombus formation can be induced not only by the GPVI-collagen interaction, but also by activation of other signaling pathways.

Thus, it is both fundamentally and clinically important to reveal the GPVI-dependent molecular mechanisms of thrombus formation under physiological conditions. Analysis of the platelet adhesion to an immobilized-collagen surface under blood flow conditions is the ex vivo assay closest to the in vivo situation. This technique allows researchers to examine the physiological function of the GPVI receptor using GPVI-deficient patients' platelets, anti-GPVI antibody and GPVI-null mouse platelets. The data obtained by this method indicate that the formation of large platelet aggregates is severely impaired if GPVI is defective or inhibited. However, the reported results of studies on platelet adhesion evaluated as platelet immobilization on the collagen surface under blood flow conditions are contradictory. Some studies indicate that GPVI-deficient platelets show little or no adhesion,[50,51] whereas others report that adhesion could still be observed in such platelets.[9,52,49] When platelet adhesion is determined under blood flow conditions using denuded blood vessels instead of using a collagen-coated surface, the control blood shows the formation of large platelet aggregates on the subendothelium. In contrast, the GPVI-deficient patient's blood shows single platelets or a low number of small aggregates adhered to the subendothelium.[53] Since the residual, albeit at low level, platelet adhesion or aggregation of GPVI-deficient patient's platelets is $\alpha_2\beta_1$-dependent,[6] it is reasonable to suggest that through mediation by integrin $\alpha_2\beta_1$, GPVI-deficient platelets can adhere to collagen but can not become activated because of the lack of GPVI, thus preventing formation of large aggregates. This could be one of the reasons why GPVI-deficient patients or mice show only a mild bleeding tendency.

In vivo studies of the thrombus formation caused by laser-induced injury of blood vessels show that this process depends upon P-selectin, tissue factor and thrombin, but it does not depend on GPVI. However, when blood vessels are chemically injured by $FeCl_3$, an agent that damages the endothelial cells and thereby exposes the subendothelium, the thrombus formation is GPVI-dependent.[54,55] Thus, the mechanism of thrombus formation in vivo differs depending on the pathological condition. GPVI might be involved in this process in cases where collagen fibrils are exposed to blood flow, as it is observed in atherosclerotic plaque rupture.[56] On the other hand, GPVI is not likely to be a major factor of the thrombus formation after mechanical injury of vessels that is used, for example, in measurements of bleeding time.

Several GPVI-deficient patients have been reported,[3-5] but the details of their genetic defects have not yet been confirmed. In addition, one patient with autoimmune thrombocytopenia[2] and one with systemic lupus erythematosus[6] developed autoantibodies against GPVI. The platelets of both patients became GPVI-negative because of the presence of autoantibody against this glyco-protein. Such acquired GPVI deficiency can be experimentally induced in mice by injection of anti-GPVI. In these mice, the platelets have been found to be depleted of GPVI as soon as 6 hours after antibody injection.[48,57] Studies on mice also show that antibodies to different GPVI epitopes and their Fab fragments also have the GPVI-depleting effect. Although the molecular mechanism responsible for the GPVI depletion is not clear, the same mechanism is likely to operate in patients with autoantibodies against GPVI. Recently, patients with idiopathic thrombocytopenic purpura (ITP) have been reported to be GPVI-deficient but no antibodies against GPVI were detected in these individuals.[58] Platelets in these patients might have been initially exposed to antibody against GPVI and thereby became depleted of this receptor and remain depleted of GPVI even after the antibody was cleared from their blood. Thus, although a small amount of antibody (below detectable levels) might still be present in the patients, exposure to the antibody seems to induce a long-lasting depletion of GPVI from the platelets. Thus, considering the potential use of anti-GPVI antibody as an antithrombotic agent in clinical practice, knowledge of the molecular mechanism of GPVI depletion can be both fundamentally and clinically important.

Since platelets contain two physiological significant collagen receptors, GPVI and integrin $\alpha_2\beta_1$, the relative contributions of both receptors to thrombus formation in vivo need to be discussed. Are they redundant collagen receptors necessary for thrombosis formation or do they have com-plimentary roles? Although $\alpha_2\beta_1$ itself does not induce platelet activation, its role in thrombus formation under physiological conditions has been supported by a number of studies. Notably, $\alpha_2\beta_1$ polymorphism with high-level expression of this integrin is associated with myocardial in-farction in younger individuals.[59,60] Integrin $\alpha_2\beta_1$ is involved in the firm adhesion of platelets on the collagen surface under flow conditions[61] and we found that the GPVI agonist CRP activates $\alpha_2\beta_1$, increasing it's affinity to collagen.[25] Thus, a coordinated contribution of GPVI and $\alpha_2\beta_1$ could enhance the interaction of platelets with collagen and stimulate their activation. In fact, a number of reports have indicated these complementary roles of GPVI and $\alpha_2\beta_1$ in collagen-induced thrombus formation.[52,62,63]

Summary and Perspectives

GPVI is a newly identified platelet-specific glycoprotein whose unique function as a collagen re-ceptor has attracted the attention of both platelet researchers and clinicians, particularly because of its important role in thrombogenesis. The structure of GPVI has been recently elucidated, showing that GPVI forms a back-to-back dimer and has a collagen groove on its D1 domain. This structure provides a foundation for a possible mechanism of interaction of GPVI with collagen fibrils.

The unique and most important characteristic of GPVI is its ability to bind the physiological agonist collagen and collagen-like molecules in vitro, initiating a signaling cascade that results in platelet activation. The GPVI/FcRγ-initiated signaling pathway is mainly mediated by tyrosine phosphorylation and formation of a large complex, the LAT signalosome. Similar mechanisms of activation are also observed in lymphocytes. From studies on knockout mice, Fyn, Lyn, Syk, LAT, SLP-76 and PLCγ2 have been identified as the main protein factors contributing to this pathway. However, how these proteins interact with each other and how other factors contribute to the signaling and activation pathway are still not fully clear. Recently, proteomics techniques of protein analysis and genetic approaches using megakaryocytes have been applied to study this complicated multi-factor and multi-interaction mechanism.

GPVI is the main collagen receptor activating platelets and its inhibition seems not to induce bleeding. Thus, an inhibitor of the GPVI-collagen interaction might have a great clinical potential as an anti-thrombotic drug. To date, the reported GPVI inhibitors are all antibodies, with the exception of triplatin, a protein from the salivary glands of the assassin bug, *Triatoma infestans*.[64] The injection of the Fab fragment of an anti-GPVI antibody into *Cynomolgus* monkeys has been recently reported to inhibit collagen-induced platelet aggregation for a longer time compared to an anti-integrin $\alpha_{IIb}\beta_3$ antibody, Abciximab and cause no bleeding tendency.[65] These findings suggest that GPVI inhibitors have potential clinical application; however, further studies are needed to determine their function under different pathological conditions. Furthermore, as discussed above, autoantibodies against GPVI appear to have a long-lasting GPVI-depletion effect on platelets. Whether this is true for exogenously added anti-GPVI antibodies needs to be addressed. In addition, small molecules seem to be more suitable for anti-thrombotic drug design and development of such compounds has been recently facilitated by the reported crystal structure of GPVI.

References

1. Horii K, Kahn ML, Herr AB. Structural basis for platelet collagen responses by immune-type receptor glycoprotein VI. Blood 2006; 108:936-942.
2. Sugiyama T, Okuma M, Ushikubi F et al. A novel platelet aggregating factor found in a patient with defective collagen-induced platelet aggregation and autoimmune thrombocytopenia. Blood 1987; 69:1712-1720.
3. Moroi M, Jung SM, Okuma M et al. A patient with platelets deficient in glycoprotein VI that lack both collagen-induced aggregation and adhesion. J Clin Invest 1989; 84:1440-1445.
4. Ryo R, Yoshida A, Sugano W et al. Deficiency of P62, a putative collagen receptor, in platelets from a patient with defective collagen-induced platelet aggregation. Am J Hematol 1992; 39:25-31.
5. Arai M, Yamamoto N, Moroi M et al. Platelets with 10% of the normal amount of glycoprotein VI have an impaired response to collagen that results in a mild bleeding tendency. Br J Haematol 1995; 89:124-130.
6. Takahashi H, Moroi M. Antibody against platelet membrane glycoprotein VI in a patient with systemic lupus erythematosus. Am J Hematol 2001; 67:262-267.
7. Ichinohe T, Takayama H, Ezumi Y et al. Cyclic AMP-insensitive activation of c-Src and Syk protein-tyrosine kinases through platelet membrane glycoprotein VI. J Biol Chem 1995; 270:28029-28036.
8. Ichinohe T, Takayama H, Ezumi Y et al. Collagen-stimulated activation of Syk but not c-Src is severely compromised in human platelets lacking membrane glycoprotein VI. J Biol Chem 1997; 272:63-68.
9. Moroi M, Jung SM, Shinmyozu K et al. Analysis of platelet adhesion to a collagen-coated surface under flow conditions: The involvement of glycoprotein VI in the platelet adhesion. Blood 1996; 88:2081-2092.
10. Morton LF, Hargreaves PG, Farndale RW et al. Integrin $\alpha_2\beta_1$-independent activation of platelets by simple collagen-like peptides: collagen tertiary (triple-helical) and quaternary (polymeric) structures are sufficient alone for $\alpha_2\beta_1$-independent platelet reactivity. Biochem J 1995; 306:337-344.
11. Knight CG, Morton LF, Onley DJ et al. Collagen-platelet interaction: Gly-Pro-Hyp is uniquely specific for platelet GP VI and mediates platelet activation by collagen. Cardiovasc Res 1999; 41:450-457.
12. Polgar J, Clemetson JM, Kehrel BE et al. Platelet activation and signal transduction by convulxin, a C-type lectin from Crotalus durussus terrificus (tropical rattlesnake) venom via the p62/GPVI collagen receptor. J Biol Chem 1997; 272:13576-13583.
13. Jandrot-Perrus M, Lagrue A-H, Okuma M et al. Adhesion and activation of human platelets induced by convulxin involve glycoprotein VI and integrin $\alpha_2\beta_1$. J Biol Chem 1997; 272:27035-27041.

14. Murakami MT, Zela SP, Gava LM et al. Crystal structure of the platelet activator convulxin, a disulfide-linked $\alpha_4\beta_4$ cyclic tetramer from the venom of Crotalus durissus terrificus. Biochem Biophys Res Commun 2003; 310:478-482.
15. Clemetson JM, Polgar J, Magnenat E et al. The platelet collagen receptor glycoprotein VI is a member of the immunoglobulin superfamily closely related to FR and the natural killer receptors. J Biol Chem 1999; 274:29019-29024.
16. Berlanga O, Tulasne D, Bori T et al. The Fc receptor γ-chain is necessary and sufficient to initiate signalling through glycoprotein VI in transfected cells by the snake C-type lectin, convulxin. Eur J Biochem 2002; 269:2951-2960.
17. Zheng Y-M, Liu C, Chen H et al. Expression of the platelet receptor GPVI confers signaling via the Fc receptor γ-chain in response to the snake venom convulxin but not to collagen. J Biol Chem 2001; 276:12999-13006.
18. Miura Y, Ohnuma M, Jung SM et al. Cloning and expression of the platelet-specific collagen receptor glycoprotein VI. Thromb Res 2000; 98:301-309.
19. Jandrot-Perrus M, Busfield S, Lagrue A-H et al. Cloning, characterization and functional studies of human and mouse glycoprotein VI: A platelet-specific collagen receptor from the immunoglobulin superfamily. Blood 2000; 96:1798-1807.
20. Ezumi Y, Uchiyama T, Takayama H. Molecular cloning, genomic structure, chromosomal localization and alternative splice forms of the platelet collagen receptor glycoprotein VI. Biochem Biophys Res Commun 2000; 277:27-36.
21. Andrews RK, Suzuki-Inoue K, Shen Y et al. Interaction of calmodulin with the cytoplasmic domain of platelet glycoprotein VI. Blood 2002; 99:4219-4221.
22. Bori-Sanz T, Suzuki-Inoue K, Berndt MC et al. Delineation of the region in the glycoprotein VI tail required for association with the Fc receptor γ-chain. J Biol Chem 2003; 278:35914-35922.
23. Suzuki-Inoue K, Tulasne D, Shen Y et al. Association of Fyn and Lyn with the proline-rich domain of glycoprotein VI regulates intracellular signaling. J Biol Chem 2002; 277:21561-21566.
24. Jung SM, Moroi M. Signal-transducing mechanisms involved in activation of the platelet collagen receptor integrin $\alpha_2\beta_1$. J Biol Chem 2000; 275:8016-8026.
25. Jung SM, Moroi M. Platelets interact with soluble and insoluble collagens through characteristically different reactions. J Biol Chem 1998; 273:14827-14837.
26. Miura Y, Takahashi T, Jung SM et al. Analysis of the interaction of platelet collagen receptor glycoprotein VI (GPVI) with collagen A dimeric form of GPVI, but not the monomeric form, shows affinity to fibrous collagen. J Biol Chem 2002; 277:46197-46204.
27. Chen H, Locke D, Liu Y et al. The platelet receptor GPVI mediates both adhesion and signaling responses to collagen in a receptor density-dependent fashion. J Biol Chem 2002; 277:3011-3019.
28. Smethurst PA, Joutsi-Korhonen L, O'Connor MN et al. Identification of the primary collagen-binding surface on human glycoprotein VI by site-directed mutagenesis and by a blocking phage antibody. Blood 2004; 103:903-911.
29. O'Connor MN, Smethurst PA, Farndale RW et al. Gain- and loss-of-function mutants confirm the importance of apical residues to the primary interaction of human glycoprotein VI with collagen. J Thromb Haemost 2006; 4:869-873.
30. Lecut C, Arocas V, Ulrichts H et al. Identification of residues within human glycoprotein VI involved in the binding to collagen. Evidence for the existence of distinct binding sites. J Biol Chem 2004; 279:52293-52299.
31. Foster CE, Colonna M, Sun PD. Crystal structure of the human natural killer (NK) cell activating receptor NKp46 reveals structural relationship to other leukocyte receptor complex immunoreceptors. J Biol Chem 2003; 278:46081-46086.
32. Emsley J, Knight CG, Farndale RW et al. Structural basis of collagen recognition by integrin $\alpha_2\beta_1$. Cell 2000; 100:47-56.
33. Watson SP, Auger JM, McCarty OJT et al. GPVI and integrin αIIbβ3 signaling in platelets. J Thromb Haemost 2005; 3:1752-1762.
34. Nieswandt B, Watson SP. Platelet-collagen interaction: Is GPVI the central receptor? Blood 2003; 102:449-461.
35. Watson SP, Asazuma N, Atkinson B et al. The role of ITAM- and ITIM-coupled receptors in platelet activation by collagen. Thromb Haemostas 2001; 86:276-288.
36. Ezumi Y, Shindoh K, Tsuji M et al. Physical and functional association of the Src family kinases Fyn and Lyn with the collagen receptor glycoprotein VI-Fc receptor γ-chain complex on human platelets. J Exp Med 1998;188:267-276.
37. Briddon SJ, Watson SP. Evidence for the involvement of p59^{fyn} and p53/56^{lyn} in collagen receptor signaling in human platelets. Biochem J 1999; 338:203-209.

38. Quek LS, Pasquet J-M, Hers I et al. Fyn and Lyn phosphorylate the Fc receptor γ chain downstream of glycoprotein VI in murine platelets and Lyn regulates a novel feedback pathway. Blood 2000; 96:4246-4253.
39. Gross BS, Melford SK, Watson SP. Evidence that phospholipase C-γ2 interacts with SLP-76, Syk, Lyn, LAT and the Fc receptor γ-chain after stimulation of the collagen receptor glycoprotein VI in human platelets. Eur J Biochem 1999; 263:612-623.
40. Asazuma N, Wilde JI, Berlanga O et al. Interaction of linker for activation of T-cells with multiple adapter proteins in platelets activated by the glycoprotein VI-selective ligand, convulxin. J Biol Chem 2000; 275:33427-33434.
41. Poole A, Gibbins JM, Turner M et al. The Fc receptor γ-chain and the tyrosine kinase Syk are essential for activation of mouse platelets by collagen. EMBO J 1997; 16:2333-2341.
42. Suzuki-Inoue K, Inoue O, Frampton J et al. Murine GPVI stimulates weak integrin activation in PLCγ2$^{-/-}$ platelets: involvement of PLCγ1 and PI3-kinase. Blood 2003; 102:1367-1373.
43. Gross BS, Lee JR, Clements JL et al. Tyrosine phosphorylation of SLP-76 is downstream of Syk following stimulation of the collagen receptor in platelets. J Biol Chem 1999; 274:5963-5971.
44. Pasquet J-M, Gross B, Quek L et al. LAT is required for tyrosine phosphorylation of phospholipase Cγ2 and platelet activation by the collagen receptor GPVI. Mol Cell Biol 1999; 19:8326-8334.
45. Judd BA, Myung PS, Obergfell A et al. Differential requirement for LAT and SLP-76 in GPVI versus T-cell receptor signaling. J Exp Med 2002; 195:705-717.
46. Watanabe N, Nakajima H, Suzuki H et al. Functional phenotype of phosphoinositide 3-kinase p85a-null platelets characterized by an impaired response to GP VI stimulation. Blood 2003; 102:541-548.
47. Hirsch E, Bosco O, Tropel P et al. Resistance to thromboembolism in PI3Kγ-dificient mice. FASEB J 2001; 15:2019-2021.
48. Nieswandt B, Schulte V, Bergmeier W et al. Lonγ-term antithrombotic protection by in vivo depletion of platelet glycoprotein VI in mice. J Exp Med 2001; 193:459-469.
49. Kato K, Kanaji T, Russell S et al. The contribution of glycoprotein VI to stable platelet adhesion and thrombus formation illustrated by targeted gene deletion. Blood 2003; 102:1701-1707.
50. Nieswandt B, Brakebusch C, Bergmeier W et al. Glycoprotein VI but not $α_2β_1$ integrin is essential for platelet interaction with collagen. EMBO J 2001; 20:2120-2130.
51. Goto S, Tamura N, Handa S et al. Involvement of glycoprotein VI in platelet thrombus formation on both collagen and von Willebrand factor surfaces under flow conditions. Circulation 2002; 106:266-272.
52. Siljander PRM, Munnix ICA, Smethurst PA et al. Platelet receptor interplay regulates collagen-induced thrombus formation in flowing human blood. Blood 2004; 103:1333-1341.
53. Moroi M, Jung SM. Platelet glycoprotein VI: Its structure and function. Thromb Res 2004; 114:221-233.
54. Dubois C, Panicot-Dubois L, Merrill-Skoloff G et al. Glycoprotein VI-dependent and -independent pathways of thrombus formation in vivo. Blood 2006; 107:3902-3906.
55. Furie B, Furie BC. Thrombus formation in vivo. J Clin Invest 2005; 115:3355-3362.
56. Penz S, Reininger AJ, Brandl R et al. Human atheromatous plaques stimulate thrombus formation by activating platelet glycoprotein VI. FASEB J 2005; 19:898-909.
57. Schulte V, Rabie T, Prostredna M et al. Targeting of the collagen-binding site on glycoprotein VI is not essential for in vivo depletion of the receptor. Blood 2003; 101:3948-3952.
58. Boylan B, Chen H, Rathore V et al. Anti-GPVI-associated ITP: An acquired platelet disorder caused by autoantibody-mediated clearance of the GPVI/FcRγ-chain complex from the human platelet surface. Blood 2004; 104:1350-1355.
59. Santoso S, Kunicki TJ, Kroll H et al. Association of the platelet glycoprotein Ia C_{807}T gene polymorphism with myocardial infarction in younger patients. Blood 1999; 93:2449-2453.
60. Moshfeghb K, Wuillemin WA, Redondo M et al. Association of two silent polymorphisms of platelet glycoprotein la/lla receptor with risk of myocardial infarction: A case-control study. Lancet 1999; 353:351-354.
61. Moroi M, Onitsuka I, Imaizumi T et al. Involvement of activated integrin $α_2β_1$ in the firm adhesion of platelets onto a surface of immobilized collagen under flow conditions. Thromb Haemostas 2000; 83:769-776.
62. Kuijpers MJE, Schulte V, Bergmeier W et al. Complementary roles of glycoprotein VI and $α_2β_1$ integrin in collagen-induced thrombus formation in flowing whole blood ex vivo. FASEB J 2003; 17:685-687.
63. Mangin P, Nonne C, Eckly A et al. A PLCγ2-independent platelet collagen aggregation requiring functional association of GPVI and integrin $α_2β_1$. FEBS Lett 2003; 542:53-59.
64. Morita A, Isawa H, Orito Y et al. Identification and characterization of a collagen-induced platelet aggregation inhibitor, triplatin, from salivary glands of the assassin bug, Triatoma infestans. FEBS J 2006; 273:2955-2962.
65. Matsumoto Y, Takizawa H, Nakama K et al. Ex vivo evaluation of anti-GPVI antibody in Cynomolgus monkeys: Dissociation between anti-platelet aggregatory effect and bleeding time. Thromb Haemostas 2006; 96:167-175.

CHAPTER 6

Clustering Models

Wolfgang W.A. Schamel* and Michael Reth

Abstract

Ligand binding to the multichain immune recognition receptors (MIRRs) leads to receptor triggering and subsequent lymphocyte activation. MIRR signal transduction pathways have been extensively studied, but it is still not clear how binding of the ligand to the receptor is initially communicated across the plasma membrane to the cells interior. Models proposed for MIRR triggering can be grouped into three categories. Firstly, ligand binding invokes receptor clustering, resulting in the approximation of kinases to the MIRR and receptor phosphorylation. Secondly, ligand binding induces a conformational change of the receptor. Thirdly, upon ligand-binding, receptors and kinases are segregated from phosphatases, leading to a net phosphorylation of the receptor. In this review, we focus on the homoclustering induced by multivalent ligands, the heteroclustering induced by simultaneous binding of the ligand to the MIRR and a coreceptor and the pseudodimer model.

Introduction

Multichain immune recognition receptor (MIRR) family members are transmembrane multiprotein complexes that are activated by binding to their appropriate antigens (or ligands). This binding event transmits information across the plasma membrane to the cytoplasm in a process termed signal transduction. The first measurable biochemical change that takes place upon antigen recognition is phosphorylation of tyrosine residues in the cytoplasmic portions of the receptor itself. These tyrosines are part of the immunoreceptor tyrosine-based activation motif (ITAM).[1] Depending on the receptor type, ITAM tyrosines are phosphorylated by kinases of the Src-family[2] and Syk/ZAP70-family.[3] Phosphorylation of the MIRR is the critical event in initiating the signaling cascades, since phosphotyrosines serve as binding sites for proteins with src homology 2 (SH2) domains. Consequently, these proteins are recruited to the receptor and activate signaling pathways.

Ligand binding to the MIRRs induces molecular changes in the receptor structure or in the distribution of the receptor within the membrane, leading to receptor activation. Although crucial for lymphocyte activation, these events still remain an enigma. One experimental possibility to decipher the molecular mechanisms underlying receptor activation is to study the structural requirements of the ligand. Ligands of interest are altered ligands that still bind to the receptor, but are unable to activate the receptor. It was demonstrated that the T-cell antigen receptor (TCR) and the high-affinity IgE receptor (FcεRI) are only activated by bi- or multivalent ligands.[4-7] This has led to the clustering models of signal initiation by the MIRRs. In case of the B-cell antigen receptor (BCR) however, it is still controversial whether monovalent antigen can induce BCR triggering.[8]

*Corresponding Author: Wolfgang W.A. Schamel—Department of Molecular Immunology, Max Planck-Institute for Immunobiology and Faculty of Biology, University of Freiburg, Stuebeweg 51, 79108 Freiburg, Germany. Email: schamel@immunbio.mpg.de

Multichain Immune Recognition Receptor Signaling: From Spatiotemporal Organization to Human Disease, edited by Alexander B. Sigalov. ©2008 Landes Bioscience and Springer Science+Business Media.

In this chapter, emphasis will be taken on the TCR, since it is the best studied MIRR. The ligand for the TCR is a peptide bound to a major histocompatibility complex molecule (MHCp) on the surface of antigen-presenting cells (APCs). One single TCR can bind to several distinct, but related, MHCp that differ in the exact sequence of the peptide. MHC presenting antigenic peptides (aMHCp) bind with moderate affinity and lead to activation of the mature T-cell. MHC molecules also present self-peptides (sMHCp) derived from endogenous proteins. sMHCp bind with weak affinity to the TCR and cannot activate the cells by themselves. TCR engagement to sMHCp, however, leads to positive selection of immature T-cells in the thymus and T-cell survival in the periphery. In the context of the immune response, both types of peptides are copresented by the APC. In this case, self-peptides aid in the recognition of agonistic peptides by the T-cell.[9-12] These features of the TCR allow the development of sophisticated MIRR triggering models (pseudodimer and permissive geometry models), which have not been extended to the BCR or FcεRI.

Homoclustering

Early stimulation experiments have shown that bivalent anti-TCR-CD3 antibodies can activate T-cells, whereas monovalent Fab fragments of these antibodies fail to do so, although they do bind to the TCR-CD3 complex.[13,14] The same holds true for the FcεRI.[6,7] Later on, soluble MHCp monomers or multimers were prepared and, as expected, only the multimeric forms could trigger the TCR.[4,5] These experiments indicate that the ligands for the TCR and FcεRI have to be multivalent, in order to be functional. This implies that two or more receptors have to bind simultaneously to one ligand molecule in order to be activated. In conjunction with the hypothesis that individual receptor molecules are distributed equally on the cell surface, these findings led to the homoclustering model of MIRR activation[15] (Fig. 1). This model has also been named cross-linking, aggregation or multimerization model. Support for this model was provided by imaging approaches that have shown that MIRRs cluster on the cell surface upon multivalent ligand binding and that TCR microclusters are formed upon interaction with an APC.[16] Indeed,

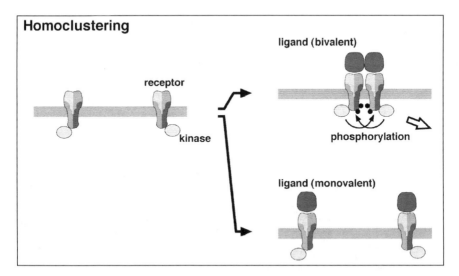

Figure 1. Homoclustering model. Monovalent receptors with associated kinases are individually expressed on the cell surface and randomly distributed. Stimulation by bi- or multivalent ligand leads to clustering of the receptors, since the ligand simultaneously binds two or more receptors. In these clusters, the kinases can transphosphorylate each other and the receptors. This represents the first step in the activation of the cell. Monovalent ligands do not cluster the receptors and thus cannot induce phosphorylation and activation. Black dots represent phosphorylated tyrosine residues. The open arrow shows activation of downstream signaling cascades.

these TCR microclusters have been shown to be the structures that initiate and sustain TCR signaling (see Chapter by Dustin).

Within the homoclustering model, each monomeric receptor is supposed to be associated with a kinase, that needs cross-wise phosphorylation for its full activity. In addition, it is possible that due to sterical reasons the kinase cannot phosphorylate the receptor that it is bound to. As random encounters of two receptors by diffusion are infrequent, the amount of phosphorylated receptor stay below the activation threshold of the cell. If a bi- or multivalent ligand simultaneously binds to at least two MIRRs and brings them into close proximity, the kinases are enabled to transphosphorylate each other and the neighboring MIRR. Thus, the cell becomes activated. A variant of the homoclustering model is the signaling chain homooligomerization (SCHOOL) model,[17] which is subject of chapters by Sigalov.

Concerning B-cell activation, studies in the 80th have shown that the higher the valency of an antigen is, the better is the B-cell response (the immunon model).[18] In addition, monovalent anti-BCR antibody Fab fragments do not trigger the BCR, whereas bivalent antibodies do.[8] Therefore, the homoclustering model also seems to apply for the BCR. This assumption was recently challenged by the finding that a monovalent antigen could activate the BCR.[8] Therefore, the triggering mechanism for the BCR is still a mystery.

The homoclustering model is widely accepted, but has several caveats. First, recent publications have questioned that the receptors are equally distributed on the cell surface.[19-22] Second, there exist antireceptor antibodies that do dimerize the receptor but do not elicit a full activation response.[6,23,24] Therefore, an alternative model has been put forward, known as the permissive geometry model[25] (Chapter by Minguet and Schamel).

The physiological ligand for the TCR is not a soluble but membrane-bound MHCp molecule. The MHC molecules of class I and II are most likely multimeric on the surface of the APCs.[26,27] Therefore, the requirement of the homoclustering model for multivalent ligand seems to hold true. In addition to a possible aMHCp, an APC also presents sMHCp. The affinity (or half-life) of an sMHCp-TCR interaction is too low (or short) to yield an effective sMHCp-TCR complex. This is the reason why sMHCp do not activate T-cells. Interestingly, an APC can activate the T-cell if it presents only around 10 aMHCp molecules that are intermingled within 10,000-100,000 sMHCp.[9] In these situations, the probability of two antigenic peptides being present on two adjacent MHC molecules is negligible. Thus, one has to postulate that TCRs can be activated by MHCp heterodimers in which one MHC molecule is loaded with the antigenic peptide and the second with a self-peptide.[9-11] Indeed, recent work has shown that it is possible to activate T-cells with those MHCp heterodimers.[12] The homoclustering model cannot explain the activating effect of the MHC heterodimers (aMHCp-sMHCp).

Heteroclustering

Variable portions of the MHCp complex, representing the peptide sequences and the adjacent regions of the MHC molecules, bind to the variable immunoglobulin domains of the TCRα/β.[28,29] In addition, constant regions of the MHC class I and class II molecules bind to the CD8 and CD4 coreceptors, respectively.[30] The simultaneous binding of MHCp to the TCR and the coreceptor leads to a heteroclustering of TCR with CD4 or CD8. Interestingly, the cytoplasmic tails of CD4 and CD8 have been shown to interact constitutively with the Src-family kinase Lck[31] that when recruited to the TCR, could phosphorylate the receptor (heteroclustering model, Fig. 2).

Formation of TCR-CD4 heterodimers by chimeric antibodies leads to TCR triggering,[32] enforcing the heteroclustering model. In favour of this model are also findings that demonstrate a requirement for the coreceptor in T-cell activation.[33] However, in other experimental systems the coreceptor is not needed.[34] The reason for this discrepancy is not clear but might be related to the affinity and amount of ligand used. The findings that a substantial proportion of TCRs is constitutively associated with coreceptors[35] and that monomeric MHCp does not activate T-cells,[4,5] argue against the heteroclustering model. Also, this model cannot explain why T-cells are triggered by anti-TCR-CD3 antibodies.[13,14] However, it seems likely that co-engagement of

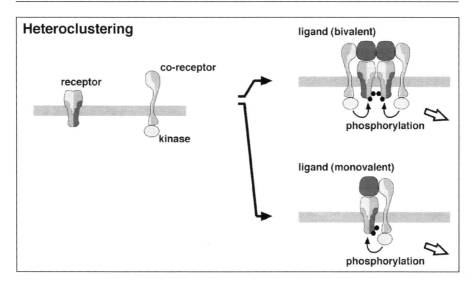

Figure 2. Heteroclustering model. TCRs are not constitutively associated to kinases, but the coreceptors CD4 and CD8 are. Multimeric and monomeric MHCp ligands simultaneously engage the TCR and the coreceptor. Hence, the kinase is brought into the vicinity of the TCR allowing phosphorylation with subsequent activation of the cell. Notably, the heteroclustering model does not apply to the BCR or the FcεRI.

the TCR-CD3 with CD4 or CD8 lowers the threshold for T-cell activation by augmenting the avidity of MHCp binding to the T-cell and/or by recruiting additional kinases (and possibly other signaling proteins) to the receptor.

The antigens for the BCR and the FcεRI can be multimeric substances of nearly any structure, shape and geometry. Therefore, signaling cannot be initiated by clustering of these MIRRs with a second receptor on the cell surface. Nevertheless, there exist coreceptors for these MIRRs that can either suppress or augment the signal emitted by the BCR or the FcεRI. Antigen that is already bound to an antibody can promote heteroclustering with FcγRs which bind to the constant regions of the antibody. FcγRs associate with the SH2 domain-containing protein tyrosine phosphatase SHP-1 and therefore weaken the signal emitted by the BCR or the FcεRI. In contrast, antigen that is bound by complement can simultaneously bind to the antigen receptor and the complement receptor CD19/CD21, resulting in an augmentation of the signal.

Pseudodimer Model

With the beginning of the new millennium it became clear that the sMHCp complexes play a role not only in positively selecting T-cells in the thymus but also in T-cell activation.[9-11] They are accumulated in the center of the immune synapse and enhance recognition of aMHCp. Final proof of their stimulating activity came from the finding that aMHCp-sMHCp heterodimers can trigger the TCR.[12] This cooperativity between antigenic and self-peptides in T-cell stimulation together with the data concerning the role of the coreceptor CD4 lead to the development of the pseudodimer model (Fig. 3).[9] In this model, one TCR binds to aMHCp, whereas the second TCR binds to sMHCp. Importantly, the low-affinity sMHCp-TCR interaction is strengthened by the binding of MHC to CD4. At the same, time this CD4 molecule interacts with the first TCR. Indeed, crystallography studies suggested that in a trimeric CD4-pMHC-TCR complex the cytoplasmic tail CD4 cannot come close to the TCR.[30] Thus, the Src-family kinase that is associated to the CD4 cytoplasmic tail can phosphorylate only the second TCR. The strength of this model is that it takes into account the effects of the self-peptides. However, the model cannot

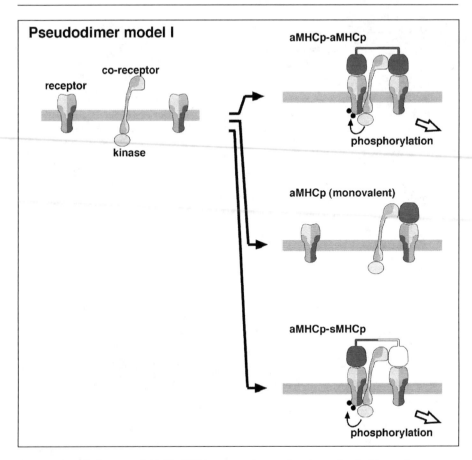

Figure 3. Pseudodimer model I. The TCR is monovalent and not bound to the kinase. In contrast, the coreceptor CD4 is associated to a Src-family kinase. Upon bi- or multivalent MHCp-binding TCRs become clustered and CD4 that binds to both the first TCR and the second MHCp, is recruited, allowing TCR phosphorylation (upper panel). Monovalent MHCp would be inactive (middle panel). At low antigenic peptide concentrations, aMHCp-sMHCp heterodimers are presented by the APC. This allows formation of a pseudodimer:one TCR binds aMHCp and the second TCR binds sMHCp (lower panel). The weak sMHC-TCR interaction is enforced by CD4 that binds to sMHCp and the first TCR. Abbreviations: antigenic peptide-MHC complex, aMHCp; self-peptide-MHC complex, sMHCp.

explain the coreceptor-independent stimulation of T-cells. In a new version of the pseudodimer model, CD4 interacts with aMHCp and the TCR bound to sMHCp (Fig. 4).[12]

Homo- and Heteroclustering and Lipid Rafts

Ligand-induced clustering of the MIRRs could also play a role in changing the lipid environment of the receptor within the plasma membrane, thereby initiating signal transduction. Central to this concept are the so-called lipid rafts (see Chapter by Dopfer et al). They represent microdomains that are enriched in sphingolipids and cholesterol and devoid of phospholipids.[36] Membrane proteins are thought to be located either inside (kinases of the Src-family) or outside (e.g., CD45) the lipid rafts. Early studies have suggested that most of the resting TCR, BCR and FcεRI receptor complexes are located outside lipid rafts. Upon ligand binding, some of them move into the rafts and become phosphorylated (Fig. 5A).[37-40] When lipid rafts are disrupted by

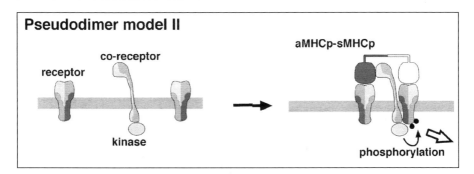

Figure 4. Pseudodimer model II. In the latest version of the pseudodimer model, the coreceptor CD4 binds to aMHCp but not to sMHCp. At the same time, CD4 binds to the second TCR that also associates to the sMHCp. Consequently, kinases bound to the cytoplasmic tail of CD4 phosphorylate the second TCR.

extracting cholesterol from the membrane, stimulation of the MIRRs does not lead to an increase in tyrosine phosphorylation, suggesting that activation of the treated cells is blocked.[41] Likewise, mutants of Src-kinase family members unable to localize to lipid rafts do not initiate signaling upon receptor triggering.[42]

The fundamental question of this model is the molecular change that ligand binding exerts on the receptor, making it move to the lipid rafts. Since signal processes are not required, as shown by inhibiting Src-kinases or disrupting the actin cytoskeleton.[39,43,44] This molecular change might be one of the crucial steps of signal initiation. For the FcεRI, an important role of its transmembrane region has been shown, indicating that interactions with membrane lipids might be critical.[45] Therefore, it can be postulated that monomeric receptors have a low affinity for lipid rafts. Homoclustering of the receptors by multivalent ligands increases their avidity for rafts so that the receptor oligomers move to lipid rafts.[46]

In contrast, other research groups have found that the receptors are already present in lipid rafts prior to stimulation.[43,47] In this case, it is postulated that receptor and kinase are in different separated rafts and that homoclustering of the receptors leads to raft coalescence that also includes kinase-possessing rafts (Fig. 5B).[48] Thus, receptors and Src-kinases come into contact only upon multivalent ligand binding.

Alternatively, CD4 and CD8 might be present in the same lipid rafts as the Src-kinases[49] and therefore heteroclustering with the TCR would bring the MIRR in contact with the kinase (Fig. 6). As discussed above, monomeric ligand cannot activate the MIRRs making the heteroclustering model rather unlikely.

Proving or disproving the "lipid raft localization" models is hampered by the current technical difficulties in studying lipid rafts[50] that might be more dynamic than previously appreciated.

The PreTCR and PreBCR

The pre-antigen receptors (preBCR and preTCR) are expressed on immature lymphocytes and play a role in their development to mature cells. Most likely, these receptors signal in the absence of ligand. Studies of this constitutive signal have shown that both receptors localize to lipid rafts without the need for ligand binding.[51,52] In addition, they might be present in preformed oligomeric structures.[53,54] These findings provide further support for the clustering models of MIRR triggering.

Acknowledgements

M.R. was supported by the European Union founded grant EPI-PEP-VAC and by the Deutsche Forschungsgemeinschaft (DFG) through the SFB620 and SFB388. W.W.A.S was supported by the DFG through the Emmy Noether program and the SFB620.

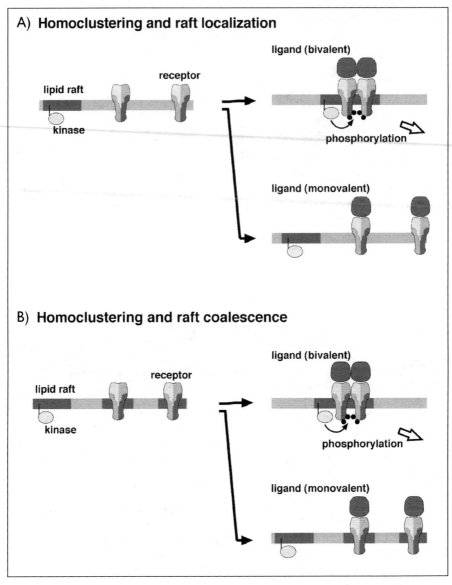

Figure 5. Homoclustering and lipid rafts. A) In the absence of ligand binding, monomeric MIRRs are located outside the cholesterol-rich microdomains, the so-called lipid rafts, due to low affinity of the interaction between MIRR and raft. Because of their fatty acid modification, the Src-family kinases are located inside the rafts. Homoclustering of the receptor by bi- or multivalent ligand increases the avidity of the receptor-raft interaction. Therefore, clustered MIRRs move to the rafts where they become phosphorylated. Monovalent ligands do not induce MIRR clustering and do not lead to colocalization of the receptor with the kinases. B) Alternatively, receptors and kinases are constitutively raft-associated but partitioned into distinct rafts. Homoclustering of the MIRR-associated rafts by bi- or multivalent ligand binding induces raft coalescence including nonreceptor rafts. Thus, the kinases come into proximity to the MIRRs. Again, monomeric ligands are inactive.

Heteroclustering and raft localization

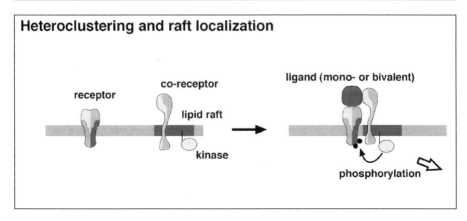

Figure 6. Heteroclustering and lipid rafts. Similar to that suggested for the heteroclustering model (Fig. 2), the ligand (MHCp) binds to the TCR and the coreceptor and simultaneously brings the TCR into the vicinity of the coreceptor CD4 or CD8 . In the "heteroclustering model for raft localization", the coreceptors are localized to the same rafts as the Src-kinases but do not directly bind to the kinases. If TCRs and kinases become coclustered by mono- or multivalent MHCp (not shown), the receptor becomes phosphorylated, activating signaling pathways.

References

1. Reth M. Antigen receptor tail clue. Nature 1989; 338:383.
2. Iwashima M, Irving BA, van Oers NS et al. Sequential interactions of the TCR with two distinct cytoplasmic tyrosine kinases. Science 1994; 263(5150):1136-1139.
3. Rolli V, Gallwitz M, Wossning T et al. Amplification of B-cell antigen receptor signaling by a Syk/ITAM positive feedback loop. Mol Cell 2002; 10(5):1057-1069.
4. Boniface JJ, Rabinowitz JD, Wülfing C et al. Initiation of signal transduction through the T-cell receptor requires the peptide multivalent engagement of MHC ligands. Immunity 1998; 9:459-466.
5. Cochran JR, Cameron TO, Stern LJ. The relationship of MHC-peptide binding and T-cell activation probed using chemically defined MHC class II oligomers. Immunity 2000; 12(3):241-250.
6. Ortega E, Schweitzer-Stenner R, Pecht I. Possible orientational constraints determine secretory signals induced by aggregation of IgE receptors on masT-cells. EMBO J 1988; 7(13):4101-4109.
7. Erickson J, Kane P, Goldstein B et al. Cross-linking of IgE-receptor complexes at the cell surface:a fluorescence method for studying the binding of monovalent and bivalent haptens to IgE. Mol Immunol 1986; 23(7):769-781.
8. Kim YM, Pan JY, Korbel GA et al. Monovalent ligation of the B-cell receptor induces receptor activation but fails to promote antigen presentation. Proc Natl Acad Sci USA 2006; 103(9):3327-3332.
9. Irvine DJ, Purbhoo MA, Krogsgaard M et al. Direct observation of ligand recognition by T-cells. Nature 2002; 419:845-849.
10. Wülfing C, Sumen C, Sjaastad MD et al. Costimulation and endogenous MHC ligands contribute to T-cell recognition. Nat Immunol 2002; 3:42-47.
11. Stefanova I, Dorfman JR, Germain RN. Self-recognition promotes the foreign antigen sensitivity of naive T-lymphocytes. Nature 2002; 420(6914):429-434.
12. Krogsgaard M, Li QJ, Sumen C et al. Agonist/endogenous peptide-MHC heterodimers drive T-cell activation and sensitivity. Nature 2005; 434(7030):238-243.
13. Chang TW, Kung PC, Gingras SP et al. Does OKT3 monoclonal antibody react with an antigen-recognition structure on human T-cells? Proc Natl Acad Sci USA 1981; 78(3):1805-1808.
14. Kaye J, Janeway CA Jr. The Fab fragment of a directly activating monoclonal antibody that precipitates a disulfide-linked heterodimer from a helper T-cell clone blocks activation by either allogeneic Ia or antigen and self-Ia. J Exp Med 1984; 159(5):1397-1412.
15. Ashwell JD, Klausner RD. Genetic and mutational analysis of the T-cell antigen receptor. Annu Rev Immunol 1990; 8:139-167.

16. Yokosuka T, Sakata-Sogawa K, Kobayashi W et al. Newly generated T-cell receptor microclusters initiate and sustain T-cell activation by recruitment of Zap70 and SLP-76. Nat Immunol 2005; 6(12):1253-1262.
17. Sigalov A. Multi-chain immune recognition receptors:spatial organization and signal transduction. Semin Immunol 2005; 17(1):51-64.
18. Dintzis HM, Dintzis RZ, Vogelstein B. Molecular determinants of immunogenicity:the immunon model of immune response. Proc Natl Acad Sci USA 1976; 73(10):3671-3675.
19. Schamel WW, Reth M. Monomeric and oligomeric complexes of the B-cell antigen receptor. Immunity 2000; 13(1):5-14.
20. Wilson BS, Pfeiffer JR, Oliver JM. Observing FcepsilonRI signaling from the inside of the masT-cell membrane. J Cell Biol 2000; 149(5):1131-1142.
21. Schamel WW, Arechaga I, Risueno RM et al. Coexistence of multivalent and monovalent TCRs explains high sensitivity and wide range of response. J Exp Med 2005; 202:493-503.
22. Alarcon B, Swamy M, van Santen HM et al. T-cell antigen-receptor stoichiometry: Preclustering for sensitivity. EMBO Rep 2006; 7(5):490-495.
23. Rojo JM, Janeway CA Jr. The biological activity of anti-T-cell receptor variable region monoclonal antibodies is determined by the epitope recognized. J Immunol 1988; 140:1081-1088.
24. Gil D, Schamel WW, Montoya M et al. Recruitment of Nck by CD3 epsilon reveals a ligand-induced conformational change essential for T-cell receptor signaling and synapse formation. Cell 2002; 109(7):901-912.
25. Minguet S, Swamy M, Alarcon B et al. Full activation of the T-cell antigen receptor requires both clustering and conformational changes at CD3. Immunity 2007; 26(1):43-54.
26. Schafer PH, Pierce SK, Jardetzky TS. The structure of MHC class II: A role for dimer of dimers. Semin Immunol 1995; 7(6):389-398.
27. Krishna S, Benaroch P, Pillai S. Tetrameric cell-surface MHC class I molecules. Nature 1992; 357(6374):164-167.
28. Garboczi DN, Ghosh P, Utz U et al. Structure of the complex between human T-cell receptor, viral peptide and HLA-A2. Nature 1996; 384(6605):134-141.
29. Garcia KC, Degano M, Stanfield RL et al. An alphabeta T-cell receptor structure at 2.5 A and its orientation in the TCR-MHC complex. Science 1996; 274(5285):209-219.
30. Wang JH, Meijers R, Xiong Y et al. Crystal structure of the human CD4 N-terminal two-domain fragment complexed to a class II MHC molecule. Proc Natl Acad Sci USA 2001; 98(19):10799-10804.
31. Kim PW, Sun ZY, Blacklow SC et al. A zinc clasp structure tethers Lck to T-cell coreceptors CD4 and CD8. Science 2003; 301(5640):1725-1728.
32. Madrenas J, Chau LA, Smith J et al. The efficiency of CD4 recruitment to ligand-engaged TCR controls the agonist/partial agonist properties of peptide-MHC molecule ligands. J Exp Med 1997; 185(2):219-229.
33. Parnes JR. Molecular biology and function of CD4 and CD8. Adv Immunol 1989; 44:265-311.
34. Viola A, Salio M, Tuosto L et al. Quantitative contribution of CD4 and CD8 to T-cell antigen receptor serial triggering. J Exp Med 1997; 186(10):1775-1779.
35. Suzuki S, Kupsch J, Eichmann K et al. Biochemical evidence of the physical association of the majority of CD3 delta chains with the accessory/coreceptor molecules CD4 and CD8 on nonactivated T-lymphocytes. Eur J Immunol 1992; 22(10):2475-2479.
36. Simons K, Toomre D. Lipid rafts and signal transduction. Nat Rev Mol Cell Biol 2000; 1(1):31-39.
37. Xavier R, Brennan T, Li Q et al. Membrane compartmentation is required for efficient T-cell activation. Immunity 1998; 8:723-732.
38. Montixi C, Langlet C, Bernard AM et al. Engagement of T-cell receptor triggers its recruitment to low-density detergent-insoluble membrane domains. EMBO J 1998; 17(18):5334-5348.
39. Field KA, Holowka D, Baird B. Compartmentalized activation of the high affinity immunoglobulin E receptor within membrane domains. J Biol Chem 1997; 272(7):4276-4280.
40. Cheng PC, Dykstra ML, Mitchell RN et al. A role for lipid rafts in B-cell antigen receptor signaling and antigen targeting. J Exp Med 1999; 190(11):1549-1560.
41. Langlet C, Bernard AM, Drevot P et al. Membrane rafts and signaling by the multichain immune recognition receptors. Curr Opin Immunol 2000; 12(3):250-255.
42. Kabouridis PS, Magee AI, Ley SC. S-acylation of LCK protein tyrosine kinase is essential for its signalling function in T-lymphocytes. EMBO J 1997; 16(16):4983-4998.
43. Janes PW, Ley SC, Magee AI. Aggregation of lipid rafts accompanies signaling via the T-cell antigen receptor. J Cell Biol 1999; 147(2):447-461.
44. Cheng PC, Brown BK, Song W et al. Translocation of the B-cell antigen receptor into lipid rafts reveals a novel step in signaling. J Immunol 2001; 166(6):3693-3701.

45. Field KA, Holowka D, Baird B. Structural aspects of the association of FcepsilonRI with detergent-resistant membranes. J Biol Chem 1999; 274(3):1753-1758.
46. Pierce SK. Lipid rafts and B-cell activation. Nat Rev Immunol 2002; 2(2):96-105.
47. Drevot P, Langlet C, Guo XJ et al. TCR signal initiation machinery is pre-assembled and activated in a subset of membrane rafts. EMBO J 2002; 21(8):1899-1908.
48. Janes PW, Ley SC, Magee AI et al. The role of lipid rafts in T-cell antigen receptor (TCR) signalling. Semin Immunol 2000; 12(1):23-34.
49. Cinek T, Hilgert I, Horejsi V. An alternative way of CD4 and CD8 association with protein kinases of the Src family. Immunogenetics 1995; 41(2-3):110-116.
50. Munro S. Lipid rafts: Elusive or illusive? Cell 2003; 115(4):377-388.
51. Guo B, Kato RM, Garcia-Lloret M et al. Engagement of the human preB-cell receptor generates a lipid raft-dependent calcium signaling complex. Immunity 2000; 13(2):243-253.
52. Saint-Ruf C, Panigada M, Azogui O et al. Different initiation of preTCR and gammadeltaTCR signalling. Nature 2000; 406(6795):524-527.
53. Ohnishi K, Melchers F. The nonimmunoglobulin portion of lambda5 mediates cell-autonomous preB-cell receptor signaling. Nat Immunol 2003; 4(9):849-856.
54. Yamasaki S, Ishikawa E, Sakuma M et al. Mechanistic basis of preT-cell receptor-mediated autonomous signaling critical for thymocyte development. Nat Immunol 2006; 7(1):67-75.

CHAPTER 7

Segregation Models

Elaine P. Dopfer, Mahima Swamy, Gabrielle M. Siegers, Eszter Molnar,
Jianying Yang and Wolfgang W.A. Schamel*

Abstract

Many antigen receptors of the immune system belong to the family of multichain immune recognition receptors (MIRRs). Binding of ligand (antigen) to MIRR results in receptor phosphorylation, triggering downstream signaling pathways and cellular activation. How ligand binding induces this phosphorylation is not yet understood. In this Chapter, we discuss two models exploring the possibility that kinases and phosphatases are intermingled on the cell surface. Thus, in resting state, MIRR phosphorylation is counteracted by dephosphorylation. Upon ligand binding, phosphatases are removed from the vicinity of the MIRR and kinases, such that phosphorylated MIRRs can accumulate (segregation models). In the first model, clustering of MIRRs by multivalent ligand leads to their concentration in lipid rafts where kinases, but not phosphatases, are localized. The second model takes into account that the MIRR-ligand pair needs close apposition of the two cell membranes, in cases where ligand is presented by an antigen-presenting cell. The intermembrane distance is too small to accommodate transmembrane phosphatases, which possess large ectodomains. Thus, phosphatases become spatially separated from the MIRRs and kinases (kinetic-segregation model).

Introduction

The B- and T-cell antigen receptors (BCR and TCR, respectively) and the high-affinity IgE receptor on mast cells (FcεRI) are transmembrane multiprotein complexes. They belong to the family of multichain immune recognition receptors (MIRRs). Upon binding to their appropriate antigens (or ligands), information is transmitted across the plasma membrane to the cytoplasm where signaling cascades are activated. These signal transduction events can lead to cell activation, proliferation, differentiation, survival or apoptosis. Cell fate depends on the nature, strength and context of the stimulus as well as on the developmental stage of the cell.

Ligand binding induces phosphorylation of the MIRR on cytoplasmic tyrosine residues that are part of the immunoreceptor tyrosine-based activation motif (ITAM), which has the consensus sequence $YxxI/Lx_{6-7}YxxI/L$.[1] These residues can be phosphorylated by Src-[2] and Syk/ZAP70-family kinases.[3] Phosphorylation of the MIRR ITAMs is the critical event initiating signaling cascades. Phosphotyrosines serve as binding sites for signaling proteins with Src-homology 2 (SH2) domains. These proteins are thereby recruited to the plasma membrane where they become activated and interact with other signaling proteins. Subsequent signaling cascades involve lymphocyte-specific signaling elements, such as Syk and ZAP70 kinases, cytoplasmic adaptors SLP-65 and SLP-76 and transmembrane adaptors NTAL and LAT as well as ubiquitously expressed signaling proteins, e.g., molecules of the Ras-Erk and PLCγ-Ca^{2+} pathways.

*Corresponding Author: Wolfgang W.A. Schamel—Department of Molecular Immunology, Max Planck-Institute for Immunobiology and Faculty of Biology, University of Freiburg, Stuebeweg 51, 79108 Freiburg, Germany. Email: schamel@immunbio.mpg.de

Multichain Immune Recognition Receptor Signaling: From Spatiotemporal Organization to Human Disease, edited by Alexander B. Sigalov. ©2008 Landes Bioscience and Springer Science+Business Media.

The molecular mechanism as to how ligand binding induces MIRR phosphorylation is still a matter of debate. In the conformational change and permissive geometry models[4,5] (Chapters 10 and 11), ligand binding induces a structural change in the receptor (or within receptor oligomers) that enables phosphorylation of cytoplasmic domains. In the homo- or hetero-clustering models of MIRR activation[6] (Chapter 6), kinases are pre-associated with the MIRR or coreceptor, respectively. In the resting state, the receptor is not phosphorylated by the associated kinase. This requires multivalent ligand binding, which leads to the approximation of receptors and kinases, resulting in receptor transphosphorylation.

The clustering models seem to be at odds with the finding that artificial inhibition of phosphatases by phosphatase inhibitors, such as H_2O_2 or pervanadate, results in massive MIRR phosphorylation independent of ligand engagement. This has been shown for both the TCR and the BCR.[7,8] The finding that inhibition of phosphatases mimics ligand engagement, suggests that kinase activity is counterbalanced by phosphatases in resting cells.[9] The main phosphatase in the plasma membrane of lymphocytes is CD45.[10] This molecule occupies as much as 10% of the cell surface, has high constitutive enzymatic activity and broad specificity,[10] allowing it to suppress MIRR signaling in unstimulated cells. Therefore, antigen binding could result in phosphatase inhibition[11] or removal of the phosphatase from the vicinity of the MIRR,[12,13] thus initiating signal transduction. How ligand binding could induce segregation of receptors and their associated kinases from phosphatases is the subject of this chapter.

Lipid Rafts

MIRRs transduce the signal ensuing from ligand binding across the plasma membrane. To understand this process, it is necessary to consider the structure and organization of biological membranes. According to the fluid-mosaic model of Singer and Nicolson (1972), lipid molecules are distributed homogeneously in each leaflet of the membrane bilayer.[14] Research done over the past two decades has shown that this model is incorrect.[15] In biological membranes, lipids are asymmetrically distributed over the exoplasmic and cytoplasmic leaflets. Eukaryotic cells possess up to 1,000 different lipids with various chemical properties[16] and distinct affinities towards each other. Therefore, it makes sense that certain lipids tend to aggregate with one another, excluding others. The most extensively studied example is that of the so-called lipid rafts.

Lipid rafts are microdomains of the plasma membrane. Their outer leaflet is enriched in glycosphingolipids and cholesterol. Glycosphingolipids consist mostly of long and saturated hydrocarbon fatty acid chains and large hydrophilic head groups. Their composition allows cholesterol to be tightly intercalated below the head groups (Fig. 1) and together these molecules form distinct liquid-ordered phases in the lipid bilayer. These microdomains are dispersed in a liquid-disordered matrix of unsaturated glycerolipids that have kinked fatty acid chains. The cytoplasmic leaflet of lipid rafts is less well characterised but it is probably enriched in phospholipids with saturated fatty acids and cholesterol.[17] To date, it is not completely understood how the two leaflets are coupled. The glycosphingolipid and cholesterol-enriched microdomains are called lipid rafts since they float up in a sucrose gradient after solubilization in certain non-ionic detergents.

Importantly, membrane segments of proteins can have variable affinities for the lipid raft environment. Dual fatty acid protein modifications, especially palmitoylation, or glycosylphosphatidylinositol (GPI)-anchoring, lead to raft localization (Fig. 1). Examples of such modified proteins are Src-family kinases and the transmembrane adaptor protein LAT. Most proteins with unmodified transmembrane regions are excluded from rafts, for example, the phosphatase CD45. It has been suggested that MIRRs are present in rafts prior to stimulation.[18,19] Others who reported that MIRRs were located outside lipid rafts[20-23] were probably unable to detect the raft localization due to harsher lysis conditions.[19]

Lipid rafts are thought to be platforms for MIRR signaling.[24] Firstly, upon stimulation, lipid raft fractions are enriched by phosphorylated proteins, including MIRRs. This is the case for the TCR, the BCR and the FcεRI.[20-23] Secondly, disruption of lipid rafts by extraction of cholesterol from the membrane abolishes MIRR phosphorylation and activation upon ligand binding.[25]

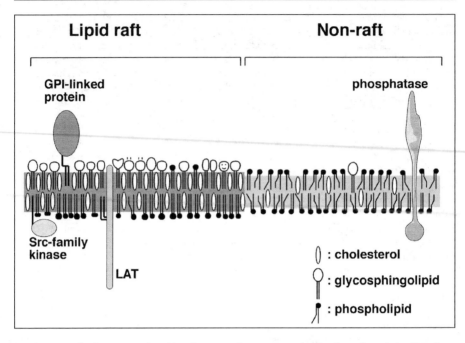

Figure 1. Lipid rafts. Proteins and lipids are not homogenously distributed on the cell surface but rather organized into membrane microdomains called lipid rafts. Lipid rafts are enriched in glycosphingolipids and cholesterol. Many proteins with fatty acid modifications are localized inside these microdomains. Src-family kinases and transmembrane adaptor proteins, e.g., LAT, are raft-localized due to their covalent attachment to fatty acids such as palmitic acid. In lipid rafts, these signaling proteins form "signaling platforms". Phospholipids with unsaturated fatty acid chains are the major constituents of the nonraft plasma membrane. Most unmodified transmembrane proteins are localized outside the rafts, e.g., phosphatases CD45 and CD148.

Thirdly, mutants of Src-kinase family members, which do not localize to lipid rafts, cannot initiate signaling upon receptor triggering.[26]

Analysis of lipid rafts is currently hampered by technical difficulties. The size, heterogeneity and dynamics of these microdomains are still controversial. Some researchers even argue that lipid rafts are merely experimental artefacts and that they might not exist on the surface of living cells at 37°C.[27]

Segregation by Raft Clustering

MIRRs are thought to be constitutively localised to lipid rafts that also contain Src-family kinases. In contrast, the phosphatase CD45 is excluded from these microdomains.[28] Due to the small size of rafts containing only few proteins, receptors might be constantly surrounded by phosphatases that suppress MIRR phosphorylation and auto-phosphorylation of the Src-family kinases at their activatory tyrosine within the kinase domain. According to the segregation-by-raft-clustering model, stimulation with a multivalent ligand results in homoclustering of MIRRs and subsequently, rafts. This leads to clustering of the MIRRs and kinases but excludes phosphatases[29] (Fig. 2). The kinase-phosphatase equilibrium in the vicinity of receptors is shifted towards phosphorylation, thereby initiating signal transduction. Importantly, many downstream signaling molecules (e.g., LAT, phospholipase C and phosphoinositides) are also present in lipid rafts.

This model is supported by the findings that clustering of lipid rafts in the absence of a TCR ligand induces signaling events very similar to those following TCR triggering.[18] CD45 deficiency

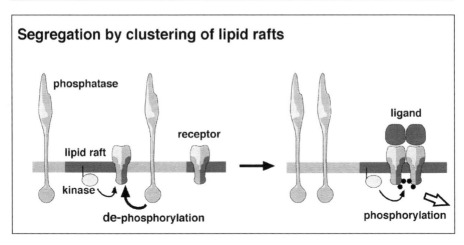

Figure 2. Segregation of rafts from nonrafts. Receptors and lipid-raft localized Src-family kinases are associated in the rafts or at their boundaries. Constitutive receptor phosphorylation is counterbalanced by massive dephosphorylation through phosphatases surrounding the rafts. Importantly, despite phosphatases are not raft-associated, dephosphorylation still dominates and MIRRs are not activated. Upon bi- or multivalent ligand binding, rafts coalesce and/or MIRRs move into the rafts. Spatial separation of phosphatases from the MIRRs leads to rapid accumulation of phosphorylated receptors. This event is the first step in cell activation. Black dots represent phosphorylated tyrosine residues and the open arrow indicates activation of downstream signaling cascades.

is not informative in this context, since this phosphatase is also required to maintain the Src-family kinases in an active state.[13]

Kinetic-Segregation Model

The TCR and the peptide-bound major histocompatibility complex (MHCp) are small molecules with an extracellular size of 7 nm,[30,31] whereas the phosphatases CD45 and CD148 each have a bulky ectodomain sizing 40 and 50 nm, respectively[13,32] (Fig. 3). If the TCR and MHCp are to interact, then the distance between the two opposing cell membranes has to be around 14 nm. Since the TCR-MHCp interaction is of low affinity,[33] apposition of the membranes is likely accomplished by other proteins. Thus, the interaction of CD2 and CD48/58 (mouse/human) proteins expressed on T-cells and antigen presenting cells (APCs), respectively, has been proposed to generate an intermembrane distance of 14 nm.[12] Apposition of the two membranes in so-called "close-contact zones"[32] or "microclusters"[34] would exclude large proteins, e.g., the phosphatases CD45 and CD148 (Fig. 4). Therefore, constitutive phosphorylation of the MIRR and the activatory tyrosine residue of the Src-kinases is not counteracted by membrane-bound phosphatases and active MIRRs can accumulate. Receptors that do not engage the correct ligand rapidly diffuse out of the close-contact zones, resulting in receptor dephosphorylation. Receptors that bind to their ligand with a sufficient half-life stay in these phosphatase-free zones long enough to initiate downstream signaling events. The length of time an individual receptor is in the phosphorylated state determines the signaling outcome. This concept is called kinetic proofreading (Chapter 8).

In summary, the kinetic-segregation model suggests that ligand binding to the TCR shifts the critical kinase-phosphatase equilibrium to support kinases and signal initiation in the microenvironment of the engaged TCR. TCR-ligand-binding kinetics is thereby responsible for the molecular segregation of kinases from phosphatases.

This kinetic-segregation model is supported by the findings that shortening of the extracellular domain of CD148 inhibits TCR triggering, possibly due to misallocation of the phosphatase to the close-contact zones.[35] Likewise, elongation of MHCp, enlarging the TCR-MHCp complex,

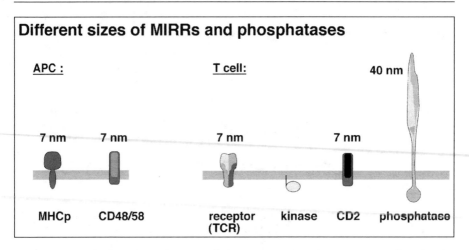

Figure 3. Sizes of selected ectodomains. The TCR and its ligand MHCp are small molecules with an extracellular length of 7 nm. Likewise, CD2, CD48 and CD58 have very similar ectodomain dimensions of approximately 7 nm. In contrast, the transmembrane phosphatases, CD45 and CD148 (not shown), have large 30-50 nm ectodomains. Src-family kinases are attached to the plasma membrane from the cytoplasmic side and do not contain extracellular sequences.

significantly increases the intermembrane distance.[36] This results in inhibition of TCR triggering, probably due to accessibility of phosphatases to the sites of TCR engagement.[36]

The role of CD45 in this context is controversial since this phosphatase also dephosphorylates Src-family kinases on inhibitory tyrosine residues. Thus, CD45 plays both negative (MIRR dephosphorylation) and positive (kinase activation) roles. Signaling abrogation in the presence of CD45 mutants in which the ectodomain is shortened, is possibly due to the lack of the CD45 activation function[37] and not necessarily due to its misallocation to the close-contact zones. On the other hand, recent fluorescent imaging experiments showed that CD45 is indeed excluded from the close-contact zones.[36,38] The kinetic-segregation model is also in line with reported CD4 or CD8 coreceptor-independent TCR activation (Chapter 6).

The kinetic-segregation model could also apply to the BCR and FcεRI, when soluble antigen is presented via antibody-antigen complexes bound to APCs. Alternatively, the antigen could be a membrane protein present on the APC surface. It has been demonstrated that during B-cell recognition CD45 is excluded from sites where the BCR accumulates,[39] although involvement of this step in the initiation of BCR triggering was not addressed.

The kinetic-segregation model fails to explain T-cell activation via soluble anti-TCR antibodies,[40,41] soluble bi- or multivalent MHCp,[42,43] or heterodimeric MHCp with one agonist and one self-peptide.[44]

Immune Synapse and Microclusters

TCR recognition of its antigen (MHCp) on an APC results in development of an immune synapse which concentrates the TCRs in its center[45,46] (Chapter 13). B-cells that recognize antigen on the surface of APC also form a synapse.[39,47] Synapse formation depends on the cytoskeleton and signal transduction events in B- and T-cells. Thus, synapse formation cannot be the molecular mechanism that initiates signaling. Recent experiments have shown that the TCRs in the center of the synapse (cSMAC) are in a nonsignaling state,[48,49] whereas those that are in the process of microclustering at the synapse periphery (during transport to the center) are the active, phosphorylated receptors[34,38] (Chapter 13). This is in line with data showing that the CD45 phosphatase is excluded from early TCR microclusters but not from the cSMAC.[38] Probably microclusters are synonymous

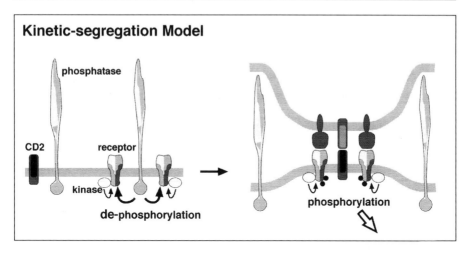

Kinetic-segregation Model

Figure 4. Kinetic-segregation model. Receptors, kinases and transmembrane phosphatases intermingle on the cell surface, resulting in constitutive phosphorylation and dephosphorylation of the MIRR. When the MIRR interacts with its ligand presented by an APC, the two opposing membranes come into close contact. The intermembrane distance in the vicinity of the MIRRs does not provide enough space for the large phosphatases. Therefore, phosphatases are segregated from receptors, resulting in accumulation of phosphorylated MIRRs. Receptors that do not encounter their cognate ligands rapidly move out of these close-contact zones. In contrast, ligand binding keeps the receptor inside phosphatase-free areas for a sufficient amount of time to allow effective initiation of signaling cascades.

with close-contact sites. Since the cSMAC is not required for TCR signaling, it is thought to be a site of TCR degradation.[38,50] In conclusion, the immune synapse concentrates and segregates signaling molecules, representing a consequence rather than the cause of MIRR triggering.

Acknowledgements

G.M.S. was supported by a University of Freiburg Wiedereinstiegsstipendium. W.W.A.S was supported by the Deutsche Forschungsgemeinschaft (DFG) through the Emmy Noether program and the SFB620.

References

1. Reth M. Antigen receptor tail clue. Nature 1989; 338:383.
2. Iwashima M, Irving BA, van Oers NS et al. Sequential interactions of the TCR with two distinct cytoplasmic tyrosine kinases. Science 1994; 263(5150):1136-1139.
3. Rolli V, Gallwitz M, Wossning T et al. Amplification of B-cell antigen receptor signaling by a Syk/ITAM positive feedback loop. Mol Cell 2002; 10(5):1057-1069.
4. Gil D, Schamel WW, Montoya M et al. Recruitment of Nck by CD3 epsilon reveals a ligand-induced conformational change essential for T-cell receptor signaling and synapse formation. Cell 2002; 109(7):901-912.
5. Schamel WW, Risueno RM, Minguet S et al. A conformation- and avidity-based proofreading mechanism for the TCR-CD3 complex. Trends Immunol 2006; 27:176-182.
6. Cochran JR, Aivazian D, Cameron TO et al. Receptor clustering and transmembrane signaling in T-cells. Trends Biochem Sci 2001; 26(5):304-310.
7. Secrist JP, Burns LA, Karnitz L et al. Stimulatory effects of the protein tyrosine phosphatase inhibitor, pervanadate, on T-cell activation events. J Biol Chem 1993; 268(8):5886-5893.
8. Wienands J, Larbolette O, Reth M. Evidence for a preformed transducer complex organized by the B-cell antigen receptor. Proc Natl Acad Sci USA 1996; 93(15):7865-7870.
9. Cooper JA, MacAuley A. Potential positive and negative autoregulation of p60c-src by intermolecular autophosphorylation. Proc Natl Acad Sci USA 1988; 85(12):4232-4236.

10. Trowbridge IS, Thomas ML. CD45: An emerging role as a protein tyrosine phosphatase required for lymphocyte activation and development. Annu Rev Immunol 1994; 12:85-116.
11. Reth M. Hydrogen peroxide as second messenger in lymphocyte activation. Nat Immunol 2002; 3(12):1129-1134.
12. Davis SJ, van der Merwe PA. The structure and ligand interactions of CD2: Implications for T-cell function. Immunol Today 1996; 17(4):177-187.
13. Shaw AS, Dustin ML. Making the T-cell receptor go the distance: A topological view of T-cell activation. Immunity 1997; 6(4):361-369.
14. Singer SJ, Nicolson GL. The fluid mosaic model of the structure of cell membranes. Science 1972; 175(23):720-731.
15. Simons K, Ikonen E. Functional rafts in cell membranes. Nature 1997; 387(6633):569-572.
16. Wenk MR. The emerging field of lipidomics. Nat Rev Drug Discov 2005; 4(7):594-610.
17. Fridriksson EK, Shipkova PA, Sheets ED et al. Quantitative analysis of phospholipids in functionally important membrane domains from RBL-2H3 mast cells using tandem high-resolution mass spectrometry. Biochemistry 1999; 38(25):8056-8063.
18. Janes PW, Ley SC, Magee AI. Aggregation of lipid rafts accompanies signaling via the T-cell antigen receptor. J Cell Biol 1999; 147(2):447-461.
19. Drevot P, Langlet C, Guo XJ et al. TCR signal initiation machinery is pre-assembled and activated in a subset of membrane rafts. EMBO J 2002; 21(8):1899-1908.
20. Xavier R, Brennan T, Li Q et al. Membrane compartmentation is required for efficient T-cell activation. Immunity 1998; 8:723-732.
21. Montixi C, Langlet C, Bernard AM et al. Engagement of T-cell receptor triggers its recruitment to low-density detergent-insoluble membrane domains. EMBO J 1998; 17(18):5334-5348.
22. Field KA, Holowka D, Baird B. Compartmentalized activation of the high affinity immunoglobulin E receptor within membrane domains. J Biol Chem 1997; 272(7):4276-4280.
23. Cheng PC, Dykstra ML, Mitchell RN et al. A role for lipid rafts in B-cell antigen receptor signaling and antigen targeting. J Exp Med 1999; 190(11):1549-1560.
24. Simons K, Toomre D. Lipid rafts and signal transduction. Nat Rev Mol Cell Biol 2000; 1(1):31-39.
25. Langlet C, Bernard AM, Drevot P et al. Membrane rafts and signaling by the multichain immune recognition receptors. Curr Opin Immunol 2000; 12(3):250-255.
26. Kabouridis PS, Magee AI, Ley SC. S-acylation of LCK protein tyrosine kinase is essential for its signalling function in T-lymphocytes. EMBO J 1997; 16(16):4983-4998.
27. Munro S. Lipid rafts:elusive or illusive? Cell 2003; 115(4):377-388.
28. Janes PW, Ley SC, Magee AI et al. The role of lipid rafts in T-cell antigen receptor (TCR) signalling. Semin Immunol 2000; 12(1):23-34.
29. Lanzavecchia A, Lezzi G, Viola A. From TCR engagement to T-cell activation: A kinetic view of T-cell behavior. Cell 1999; 96(1):1-4.
30. Garboczi DN, Ghosh P, Utz U et al. Structure of the complex between human T-cell receptor, viral peptide and HLA-A2. Nature 1996; 384(6605):134-141.
31. Garcia KC, Degano M, Stanfield RL et al. An αβ T-cell receptor structure at 2.5 A and its orientation in the TCR-MHC complex. Science 1996; 274(5285):209-219.
32. Davis SJ, van der Merwe PA. The kinetic-segregation model: TCR triggering and beyond. Nat Immunol 2006; 7(8):803-809.
33. Davis MM, Boniface JJ, Reich Z et al. Ligand recognition by αβ T-cell receptors. Annu Rev Immunol 1998; 16:523-544.
34. Yokosuka T, Sakata-Sogawa K, Kobayashi W et al. Newly generated T-cell receptor microclusters initiate and sustain T-cell activation by recruitment of Zap70 and SLP-76. Nat Immunol 2005; 6(12):1253-1262.
35. Lin J, Weiss A. The tyrosine phosphatase CD148 is excluded from the immunologic synapse and down-regulates prolonged T-cell signaling. J Cell Biol 18 2003; 162(4):673-682.
36. Choudhuri K, Wiseman D, Brown MH et al. T-cell receptor triggering is critically dependent on the dimensions of its peptide-MHC ligand. Nature 2005; 436(7050):578-582.
37. Irles C, Symons A, Michel F et al. CD45 ectodomain controls interaction with GEMs and Lck activity for optimal TCR signaling. Nat Immunol 2003; 4(2):189-197.
38. Varma R, Campi G, Yokosuka T et al. T-cell receptor-proximal signals are sustained in peripheral microclusters and terminated in the central supramolecular activation cluster. Immunity 2006; 25(1):117-127.
39. Batista FD, Iber D, Neuberger MS. B-cells acquire antigen from target cells after synapse formation. Nature 2001; 411(6836):489-494.
40. Chang TW, Kung PC, Gingras SP et al. Does OKT3 monoclonal antibody react with an antigen-recognition structure on human T-cells? Proc Natl Acad Sci USA 1981; 78(3):1805-1808.

41. Kaye J, Janeway CA Jr. The Fab fragment of a directly activating monoclonal antibody that precipitates a disulfide-linked heterodimer from a helper T-cell clone blocks activation by either allogeneic Ia or antigen and self-Ia. J Exp Med 1984; 159(5):1397-1412.
42. Boniface JJ, Rabinowitz JD, Wülfing C et al. Initiation of signal transduction through the T-cell receptor requires the peptide multivalent engagement of MHC ligands. Immunity 1998; 9:459-466.
43. Cochran JR, Cameron TO, Stern LJ. The relationship of MHC-peptide binding and T-cell activation probed using chemically defined MHC class II oligomers. Immunity 2000; 12(3):241-250.
44. Krogsgaard M, Li QJ, Sumen C et al. Agonist/endogenous peptide-MHC heterodimers drive T-cell activation and sensitivity. Nature 2005; 434(7030):238-243.
45. Monks CR, Freiberg BA, Kupfer H et al. Three-dimensional segregation of supramolecular activation clusters in T-cells. Nature 1998; 395(6697):82-86.
46. Grakoui A, Bromley SK, Sumen C et al. The immunological synapse: A molecular machine controlling T-cell activation. Science 1999; 285(5425):221-227.
47. Carrasco YR, Batista FD. B-cell recognition of membrane-bound antigen: An exquisite way of sensing ligands. Curr Opin Immunol 2006; 18(3):286-291.
48. Lee KH, Holdorf AD, Dustin ML et al. T-cell receptor signaling precedes immunological synapse formation. Science 2002; 295(5559):1539-1542.
49. Freiberg BA, Kupfer H, Maslanik W et al. Staging and resetting T-cell activation in SMACs. Nat Immunol 2002; 3(10):911-917.
50. Lee KH, Dinner AR, Tu C et al. The immunological synapse balances T-cell receptor signaling and degradation. Science 2003; 302(5648):1218-1222.

CHAPTER 8

Kinetic Proofreading Model

Byron Goldstein,* Daniel Coombs, James R. Faeder and William S. Hlavacek

Abstract

K inetic proofreading is an intrinsic property of the cell signaling process. It arises as a consequence of the multiple interactions that occur after a ligand triggers a receptor to initiate a signaling cascade and it ensures that false signals do not propagate to completion. In order for an active signaling complex to form after a ligand binds to a cell surface receptor, a sequence of binding and phosphorylation events must occur that are rapidly reversed if the ligand dissociates from the receptor. This gives rise to a mechanism by which cells can discriminate among ligands that bind to the same receptor but form ligand-receptor complexes with different lifetimes. We review experiments designed to test for kinetic proofreading and models that exhibit kinetic proofreading.

Introduction

For many receptors the occupancy of their binding sites by the appropriate ligand is insufficient to initiate a cellular response; rather these receptors must aggregate with additional membrane proteins or among themselves to initiate a signal. As exemplified by the multichain immune recognition receptors (MIRRs), transmitting information across the cell membrane by juxtaposing the cytoplasmic tails of receptors is a common signaling mechanism used in every facet of immune system function. The subunits of the MIRRs can be divided into those that participate in binding and those that participate in signaling. All the signaling subunits, but none of the binding subunits, have at least one copy of a common sequence motif, an immunoreceptor tyrosine-based activation motif (ITAM) in their cytoplasmic domains. Each ITAM is composed of a pair of YXXL/I sequences usually separated by seven or eight amino acid residues.[1,2] Upon receptor aggregation, ITAM tyrosines become phosphorylated. It is by converting cytoplasmic domains of the receptor to phosphorylated forms that the cell first "senses" the external ligand and a signaling cascade is initiated.

The cell synthesizes information during signal transduction through chemical reactions that build and use transient molecular scaffolds. The cytoplasmic domains of the receptors and other scaffolding proteins are sites for coalescence of kinases, phosphatases and adapters. The structures formed are ephemeral with components going on and off rapidly. Completion of construction depends on the lifetime of the receptor-ligand complex. If the lifetime of the complex is too short, most of the chemical cascades that are initiated will fail to go to completion, the signaling will be aborted and no response will be produced. This is the essence of kinetic proofreading, an idea introduced by Hopfield[3,4] to explain how high specificity arises in biosynthetic pathways and resurrected in the context of cell signaling by McKeithan,[5] who proposed it as a mechanism to explain how T-cell antigen receptors (TCRs) discriminate between foreign and self-antigens.

*Corresponding Author: Byron Goldstein—Theoretical Biology and Biophysics Group, T-10 MS K710, Theoretical Division, Los Alamos National Laboratory, Los Alamos, NM 875435, USA. Email: bxg@lanl.gov

Multichain Immune Recognition Receptor Signaling: From Spatiotemporal Organization to Human Disease, edited by Alexander B. Sigalov. ©2008 Landes Bioscience and Springer Science+Business Media.

Kinetic Proofreading Illustrated through FcεRI Signaling

The clearest demonstration of kinetic proofreading in cell signaling used ligands of low valence with different dissociation rate constants to aggregate high-affinity IgE receptor FcεRI on mast cells and trigger downstream responses.[6] We will review this experimental work but first we will illustrate kinetic proofreading using a detailed mathematical model of the initial signaling cascade mediated by FcεRI.[7] The beauty of using such a model is that we know exactly what is in the model, which molecules make up the chemical network and how they interact and we can pick the perfect ligands and concentrations to test our ideas.

In Figure 1, we review how FcεRI ITAMs become phosphorylated, how this leads to the recruitment of the protein tyrosine kinase (PTK) Syk from the cytosol and how Syk then becomes

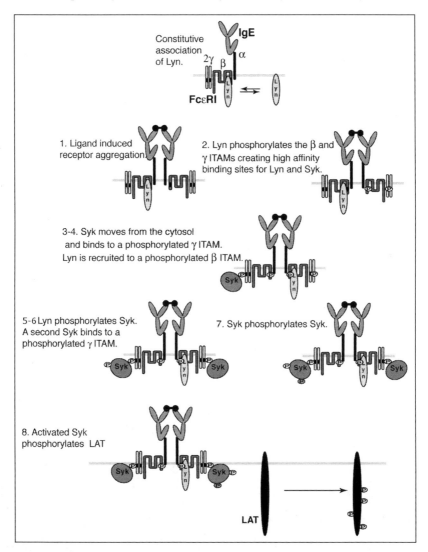

Figure 1. Initial steps in FcεRI signaling. FcεRI is a four chain MIRR. The α chain binds IgE with high affinity. The β and two γ chains participate in signaling and each contain a single ITAM. In the basal state, the Src kinase Lyn constitutively associates with the unphosphorylated β chain.[11,12]

activated. All the reactions shown in Figure 1 are included in the mathematical model. Briefly, the initiating Src family kinase Lyn constitutively associates with the β chain of the receptor. Upon receptor aggregation, if Lyn is present in an aggregate, it can transphosphorylate both the β and γ ITAM tyrosines. The amount of Lyn available to the receptor in rat basophilic leukemia (RBL) cells is limiting, so that often (at least in this cell line) receptor aggregates initially contain no Lyn.[8-10] In the basal state, the Src kinase Lyn constitutively associates with the unphosphorylated β chain.[11,12] Lyn can bind through its single SH2 domain to a phosphorylated β ITAM while Syk, with two SH2 domains, can form a stable complex with a γ chain ITAM only when both of the ITAM tyrosines are phosphorylated.[13,14] Syk is phosphorylated on multiple tyrosines by both Lyn and Syk.[15,16] Syk is fully activated when the two tyrosines in its activation loop, Tyr519 and Tyr 520, are phosphorylated.[17,18] The roles of the specific tyrosines of Syk are reviewed in ref. 18. In the model, all the phosphorylation reactions that take place at the receptor are trans, i.e., Lyn and Syk cannot phosphorylate substrates that are associated with any of the chains of the receptors they are associated with. Transphosphorylation of the receptor by Lyn has been demonstrated[19] and there is indirect evidence that Syk phosphorylation of Syk is trans as well.[15] In the model, this is assumed and thus two Syk molecules must be simultaneously associated with a receptor aggregate for a Syk to transphosphorylate the activation loop of an adjacent Syk. Whether signaling proceeds depends to a large extent on the competition between kinases and phosphatases. Even while receptors are held in an aggregate, they are constantly undergoing phosphorylation and dephosphorylation with kinases and phosphatases moving in and out of the aggregate.[20,21] When a receptor leaves an aggregate and is separated from the kinase that is phosphorylating its ITAMs, it undergoes rapid dephosphorylation.[22] In the mathematical model, a pool of unspecified protein tyrosine phosphatases account for the dephosphorylation reactions.

A detailed description of the model along with all the parameters and how they were obtained is given in Faeder et al[7] The model consists of the receptor, FcεRI, a bivalent ligand that can only aggregate receptors into dimers (higher aggregates can't form), the PTKs Lyn and Syk, and a background pool of phosphatases. The model describes the association, dissociation, phosphorylation and dephosphorylation among the components. In the model, these reactions can lead to 354 distinct chemical species. Using software called BioNetGen (http://bionetgen.lanl.gov/) that was developed to build cell signaling models,[23,24] 354 ordinary differential equations can be generated whose solutions give the concentrations of each chemical species as a function of time. Although we can look at the time courses of all 354 chemical species,[25] the most useful outputs of the model usually correspond to experimentally determined quantities. For example, if we are interested in how the phosphorylation of the β ITAM changes in time, we add up the time courses of all the concentrations of the chemical species that have the β ITAM phosphorylated, a few of which are shown in Figure 2. We can then predict how the phosphorylation of the β ITAM changes in time after the addition of the bivalent ligand.

To see if this network model of the early events of FcεRI-mediated cell signaling exhibits kinetic proofreading, we take as our ligands three monoclonal anti-FcεRI that aggregate FcεRI into dimers (Fig. 3). We take these ligands to have the same forward rate constants, k_{+1} and k_{+2}, but to differ in their reverse rate constants, k_{-1} and k_{-2}. Note that the mean lifetime of a receptor

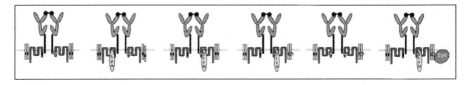

Figure 2. Six of the 170 chemical species in the model that have at least one β ITAM phosphorylated. The model lumps the two ITAM tyrosines together so that an ITAM is either phosphorylated or not phosphorylated.

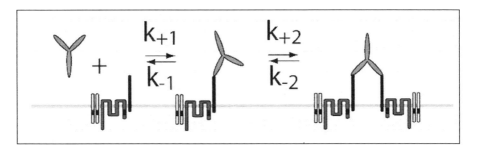

Figure 3. Kinetic scheme for the binding of an anti-IgEα monoclonal antibody to an FcεRI.

in a dimer formed by a monoclonal antibody bridging two FcεRI is $1/k_{-2}$. In the simulations, we take the value of k_{-2} for the three ligands to be 0.05 s^{-1}, 0.5 s^{-1} and 5 s^{-1}.

Shown in Figure 4 are the predicted levels of β and γ ITAM phosphorylation and Syk autophosphorylation (Syk is phosphorylated in its activation loop by another Syk) at long times when the model has reached a steady state. Because none of the components in the model are downregulated, the model goes to a steady state with a distribution of chemical species populated. These quantities are plotted as a function of the number of FcεRI in aggregates. From Figure 4, we can compare the responses induced by the three ligands when they form the same number of aggregates on the cell surface. For example, for the vertical dotted line in Figure 4, all the ligands maintain 3000 dimers on the cell surface in the steady state. To achieve this number of dimers, the ligand concentration for the most rapidly dissociating ligand is about 100-fold higher than the ligand with the intermediate dissociation rate, whose concentration is in turn about 100-fold higher than the slowest dissociating ligand (see the three horizontal dotted lines in Fig. 4). As can be seen, there is little difference in the amount of β and γ ITAM phosphorylation induced by the ligand but for the response furthest downstream in our model, the autophosphorylation of Syk, there is a dramatic difference in the response to the three ligands. This is a manifestation of kinetic proofreading, where the lifetime of the receptor in an aggregate strongly influences downstream responses. In summary, the model shows that a rapidly dissociating ligand will often be less effective in generating a response than a slowly dissociating ligand, even when the ligand concentrations are chosen to give the same number of receptors in aggregates.

As illustrated in Figure 5, higher ligand concentrations can compensate for faster rates of dissociation up to a point. For the slowest and intermediate dissociating ligands one can choose concentrations for each, 3×10^{-8} M and 2×10^{-11} M, respectively, that induce 1000 molecules per cell of autophosphorylated Syk. If there were no further proofreading downstream of Syk, then we would expect these ligands at the two different concentrations to produce similar, but probably not identical, cellular responses. Since the time course of binding and formation of receptor aggregates differ for the two ligands, the time courses of the responses and possibly their magnitudes might differ as well. We will address the question of whether there is further proofreading beyond Syk shortly.

To test for kinetic proofreading experimentally, Torigoe et al[6] compared the time courses of cellular responses triggered by rapidly and slowly dissociating multivalent ligands that bound to and aggregated a monoclonal anti-2,4-dinitrophenyl (DNP) IgE. RBL cells were sensitized with anti-DNP IgE, creating long-lived complexes of anti-DNP IgE and FcεRI on the cell surface. The concentrations of the two ligands were adjusted so that the rapidly dissociating ligand induced a maximal receptor phosphorylation (the sum of β and γ ITAM phosphorylation) that was twice that of the slowly dissociating ligand. Figure 6 shows that the rapidly dissociating ligand was progressively less effective in stimulating downstream events. The experiments permit comparison between the relative phosphorylation levels of each protein induced by the two ligands, but do not permit comparison between the phosphorylation levels of different proteins. As seen in Figure 6, the maximum phosphorylation of Syk induced by the rapidly dissociating ligand was less than

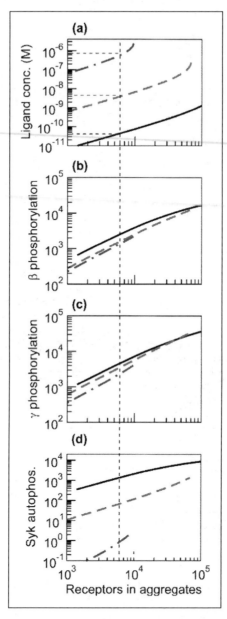

Figure 4. Simulation using a network model for three monoclonal anti-FcεRIα that aggregate receptors only into dimers. The figure is modified from Figure 8 of Faeder et al (2003) where all the parameters values for the simulation are given. The values plotted are obtained at steady state. For the three ligands, the forward rate constants are taken to be the same at $k_{+1} = 10^6$ M^{-1} s^{-1} and $k_{+2}R_T = 0.5$ s^{-1} where R_T is the RBL surface concentration of FcεRI. The reverse rate constants, $k_{-1} = k_{-2}$, have the following values: 0.05 s^{-1} (black solid lines), 0.5 s^{-1} (red dashed lines) and 5.0 s^{-1} (blue dash-dot lines). The x axis is in number of receptors in aggregates. (a) Ligand concentration required to achieve given levels of aggregation, (b) number of receptors per cell with the β ITAM phosphorylated, (c) number of receptors per cell with the γ ITAM phosphorylated and (d) number of autophosphorylated Syk per cell. Modified from a figure in J Immunol.

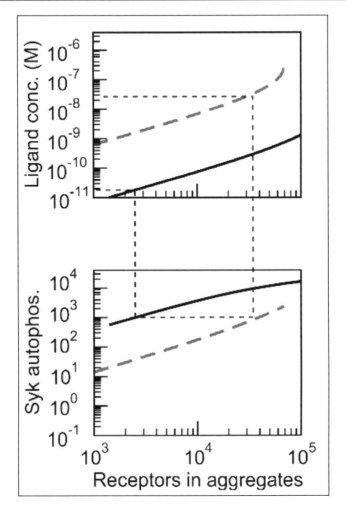

Figure 5. The two panels from Figure 4 with the results for the rapidly dissociating ligand, $k_{-1} =$ 5.0 s^{-1}, omitted. Over a limited range higher ligand concentrations can compensate for faster rates of dissociation. Concentrations for the two ligands (top panel) can be found so that both activate 1000 Syk molecules per cell (bottom panel). Modified from a figure in J Immunol.

one-third of that induced by the slowly dissociating ligand. Follow up experiments confirmed these results for Syk.[26] For Erk phosphorylation, which is known to be downstream of Syk, the ratio of the maximal responses was about a tenth. From these experiments, Torigoe et al[6] concluded that "these findings are consistent with a kinetic proofreading regime."

Although the experiments are consistent with the predictions of kinetic proofreading, a feature of cell signaling not controlled for in these experiments was the level of phosphorylation of the individual tyrosines in the β and γ ITAMs of the receptors. Even though the receptors on RBL cells exposed to the rapidly dissociating ligand had maximum levels of total receptor phosphorylation that were higher than for the receptors on cells exposed to the slowly dissociating ligand, the distribution of phosphorylated tyrosines for these two cases may have been quite different. If that were the case, the differences in downstream signaling could result from different patterns of tyrosine phosphorylation rather than the different rates of dissociation. Simulations of the model indeed show that γ and β phosphorylation exhibits different behaviors as the ligand dissociation rate is varied,[38] but this relatively small difference does not account for the dramatic reduction of Syk phosphorylation at higher dissociation rates (Fig. 4d). Experiments also show that the ratio of γ/β phosphorylation changes as the ligand dissociation rate increases,[51] but these differences seem unlikely to account for the full extent of the decrease in downstream activation events. Because the β and γ ITAMs contain 3 and 2 tyrosines respectively, it would be necessary to repeat these studies using site-specific anti-phosphotyrosine antibodies to fully resolve the effects of differential phosphorylation of individual tyrosines and ITAMs.

The Extent of Kinetic Proofreading in FcεRI Signaling

A major function of activated Syk associated with FcεRI is to phosphorylate the transmembrane adaptor protein linker for activation of T-cells (LAT).[27] When receptor aggregates are broken up by adding large amounts of hapten that compete for IgE-binding sites with the ligand responsible for receptor aggregation, LAT is rapidly dephosphorylated.[28] The present view is that LAT phosphorylation is maintained through enzyme-substrate reactions involving transient associations of LAT with activated Syk- FcεRI complexes. These reactions are thought to occur predominantly in specialized lipid domains where LAT is preferentially located[29] and where aggregated receptors tend to cluster,[30-32] although electron microscopy suggests a more complex topographical organization of membrane microdomains.[33] If the lifetime of the association of activated Syk with LAT is short compared to the lifetime of a receptor in an aggregate, then we expect events that stem from LAT phosphorylation not to be subject to kinetic proofreading. The experiments of Torigoe et al[6] suggest a way to test this idea—use the same rapidly and slowly dissociating ligands but choose concentrations such that the two ligands induce the same level of LAT phosphorylation, then see if downstream events still show signs of kinetic proofreading. Counter to what we anticipated, our results were consistent with kinetic proofreading beyond LAT phosphorylation.[52] In these experiments, however, although the overall level of LAT phosphorylation was the same, the pattern of LAT phosphorylation was not determined, leaving open the possibility that differences in the phosphorylation pattern could be responsible for the apparent proofreading.

Some Responses May Escape Kinetic Proofreading

Although many cellular responses are subject to kinetic proofreading some responses have been observed that appear to "escape" kinetic proofreading.[26,34,35] The same rapidly and slowly dissociating ligands used by Torigoe et al[6] to detect kinetic proofreading were subsequently found to

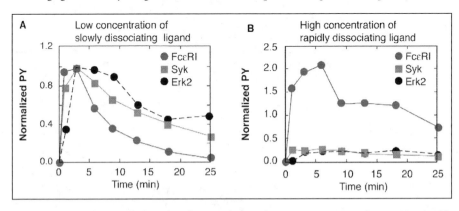

Figure 6. Time course of phosphorylation of several proteins in RBL cells sensitized with anti-DNP IgE after stimulation with (A) a slowly dissociating (high affinity) DNP-conjugated antigen (50 ng/ml) or (B) a rapidly dissociating (low affinity) 2NP-conjugated antigen (375 ng/ml). Adapted from Science: Torigoe C, Inman JK and Metzger H. An unusual mechanism for ligand antagonism. Science 1998; 281:568-72.

stimulate the transcription of the gene for monocyte chemotactic protein-1 (MCP-1) to about the same extent. Even though this is a late response and one might think it should be subject to many proofreading steps, it appears that that is not the case. To explain how a cellular response might avoid kinetic proofreading, a mechanism was proposed in which a soluble "messenger" generated early in the chain of chemical modifications that follow receptor aggregation determines the response. To investigate this idea, McKeithan's[5] mathematical formulation of kinetic proofreading (see below) was extended to allow a branch reaction in which a modified receptor, one that has associated with appropriate signaling molecules, acts as an enzyme on an intracellular substrate to generate a messenger.[36,37] If the response saturates as a function of the messenger concentration, then even if the rapidly dissociating ligand leads to the production of less messenger than the slowly dissociating ligand, both amounts may be sufficient to produce the same response. If this is the mechanism, then at lower concentrations of both ligands, when the rapidly dissociating ligand produces message below the saturating level, kinetic proofreading should once again be observed as it is for the production of mRNA for MCP-1.[26] In general, if a response saturates with respect to the level of an intermediate messenger, kinetic proofreading may be masked at high ligand concentrations but be revealed at low concentrations.

McKeithan's Mathematical Formulation

McKeithan[5] introduced a simple mathematical model to explain how the binding properties of peptide-MHC (pMHC) for the TCR influence the activation of the TCR (reviewed in refs. 38 and 39). According to the model, a bound receptor must complete a series of modifications to generate a cellular response. The model replaces the complex chemistry of the signaling cascade by a linear sequence of reactions but tries to capture a basic feature of the signaling process, that multiple events unspecified in the model must happen for the TCR to become activated. The series of modifications can be thought of as representing intermediate steps, such as the steps required for the formation of the scaffolding about the receptor and LAT. If the ligand dissociates before all the modifications have occurred, the receptor reverts to its basal state and no productive signal results as, for example, when FcεRI dissociates from aggregates and rapidly undergoes dephosphorylation.[22] An attractive feature of the model, illustrated in Figure 7, is that minor differences in the lifetime of a TCR-pMHC complex lead to huge differences in TCR-mediated signaling.

Since McKeithan[5] introduced his model, the view of the surface events that drive T-cell activation has changed. At low agonist pMHC surface concentrations, it appears that agonist/endogenous pMHC heterodimers act as the signaling unit that initiates T-cell signaling.[40] Thus, it might be useful to generalize the model to include endogenous pMHC and heterodimer formation, but as we discuss below, the main weakness of the McKeithan model and extended versions considered so far is the overly simplistic way in which signaling is treated.

There are four parameters in the McKeithan model, the forward and reverse rate constants for binding and dissociation of the pMHC to and from the TCR in the immunological synapse, k_{+1} and k_{-1} and the parameters that characterize the signaling events, k_p and N. From Figure 7, we see that in the model, N is the number of successive, irreversible modifications a bound TCR must complete to become activated and k_p is the rate constant for each modification. The difficulty is that the model does not specify how to determine these parameters from experiment and as a result, the model cannot make testable quantitative predictions. At the heart of the problem is the replacement of a highly branched biochemical network with a linear chain of reactions. Not surprisingly, there is no clear correspondence between the model parameter k_p and the set of reaction rates in the chemical cascade nor between N and the average number of reactions that occur for receptor activation. In the model N/k_p is the mean time for a receptor to become activated so we can put an estimate on this quantity, N/k_p. If we are interested in a rapid response, such as the activation of ZAP-70, it is probably of order seconds, whereas if we are interested in late response, such as IL-2 production, it might be hours.

T-Cell Activation and the Competition between Kinetic Proofreading and Serial Engagement

Despite its shortcomings, the McKeithan model has been used to make interesting qualitative predictions. The most investigated of these concerns the implications of the competition between kinetic proofreading and serial engagement (Chapter 9 and Valitutti et al[41]). To achieve a robust T-cell response, the activation of many TCRs is required. Kinetic proofreading works at the level of the individual receptor and to become activated, a TCR bound to a pMHC must complete a series of biochemical modifications before dissociating from the pMHC. Thus, the half-life of the pMHC-TCR complex must be long enough to allow completion of the signaling events required for TCR activation. Under physiological conditions, the density of cognate pMHC on antigen-presenting cells (APCs) is low. Thus, the half-life of the pMHC-TCR complex must be short enough to permit a single peptide to bind and dissociate many times, i.e., to serially engage multiple TCRs. The recognition that there is a trade off between kinetic proofreading and serial engagement led to the proposal that there is an optimal pMHC-TCR half-life for T-cell activation.[41-44] One way to see this is to look at the initial rate of TCR activation after synapse formation.

$$\text{activation rate} = (\text{hits/s}) \times (\text{fraction activated}) \tag{1}$$

By hitting rate (hits/s) we mean the rate at which a single pMHC engages TCRs. In the absence of TCR internalization, or if internalization of TCR only occurs when a TCR is not bound to pMHC, the rate of serial engagement per pMHC is[44,45]

$$\text{hits/s} = 1/(1/k_{-1}+1/(k_{+1}T)) = k_{-1}KT/(1+KT) \tag{2}$$

where $K = k_{+1}/k_{-1}$ is the two-dimensional equilibrium constant K for binding in the immunological synapse of pMHC on an APC to a TCR on a T-cell and T is the concentration of unbound TCR in the synapse. An assumption underlying Eq. (2) is that the density of cognate pMHC is small relative to the density of TCR so that individual pMHC do not compete for TCRs.

For the model in Figure 7, the fraction of pMHC-bound TCRs that go through all the steps leading to activation is

$$\text{fraction activated} = (k_p/(k_p + k_{-1}))^N \tag{3}$$

The initial activation rate, Eq. (1), therefore becomes

$$\text{activation rate} = (k_{-1}KT/(1+KT)) (k_p/(k_p + k_{-1}))^N \tag{4}$$

For even modest values of N the fraction of activated TCR can be quite sensitive to small variations in k_{-1}. For example, consider two pMHC-TCR complexes whose half-lives differ by a factor of 2. For N = 10 and k_{-1}/k_p = 1 and 2, respectively, the fractions activated are 9.8×10^{-4} and 1.7×10^{-5}, i.e., 57 times more TCRs are activated by the pMHC with the longer complex half-life. However, in the model this power of discrimination comes at the expense of the overall sensitivity to specific pMHC. To continue with our example, if k_{-1} = 0.05 s^{-1} and 0.1 s^{-1} for the two pMHC and KT>>1 for both, the initial activation rates per pMHC would be 4.9×10^{-5} s^{-1} and 1.7×10^{-6} s^{-1}, respectively. Even if there were a thousand peptides per APC, the agonist pMHC, k_{-1} = 0.05 s^{-1}, would only activate on average one TCR every 20s. To get around this problem, McKeithan proposed (not shown in Fig. 6) that once the TCR had gone through N modifications and becomes activated, its half-life increases. This would allow the fully modified TCR to build up over time and achieve higher sensitivity. One problem with this proposal is that such a mechanism would inhibit serial engagement. In what follows we ignore the possibility of a change in the half-life of a pMHC-TCR complex when a TCR is fully modified since we know of no evidence that indicates this occurs.

By differentiating Eq. (4) with respect to k_{-1} and assuming that KT>>1, we find the value of k_{-1} where the rate of activation is maximal, is given approximately by $k_{-1}^{max} = k_p/(N-1)$. We expect k_{-1}^{max} to correspond to a dissociation rate for strong agonist peptides, about 0.01 s^{-1}. For the expression

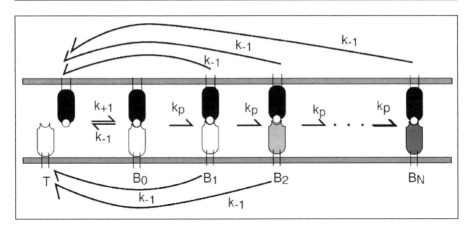

Figure 7. Kinetic proofreading model. T is the concentration of unbound TCR, B_0 is the concentration of bound TCRs that have not been modified and B_i is the concentration of TCRs that have undergone i modifications. Each modification occurs with the same rate constant k_p. When a TCR has gone through N modifications, it is activated.

for k_{-1}^{max} to be a reasonable approximation, we need KT > 1. For T-cells with 50,000 TCR per cell and a surface area of 800 μm^2 this requires that K > 1.6×10^{-9} cm.^{2} For some but not all agonist peptide-MHC-TCR complexes, this inequality is satisfied.[45] From Eq. (2) we see that in this limit hits/s = k_{-1} and the rate limiting step in serial engagement is the breaking of the pMHC-TCR complex. Once a pMHC is free it rapidly finds a new TCR and binds to it.

To show that there is an optimal range of half-lives for cytotoxic T-lymphocyte (CTL) activation, Kalergis et al[46] used a series of CD8$^+$ T-cell hybridomas expressing the wild type TCR and TCRs that had point mutations in their CDR3β domain and bound a VSV peptide-MHC with differing affinities. Using the same system, similar results were observed for the downregulation of TCR.[44] If these results arise from competition between kinetic proofreading and serial engagement, then the model predicts that going to high pMHC densities on the APC surface should remove the need for serial engagement in T-cell activation and an optimal range of half-lives for activation should no longer be observed. Consistent with this prediction, it was observed that when pMHC was present at high density, T-cell activation no longer went through a maximum as a function of the pMHC-TCR half-life but became a monotonic increasing function that reached a plateau for long half-lives.[47]

A problem with these studies is that the half-lives were not determined directly or at the same temperature that activation was measured. Rather the half-lives of pMHC-tetramers were measured at 4°C and it was assumed that the order of the half-lives was unchanged for the pMHC-monomers at 25°C. Whether the same ordering of half-lives occurs at 25°C is unknown. Holler and Kranz[48] used T-cells transfected with normal and engineered TCRs that had a wide range of pMHC-TCR half-lives and determined their half-lives and affinities at 25°C. They observed that T-cell activation was a monotonic increasing function of affinity that reached a plateau as a function of the pMHC-TCR halflife. They detected no optimal lifetime for the pMHC-TCR complex. Consistent with their results, Weber et al[49] engineered high affinity TCRs with slower off rates (also faster on rates) and found that the cognate peptide acted as a strong agonist for T-cells that had been transfected with these TCRs.

The evidence for pMHC serially engaging multiple TCRs at low peptide densities is compelling[41,45,50] as is the evidence for kinetic proofreading occurring in MIRR signaling.[6] As discussed, it has been proposed that at low pMHC densities TCR activation is a result of competition between these two effects and should result in TCR activation increasing to a maximum value and declining as a function of the pMHC-TCR complex lifetime. For interleukin 2 (IL-2) production, Kalergis et al[46] observed an optimal range of half-lives for TCR activation while Holler and Kranz[48] did not. IL-2 production is a late response with many signal events occurring between pMHC-TCR binding and IL-2 production. To help resolve the question of whether competition occurs during signaling and leads to an optimal half-life for pMHC-TCR complex-triggered responses, early responses such as the activation of ZAP70 or the phosphorylation of LAT would make a much better target for study. As yet these studies have not been done.

Concluding Remarks

Kinetic proofreading is an intrinsic property of the cell signaling process. It arises as a consequence of the multiple interactions that occur after a ligand triggers a receptor to initiate a signaling cascade and it ensures that false signals do not propagate to completion. An important consequence of kinetic proofreading is that it prevents signal propagation in the basal state due to fluctuations in the density of receptors (Faeder and Goldstein, unpublished results). In its basal state, a cell with receptors diffusing over its surface will always have some receptors close enough together that if one of the receptors is a PTK or associated with a PTK it can transphosphorylate its neighbor. Low levels of receptor phosphorylation occur in the absence of cognate ligands, but this phosphorylation does not generate a cellular response unless the receptors are overexpressed. Similarly, kinetic proofreading allows quality-controlled responses to the various ligands that a receptor can bind. Signals initiated by ligands with short binding half-lives are arrested at an early stage. This is particularly important for the MIRRs because either directly (the T-cell and B-cell receptors) or indirectly (the Fc receptors) they bind ligands with widely varying half-lives. It is therefore not surprising that kinetic proofreading was first introduced to cell-surface receptor signaling in this context and that the MIRRs have become the testing-ground for understanding the details of the manifestation of kinetic proofreading. There is considerable work still to be done.

Acknowledgements

The work was supported by grants R37GM35556 and RR18754 from the National Institutes of Health and by the Department of Energy through contract DE-AC52-06NA25396.

References

1. Reth M. Antigen receptor tail clue. Nature 1989; 338:383-384.
2. Cambier JC. Antigen and Fc receptor signaling: The awesome power of the immunoreceptor tyrosine-based activation motif (ITAM). J Immunol 1995; 155:3281-2185.
3. Hopfield JJ. Kinetic proofreading: A new mechanism for reducing errors in biosynthetic processes requiring high specificity. Proc Natl Acad Sci USA 1974; 71:4135-4139.
4. Hopfield JJ, Yamane T, Yue V et al. Direct experimental evidence for kinetic proofreading in amino acylation of tRNAIle. Proc Natl Acad Sci USA 1976; 73:1164-1168.
5. McKeithan TW. Kinetic Proofreading in T-cell receptor signal-transduction. Proc Natl Acad Sci USA 1995; 92:5042-5046.
6. Torigoe C, Inman JK, Metzger H. An unusual mechanism for ligand antagonism. Science 1998; 281:568-572.
7. Faeder JR, Hlavacek WS, Reischl I et al. Investigation of early events in FcεRI-mediated signaling using a detailed mathematical model. J Immunol 2003; 170:3769-3781.
8. Wofsy C, Kent UM, Mao SY et al. Kinetics of tyrosine phosphorylation when IgE dimers bind to Fcε Receptors on rat basophilic leukemia-cells. J Biol Chem 1995; 270:20264-20272.
9. Wofsy C, Torigoe C, Kent UM et al. Exploiting the differece between intrinsic and extrinsic kinases: Implications for regulation of signaling by immunoreceptors. J Immunol 1997; 259:5984-5992.
10. Torigoe C, Goldstein B, Wofsy C et al. Shuttling of initiating kinase between discrete aggregates of the high affinity receptor for IgE regulates the cellular response. Proc Natl Acad Sci USA 1997; 94:1372-1377.
11. Yamashita T, Mao SY, Metzger H. Aggregation of the high-affinity IgE receptor and enhanced activity of P53/56(Lyn) protein-tyrosine kinase. Proc Natl Acad of Sci USA 1994; 91:11251-11255.
12. Vonakis BM, Haleem-Smith H, Benjamin P et al. Interaction between the unphosphorylated receptor with high affinity for IgE and Lyn kinase. J Biol Chem 2001; 276:1041-1050.
13. Kihara H, Siraganian RP. Src homology 2 domains of Syk and Lyn bind to tyrosine-phosphorylated subunits of the high affinity IgE receptor. J Biol Chem 1994; 269:22427-22432.

14. Chen T, Repetto B, Chizzonite R et al. Interaction of phosphorylated FcεRIγ immunoglobulin receptor tyrosine activation motif-based peptides with dual and single SH2 domains of p72syk: Assessment of binding parameters and real time binding kinetics. J Biol Chem 1996; 271:25308-25315.
15. El-Hillal O, Kurosaki T, Yamamura H et al. Syk kinase activation by a src kinase-initiated activation loop phosphorylation chain reaction. Proc Natl Acad Sci USA 1997; 94:1919-1924.
16. Keshvara LM, I saacson CC, Yankee TM et al. Syk- and Lyn-dependent phosphorylation of Syk on multiple tyrosines following B-cell activation includes a site that negatively regulates signaling. J Immunol 1998; 161:5276-5283.
17. Zhang J, Kimura T, Siraganian RP. Mutations in the activation loop tyrosines of protein tyrosine kinase Syk abrogate intracellular signaling but not kinase activity. J Immunol 1998; 161:4366-4374.
18. Siraganian RP, Zhang J, Suzuki K et al. Protein tyrosine kinase Syk in mast cell signaling. Mol Immunol 2002; 38:1229-1233.
19. Pribluda VS, Pribluda C, Metzger H. Transphosphorylation as the mechanism by which the high-affinity receptor for IgE is phosphorylated upon aggregation. Proc Natl Acad Sci USA 1994; 91:11246-11250.
20. Kent UM, Mao S-Y, Wofsy C et al. Dynamics of signal transduction after aggregation of cell-surface receptors: Studies on the type I receptor for IgE. Proc Natl Acad Sci USA 1994; 91:3087-3091.
21. Bunnell SC, Hong DI, Kardon JR et al. T-cell receptor ligation induces the formation of dynamically regulated signaling assemblies. J Cell Biol 2002; 158:1263-1275.
22. Mao SY, Metzger H. Characterization of protein-tyrosine phosphatases that dephosphorylate the high affinity IgE receptor. J Biol Chem 1997; 272:14067-14073.
23. Blinov ML, Faeder JR, Goldstein B et al. BioNetGen: Software for rule-based modeling of signaling transduction based on the interactions of molecular domains. Bioinformatics 2004; 20:289-291.
24. Faeder JR, Blinov ML, Goldstein B et al. Ruled-based modeling of biochemical networks. Complexity 2005; 10:22-41.
25. Faeder JR, Blinov ML, Goldstein B et al. Combinatorial complexity and dynamical restriction of network flows in signal transduction. Syst Biol 2005; 2:5-15.
26. Liu Z-J, Haleem-Smith H, Chen H et al. Unexpected signals in a system subject to kinetic proofreading. Proc Natl Acad Sci USA 2001; 98:7289-7294.
27. Saitoh S, Arudchandran R, Manetz TS et al. LAT is essential for FcεRI-mediated mast cell activation. Immunity 2000; 12:525-535.
28. Pierce M, Metzger H. Detergent-resistant microdomains offer no refuge for proteins phophorylated by the IgE receptor. J Biol Chem 2000; 275:34976-34982.
29. Zhang W, Trible RP, Samelson LE. LAT palmitoylation: Its essential role in membrane microdomain targeting and tyrosine phosphorylation during T-cell activation. Immunity 1988; 9:239-246.
30. Field KA, Holowka D, Baird B. FcεRI-mediated recruitment of p53/56-lyn to detergent-resistant membrane domains accompanies cellular signaling. Proc Natl Acad Sci USA 1995; 92:9201-9205.
31. Field KA, Holowka D, Baird B. Compartmentalized activation of the high affinity immunoglobulin E receptor within membrane domains. J Biol Chem 1997; 272:4276-4280.
32. Rivera J, Cordero JR, Furumoto Y et al. Macromolecular protein signaling complexes and mast cell responses: A view of the organization of IgE-dependent mast cell signaling. Mol Immunol 2002; 38:1253-1258.
33. Wilson BS, Steinberg SL, Liederman K et al. Markers for detergent-resistant lipid rafts occupy distinct and dynamic domains in native membranes. Mol Biol Cell 2004; 15:2580-2592.
34. Rosette C, Werlen G, Daniels MA et al. The impact of duration versus extent of TCR occupancy on T-cell activation: A revision of the kinetic proofreading model. Immunity 2001; 15:59-70.
35. Eglite S, Morin JM, Metzger H. Sythesis and Secretion of monocyte chemotactic protein-1 stimulated by high affinity receptor for IgE. J Immunol. 2003; 170:2680-2687.
36. Hlavacek WS, Redondo A, Metzger H et al. Kinetic proofreading models for cell signaling predict ways to escape kinetic proofreading. Proc Natl Acad Sci USA 2001; 98:7295-7200.
37. Hlavacek WS, Redondo A, Wofsy C et al. Kinetic proofreading in receptor-mediated transduction of cellular signals. Bull Math Biol 2002; 64:887-811.
38. Goldstein B, Faeder JR, Hlavacek WS. Mathematical models of immune receptor signaling. Nat Rev Immunol 2004; 4:445-456.
39. Coombs D, Goldstein B. T-cell activation: Kinetic proofreading, serial engagement and cell adhesion. J of Computation and Appl Math 2005; 184:121-139.
40. Krogsgaard M, Li Q.-J, Sumen C et al. Agonist/endogenous peptide-MHC heterodimers drive T-cell activation and sensitivity. Nature 2005; 434:238-243.
41. Valitutti S, Muller S, Cella M et al. Serial triggering of many T-cell receptors by a few peptide-MHC complexes. Nature 1995; 375:148-151.

42. Lanzavecchia A, Lezzi G, Viola A. From TCR engagement to T-cell activation: A kinetic view of T-cell behavior. Cell 1999; 96:1-4.
43. van den Berg HA, Rand DA, Burroughs NJ. A reliable and safe T-cell repertoire based on low-affinity T-cell receptors. J Theor Biol 2001; 209:465-486.
44. Coombs D, Kalergis AM, Nathenson SG et al. Activated TCR remain marked for internalization after dissociation from peptide-MHC. Nature Immunol 2002; 3:926-931.
45. Wofsy C, Coombs D, Goldstein B. Calculations show substantial serial engagement of T-cell receptors. Biophys J 2001; 80:606-612.
46. Kalergis AM, Boucheron N, Doucey M-A et al. Efficient T-cell activation requires an optimal dwell-time of interaction between the TCR and the pMHC complex. Nature Immunol 2001; 2:229-234.
47. Gonzalez PA, Carreno LJ, Coombs D et al. Effects of pMHC and TCR dwell time on T-cell activation. Proc Natl Acad Sci USA 2005; 102:4824-4829.
48. Holler PD, Kranz DM. Quantitative analysis of the contribution of TCR/pepMHC affinity and CD8 to T-cell activation. Immunity 2003; 18:255-264.
49. Weber KS, Donermeyer DL, Allen PM et al. Class II-restricted T-cell receptor engineered in vitro for higher affinity retains peptide specificity and function. Proc Natl Acad of Sci USA 2005; 102:19033-19038.
50. Itoh Y, Hemmer B, Martin R et al. Serial TCR engagement and down-modulation by peptide: MHC molecule ligands: Relationship to the quality of individual TCR signaling events. J Immunol 1999; 162:2073-2080.
51. Torigoe C, Song JM, Barisas BG et al. The influence of actin microfilaments on signaling by the receptor with high-affinity for IgE. Mol Immunol 2004; 41:817-829.
52. Torigoe C, Faeder JR, Oliver JM et al. Kinetic proofreading of ligand-FceRI interactions may persist beyond LAT phosphorylation, J Immunol 2007; 178:3530-3535.

CHAPTER 9

Serial Triggering Model

Jacob Rachmilewitz*

Abstract

T-cells recognize a foreign antigen when presented on antigen-presenting cells (APCs) in the context of a peptide bound to major histocompatibility complex (MHC). The recognition of an antigen takes place at the T-cell:APC contact site where an "immune synapse" is formed and the multichain T-cell antigen receptor (TCR) is triggered. This initiates a signal transduction cascade that involves activation of tyrosine kinases, which in turn activate downstream events that elicit a diverse array of effector functions. T-cell activation requires a sustained signal that lasts for several hours. However, TCR affinity to its antigen is low and activation of TCR induces only a brief spike of intracellular signals. The serial triggering model resolves these seemingly paradoxical requirements for T-cell activation. The model states that sustained signaling is accomplished by the concerted action of multiple T-cell receptors that are sequentially engaged with and triggered by the peptide:MHC complex. In this chapter, we review the serial triggering model and two other models that expand this model. These models describe kinetic aspects of T-cell activation such as the pivotal question of how the T-cell "counts" the number of serially triggered receptors over time and how it determines that a threshold level has been reached for the activation of T-cell response.

T-Cell Receptor Signaling Cascade

A great deal of progress has been made in the past few years towards elucidating the cascade of signaling events and molecular mechanisms involved in the control of T-cell activation. T-cell activation is initiated in secondary lymphoid organs where T-cells encounter antigen-presenting cell (APC). T-cell activation leading to a productive response (i.e., cytokine secretion and proliferation) requires interactions of T-cell antigen receptors (TCRs) and peptides presented by major histocompatibility complex (MHC) molecules in an adhesive junction between the T-cell and APC. The main signaling pathways elicited by binding the TCR and some of the coreceptors are briefly described below.

The interaction of TCR with peptide:MHC complex (pMHC) leads to the phosphorylation of tyrosines of immunoreceptor tyrosine-based activation motifs (ITAMs) presented on the ζ and CD3 signaling subunits of the multichain TCR complex. Phosphorylation of these tyrosines by the Src family protein tyrosine kinase, Lck, promotes recruitment and activation of the nonreceptor tyrosine kinase zeta-associated protein 70 (ZAP-70) which in turn activates several target proteins including the adaptors, linker for activation of T-cells (LAT) and SH2 domain-containing leukocyte protein of 76 kDa (SLP-76). Phosphorylation of LAT and SLP-76 serves as a docking site, among others, for the SH2 domain of the phospholipase Cγ1 (PLCγ1). PLCγ1 catalyzes the formation of the second messengers, inositol 1, 4, 5-triphosphate and diacylglycerol, which trigger calcium flux and contribute to protein kinase C (PKC) and Ras activation, respectively.

*Jacob Rachmilewitz—Goldyne Savad Institute of Gene Therapy, Hadassah University Hospital, P.O.B. 12000, Jerusalem, 91120, Israel. Email: rjacob@hadassah.org.il

Multichain Immune Recognition Receptor Signaling: From Spatiotemporal Organization to Human Disease, edited by Alexander B. Sigalov. ©2008 Landes Bioscience and Springer Science+Business Media.

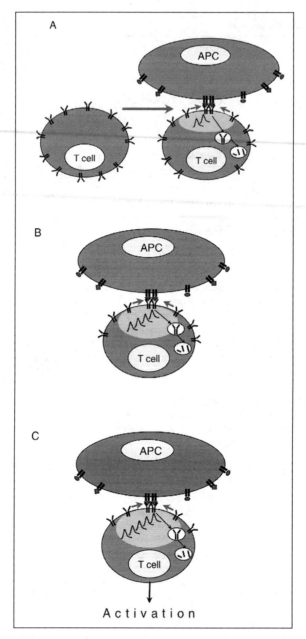

Figure 1. Naïve T-cell encounters a specific antigen on APC that presents on its surface multiple peptide antigens in association with MHC molecules (A). Following TCR triggering, signaling is initiated (red) and the triggered TCRs are internalized and degraded. New TCRs are moving into the T-cell:APC contact site where they are serially triggered by the pMHC complexes (A and B). Transient signal spike elicited by an individual TCR is represented by a curve with each spike adding to the falling phase of the one before. Eventually successive TCR-mediated transient signals are integrated up to the threshold of activation (C). Curves represent individual signal spikes.

PKC and Ras activation results in the activation of several members of the mitogen-activated protein kinase (MAPK) superfamily. The MAPKs are serine/threonine kinases that activate signaling cascades, resulting in the activation of the transcription factor NF-κB and AP-1. The transcription of several key genes involved in the T-cell response require both of these factors. Hence, the tyrosine phosphorylation cascade ultimately activates several transcription factors, such as NF-AT and AP-1, that in turn direct the transcription of new genes needed for T-cell response.

Vav1 protein expressed exclusively in the hematopoietic system is a signaling molecule that couples TCR to its effector function. Studies of Vav-deficient mice indicated that Vav is required for signaling through the TCR.[1,2] The block in Vav$^{-/-}$ T-cells appears to be in a proximal component of the TCR signaling pathway since the defect could be rescued by treatment with phorbol ester and ionomycin, pharmacological agents that mimic the downstream events induced by TCR stimulation.[1-4] Studies using Vav-deficient T-cells have indicated that Vav is the regulator of TCR-induced actin polymerization.[1,3] Defects in TCR signaling in Vav-deficient T-cells can be mimicked by treating T-cells with cytochalasin D (CD) that blocks actin polymerization. CD inhibits interleukin 2 (IL-2) production, cell proliferation and TCR capping[3,5,6] without impairing the initiation of early receptor events such as activation of tyrosine kinase, MAPK and JNK. The Wiskott-Aldrich syndrome protein (WASP) also plays an important role in the cascade linking TCR stimulation and actin polymerization. Upon TCR stimulation, both human[7,8] and mouse[4] WASP-deficient T-cells share defects in actin polymerization and T-cell proliferation with Vav-deficient cells. Thus, both Vav and WASP proteins appear to be crucial in linking the TCR signaling to cytoskeleton reorganization.

Due to the changes in the cytoskeleton, T-cells interacting with APCs undergo rearrangement of cytoskeletal elements that in turn drives the reorganization of specialized lipid microdomains and surface receptors to face the zone of contact with the APC. This active accumulation of signaling complexes and TCRs at the T-cell: APC interface results in formation of a supramolecular structure, termed "the immune synapse", that serves to increase the amplitude and duration of TCR signaling (reviewed by ref. 9). This process requires the engagement of costimulatory receptors, such as CD28.[10,11] Interestingly, in T-cells, the molecular adaptor Cbl-b has been shown to selectively inhibit TCR-mediated Vav activation, receptor clustering and raft aggregation, whereas CD28 costimulation has been found to be necessary to overcome this inhibitory effect.[12-14]

Hence, T-cell activation involves a series of signal transduction events, cytoskeleton polarization and dynamic accumulation of accessory molecules at APC: T-cell contact sites. The overall integration of these events determines the functional outcome of TCR engagement.

Serial Triggering Model

Despite knowing the major players in the signaling pathway triggered by TCR recognition of pMHC, full understanding of T-cell activation should take into account the time scale and kinetic aspects of the process.

Blocking the pMHC:TCR interaction, actin polymerization[15] or inhibiting Lck kinase activity[10] aborts TCR signaling as well as T-cell effector function. Hence, activation of TCR induces a very transient increase in tyrosine-phosphorylated intermediates and a brief spike in intracellular calcium concentration that is rapidly lost. However, a sustained elevation of signaling intermediates and intracellular calcium is an indispensable step for a productive T-cell activation.[15,16] In fact, elevated intracellular calcium levels need to be maintained for several hours to permit downstream signaling events such as NF-AT translocation to the nucleus.[17] Therefore, prolonged TCR occupancy is required to maintain sustained signaling essential for full T-cell activation.

While prolonged TCR occupancy is required for full T-cell activation, a single APC can simultaneously present multiple MHC-bound antigens, many of which are autologous peptides. Therefore, on the APC surface, the fraction of MHC molecules occupied with a specific antigen recognized by an individual T-cell is very low. It is estimated that as few as 100 pMHC complexes displayed on the APCs are sufficient to trigger T-cell activation.[18,19] This high sensitivity of TCR recognition for the low-frequency pMHC molecules and the requirement for prolonged TCR occupancy seems to be in sharp contrast to the low affinity and the high off rate of TCR:pMHC interaction, the half-life of which is estimated to be in the range of seconds to a few minutes.[20-24] Any molecular model explaining activation by triggered receptors must reconcile the two opposing kinetic and seemingly paradoxical features of T-cell activation. On the one hand, the signal emanating from each triggered TCR is short-lived and decays in the absence of a continuous TCR triggering.[15,25-27] Yet on the other hand, T-cell activation requires intracellular signaling prolonged over several hours.

The serial triggering model proposed by Lanzavecchia and coworkers[29] reconciles these apparently paradoxical findings and explains how so few pMHC complexes can over time engage enough TCRs to transduce an activation signal. In fact, they have shown that it is the low affinity of the TCR:pMHC interaction that allows for the high sensitivity of T-cells scanning for the rare antigen bound to the MHC molecule. According to the serial triggering model, a single pMHC complex can engage and trigger up to 200 TCRs one after the other and thus enables a small number of pMHC complexes to achieve a high and prolonged TCR occupancy. This process of TCR serial triggering takes place in the immune synapse where a pMHC complex presented on the APC surface triggers a TCR on the T-cell surface. The latter dissociates shortly after its activation, due to its low affinity and high off-rate and gets internalized and degraded (Fig. 1A). The pMHC complex is now available to engage with a new TCR that has transited to the APC:T-cell contact site where TCR triggering transpires (Fig. 1B). The continuous supply of new TCRs to replace those dissociated and consumed is derived by the cytoskeleton rearrangement. These dynamic changes in surface molecular topology and immune synapse organization are important in sustaining TCR signaling. Hence, the low affinity binding of the TCR to pMHC complex plays an important part in achieving serial triggering as it allows optimal turnover of the TCR:pMHC interaction (reviewed by refs. 30,31).

Flexible and Hierarchical T-Cell Activation Thresholds

Commitment of T-cells to cytokine production and proliferation requires TCR signaling sustained for as long as several hours, achieved through serial engagements. Upon productive interaction with the pMHC molecule, the triggered TCR initiates a signaling cascade and is subsequently internalized and degraded.[32] Experimentally, the extent of TCR downmodulation from the T-cell surface enables researchers to measure the number of receptors triggered before the onset of activation.[33] It has been shown that T-cell activation is elicited when a defined number of TCRs are triggered and internalized. However, it was further shown that different responses are elicited at different activation thresholds, as measured by the number of triggered and internalized TCRs. Cytotoxic T-cells (CTLs) can kill a target cell displaying as few as a single antigenic pMHC complex that triggers only a very small number of TCRs.[34] On the other hand, cytokine response requires significantly higher levels of antigen concentrations and TCR occupancy.[29] Single-cell analysis established the existence of a hierarchy in TCR signaling threshold for induction of distinct cytokine responses.[35] For example, IL-2 production requires higher TCR occupancy than interferon-gamma (IFN-γ). These single-cell studies further demonstrate that the hierarchy in activation thresholds observed in cell populations reflects the hierarchy that takes place at the individual cell level. As a result of this mechanism, antigen dosage may dictate not only quantitative but also qualitative aspects of the immune response due to changes in the relative levels of cytokine produced. Interestingly, the induction of IL-4, a Th2-like cytokine, has been reported to require a lower antigen dose and receptor ligation as compared to IFN-γ, showing that antigen dose and the extent of TCR triggering may effect the development of distinct T helper cell phenotypes.[36,37]

It is important to note that the threshold number is flexible. Costimulation can significantly lower the number of TCRs required for individual responses[29] without affecting TCR occupancy or triggering as evident by TCR down-regulation.[10] Studies of CD28-mediated costimulation suggest that the stability of the TCR-induced phosphorylation signal is flexible and can be regulated.[10] The changes in the stability of TCR-evoked signaling mediated by CD28 result in a decreased threshold number of triggered TCRs required for activation. Taken together these findings suggest that the stability of phosphorylation induced by an individual TCR is critical for setting the threshold number of TCRs required for activation. The effect of costimulation on TCR signal duration may result from costimulator-promoted reorganization of membrane micro-domains and immune synapse maturation that stabilizes signal spikes.[10,11,38,39] On the other hand, the pregnancy-associated immunoregulatory protein, placental protein 14 (PP14; also known as glycodelin), was shown to decrease the stability of TCR-induced phosphorylated proteins[40] and to elevate the T-cell activation threshold, thereby rendering T-cells less sensitive to a given level of TCR stimulation.[41]

Temporal Summation as a Mechanism for Signal Integration

As indicated above, to fully activate T-cells, antigen-stimulated TCR signaling needs to be sustained for as long as several hours, thus requiring ongoing TCR triggering. The magnitude of the T-cell response correlates with the level of TCR occupancy,[25,26] thus pegging distinct cytokine and other functional responses to a different level of TCR occupancy at the single cell level. Furthermore, the threshold number of triggered TCRs required for T-cell activation is flexible and costimulation can significantly lower the number of TCRs required for individual responses. Given the transient nature of the TCR response, it is not clear how a T-cell integrates signaling events from multiple triggered TCRs over several hours. Thus, pivotal questions arise as to: (1) how serially triggered receptors that are promptly internalized (reviewed by ref. 32) can be "counted" by the T-cell and (2) how their transient signaling events are accumulated over time and integrated to yield a threshold level sufficient to induce a corresponding biological response.

Most kinetic models for T-cell activation do not address specifically the path followed by T-cells from serial engagements to signal strength, assuming simply that prolonged TCR occupancy is the critical event. A more recently proposed model for T-cell activation is based on temporal summation of successive signals from individual TCRs,[42] similar to temporal summation in neural cells, which translate the frequency of presynaptic signals into the size of a postsynaptic potential. Within this model, it is possible to explain how T-cells combine the effects of signals received from successively triggered TCRs. Thus, signaling intermediates produced by serially triggered TCR incrementally and locally build up, with each signaling event adding to the falling phase of the one before (Fig. 1). In this way, small and short signals that are unable to trigger a response by themselves can be summed up over time to reach a threshold level. The magnitude of the signal reflects the rate of TCRs triggering and the life-span of the signal induced by each triggered TCR. Thus, the T-cell "counts" the number of occupied TCRs by summing TCR-induced signals. When threshold amplitude is reached, activation ensues (Fig. 1C). This model suggests that there is more to activating T-cells than simple receptor occupancy. Several phenomena associated with T-cell activation, such as the flexibility in the threshold number of activated receptors required for T-cell activation, are apparent by the model.

According to this model, costimulation extends the duration of individual signals by stabilizing TCR-induced tyrosine phosphorylation,[10] thus extending the time course of an individual signal and driving signal summation. Costimulation not only reduces the level of TCR occupancy required for activation and increases the amplitude of the response, but also decreases the duration of the signaling. In contrast, factors that shorten the duration of the signal, such as PP14,[40,41] will impede signal summation. Furthermore, the temporal summation model posits that summation may occur more effectively if the triggered receptors are in close proximity. Indeed, proximity is an intrinsic characteristic of serial triggering of many receptors by a single pMHC complex that takes place in the centre of the immune synapse.[43] Nevertheless, this model can also explain how a T-cell can integrate and sum up discontinuous signals that are generated over time in a mobile T-cell moving from one APC to another.[44]

An essential feature of the temporal summation model is that signaling events originating from successively triggered TCRs are gradually accumulating over time. Using experimental conditions that provide T-cell-APC or antibody-coated bead contacts and allow for serial TCR triggering at both bulk cell population and at a single T-cell level, several studies have demonstrated that signaling events originating from serially triggered TCR are not simply sustained but are gradually accumulated and integrated in order to generate a corresponding response.[45-48] The rate and extent of accumulation of signaling intermediates are essentially determined by the level of TCR engagement and augmented by costimulation. As predicted by the temporal summation model, a good correlation exists between the strength of the stimulus and the delay duration for the onset of the response.

Previous studies have shown that the amplitude and duration of calcium signals differentially control transcription factors shuttling to the nucleus. Consequently, the nature of the calcium signal determines the specificity of gene expression.[17,49-52] Hence, the different oscillations patterns and amplitudes of signaling intermediates (such as calcium) generated in the process of serial triggering can be translated to the shuttling of transcription factors into the nucleus at various combinations, consequently producing different biological responses. This may explain how the various threshold levels of TCR occupancy generate different T-cell responses over time.

Summary

Antigen recognition by T-cells is a very sensitive process since only a very small number of peptides presented on MHC molecules and displayed on the surface of APCs are capable of activating T-cells. Receptor engagement leads only to a transient response, while full T-cell activation requires antigen-stimulated TCR signaling to be sustained for as long as several hours. Therefore, activation of T-cells leading to cytokine production and proliferation requires ongoing TCR triggering. This triggering is achieved through serial engagement at the T-cell: APC interface in a structure termed "the immune synapse". However, signaling events originating from serially triggered TCR are not simply sustained but gradually accumulated and integrated over time to reach threshold levels required to elicit a response. Distinct responses have different activation thresholds. Thus, the nature of the antigen, its dose and the presence of costimulatory signals determine T-cell fate and the quantitative and qualitative aspects of the biological outcome. This mechanistic insight into the kinetics of signaling events during TCR serial triggering provides a useful framework to further explore the TCR-triggered intracellular pathways and the impact of TCR serial triggering on T-cell fate in particular and the immune response in general.

References

1. Fischer KD, Kong YY, Nishina H et al. Vav is a regulator of cytoskeletal reorganization mediated by the T-cell receptor. Curr Biol 1998; 8(10):554-62.
2. Zhang R, Alt FW, Davidson L et al. Defective signalling through the T- and B-cell antigen receptors in lymphoid cells lacking the vav proto-oncogene. Nature 1995; 374(6521):470-3.
3. Holsinger LJ, Graef IA, Swat W et al. Defects in actin-cap formation in Vav-deficient mice implicate an actin requirement for lymphocyte signal transduction. Curr Biol 1998; 8(10):563-72.
4. Snapper SB, Rosen FS, Mizoguchi E et al. Wiskott-Aldrich syndrome protein-deficient mice reveal a role for WASP in T but not B cell activation. Immunity 1998; 9(1):81-91.
5. Berg NN, Puente LG, Dawicki W et al. Sustained TCR signaling is required for mitogen-activated protein kinase activation and degranulation by cytotoxic T-lymphocytes. J Immunol 1998; 161(6):2919-24.
6. Kong YY, Fischer KD, Bachmann MF et al. Vav regulates peptide-specific apoptosis in thymocytes. J Exp Med 1998; 188(11):2099-111.
7. Gallego MD, Santamaria M, Pena J et al. Defective actin reorganization and polymerization of Wiskott-Aldrich T-cells in response to CD3-mediated stimulation. Blood 1997; 90(8):3089-97.
8. Zhang J, Shehabeldin A, da Cruz LA et al. Antigen receptor-induced activation and cytoskeletal rearrangement are impaired in Wiskott-Aldrich syndrome protein-deficient lymphocytes. J Exp Med 1999; 190(9):1329-42.
9. Dustin ML, Shaw AS. Costimulation: Building an immunological synapse. Science 1999; 283(5402):649-50.
10. Viola A, Schroeder S, Sakakibara Y et al. T-lymphocyte costimulation mediated by reorganization of membrane microdomains. Science 1999; 283(5402):680-2.
11. Wulfing C, Davis MM. A receptor/cytoskeletal movement triggered by costimulation during T-cell activation. Science 1998; 282(5397):2266-9.
12. Bachmaier K, Krawczyk C, Kozieradzki I et al. Negative regulation of lymphocyte activation and autoimmunity by the molecular adaptor Cbl-b. Nature 2000; 403(6766):211-6.

13. Chiang YJ, Kole HK, Brown K et al. Cbl-b regulates the CD28 dependence of T-cell activation. Nature 2000; 403(6766):216-20.
14. Krawczyk C, Bachmaier K, Sasaki T et al. Cbl-b is a negative regulator of receptor clustering and raft aggregation in T-cells. Immunity 2000; 13(4):463-73.
15. Valitutti S, Dessing M, Aktories K et al. Sustained signaling leading to T-cell activation results from prolonged T-cell receptor occupancy. Role of T-cell actin cytoskeleton. J Exp Med 1995; 181(2):577-84.
16. Goldsmith MA, Weiss A. Early signal transduction by the antigen receptor without commitment to T-cell activation. Science 1988; 240(4855):1029-31.
17. Timmerman LA, Clipstone NA, Ho SN et al. Rapid shuttling of NF-AT in discrimination of Ca2+ signals and immunosuppression. Nature 1996; 383(6603):837-40.
18. Demotz S, Grey HM, Sette A. The minimal number of class II MHC-antigen complexes needed for T-cell activation. Science 1990; 249(4972):1028-30.
19. Harding CV, Unanue ER. Quantitation of antigen-presenting cell MHC class II/peptide complexes necessary for T-cell stimulation. Nature 1990; 346(6284):574-6.
20. Corr M, Slanetz AE, Boyd LF et al. T-cell receptor-MHC class I peptide interactions:Affinity, kinetics and specificity. Science 1994; 265(5174):946-9.
21. Lyons DS, Lieberman SA, Hampl J et al. A TCR binds to antagonist ligands with lower affinities and faster dissociation rates than to agonists. Immunity 1996; 5(1):53-61.
22. Matsui K, Boniface JJ, Reay PA et al. Low affinity interaction of peptide-MHC complexes with T-cell receptors. Science 1991; 254(5039):1788-91.
23. Sykulev Y, Brunmark A, Jackson M et al. Kinetics and affinity of reactions between an antigen-specific T-cell receptor and peptide-MHC complexes. Immunity 1994; 1(1):15-22.
24. Weber S, Traunecker A, Oliveri F et al. Specific low-affinity recognition of major histocompatibility complex plus peptide by soluble T-cell receptor. Nature 1992; 356(6372):793-6.
25. Beeson C, Rabinowitz J, Tate K et al. Early biochemical signals arise from low affinity TCR-ligand reactions at the cell-cell interface. J Exp Med 1996; 184(2):777-82.
26. Muller S, Demotz S, Bulliard C et al. Kinetics and extent of protein tyrosine kinase activation in individual T-cells upon antigenic stimulation. Immunology 1999; 97(2):287-293.
27. Huppa JB, Gleimer M, Sumen C et al. Continuous T-cell receptor signaling required for synapse maintenance and full effector potential. Nat Immunol 2003; 4(8):749-55.
28. Kalergis AM, Boucheron N, Doucey MA et al. Efficient T-cell activation requires an optimal dwell-time of interaction between the TCR and the pMHC complex. Nat Immunol 2001; 2(3):229-34.
29. Valitutti S, Muller S, Cella M et al. Serial triggering of many T-cell receptors by a few peptide-MHC complexes. Nature 1995; 375(6527):148-51.
30. Valitutti S, Lanzavecchia A. Serial triggering of TCRs: A basis for the sensitivity and specificity of antigen recognition. Immunol Today 1997; 18(6):299-304.
31. Lanzavecchia A, Sallusto F. Antigen decoding by T-lymphocytes: From synapses to fate determination. Nat Immunol 2001; 2(6):487-92.
32. Alcover A, Alarcon B. Internalization and intracellular fate of TCR-CD3 complexes. Crit Rev Immunol 2000; 20(4):325-46.
33. Padovan E, Casorati G, Dellabona P et al. Expression of two T-cell receptor alpha chains: Dual receptor T-cells. Science 1993; 262(5132):422-4.
34. Valitutti S, Muller S, Dessing M et al. Different responses are elicited in cytotoxic T-lymphocytes by different levels of T-cell receptor occupancy. J Exp Med 1996; 183(4):1917-21.
35. Itoh Y, Germain RN. Single cell analysis reveals regulated hierarchical T-cell antigen receptor signaling thresholds and intraclonal heterogeneity for individual cytokine responses of CD4+ T-cells. J Exp Med 1997; 186(5):757-66.
36. Constant S, Pfeiffer C, Woodard A et al. Extent of T-cell receptor ligation can determine the functional differentiation of naive CD4+ T-cells. J Exp Med 1995; 182(5):1591-6.
37. Hosken NA, Shibuya K, Heath AW et al. The effect of antigen dose on CD4+ T helper cell phenotype development in a T-cell receptor-alpha beta-transgenic model. J Exp Med 1995; 182(5):1579-84.
38. Das V, Nal B, Roumier A et al. Membrane-cytoskeleton interactions during the formation of the immunological synapse and subsequent T-cell activation. Immunol Rev 2002; 189:123-35.
39. Huppa JB, Davis MM. T-cell-antigen recognition and the immunological synapse. Nat Rev Immunol 2003; 3(12):973-83.
40. Rachmilewitz J, Borovsky Z, Mishan-Eisenberg G et al. Focal localization of placental protein 14 toward sites of TCR engagement. J Immunol 2002; 168(6):2745-50.
41. Rachmilewitz J, Riely GJ, Huang JH et al. A rheostatic mechanism for T-cell inhibition based on elevation of activation thresholds. Blood 2001; 98(13):3727-32.
42. Rachmilewitz J, Lanzavecchia A. A temporal and spatial summation model for T-cell activation: Signal integration and antigen decoding. Trends Immunol 2002; 23(12):592-5.

43. Monks CR, Freiberg BA, Kupfer H et al. Three-dimensional segregation of supramolecular activation clusters in T-cells. Nature 1998; 395(6697):82-6.
44. Gunzer M, Schafer A, Borgmann S et al. Antigen presentation in extracellular matrix: Interactions of T-cells with dendritic cells are dynamic, short lived and sequential. Immunity 2000; 13(3):323-32.
45. Borovsky Z, Mishan-Eisenberg G, Yaniv E et al. Serial triggering of T-cell receptors results in incremental accumulation of signaling intermediates. J Biol Chem 2002; 277(24):21529-36.
46. Rosette C, Werlen G, Daniels MA et al. The impact of duration versus extent of TCR occupancy on T-cell activation: A revision of the kinetic proofreading model. Immunity 2001; 15(1):59-70.
47. Wei X, Tromberg BJ, Cahalan MD. Mapping the sensitivity of T-cells with an optical trap: polarity and minimal number of receptors for Ca(2+) signaling. Proc Natl Acad Sci USA 1999; 96(15):8471-6.
48. Wulfing C, Rabinowitz JD, Beeson C et al. Kinetics and extent of T-cell activation as measured with the calcium signal. J Exp Med 1997; 185(10):1815-25.
49. De Koninck P, Schulman H. Sensitivity of CaM kinase II to the frequency of Ca2+ oscillations. Science 1998; 279(5348):227-30.
50. Dolmetsch RE, Lewis RS, Goodnow CC et al. Differential activation of transcription factors induced by Ca2+ response amplitude and duration. Nature 1997; 386(6627):855-8.
51. Dolmetsch RE, Xu K, Lewis RS. Calcium oscillations increase the efficiency and specificity of gene expression. Nature 1998; 392(6679):933-6.
52. Li W, Llopis J, Whitney M et al. Cell-permeant caged InsP3 ester shows that Ca2+ spike frequency can optimize gene expression. Nature 1998; 392(6679):936-41.

CHAPTER 10

Conformational Model

Ruth M. Risueño, Angel R. Ortiz and Balbino Alarcón*

Abstract

The failure to identify changes in the crystal structure of the T-cell antigen receptor (TCR) α/β ectodomains beyond the ligand-binding complementarity-determining region loops is most probably responsible for conformational changes having been relegated to a second plane as a mechanism of signal transduction. However, there is strong biochemical and spectroscopic evidence that the cytoplasmic tails of the tcr and the B-cell antigen receptor undergo conformational changes upon stimulation. This suggests that in the context of the whole TCR complex, including both the TCRα/β ectodomains and the complete CD3 subunits with their transmembrane and cytoplasmic tails, the conformational change has to be transmitted from the ectodomains to the cytoplasmic tails upon ligand binding. While the mechanism of transmission and the importance of conformational changes in T- and B-cell activation are still being elucidated, there are already functional correlates that establish a link between full T-cell activation and this conformational change.

Introduction

Upon ligand binding, membrane receptors must transmit information from their ectodomains to their cytoplasmic tails. It is unclear in almost all membrane receptors systems how this is achieved and the multichain immune recognition receptors (MIRRs) are no exception. One obvious manner by which this transmission can be achieved that is compatible with all other mechanisms proposed in this chapter is via conformational changes. These changes can be initiated in the ectodomains of the receptors and are somehow transmitted to the cytoplasmic tails where they may lead to the unmasking of recognition sites for intracellular effectors. The G protein-coupled receptors are a good example of this as they contain a G-protein binding site situated within an intracellular loop.[1,2] However, the mechanisms of aggregation and conformational change are not necessarily independent. Indeed, it is now clear that many receptors with intrinsic tyrosine kinase activity initiate signaling through clustering and transphosphorylation and that they also undergo conformational changes during this process. To date, a considerable body of evidence has accumulated indicating that several receptors of immunological interest undergo a conformational change upon ligand binding, including: integrins, the erythropoietin (EPO) receptor, the tumor necrosis factor (TNF) receptors, Fas and the interleukin 6 (IL6) receptor.[4-11] In MIRRs, the conformational change induced by ligand binding could be a result of the rearrangement of the quaternary structure of the complex. However, the evidence that MIRRs undergo ligand-induced conformational changes that are responsible for the initiation of signal transduction remains scarce. Several lines of evidence support the existence of conformational changes in the T-cell antigen

*Corresponding Author: Balbino Alarcón—Centro de Biología Molecular Severo Ochoa, CSIC-Universidad Autónoma de Madrid, Cantoblanco, Madrid 28049, Spain. Email: balarcon@cbm.uam.es

Multichain Immune Recognition Receptor Signaling: From Spatiotemporal Organization to Human Disease, edited by Alexander B. Sigalov. ©2008 Landes Bioscience and Springer Science+Business Media.

receptor (TCR), mostly based on the exposure of a cytoplasmic sequence in CD3ε. In contrast, there is much less evidence of a conformational change in either the B-cell antigen receptor (BCR) or the immunoglobulin (Ig) Fc receptors.

The existence of ligand-induced conformational changes in the complementarity-determining regions (CDRs) of the TCRα and TCRβ is now widely accepted (reviewed in ref. 12). Nevertheless, the idea that conformational changes act as a mechanism to initiate the signaling cascade that transmits information from the ligand-binding TCRα/β ectodomains to the CD3 cytoplasmic tails has faced resistance from immunologists. This resistance is mostly based on the virtual absence of structural information supporting the existence of signal-transmitting conformational changes. Except for the LC13 TCR, where a ligand-induced shift of the Cα AB loop has been demonstrated, no other crystals of TCRα/β ectodomains have shown any evidence of structural changes beyond the adjustments observed in the CDRs.[13] Furthermore, the structure of the A-6 TCR α/β ectodomains remains unaltered when bound to agonist, partial agonist and antagonist peptide-major histocompatibility complex (pMHC) complexes, suggesting that these conformational changes are not transmitted from the ligand-binding sites across the TCRα/β ectodomains.[14] However, when analyzing conformational changes one must consider the TCR complex as a whole rather than just the TCRα/β ectodomains. Indeed, the structural information available regarding the TCR complex is incomplete. Not only do we lack essential details regarding how the TCRα/β and CD3 ectodomains interact, but it is also unclear how the transmembrane domains of the six subunits are interrelated. This is especially relevant when considering that the TCR forms pre-existing multivalent arrays,[15] and that a multivalent engagement of the TCR is required for transmission of the conformational change.[16]

Evidence in Favour of Conformational Changes in MIRRs

The study of conformational changes in MIRRs is still at a preliminary stage. Indeed, only a few publications have shed any light on the mechanisms underlying the structural rearrangements and their consequences in signal transmission (Table 1).

Table 1. Experimental evidences in favour of conformational changes in MIRRs

	Structural	Co-Capping	FRET	Neo-Epitopes	Induced Recruitment
TCR	+/−[1]	+[2]	−	+[3]	+[4]
BCR	−	−	+[5]	−	−
IG FCR	+/−[6]	−	−	−	−

[1]Kjer-Nielsen L, Clements CS, Purcell AW et al. A Structural Basis for the Selection of Dominant alphabeta T-Cell Receptors in Antiviral Immunity. Immunity 2003; 18: 53-64.
[2]Janeway CA Jr. Ligands for the T-cell receptor: hard times for avidity models. Immunol Today 1995; 16:223-225.
[3]Risueno RM, Gil D, Fernandez E et al. Ligand-induced conformational change in the T-cell receptor associated with productive immune synapses. Blood 2005; 106:601-608.
[4]Gil D, Schamel WW, Montoya M et al. Recruitment of Nck by CD3 epsilon reveals a ligand-induced conformational change essential for T-cell receptor signaling and synapse formation. Cell 2002; 109:901-912.
[5]Tolar P, Sohn HW, Pierce SK. The initiation of antigen-induced B-cell antigen receptor signaling viewed in living cells by fluorescence resonance energy transfer. Nat Immunol 2005; 6:1168-1176.
[6]Kato K, Fridman WH, Arata Y et al. A conformational change in the Fc precludes the binding of two Fcgamma receptor molecules to one IgG. Immunol Today. 2000; 21:310-312.

T-Cell Antigen Receptor

The initial evidence for the existence of ligand-induced conformational changes in the TCR complex came from cocapping experiments performed by Janeway and colleagues.[17,18] They observed that stimulation of T-cells with monovalent Fab fragments of anti-TCR antibodies promoted cocapping with the CD4 coreceptor, suggesting that anti-TCR antibodies may promote changes in the T-cell in the absence of TCR crosslinking.[18] More recently, it was shown that the nonphosphorylated form of the cytoplasmic tail of TCRζ, when studied in isolation, adopts helical conformation in the presence of micelles of lysomyristoylphosphatidylglycerol and small unilamellar vesicles of dimyristoylphosphatidylglycerol, whereas the phosphorylated form of this protein does not.[19] It is also known that a synthetic peptide corresponding to the third immunoreceptor tyrosine-based activation motif (ITAM) of TCRζ adopts an α-helical structure in the nonphosphorylated form and a β-strand conformation when phosphorylated.[20] However, both these studies were performed in cell-free systems with synthetic peptides and recombinant proteins in the absence of other TCR subunits. So far, the strongest evidence for a conformational change in the TCR has emerged from studies showing ligand-induced exposure of a polyproline sequence in the cytoplasmic tail of CD3ε.[21] The exposure of this sequence facilitates the binding of the TCR to the N-terminal SH3 (SH3.1) domain of the adapter Nck as demonstrated in a pull-down assay using immobilized GST-SH3.1 fusion protein. This biochemical assay has served to demonstrate that certain stimulatory anti-CD3 antibodies[21] and a panel of pMHC ligands[22] induce a conformational change in the TCR. The utility of the biochemical assay is, however, hampered by the fact that it involves the disruption of T-cells. A second assay to detect these conformation changes has been developed based on the exposure of a neo-epitope recognized by the monoclonal antibody APA1/1. This epitope coincides with the polyproline sequence in CD3ε and reveals, like the pull-down assay, a conformational change transmitted to the cytoplasmic tail of CD3ε.[23] Through the use of APA1/1, it has been possible to demonstrate that the TCR undergoes conformational changes during antigen recognition in vivo and that the TCR complexes undergoing the conformational change are located in the immune synapse. Moreover, it has been shown that the conformational change is elicited by full but not by partial agonist/antagonist peptides.[23] Unfortunately, both the pull-down assay and APA1/1 recognition reflect the same molecular event (i.e., a re-arrangement of the polyproline sequence in the tail of CD3ε). Thus, there is still no clear evidence that conformational changes affect TCR subunits other than CD3ε.

B-Cell Antigen Receptor

An allosteric activation model for the BCR (reviewed in ref. 24) was first suggested by Cambier et al.[25] In this model, it was proposed that antigen binding could promote a conformational change in the structure of the cytoplasmic portion of the BCR, resulting in the activation of a prebound tyrosine kinase. In addition, it was proposed by Reth that a conformational change could activate proximal tyrosine kinases by inhibiting phosphatases through the production of hydrogen peroxide.[26] Accordingly, the existence of ligand-induced conformational changes in the BCR was recently addressed using fluorescence energy transfer (FRET). In an elegant study, Pierce and colleagues showed that antigen binding induces a more complex pattern of FRET changes than would be expected from simple clustering.[27] Indeed, an initial rise in the FRET signal was followed by a rapid decline. This pattern was thought to reflect the initial clustering of the BCR that was followed by a conformational change which resulted in the opening of the Igα and Igβ cytoplasmic tails, resembling the opening of an umbrella. This pattern of fluctuating FRET signals was observed irrespective of the pairs of BCR subunits that were considered (Igα-Igβ, IgH-Igα, IgH-Igβ) and independent of the membrane IgH chain studied (μ or γ). However, the decline in the FRET signal was not only completely abolished by the presence of the src family kinase inhibitor PP2 but also was dependent on the presence of a phosphorylatable ITAM. These results suggest that the BCR undergoes a conformational change secondary to tyrosine phosphorylation of the ITAMs in Igα and Igβ. Therefore, it would appear that a conformational change in the BCR tails could participate in BCR signaling. However, these data do not demonstrate that the conformational change is the cause for the transmission of the initial signaling cues from the extracellular to the cytoplasmic tails of the BCR.

Immunoglobulin Fc Receptors

Upon binding to FcγRIII, a conformational change in the Fc portion of IgG has been documented as a mechanism to prevent FcR crosslinking and permanent activation by isolated IgG molecules.[28] In addition, IgG Fc binding to FcγRIII is accompanied by significant conformational adjustments at the binding site. However, these changes refer either to the ligand or to the receptor-ligand binding site and as yet there has been no demonstration that conformational changes initiate signal transduction in Ig Fc receptors.

Consequences for Ligand Recognition

The existence of ligand-induced conformational changes in MIRRs introduces a mechanism by which ligands can be discriminated based on the differences in free energy between the open and closed states of the receptor. This discrimination is particularly relevant for the TCR since the difference in affinity for agonist pMHC and self-peptide pMHCs is very small. However, proper distinction between these two types of ligand is crucial for maintenance of self-tolerance. The kinetic proofreading model proposes that the time of occupancy of the TCR forms the basis of the activation thresholds imposed,[29,30] which generally correlates well with the strength of the ligand.[31-33] However, there are numerous exceptions of strong pMHC ligands with similar or shorter half-lives than weak ligands.[34,35] It was recently demonstrated that when movement restrictions imposed by the plasma membrane are taken into account when measuring the half-life of these outliers, the fit with signal strength is much better.[36] Three of these "outliers" promote large changes in the heat capacity, indicative of conformational changes that could be correlated with the induced fit of pMHC-binding to the CDR loops in the TCRα/β heterodimer.[37] The conformational changes associated with the induced fit of the CDR loops have been proposed as a mechanism that contributes to ligand discrimination.[36] This mechanism can be generalized to all TCR-pMHC interactions if conformational changes in the whole TCR complex, not just those in the CDRs, are considered.

At the plasma membrane, the TCR is expressed as a combination of monovalent and multivalent complexes whose proportions vary.[15] Thus, multivalent engagement of the TCR and the conformational change together have been proposed to constitute a powerful mechanism for ligand discrimination.[38] This model is supported by the data showing that: a) a multivalent ligand is necessary for T-cell activation; b) the TCR is expressed as multivalent complexes; and c) induction of the conformational change requires multivalent engagement of the TCR.[16] This proposed mechanism is based on two processes that act in conjunction. Firstly, the affinity of the TCR for the agonist and non-agonist pMHC is translated into different residence times (kinetic proofreading)[30] that augment exponentially with multivalency.[38] Secondly, multivalent binding allows the TCR complex to discriminate between agonists that induce a conformational change in the receptor and non-agonists that do not. The conformational change increases the avidity of the interaction. Thus, binding of the pMHC cluster begins with a monovalent interaction between one of the agonist-loaded MHC units and one of the TCR units in the TCR cluster. This initial binding event facilitates the binding of a second agonist-loaded MHC unit. If only multivalent engagement of the TCR produces the fully active state, this equilibrium may constitute a proofreading mechanism for the quality of the antigen. Indeed, multivalent pMHC is necessary to activate T-cells.[39,40] Thus, the binding of two or more agonists to the TCR cluster introduces a cooperative interaction between sites, resulting in a conformational change in the TCR. If an affinity threshold must be reached to trigger the conformational change at any given time point, weak pMHC ligands would produce a higher proportion of TCR-pMHC complexes in which the TCR is mono- rather than multivalently bound. The result of this cooperative effect is that the apparent binding affinity of the pMHC to the TCR is increased by a factor that exponentially reflects the valency of the interaction. In this way, a small difference in the intrinsic affinity can be converted into a highly effective discriminatory mechanism (Fig. 1) since the cooperative effect results from the multivalency of the receptor and ligand. Together and through thermodynamic considerations alone, multivalency and the ensuing conformational change can explain the high specificity of antigen recognition.[38]

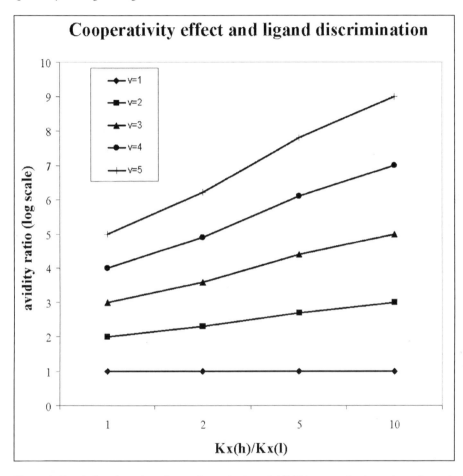

Figure 1. Simulations based on the mathematical model.[38] This example shows how the multivalency and conformational change within a multivalent TCR complex can help discriminate among ligands. The y-axis shows in logarithmic scale the ratio of avidities between an agonist (h) that is capable of eliciting the conformational change and a self-peptide (l) that is not. In the example, the intrinsic affinity (KA, the affinity for a monomeric TCR-pMHC interaction) for the agonist-TCR interaction is fixed to a value only 10-fold higher than the affinity of the self peptide, corresponding to the monovalent case (v = 1) in the plot. The effect of the conformational change is measured by the ratio of cooperativity constants between the agonist and the self peptide K × (h)/K × (l) shown in the x-axis. Note how multivalency (v = 2 to v = 5) and cooperativity interact to exponentially amplify avidity. For instance, in a situation where a pentavalent TCR is engaged by 5 pMHC complexes loaded either with agonist peptide or with self-peptide, the ratio of avidities is 100,000 if K × (h)/K × (l) equals 1, but increases by 1000-fold (i.e., to 100,000,000) if K × (h)/K × (l) equals 5.

Model for Transmission of Conformational Changes

Transmission of conformational changes across the membrane has generally been difficult to understand. While signaling by the bacterial aspartate receptor is thought to be mediated by a piston mechanism,[41] productive dimerization of the ErbB2/Neu receptor requires a precise rotational coupling of each of the monomers.[42] A similar mechanism involving rotation has also been proposed for transmission of conformational changes from the ectodomain to the transmembrane domain of the atrial natriuretic peptide receptor.[43] Indeed, rotation or sliding between transmembrane domains of the two subunits of the G protein-coupled metabotropic glutamate receptor mGlur1 is thought to be responsible for the outside-in transmission of the conformational change.[44] Finally, a switchblade model has been proposed for the inside-out and outside-in signaling of integrins.[45] In this case, ligand binding may promote separation of the TM and cytoplasmic segments that could permit novel interactions with signaling complexes.

With regard to MIRRs, the mechanism proposed for the transmission of the conformational change across the membrane involves piston-like, rotational and sliding movements. The structure of the CD3γ-CD3ε dimer ectodomains led to the proposal that the TM domain of the dimer undergoes a piston-like movement. In this way, the paired G beta strands of CD3ε and CD3γ (as well as those of CD3ε and CD3δ) form a rigid rod-like connector that could produce a displacement of the transmembrane helices.[46] On the other hand, monovalent Fab fragments of the anti-CD3 antibody OKT3 bind to CD3ε in a side-on fashion[47] and this interaction induces a conformational change in the TCR.[21] When considered in conjunction with the electrostatic properties of CD3ε, the transmission of the conformational change appears to require a rotational or scissor-like movement of the TM and the cytoplasmic tails of the CD3 subunits.[47] Similarly, an umbrella model has been proposed for the opening of the cytoplasmic tails of the BCR upon antigen-induced clustering.[27]

Recently, in a review of the structure of the TCR complex, five possible ways of transmitting the conformational change were proposed.[48] In the first model, binding of pMHC to the TCRα/β heterodimer causes a conformational change that is transmitted to the constant regions as observed in the TCR LC13 crystal structure.[13] In turn, this conformational change would promote a change in the orientation of the CD3 dimers (Fig. 2A, model 1). In the second model (Fig. 2A, model 2), the conformational change in the TCRα/β heterodimer causes a conformational change in the CD3 dimers. In the third model (Fig. 2A, model 3), pMHC binding does not alter the conformation of the TCRα/β heterodimer beyond the CDR loops, but rather repositions the CD3 dimers with respect to TCRα/β. In the fourth model (Fig. 2A, model 4), again the binding of pMHC does not affect the conformation of the constant regions of the TCRα/β, but rather promotes a vertical displacement of the CD3 dimers in a piston-like fashion. In the fifth model (Fig. 2A, model 5), pMHC binding causes the TCRα/β heterodimer to push one of the CD3 dimers while pulling the other, a modification of the piston-like movement. If we consider the TCR as a monovalent structure without the restrictions in the mobility of the complex that could result from binding of two or more ligands to a multivalent complex, models 3, 4 and 5 are difficult to understand. Thus, binding of a cluster of two agonist pMHC to two TCRα/β heterodimers within a multivalent TCR complex would generate a torque on the TCRα/β heterodimers that could be transmitted to the CD3 dimers. In turn, this binding would induce the rotation or sliding of the TM domains of the CD3 dimer with respect to those of the TCRα/β heterodimers (Fig. 2B) and the ensuing transmission of this movement to the CD3 cytoplasmic tails. Crystal structures of TCRα/β ectodomains bound to pMHC complexes clearly show that the orientation of TCRα/β is approximately diagonal to the MHC peptide-binding groove (reviewed in ref. 12). We hypothesize that in the context of a multivalent TCR, the diagonal orientation imposed on the two TCRα/β heterodimers upon pMHC binding may be responsible for the torque transmitted to the CD3 subunits (Fig. 3).

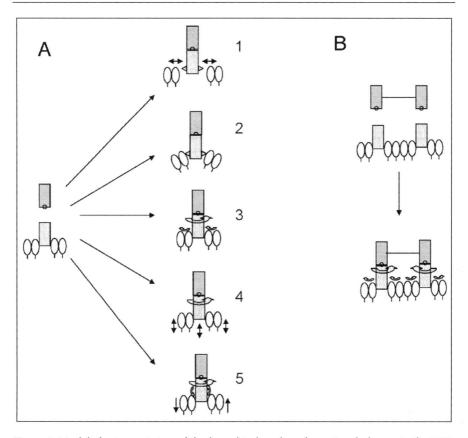

Figure 2. Models for transmission of the ligand-induced conformational change in the TCR. A) Monovalent TCR-pMHC interaction. The TCR complex is represented as the TCRα/β heterodimer (shown as a dotted rectangle) associated with two CD3 dimers (CD3ε-CD3γ and CD3ε-CD3δ). The MHC is represented as a grey rectangle bound to a little ball (peptide antigen). Upon binding of the pMHC to the TCR complex, the TCRα/β ectodomains undergo a conformational change not only in the ligand-binding CDR loops but also in the Cα and/or Cβ domains (indicated by triangles). This conformational change is transmitted to the CD3 dimers, which subsequently shift their orientation with respect to the TCRα/β heterodimer (model 1). According to model 2, in a rigid TCRα/β-CD3 interaction, the TCRα/β heterodimer undergoes a conformational change that as in model 1 is transmitted to the CD3-interacting domains which also change their conformation. According to model 3, the TCRα/β heterodimer does not change its conformation beyond the CDR loops, but binding to the pMHC generates a torque that causes the reorientation of the TCRα/β heterodimer with respect to the CD3 dimers. Model 4 is like model 3, but torque generated upon pMHC binding is transmitted to the CD3 dimers that undergo a rotation and a piston-like movement. Model 5 is a piston-like model in which pMHC binding changes the angle of TCRα/β heterodimer anchoring to the membrane and this pushes one of the CD3 dimers and pulls the other. B) Multivalent TCR-pMHC interaction. In the context of a multivalent TCR, the restrictions imposed on two TCRα/β heterodimers upon binding of the pMHC cluster results in a torque of the TCRα/β heterodimers, causing a rotation or sliding of the CD3 dimers.

Conclusions

Unlike other membrane receptors, studies on the conformational changes involved in signal transduction are still only preliminary in MIRRs. This may be because of incomplete structural

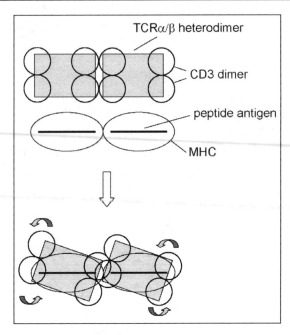

Figure 3. Model for torque generation in the TCR complex. In this cartoon, the TCRα/β heterodimers are symbolized as grey squares that are associated with CD3 dimers (illustrated as pairs of circles) while a pMHC superdimer is shown as two ellipses loaded with an agonist peptide (thick black line). The diagonal orientation that the TCRα/β heterodimer adopts upon binding to the pMHC superdimer causes a sliding or rotation of the two TCRα/β heterodimers and a torque that is transmitted to the associated CD3 dimers.

information available for these multisubunit receptors. Nevertheless, it has been fairly well demonstrated that the TCR and BCR undergo ligand-induced conformational changes that are transmitted to their cytoplasmic tails, although the functional importance of these mechanisms remains to be fully demonstrated. When considering conformational changes in MIRRs, we frequently refer to "a conformational change" as an on/off switch that leads to full cell activation. However, it remains possible that different conformational changes, or different degrees of a given conformational change, are responsible for modulating cellular responses to agonist, partial agonist and antagonist ligands.

Acknowledgements

We thank Hisse M van Santen and Mark Sefton for critically reading the manuscript. The laboratory of BA is supported by the grant SAF2005-00937 from Comisión Internacional de Ciencia y Tecnología.

References

1. Wess J. G-protein-coupled receptors: molecular mechanisms involved in receptor activation and selectivity of G-protein recognition. FASEB J 1997; 11:346-354.
2. Wess J, Liu J, Blin N et al. Structural basis of receptor/G protein coupling selectivity studied with muscarinic receptors as model systems. Life Sci 1997; 60:1007-1014.
3. Jiang G, Hunter T. Receptor signaling: when dimerization is not enough. Curr Biol 1999; 9:R568-571.
4. Banner DW, D'Arcy A, Janes W et al. Crystal structure of the soluble human 55 kd TNF receptor-human TNF beta complex: implications for TNF receptor activation. Cell 1993; 73:431-445.
5. Chan FK, Chun HJ, Zheng L et al. A domain in TNF receptors that mediates ligand-independent receptor assembly and signaling. Science 2000; 288:2351-2354.

6. Chan FK, Siegel RM, Zacharias D et al. Fluorescence resonance energy transfer analysis of cell surface receptor interactions and signaling using spectral variants of the green fluorescent protein. Cytometry 2001; 44:361-368.

7. Krause CD, Mei E, Xie J et al. Seeing the light: Preassembly and ligand-induced changes of the interferon gamma receptor complex in cells. Mol Cell Proteomics 2002; 1:805-815.

8. Murali R, Cheng X, Berezov A et al. Disabling TNF receptor signaling by induced conformational perturbation of tryptophan-107. Proc Natl Acad Sci USA 2005; 102:10970-10975.

9. Remy I, Wilson IA, Michnick SW. Erythropoietin receptor activation by a ligand-induced conformation change Science 1999; 283:990-993.

10. Siegel RM, Frederiksen JK, Zacharias DA et al. Fas preassociation required for apoptosis signaling and dominant inhibition by pathogenic mutations. Science 2000; 288:2354-2357.

11. Walter MR, Windsor WT, Nagabhushan TL et al. Crystal structure of a complex between interferon-gamma and its soluble high-affinity receptor. Nature 1995; 376:230-235.

12. Rudolph MG, Stanfield RL, Wilson IA. How TCRs bind MHCs, peptides and coreceptors. Annu Rev Immunol 2006; 24:419-466.

13. Kjer-Nielsen L, Clements CS, Purcell AW et al. A Structural basis for the selection of dominant alpha-beta T-cell receptors in antiviral immunity. Immunity 2003; 18:53-64.

14. Ding YH, Baker BM, Garboczi DN et al. Four A6-TCR/peptide/HLA-A2 structures that generate very different T-cell signals are nearly identical. Immunity 1999; 11:45-56.

15. Schamel WW, Arechaga I, Risueno RM et al. Coexistence of multivalent and monovalent TCRs explains high sensitivity and wide range of response. J Exp Med 2005; 202:493-503.

16. Minguet S, Swamy M, Alarcon B et al. Full activation of the T-cell receptor requires both clustering and conformational changes at CD3. Immunity 2007; 26:43-54.

17. Janeway CA Jr. Ligands for the T-cell receptor: Hard times for avidity models. Immunol Today 1995; 16:223-225.

18. Yoon ST, Dianzani U, Bottomly K et al. Both high and low avidity antibodies to the T-cell receptor can have agonist or antagonist activity. Immunity 1994; 1:563-569.

19. Aivazian D, Stern LJ. Phosphorylation of T-cell receptor zeta is regulated by a lipid dependent folding transition. Nat Struct Biol 2000; 7:1023-1026.

20. Laczko I, Hollosi M, Vass E et al. Conformational effect of phosphorylation on T-cell receptor/CD3 zeta-chain sequences. Biochem Biophys Res Commun 1998; 242:474-479.

21. Gil D, Schamel WW, Montoya M et al. Recruitment of Nck by CD3 epsilon reveals a ligand-induced conformational change essential for T-cell receptor signaling and synapse formation. Cell 2002; 109:901-912.

22. Gil D, Schrum AG, Alarcon B et al. T-cell receptor engagement by peptide-MHC ligands induces a conformational change in the CD3 complex of thymocytes. J Exp Med 2005; 201:517-522.

23. Risueno RM, Gil D, Fernandez E et al. Ligand-induced conformational change in the T-cell receptor associated with productive immune synapses. Blood 2005; 106:601-608.

24. Geisberger R, Crameri RA, chatz G. Models of signal transduction through the B-cell antigen receptor. Immunology. 2003; 110:401-410.

25. Cambier JC, Pleiman, CMClark, MR. Signal transduction by the B-cell antigen receptor and its coreceptors. Annu Rev Immunol 1994; 12:457-486.

26. Reth M. Hydrogen peroxide as second messenger in lymphocyte activation. Nat Immunol 2002; 3:1129-1134.

27. Tolar P, Sohn HW, Pierce SK. The initiation of antigen-induced B-cell antigen receptor signaling viewed in living cells by fluorescence resonance energy transfer. Nat Immunol 2005; 6:1168-1176.

28. Kato K, Fridman WH, Arata Y et al. A conformational change in the Fc precludes the binding of two Fcgamma receptor molecules to one IgG. Immunol Today 2000; 21:310-312.

29. Matsui K, Boniface JJ, Steffner P et al. Kinetics of T-cell receptor binding to peptide/I-Ek complexes: correlation of the dissociation rate with T-cell responsiveness. Proc Natl Acad Sci USA 1994; 91:12862-12866.

30. McKeithan, TW. Kinetic proofreading in T-cell receptor signal transduction. Proc Natl Acad Sci USA 1995; 92:5042-5046.

31. Davis MM, Boniface JJ, Reich Z et al. Ligand recognition by alpha beta T-cell receptors. Annu Rev Immunol 1998; 16:523-544.

32. Germain RN, Stefanova I. The dynamics of T-cell receptor signaling: complex orchestration and the key roles of tempo and cooperation. Annu Rev Immunol 1999; 17:467-522.

33. van der Merwe PA, Davis SJ. Molecular interactions mediating T-cell antigen recognition. Annu Rev Immunol 2003; 21:659-684.

34. Rudolph MG, Wilson IA. The specificity of TCR/pMHC interaction. Curr Opin Immunol 2002; 14:52-65.

35. van der Merwe PA. The TCR triggering puzzle. Immunity 2001; 14:665-668.
36. Qi S, Krogsgaard M, Davis MM et al. Molecular flexibility can influence the stimulatory ability of receptor-ligand interactions at cell-cell junctions. Proc Natl Acad Sci USA 2006; 103:4416-4421.
37. Krogsgaard M, Prado N, Adams EJ et al. Evidence that structural rearrangements and/or flexibility during TCR binding can contribute to T-cell activation. Mol Cell 2003; 12:1367-1378.
38. Schamel WW, Risueno RM, Minguet S et al. A conformation- and avidity-based proofreading mechanism for the TCR-CD3 complex. Trends Immunol 2006; 27:176-182.
39. Boniface JJ, Rabinowitz JD, Wulfing C et al. Initiation of signal transduction through the T-cell receptor requires the multivalent engagement of peptide/MHC ligands [corrected]. Immunity 1998; 9:459-466.
40. Cochran JR, Cameron TO, Stern LJ. The relationship of MHC-peptide binding and T-cell activation probed using chemically defined MHC class II oligomers. Immunity 2000; 12:241-250.
41. Ottemann KM, Xiao W, Shin YK et al. A piston model for transmembrane signaling of the aspartate receptor. Science 1999; 285:1751-1754.
42. Burke CL, Stern DF. Activation of Neu (ErbB-2) mediated by disulfide bond-induced dimerization reveals a receptor tyrosine kinase dimer interface. Mol Cell Biol 1998; 18:5371-5379.
43. Misono KS, Ogawa H, Qiu Y et al. Structural studies of the natriuretic peptide receptor: A novel hormone-induced rotation mechanism for transmembrane signal transduction. Peptides 2005; 26:957-968.
44. Kubo Y, Tateyama M. Towards a view of functioning dimeric metabotropic receptors. Curr Opin Neurobiol 2005; 15:289-295.
45. Arnaout MA, Mahalingam B, Xiong JP. Integrin structure, allostery and bidirectional signaling. Annu Rev Cell Dev Biol 2005; 21:381-410.
46. Sun ZJ, Kim KS, Wagner G et al. Mechanisms contributing to T-cell receptor signaling and assembly revealed by the solution structure of an ectodomain fragment of the CD3 epsilon gamma heterodimer. Cell 2001; 105:913-923.
47. Kjer-Nielsen L, Dunstone MA, Kostenko L et al. Crystal structure of the human T-cell receptor CD3 epsilon gamma heterodimer complexed to the therapeutic mAb OKT3. Proc Natl Acad Sci USA 2004; 101:7675-7680.
48. Kuhns MS, Davis MM, Garcia KC. Deconstructing the form and function of the TCR/CD3 complex. Immunity 2006; 24:133-139.

CHAPTER 11

Permissive Geometry Model

Susana Minguet and Wolfgang W.A. Schamel*

Abstract

L igand binding to the T-cell antigen receptor (TCR) evokes receptor triggering and subsequent T-lymphocyte activation. Although TCR signal transduction pathways have been extensively studied, a satisfactory mechanism that rationalizes how the information of ligand binding to the receptor is transmitted into the cell remains elusive. Models proposed for TCR triggering can be grouped into two main conceptual categories: receptor clustering by ligand binding and induction of conformational changes within the TCR. None of these models or their variations (see Chapter 6 for details) can satisfactorily account for the diverse experimental observations regarding TCR triggering. Clustering models are not compatible with the presence of preformed oligomeric receptors on the surface of resting cells. Models based on conformational changes induced as a direct effect of ligand binding, are not consistent with the requirement for multivalent ligand to initiate TCR signaling. In this chapter, we discuss the permissive geometry model. This model integrates receptor clustering and conformational change models, together with the existence of preformed oligomeric receptors, providing a mechanism to explain TCR signal initiation.

Introduction

The antigen receptors expressed on lymphocytes belong to the multichain immune recognition receptor (MIRR) family. The B- and T-cell antigen receptors (BCR and TCR) are expressed on B- and T-cells, respectively. BCRs bind to folded native antigens. These ligands can be any chemical substance and therefore can vary enormously in size, shape and geometry. In contrast, TCRs recognize antigenic peptides that are presented on the surface of antigen presenting cells (APCs). Therefore, the basic structure and geometry of the TCR ligand is constant. In both cases, antigen binding leads to MIRR triggering, which in turn activates several cytosolic signaling pathways. To date, the molecular mechanism of how antigen binding evokes MIRR activation is not very well understood and a matter of intense debate. The consensus is that receptor triggering leads to phosphorylation of tyrosine residues in the cytoplasmic portions of the receptor itself. Phosphorylation of the MIRR is the critical event in initiating the signaling cascades.

The TCR comprises the ligand-binding TCR$\alpha\beta$ heterodimer and the signal-transducing dimers CD3$\varepsilon\gamma$, CD3$\varepsilon\delta$ and $\zeta\zeta$ (Chapter 1). The cytoplasmic tyrosines are present within the immunoreceptor tyrosine-based activation motif (ITAM, YxxI/Lx$_{6-8}$YxxI/L)[1] in the CD3 and ζ subunits.

Several approaches have been undertaken to decipher the changes that the receptor undergoes upon ligand binding. Biochemical studies suggest that MIRRs are only activated by bi- or multivalent ligands, implying that one ligand molecule binds simultaneously to several receptor molecules. Based on these observations, the clustering models for MIRR activation were postulated (Chapter 6). Our

*Corresponding Author: Wolfgang W.A. Schamel—Department of Molecular Immunology, Max Planck-Institute for Immunobiology and Faculty of Biology, University of Freiburg, Stuebeweg 51, 79108 Freiburg, Germany. Email: schamel@immunbio.mpg.de

Multichain Immune Recognition Receptor Signaling: From Spatiotemporal Organization to Human Disease, edited by Alexander B. Sigalov. ©2008 Landes Bioscience and Springer Science+Business Media.

recent data on induction of conformational changes in the cytoplasmic tails of the receptor upon ligand binding and the fact that resting MIRRs can be found as preformed multimers, motivated us to assert the permissive geometry model of signal initiation by the MIRRs.[2]

On the surface of APCs, the TCR recognizes its ligand, a peptide bound to major histocompatibility complex molecule (MHCp). A given TCR can bind to several distinct, but related, MHCp that differ in the exact sequence of the peptide. In general, these peptides can be agonist, null or antagonist peptides, depending on their affinity for a given TCR and the strength of the signal generated. In the thymus, where only self-peptides are presented, strong activation of immature T-cells leads to apoptosis and, thus, to negative selection of self-reactive lymphocytes. In contrast, MHCp with weak affinity deliver a survival signal for the differentiation into mature lymphocytes. In the periphery, strong signals activate mature T-cells and weak signaling is required for survival.

In the context of an immune response, both self-peptides, derived from endogenous proteins and antigenic peptides are copresented by the APCs. Recent studies have clearly shown that self-peptides can aid in the activation of T-cells by agonist peptides.[3-6] In this Chapter, the mechanism of this phenomenon will be discussed from the perspective of the permissive geometry model.

The Clustering Model of TCR Triggering

Soluble MHCp monomers as well as monovalent Fab-fragments of anti-TCR antibodies can bind to the TCR but fail to stimulate the receptor.[7-10] In contrast, di- and multimeric soluble MHCp as well as complete anti-TCR antibodies can activate T-cells. These findings imply that two or more receptors have to bind simultaneously to one ligand molecule to be activated. In conjunction with the assumption that individual TCRs are distributed equally on the cell surface, these results led to the proposals of the homoclustering and the pseudodimer models of TCR activation[3,11] (Chapter 6).

Briefly, in the homoclustering model (Fig. 1A), monomeric MIRRs are associated with kinases that need cross-wise phosphorylation for full activity. In addition, the kinases cannot phosphorylate the receptors that they are bound to. If a bi- or multivalent ligand simultaneously binds to at least two MIRRs, the kinases are enabled to transphosphorylate each other and the neighboring MIRR, initiating the signaling cascades.

A variant of the homoclustering model is the pseudodimer model. It takes into account the cooperation between antigenic and self-peptides in T-cell stimulation, together with the role of the coreceptor CD4.[3,6] Since MHC molecules of class I and II are most likely multimeric proteins on the surface of the APCs,[12,13] at low antigen concentrations most antigenic peptides (aMHCp) are presented next to a self-peptides (sMHCp). Indeed, heterodimeric aMHCp-sMHCp were shown to activate T-cells.[6] Since sMHCp cannot stably bind to the TCR, the homoclustering model cannot account for this finding. In the pseudodimer model, dimeric MHCp brings two TCRs together in conjunction with CD4. The interaction between dimeric MHCp and two TCRs might therefore be enhanced by simultaneous binding of CD4 to MHCp and the TCR (Chapter 6).

Oligomeric MIRRs

The homoclustering as well as pseudodimer models require that the unstimulated receptor is monomeric. However, several studies have indicated the existence of oligomeric TCRs, i.e., complexes that have several ligand-binding $TCR\alpha/\beta$ subunits[14,15] (Fig. 1B). Other studies failed to detect them, possibly due to the use of the detergent called digitonin.[16,17] Digitonin disrupts all TCR oligomeric structures[15,18] (Fig. 1B). Likewise, the BCR and the FcεRI might exist as clusters on the surface of unstimulated cells.[19,20]

How can the requirement for a multivalent ligand and the presence of preformed oligomeric receptors be integrated into a unique model? One intriguing possibility is that binding of multivalent ligands disturbs the structure of the receptor oligomer.[21,22] The consequent reorientation of the receptor units might lead to conformational changes within the cytoplasmic tails of the receptor, thereby allowing phosphorylation by the associated kinases. Monovalent ligands are not capable of disturbing the structure of the receptor oligomer and thus they are inactive. Therefore,

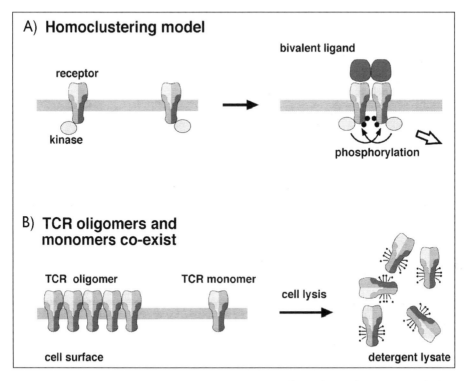

Figure 1. The clustering model is not compatible with preformed TCR oligomers. A) The homoclustering model requires that monomeric TCRs with associated kinases are individually expressed on the cell surface. Stimulation by bi- or multivalent ligand leads to clustering of the receptors, since the ligand simultaneously binds two or more receptors. In these clusters the kinases transphosphorylate each other and the receptors. This initiates the signaling cascades, resulting in activation of the cell. Black dots represent phosphorylated tyrosine residues and the open arrow represents activation of downstream signaling cascades. B) On the T-cell surface, the TCR is expressed as mixture of monomers and oligomers of different sizes. Thus, the requirement for the clustering models is not fulfilled. Note, that cell lysis with commonly used detergents (e.g., digitonin) disrupts the oligomers, hampering their detection by biochemical means.

structural changes of the receptor upon ligand binding might be a useful concept to understand MIRR triggering.

Conformational Changes within the MIRRs

Although the study of structural rearrangements within MIRRs and their consequences in activation are still at a preliminary stage, evidences of conformational changes within the TCR, BCR and FcεRI have been described[23-25] (Chapter 10). Initially, the existence of structural changes in the TCR complex upon ligand binding was proposed to explain early T-cell signaling studies where differences in receptor clustering or antibody affinities were insufficient to explain distinct activation potentials of anti-TCR antibodies.[26] However, crystallographic structures of free and MHCp-bound soluble TCRαβ have revealed induced-fit type changes only in the variable regions at the ligand-binding interface, whereas no alterations were observed at the distal parts of the heterodimer, which are in contact with the CD3 signal-transducing units. Thus, it is difficult to understand how structural changes should be transmitted to the cytoplasmic tails of CD3. Nevertheless, the group of Balbino Alarcón found that a proline-rich region in CD3ε becomes exposed upon TCR stimulation (Chapter 10).[23] Induction of conformational changes within the TCR complex has mainly been discussed as a direct consequence of ligand binding. But the lack of support from the crystal structures and the absence of a consistent mechanism have generated scepticism. The permissive geometry model integrates the main models of TCR triggering, mainly receptor clustering and conformational changes, together with the presence of preformed oligomeric receptors. It provides a mechanism that accounts for most of the experimental observations regarding TCR activation.

Likewise, the signaling chain homooligomerization (SCHOOL) model (Chapters 12 and 20) introduces the necessity of a defined orientation between two or more MIRRs cross-linked by multivalent ligand. It also explains why initially preformed MIRR oligomers on resting cells do not trigger cell activation.[27]

Permissive Geometry Model

The current clustering models for TCR activation, namely the homoclustering and the pseudodimer models, are not compatible with the presence of preformed oligomeric receptors (Chapter 6). Similarly, the conformational change model, as a direct effect of ligand binding, is not consistent with the requirement for multivalent ligand to initiate cell signaling. Our recent data have evoked a new perspective for the mechanism underlying the TCR activation. We show that conformational changes within the TCR complex can only be induced by multivalent ligands (Fig. 2).[2] The receptor oligomers can either be preformed or achieved by multivalent ligand binding, since homoclustering is necessary but not sufficient to induce conformational changes that initiate signal transduction.

When TCRs are clustered in detergent lysates, the conformational change at CD3 is not induced. This indicates that two TCRs need to be brought not only into close proximity, but also into a defined orientation. Likewise, not all anti-TCR antibodies have the same capability to induce the conformational change, even when they bind to the same number of TCRs. Since these antibodies bind to distinct regions of TCR complex, they should lead to different geometries of the clustered complex. We therefore suggested that the exact geometry determines whether a conformational change takes place or not. A permissive geometry would lead to structural reorganization within the TCR complex (Fig. 2), whereas a different inert geometry would not.[2]

Additionally, we designed an experimental approach that can separate the conformational change at CD3 from receptor clustering. By using ligands that induce conformational changes in the absence of receptor clustering and ligands that can keep two TCRs within close proximity without forcing the conformational change, we showed that both TCR clustering and the conformational change are needed for receptor activation.[2]

The necessity of the conformational change for effective TCR signaling in combination with the permissive geometry requirements, encouraged us to assert the permissive geometry model of signal initiation of the MIRRs (Fig. 2).[2] In the resting state, the cytoplasmic tails of the TCR-CD3 complex might be in a closed conformation and not accessible to kinases and/or adaptor proteins. Monovalent MHCp-binding does not lead to structural changes of $TCR\alpha/\beta$ outside the direct contact region, thus not rearranging the cytoplasmic tails of CD3 and not leading to TCR activation (Fig. 2, upper panel). Similarly, bi- and multivalent MHCp binding does not change the structure of one $TCR\alpha/\beta$ (Fig. 2, lower panel). However, since two TCR-CD3 complexes are engaged simultaneously, they have to adjust to the geometry of the preformed MHCp dimer. This results in a reorientation of two $TCR\alpha\beta$ into a permissive geometry. A mechanical force is then exerted on the extracellular and transmembrane regions of the CD3 and ζ subunits, "pushing" them away from their original positions. This rearrangement is transmitted through the membrane and affects the conformation of the cytoplasmic regions of CD3, making them accessible for activation effectors (Fig. 2, lower panel). This model is compatible with preformed receptor oligomers, suggesting that initially they are in a nonpermissive geometry, whereas binding of bi- or multivalent ligands induces the signaling-permissive geometry within the oligomer (Fig. 3).

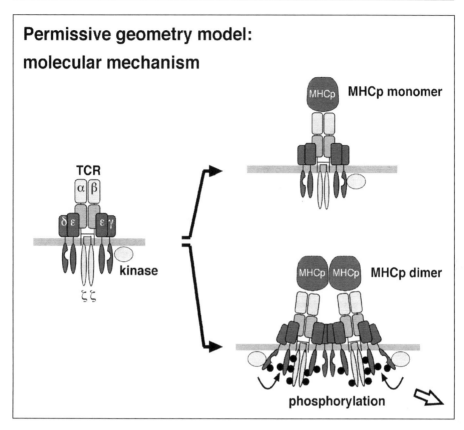

Permissive geometry model:
molecular mechanism

Figure 2. Permissive geometry model. The TCR-CD3 complex is in a closed conformation, preventing phosphorylation by the associated kinases. Monovalent MHCp does not change the structure of TCRα/β. Thus, a conformational change cannot be transmitted to CD3 and the receptor stays in an inactive, nonphosphorylated state (upper panel). In the permissive geometry model, bi- or multivalent MHCp ligands reorientate two TCRα/β without changing the structure of one TCRα/β. The reorientation exerts a mechanical force on CD3, inducing conformational changes in the cytoplasmic tails of CD3 and allowing phosphorylation. Thus, dimeric MHCp activates T-cells due to induction of conformational changes within a TCR oligomer (lower panel). Since monomeric MHCp does not engage several TCRα/β simultaneously, it does not induce conformational changes in CD3.

Similar models have been previously suggested for the TCR[28,29] (dimer conformational change model), the FcεRI[25] and for MIRRs in general (SCHOOL model, Chapters 12 and 20).[27]

Agonist/Self-Peptide-MHC Heterodimers

In the context of the immune response, both antigenic peptides and self-peptides are copresented by the APCs. Recently, it was shown that self-peptides can help agonistic peptides to activate T-cells.[3-6] The role of sMHCp in TCR triggering can be incorporated into the permissive geometry model. The heterodimeric MHCp (aMHCp-sMHCp) would have a sufficient avidity to bind a TCR oligomer stably enough to induce clustering within the permissive geometry and thereby conformational changes at CD3, leading to receptor triggering and cell activation (Fig. 4, middle panel). In contrast, the sMHCp-TCR affinity is not sufficient to cluster two monomeric receptors with a half-life that is long enough for the initiation of signaling (homoclustering model).

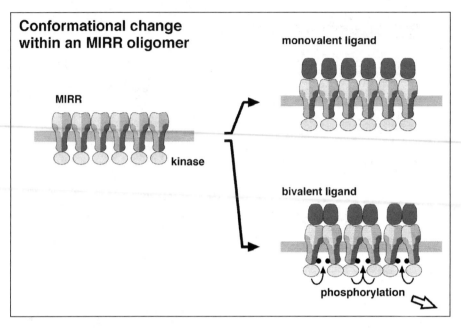

Figure 3. Rearrangement of preformed MIRR oligomers. Resting MIRR oligomers are in a nonpermissive geometry and inactive. MIRR clusters are pre-associated with kinases but due to structural constraints these kinases are not able to transphosphorylate each other and the receptor tails. Monovalent ligands do not disturb the original structure of the oligomer (upper panel) and thus, do not lead to MIRR triggering. By binding simultaneously to at least two receptor units within the cluster, bi- and multivalent ligands change the oligomeric structure into the permissive geometry (lower panel). Consequently, the kinases are able to reach each other and the receptors for their phosphorylation. Thus, this model combines the requirement for a multivalent ligand and the presence of oligomeric receptors.

Mathematical calculations using the permissive geometry model were able to explain how T-cells can respond with extreme sensitivity to low amounts of antigenic peptides without normally being activated by the self-peptides alone.[30] At high antigen doses, sufficient amounts of aMHCp homodimers might be present on the APC surface, allowing homoclustering of two monomeric TCRs and inducing cell activation (Fig. 4, lower panel). This model provides an explanation for the fact that multimeric TCRs become activated at low antigen doses, whereas multi- and monomeric TCRs are activated at high doses.[15] This permissive geometry model does not require the molecular aid of the CD4 or CD8 coreceptors but still allows the coreceptor-mediated enhancement of the signal. This could be due to increased recruitment of kinases of the Src-family or an enhancement of the avidity of TCR/co receptor towards MHCp, prolonging the duration of the TCR-ligand engagement and therefore the strength of the activation signal, as proposed by the kinetic proofreading model[31] (Chapter 8).

Acknowledgements
We thank Balbino Alarcón and Michael Reth for continuous support and Mahima Swamy for critical reading of the manuscript. S.M. was supported by the European Union grant EPI-PEP-VAC. W.W.A.S was supported by the Deutsche Forschungsgemeinschaft (DFG) through the Emmy Noether program and the SFB620.

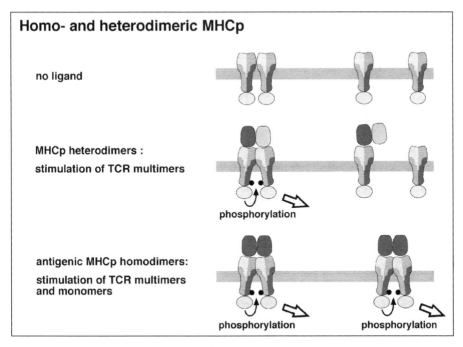

Figure 4. Role of self-peptide-MHC complexes. TCRs are co-expressed as multimers and monomers with associated kinases (upper panel). Heterodimeric aMHCp-sMHCp binds to a TCR oligomer bivalently and with sufficient avidity, inducing rearrangement of the cytoplasmic tails. This results in receptor phosphorylation (middle panel). MHCp heterodimers cannot stably cluster two monomeric TCRs, thus leaving them unphosphorylated. In contrast, homodimers of aMHCp can activate both TCR oligomers and monomers due to the moderate binding affinity of aMHCp molecules (lower panel).

References

1. Reth M. Antigen receptor tail clue. Nature 1989; 338-383.
2. Minguet S, Swamy M, Alarcon B et al. Full activation of the T-cell receptor requires both clustering and conformational changes at CD3. Immunity 2007; 26(1):43-54.
3. Irvine DJ, Purbhoo MA, Krogsgaard M et al. Direct observation of ligand recognition by T-cells. Nature 2002; 419:845-849.
4. Wülfing C, Sumen C, Sjaastad MD et al. Costimulation and endogenous MHC ligands contribute to T-cell recognition. Nat Immunol 2002; 3:42-47.
5. Stefanova I, Dorfman JR, Germain RN. Self-recognition promotes the foreign antigen sensitivity of naive T-lymphocytes. Nature 2002; 420(6914):429-434.
6. Krogsgaard M, Li QJ, Sumen C et al. Agonist/endogenous peptide-MHC heterodimers drive T-cell activation and sensitivity. Nature 2005; 434(7030):238-243.
7. Boniface JJ, Rabinowitz JD, Wülfing C et al. Initiation of signal transduction through the T-cell receptor requires the peptide multivalent engagement of MHC ligands. Immunity 1998; 9:459-466.
8. Cochran JR, Cameron TO, Stern LJ. The relationship of MHC-peptide binding and T-cell activation probed using chemically defined MHC class II oligomers. Immunity 2000; 12(3):241-250.
9. Chang TW, Kung PC, Gingras SP et al. Does OKT3 monoclonal antibody react with an antigen-recognition structure on human T-cells? Proc Natl Acad Sci USA 1981; 78(3):1805-1808.
10. Kaye J, Janeway CA, Jr. The Fab fragment of a directly activating monoclonal antibody that precipitates a disulfide-linked heterodimer from a helper T-cell clone blocks activation by either allogeneic Ia or antigen and self-Ia. J Exp Med 1984; 159(5):1397-1412.
11. Ashwell JD, Klausner RD. Genetic and mutational analysis of the T-cell antigen receptor. Annu Rev Immunol 1990; 8:139-167.

12. Schafer PH, Pierce SK, Jardetzky TS. The structure of MHC class II: A role for dimer of dimers. Semin Immunol 1995; 7(6):389-398.
13. Krishna S, Benaroch P, Pillai S. Tetrameric cell-surface MHC class I molecules. Nature 1992; 357(6374):164-167.
14. Fernandez-Miguel G, Alarcon B, Iglesias A et al. Multivalent structure of an alphabeta T-cell receptor. Proc Natl Acad Sci USA 1999; 96(4):1547-1552.
15. Schamel WW, Arechaga I, Risueno RM et al. Coexistence of multivalent and monovalent TCRs explains high sensitivity and wide range of response. J Exp Med 2005; 202:493-503.
16. Punt JA, Roberts JL, Kearse KP et al. Stoichiometry of the T-cell antigen receptor (TCR) complex: each TCR/CD3 complex contains one TCRa, one TCRb and two CD3e chains. J Exp Med 1994; 180:587-593.
17. Call ME, Pyrdol J, Wiedmann M et al. The organizing principle in the formation of the T-cell receptor- CD3 complex. Cell 2002; 111:967-979.
18. Alarcon B, Swamy M, van Santen HM et al. T-cell antigen-receptor stoichiometry: preclustering for sensitivity. EMBO Rep 2006; 7(5):490-495.
19. Schamel WW, Reth M. Monomeric and oligomeric complexes of the B-cell antigen receptor. Immunity 2000; 13(1):5-14.
20. Wilson BS, Pfeiffer JR, Oliver JM. Observing FcepsilonRI signaling from the inside of the masT-cell membrane. J Cell Biol 2000; 149(5):1131-1142.
21. Reth M, Wienands J, Schamel WW. An unsolved problem of the clonal selection theory and the model of an oligomeric B-cell antigen receptor. Immunol Rev 2000; 176:10-18.
22. Reth M. Oligomeric antigen receptors: a new view on signaling for the selection of lymphocytes. Trend Immunol 2001; 22(7):356-360.
23. Gil D, Schamel WW, Montoya M et al. Recruitment of Nck by CD3 epsilon reveals a ligand-induced conformational change essential for T-cell receptor signaling and synapse formation. Cell 2002; 109(7):901-912.
24. Tolar P, Sohn HW, Pierce SK. The initiation of antigen-induced B-cell antigen receptor signaling viewed in living cells by fluorescence resonance energy transfer. Nat Immunol 2005; 6(11):1168-1176.
25. Ortega E, Schweitzer-Stenner R, Pecht I. Possible orientational constraints determine secretory signals induced by aggregation of IgE receptors on masT-cells. EMBO J 1988; 7(13):4101-4109.
26. Janeway CAJ. Ligands for the T-cell receptor: hard times for avidity models. Immunol Today 1995; 16:223-225.
27. Sigalov AB. Multichain immune recognition receptor signaling: Different players, same game? Trends Immunol 2004; 25(11):583-589.
28. van der Merwe PA. The TCR triggering puzzle. Immunity 2001; 14(6):665-668.
29. Risueno RM, Gil D, Fernandez E et al. Ligand-induced conformational change in the T-cell receptor associated with productive immune synapses. Blood 2005; 106(2):601-608.
30. Schamel WW, Risueno RM, Minguet S et al. A conformation- and avidity-based proofreading mechanism for the TCR-CD3 complex. Trends Immunol 2006; 27:176-182.
31. McKeithan TW. Kinetic proofreading in T-cell receptor signal transduction. Proc Natl Acad Sci USA 1995; 92(11):5042-5046.

CHAPTER 12

Signaling Chain Homooligomerization (SCHOOL) Model

Alexander B. Sigalov*

Abstract

Multichain immune recognition receptors (MIRRs) represent a family of surface receptors expressed on different cells of the hematopoietic system and function to transduce signals leading to a variety of biologic responses. The most intriguing and distinct structural feature of MIRR family members is that extracellular recognition domains and intracellular signaling domains are located on separate subunits. The biochemical cascades triggered by MIRRs are understood in significant detail, however, the mechanism by which extracellular ligand binding initiates intracellular signal transduction processes is not clear and no model fully explains how MIRR signaling commences.

In this Chapter, I describe a novel mechanistic model of MIRR-mediated signal transduction, the signaling chain homooligomerization (SCHOOL) model. The basic concept of this model assumes that the structural similarity of the MIRRs provides the basis for the similarity in the mechanisms of MIRR-mediated transmembrane signaling. Within the SCHOOL model, MIRR triggering is considered to be a result of the ligand-induced interplay between (1) intrareceptor transmembrane interactions between MIRR recognition and signaling subunits that stabilize and maintain receptor integrity and (2) interreceptor homointeractions between MIRR signaling subunits that lead to the formation of oligomeric signaling structures, thus triggering the receptors and initiating the signaling cascade. Thus, the SCHOOL model is based on specific protein-protein interactions—biochemical processes that can be influenced and controlled. In this context, this plausible and easily testable model is fundamentally different from those previously suggested for particular MIRRs and has several important advantages. The basic principles of transmembrane signaling learned from the SCHOOL model may be used in different fields of immunology and cell biology to describe, explain and predict immunological phenomena and processes mediated by structurally related but functionally different membrane receptors. Important applications of the SCHOOL model in clinical immunology, molecular pharmacology and virology are described in the Chapters 20 and 22 of this book.

Introduction

Immune cells respond to the presence of foreign antigens with a wide range of responses, including the secretion of preformed and newly formed mediators, phagocytosis of particles, endocytosis, cytotoxicity against target cells, as well as cell proliferation and/or differentiation. Antigen recognition by immune cells is mediated by the interaction of soluble, particulate and cellular antigens with an array of membrane-bound signaling receptors. Key among these receptors is the family of

*Alexander B. Sigalov—Department of Pathology, University of Massachusetts Medical School, 55 Lake Avenue North, Worcester 01655, MA, USA. Email: alexander.sigalov@umassmed.edu

Multichain Immune Recognition Receptor Signaling: From Spatiotemporal Organization to Human Disease, edited by Alexander B. Sigalov. ©2008 Landes Bioscience and Springer Science+Business Media.

structurally related but functionally different multichain immune recognition receptors (MIRRs) that are expressed on many different immune cells, including T-and B-cells, natural killer (NK) cells, mast cells, macrophages, basophils, neutrophils, eosinophils, dendritic cells and platelets.[1,2] Figure 1 shows typical examples of MIRRs including the T-cell receptor (TCR) complex, the B-cell receptor (BCR) complex, Fc receptors (e.g., FcεRI, FcαRI, FcγRI and FcγRIII), NK receptors (e.g., NKG2D, CD94/NKG2C, KIR2DS, NKp30, NKp44 and NKp46), immunoglobulin (Ig)-like transcripts and leukocyte Ig-like receptors (ILTs and LIRs, respectively), signal regulatory proteins (SIRPs), dendritic cell immunoactivating receptor (DCAR), myeloid DNAX adapter protein of 12 kD (DAP12)-associating lectin 1 (MDL-1), novel immune-type receptor (NITR), triggering receptors expressed on myeloid cells (TREMs) and the platelet collagen receptor, glycoprotein VI (GPVI). For more information on the structure and function of these and other MIRRs, I refer the reader to Chapters 1-5 of this book and recent reviews.[3-22]

A distinct but common structural characteristic of MIRRs is that the extracellular recognition (or ligand-binding) domains and the intracellular signaling domains of these multisubunit complexes are intriguingly located on separate subunits (Figs. 1 and 2). The MIRR ligand-binding subunits are integral membrane proteins with small intracellular domains that are themselves inert with regard to signaling. Signaling is achieved through the association of the ligand-binding chains with signal-transducing subunits that contain in their cytoplasmic domains one or more copies of the immunoreceptor tyrosine-based activation motifs (ITAMs) with two appropriately spaced tyrosines ($YxxL/Ix_{6-8}YxxL/I$; where x denotes nonconserved residues)[23] or the YxxM motif,[24,25] found in the DAP10 cytoplasmic domain[25] (Fig. 1). The association of the MIRR subunits in resting cells is driven mostly by the noncovalent transmembrane (TM) interactions between recognition and signaling components (Fig. 2) and plays a key role in receptor assembly, integrity and surface expression.[2,9-11,13,18,21,26-37]

The MIRR-mediated activation signal can be divided into four parts: (1) the extracellular recognition of a multivalent antigen resulting in the aggregation, or clustering, of the MIRRs, (2) MIRR triggering and TM signal transduction, (3) phosphorylation of the ITAM or YxxM tyrosine residues by protein tyrosine kinases (PTKs) and activation of specific intracellular pathways and (4) the activation of genes in the nucleus. The extracellular recognition of an antigen, the MIRR-triggered biochemical cascades and the mechanisms of gene activation are understood in significant detail. However, the mechanism by which the MIRR transduces ordered information such as antigen recognition from outside the cell via receptor TM and juxtamembrane (JM) regions into intracellular biochemical events (part 2) is not well defined. In other words, the key question remains unanswered: what is the molecular mechanism by which clustering of the extracellular recognition domains of MIRRs leads to receptor triggering and tyrosine phosphorylation of the intracellular ITAMs or YxxMs, thus initiating specific pathways and resulting in immune cell functional outcomes? It is also not known how this putative mechanism can explain the intriguing ability of immune cells to discern and differentially respond to slightly different ligands.

MIRR-mediated signal transduction plays an important role in health and disease making these receptors attractive targets for rational intervention in a variety of immune disorders.[7,10,13,38-42] Thus, future therapeutic strategies depend on our detailed understanding of the molecular mechanisms underlying MIRR triggering and subsequent TM signal transduction. In addition, knowing these mechanisms would give us a new handle in dissecting the basic structural and functional aspects of the immune response.

Despite numerous models of MIRR-mediated TM signal transduction suggested for particular MIRRs (e.g., TCR, BCR, Fc receptors, NK receptors, etc.), no current model fully explains at the molecular level how ligand-induced TM signal transduction commences. As a consequence, these models are mostly descriptive, do not explain mechanistically a vast majority of the specific processes behind "outside-in" MIRR signaling and do not reveal clinically important points of therapeutic intervention. In addition, since the term "MIRR" was first introduced in 1992[1] and MIRR-triggered signaling pathways were hypothesized to be similar,[1,43-46] no general mechanistic model of MIRR-mediated immune cell activation has been suggested up to date. This impedes our

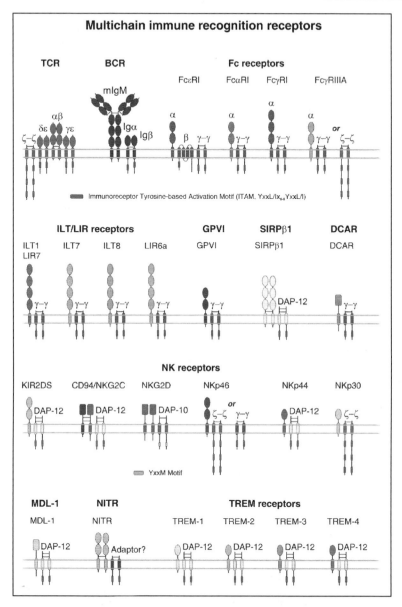

Figure 1. Multichain immune recognition receptors (MIRRs). Schematic presentation of the MIRRs expressed on many different immune cells including T- and B-cells, natural killer cells, mast cells, macrophages, basophils, neutrophils, eosinophils, dendritic cells and platelets. Abbreviations: TCR, T-cell receptor; BCR, B-cell receptor; ILT, Ig-like transcript; LIR, leukocyte Ig-like receptor; GPVI, glycoprotein VI; DNAX adapter proteins of 10 and 12 kD, DAP-10 and DAP-12, respectively; signal regulatory protein, SIRP; dendritic cell immunoactivating receptor, DCAR; NK, natural killer cells; KIR, killer cell Ig-like receptor; myeloid DAP12-associating lectin 1, MDL-1; novel immune-type receptor, NITR; TREM receptors, triggering receptors expressed on myeloid cells. Reprinted from Trends Pharmacol Sci, 27, Sigalov AB, Immune cell signaling: a novel mechanistic model reveals new therapeutic targets, 518-524, copyright 2006 with permission from Elsevier.

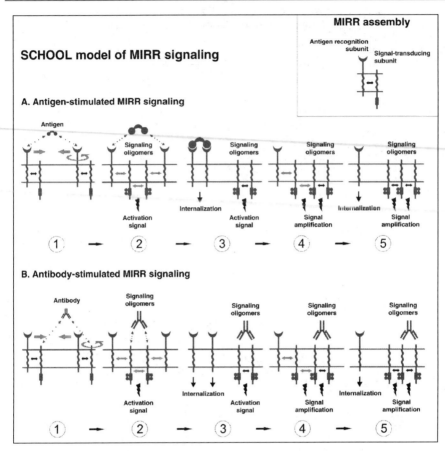

Figure 2. Structural assembly of MIRRs (the inset) and the signaling chain homooligomerization (SCHOOL) model of MIRR signaling. The model proposes that formation of competent MIRR signaling subunit oligomers driven by the homooligomerization of signaling subunits is necessary and sufficient to trigger the receptors and induce transmembrane (TM) signal transduction and downstream sequence. MIRR clustering and receptor reorientation (stage 1) induced by ligand (A) or antibodies to MIRR recognition (not shown) or signaling (B) subunits (e.g., anti-TCRα, anti-TCRβ, anti-CD3ε, anti-Igβ, etc.) lead to formation of a dimeric/oligomeric intermediate in which signaling chains from different receptor units start to trans-homointeract and form signaling oligomers (stage 2). All interchain interactions in this intermediate are shown by light gray arrows reflecting their transition state. Upon formation of signaling oligomers, protein tyrosine kinases phosphorylate the tyrosine residues in the cytoplasmic signaling motifs, the immunoreceptor tyrosine-based activation motifs (ITAMs, shown as gray rectangles), that leads to generation of the activation signal, dissociation of signaling oligomers and internalization of the engaged MIRR binding domains (stages 2 and 3). Next, the signaling oligomers sequentially homointeract with the relevant signaling subunits of nonengaged receptors resulting in formation of higher-order signaling oligomers, thus propagating and amplifying the signals (stages 4 and 5). This also leads to the release and subsequent internalization of the nonengaged ligand-binding MIRR subunits. Small solid black arrows indicate specific intersubunit hetero- and homointeractions between TM and cytoplasmic domains, respectively. The TM interactions between MIRR antigen recognition and signal-transducing subunits have a key role in receptor assembly and integrity on resting cells while cytoplasmic homointeractions represent a main driving force of MIRR triggering. Circular arrows indicate ligand-induced receptor reorientation. Phosphate groups are shown as gray circles.

advanced understanding of the immune response and even more important, prevents the potential transfer of therapeutic strategies between seemingly disparate immune disorders.

Central Hypothesis

My central hypothesis is that the similar structural architecture of the MIRRs dictates similar mechanisms of MIRR triggering and subsequent TM signal transduction and therefore suggests the existence of similar therapeutic targets in seemingly unrelated diseases. This makes possible the development of novel pharmacological approaches as well as the transfer of clinical knowledge, experience and therapeutic strategies between immune disorders. In addition, this hypothesis significantly improves our understanding of the immune modulatory activity of human viruses such as human immunodeficiency virus (HIV) and human cytomegalovirus (CMV) and assumes that the lessons learned from viral pathogenesis can be used for the development of new therapeutic approaches.

In this chapter, I describe a novel mechanistic model of MIRR triggering and subsequent TM signal transduction, the signaling chain homooligomerization (SCHOOL) model.[2,47-49] The model suggests similar mechanisms for all known MIRRs (Fig. 2) and reveals new therapeutic targets in MIRR triggering pathways[2,49] that are described in Chapter 20. Important applications of this model in basic and clinical virology are considered in Chapter 22.

SCHOOL Model of MIRR Signaling

Ligand-induced dimerization/oligomerization of cell surface receptors is frequently employed in signal transduction,[50,51] with dimerization of receptors being the most frequent. For MIRRs, binding of multivalent but not monovalent ligand and subsequent receptor clustering/oligomerization are also thought to be required for induction of the signaling cascade, with MIRR dimerization constituting a necessary and sufficient step for triggering cell activation.[43,50,52-76] Thus, the receptor dimer can be considered as an "elementary stimulatory unit" leading to an immune response. Intracellularly, the need for MIRR dimerization is consistent with the suggested structural hypothesis of cross-phosphorylation[43,77] that assumes that the kinase(s) responsible for catalyzing ITAM Tyr residue phosphorylations exist associated with the receptors, however, for steric reasons they cannot phosphorylate tyrosine residues on chains of the same receptor complex. Upon dimerization/oligomerization, these kinases phosphorylate the tyrosines of a distinct receptor complex (cross-phosphorylation, or transphosphorylation), thus triggering the receptor.[43]

Basic Concepts and Principles

The SCHOOL model suggests that formation of competent MIRR signaling subunit oligomers, rather than MIRR oligomers per se, is necessary and sufficient to trigger the receptors and induce TM signal transduction and the downstream signaling sequence.[2,47,48] Within the model, this oligomerization is driven by the specific homotypic interactions I discovered in 2001 and first reported in 2004,[78] of intrinsically disordered cytoplasmic domains of MIRR signaling subunits. Later, the natural propensity of the TCR ζ cytoplasmic domain to homodimerize has been confirmed by other investigators.[79] Surprisingly, in contrast to other unfolded proteins,[80] the homodimerization/oligomerization of the unstructured protein molecules studied is not accompanied by a structural transition to a folded form,[78,81,82] thus opposing the generally accepted view on the behavior of intrinsically disordered proteins and representing a very unique and unusual biophysical phenomenon itself. Hypothesizing a crucial physiological role of these homointeractions in MIRR triggering and cell activation, the SCHOOL model[2,47,48] indicates that MIRR engagement by multivalent antigen or anti-MIRR-signaling subunit antibodies (e.g., anti-CD3ε or anti-Igβ antibodies for TCRs and BCRs, respectively) leads to receptor clustering coupled with a multi-step structural reorganization driven by the homooligomerization of MIRR signaling subunits (Fig. 2). The model also assumes that the diversity of the immune cell response is partly provided by the combinatorial nature of MIRR-mediated signaling. Signal diversification may be achieved through different patterns of MIRR signaling subunit oligomerization[2,47,48] in combination with distinct activation signals provided by different MIRR signaling modules[83-94] and/or different ITAMs located on the same signaling module (e.g., TCR ζ chain).[95] Thus, according to the model, the more signaling subunits that are in the MIRR complex, the higher is the diversity of immune cell functional outcomes in response to different ligands.

Within the proposed model, MIRR triggering is considered to be the result of the ligand-induced interplay between (1) intrareceptor TM interactions that stabilize and maintain receptor integrity,[2,9-11,13,18,19,21,26-37,96-100] and (2) interreceptor homointeractions between the cytoplasmic domains of MIRR signaling subunits[78,81] that lead to the formation of oligomeric signaling structures and dissociation of the signaling subunits from their respective recognition subunits. Formation of these signaling oligomers triggers phosphorylation of ITAMs, thus initiating the signaling cascade.

Main Stages of MIRR Triggering/Signaling

According to the SCHOOL model, MIRR triggering and TM signaling induced by binding to multivalent antigen or anti-MIRR antibodies can be divided into five main stages (Fig. 2):

1. *Dynamic lateral clustering and rotation.* Antigen/antibody brings two or more MIRRs together in sufficient proximity and correct relative orientation toward each other to promote the interreceptor homointeractions between signaling subunits. Once initiated, these homointeractions weaken the intrareceptor TM interactions between recognition and signaling subunits.

2. *Intermediate complex formation.* A signaling-competent oligomeric intermediate complex is formed, bringing together the cytoplasmic domains of the signaling subunits, protein kinases and various adaptor/effector proteins, to create a competent, activated receptor complex. In the signaling subunit oligomers formed, the ITAM Tyr residues become phosphorylated, thus starting the signaling cascade.

3. *Dissociation and internalization.* Signaling oligomers dissociate from the engaged ligand-recognition subunits, which are internalized.

4. *Interactions with nonengaged receptors, lateral signal propagation and amplification.* Signaling oligomers interact with the signaling subunits of nonengaged receptors resulting in formation of higher-order signaling oligomers, thus propagating and amplifying the activation signal.

5. *Dissociation and internalization.* Signaling oligomers dissociate from the nonengaged ligand-recognition subunits, which later are internalized.

This plausible and easy testable model is fundamentally different from those previously suggested for particular MIRRs (TCR, BCR, FcεRI, GPVI, etc.) and has several important advantages.

First, this model is based on specific protein-protein interactions—biochemical processes that constitute the foundation for the majority of cell recognition and signal transduction processes in health and disease. Protein-protein interactions can be influenced and controlled[101] and specific inhibition and/or modulation of these interactions provides a promising novel approach for rational drug design, as revealed by the recent progress in the design of inhibitory antibodies, peptides and small molecules.[102,103] Second, assuming that the general principles underlying MIRR-mediated TM signaling mechanisms are similar, the SCHOOL model can be applied to any particular receptor of the MIRR family, including but not limiting to those shown in (Fig. 1). Third, based on specific protein-protein interactions, the model reveals new therapeutic targets for the treatment of a variety of disorders mediated by immune cells (see Chapter 20). Fourth, this model represents a powerful tool to dissect molecular mechanisms of MIRR-mediated signaling and related cell functional outcomes in response to antigen. Finally, an important application of the model is that similar therapeutic strategies targeting key protein-protein interactions involved in MIRR triggering and TM signal transduction may be used to treat diverse immune-mediated diseases. This assumes that clinical knowledge, experience and therapeutic strategies can be transferred between seemingly disparate immune disorders or used to develop novel pharmacological approaches. These and other clinically important features of the SCHOOL model will be discussed in more detail in Chapters 20 and 22.

Main Restraints of MIRR Triggering/Signaling Imposed by the SCHOOL Model

Interactions between TM helixes of recognition and signaling MIRR subunits maintain receptor integrity in unstimulated cells and determine the relative positions of these subunits in the receptor complex (angles, distances, etc.), thus dictating the overall geometry and topology of MIRRs.[2,9-11,13,18,21,26-37,99,100] Within the SCHOOL model, the overall structural architecture of MIRRs, in combination with the requirement to initiate interreceptor homointeractions between MIRR signaling subunits (Fig. 2), impose several restraints for MIRR triggering:

- sufficient interreceptor proximity in MIRR dimers/oligomers,
- correct (permissive) relative orientation of the receptors in MIRR dimers/oligomers,
- long enough duration of the MIRR-ligand interaction that generally correlates with the strength (affinity/avidity) of the ligand and
- sufficient lifetime of an individual receptor in MIRR dimers/oligomers.

The importance of these factors for productive MIRR triggering is strongly supported by a growing body of evidence.[4,32,54,56,62,67,68,73,90,104-134] Interestingly, relative receptor orientation also has been shown to be critically important for the activation of other dimeric/oligomeric TM receptors.[135-139]

Therefore, the restraints imposed by the model play an especially important role during the first stage of MIRR triggering (Fig. 2), at which point these spatial, structural and temporal requirements (correct relative orientation, sufficient proximity, long enough duration of the MIRR-ligand interaction and lifetime of MIRR dimers/oligomers) should be fulfilled to favor initiation of trans-homointeractions between MIRR signaling subunits and formation of competent signaling subunit oligomers. If these requirements are not fulfilled at this "final decision-making" point, the formed MIRR dimers/oligomers may dissociate from the ligand and remain signaling incompetent and/or break apart to its initial monomeric receptor complexes. Also, at this stage, slightly different ligands may bring two or more MIRRs in different relative orientations that favor homointeractions between different signaling subunits and result in formation of different signaling oligomers or their combinations, thus initiating distinct signaling pathways. This mechanism might explain the ability of MIRRs to differentially activate a variety of signaling pathways depending on the nature of the stimulus.

Within the proposed model, the signaling oligomers formed dissociate from ligand-binding chains, which later are internalized (Fig. 2, stage 3). This mechanism provides a structural and mechanistic basis for our improved understanding of many immunological phenomena, such as adaptive T-cell tolerance or anergy,[140-143] differential biological role of CD3 chains,[144] ligand- or antibody-induced exposure of a cryptic polyproline sequence in the cytoplasmic domain of CD3ε,[113,145-147] BCR desensitization,[148-151] cytomegaloviral (CMV) escape from NK attack[152] and others. The dissociation mechanism allows the initially formed signaling oligomers to sequentially homointeract with the signaling subunits of nonengaged receptors (Fig. 2, stages 4 and 5) resulting in formation of higher-order signaling oligomers, thus propagating and amplifying the signal. Also, this leads to dissociation and subsequent internalization of the nonengaged ligand-binding subunits. Thus, as with bacterial chemoreceptors,[153-155] the SCHOOL model-based mechanism of MIRR-mediated cell activation suggests spreading (propagation) activation signal from engaged to nonengaged receptors within receptor clusters.

Finally, it should be noted that similar spatial, structural and temporal restraints are imposed within the proposed model for MIRR triggering by not only antigen (Fig. 2A) but also the anti-MIRR (Fig. 2B) antibodies such as anti-TCRα, anti-TCRβ, anti-CD3ε, anti-Igβ and others. This may explain differential immune cell functional outcomes mediated by MIRRs depending on the specificity of the antibodies.[107,108,111-113,156-160]

Supportive Evidence

I developed the SCHOOL model as a general model for the structurally related MIRR family members, namely, for all receptors that have extracellular recognition and intracellular signaling modules located on separate receptor subunits. For this reason, in order to support the main concept and assumptions of the model, I use a rapidly growing body of evidence coming from studies of various MIRRs.

Clustering and Proximity

In order to trigger the MIRR, within the SCHOOL model, two or more receptors should be clustered/oligomerized in sufficient proximity to each other to initiate homointeractions between signaling subunits with subsequent formation of competent signaling subunit oligomers (Fig. 2).[2,47,48] To date, these spatial restraints imposed by the model on MIRR triggering and initiation of the signaling cascade are consistent with the experimental data observed.

T-Cell Receptor

There is a growing line of structural, biophysical and cellular evidence suggesting that ligand-specific TCR oligomerization is critical to generate a functional signal and that TCR dimerization constitutes a necessary and sufficient step for triggering T-cell activation (see also Chapters 6 and 11).[52,57-60,65-68,72,74,105,161-167] These findings clearly demonstrate that dimeric/oligomeric antigens are able to stimulate T-cells, whereas monomeric fail to do so. Interestingly, a correlation between antigenicity and repetitiveness of major histocompatibility complex (MHC)-bound peptides (pMHCs) has been also shown.[105] For dimeric pMHC class I and II complexes, the ability to trigger T-cells has been reported to decrease with increasing length of the connecting spacer.[168,169] Recently, by testing well-defined dimeric, tetrameric and octameric pMHC complexes containing rigid polyproline spacers of different lengths, it has been also shown that their ability to activate cytotoxic T-lymphocytes decreases as the distance between their subunit MHC complexes increases.[104] Intriguingly, the preTCR complex has been shown to form oligomers spontaneously, in a ligand-independent manner.[170,171] This oligomerization is mediated by specific charged residues in the extracellular domain of the preTCRα chain and is necessary and sufficient to induce autonomous signaling and stimulate preTCR function.[170,171] Recently, TCR-coreceptor complexes from naïve or activated CD4+ or CD8+ T-cells have been found to exist as either dimers or tetramers, whereas no monomers or multimers were detected.[167]

B-Cell Receptor

Similar to the TCR-induced signaling, the BCR activation signal is shown to be triggered by cross-linking of receptors through multivalent antigen,[4,54,56,114-116,172] thus confirming the necessity of BCR clustering for competent signaling and cell activation (see also Chapter 6). Interestingly, as it has been shown in 2007 for the preBCR, the ability of the purified recombinant receptor to dimerize indicates that accessory protein(s) are not required for dimerization and by extension, preBCR signaling through multimerization can occur in a ligand-independent fashion.[55] Showing strong similarities to the observations reported for the preTCR-mediated signaling,[170,171] these findings are well consistent with the molecular mechanisms proposed by the SCHOOL model.

Fc Receptors

Multichain Fc receptors, such as FcεRI, FcαRI, FcγRI and FcγRIII, have been known to initiate cell signaling following interactions with multivalent ligands that induce their clustering (see also Chapter 3).[29,62,63,77,88,111,117-121,173-175] FcεRI aggregates as small as dimers have been reported to be capable of providing an effective activation signal to cause mediator secretion.[111] Using a set of chemically well defined ligands of valences 1-3, the magnitude of the cellular response has been demonstrated to dramatically increase as the valency of a ligand raises from two to three.[62] Trivalent ligands with rigid double-stranded DNA spacers have been shown to effectively stimulate FcεRI-mediated degranulation responses in a length-dependent manner, providing direct evidence for receptor transphosphorylation as a key step in the mechanism of signaling by this receptor, whereas long bivalent ligands with flexible spacers has been demonstrated to be very potent inhibitors of mast cell degranulation stimulated by multivalent antigen.[122] In other studies, the spacing of receptors in ligand-specific FcεRI aggregates has been also shown to be important for generating the activation signal.[174]

NK Receptors

Multivalent ligand-induced receptor oligomerization is presumed to be a common mechanism for initiating NK receptor-mediated signaling.[123-125] Also, structural and biochemical studies of NKG2D receptor[69,176,177] have demonstrated that the receptor exists as a dimer not only in the crystal but also at the surface of unstimulated NK cells. However, in contrast to preBCR and preTCR, this ligand-independent dimerization does not trigger the receptor and initiate downstream signaling, suggesting that dimerization is necessary but not sufficient to trigger the receptor.

Glycoprotein VI

Collagen, a natural ligand of GPVI, contains the GPVI-binding GPO (glycine-proline-hydroxyproline) motifs that form about 10% of the fibrillar collagen sequence and thus represent multiple GPVI-binding sites.[178] Using a series of collagen-like model peptides containing GPO motifs of increasing length within $(GPP)_n$ sequences, Smethurst et al[179] have demonstrated that platelet aggregation and protein tyrosine phosphorylation can be induced only by cross-linked peptides that contain two or more GPO triplets. Multimeric snake venom proteins such as convulxin also strongly activate GPVI in a multimer size-dependent manner,[180,181] suggesting that clustering of GPVI receptors through multiple binding events leads to activation. Structural studies have revealed a dimeric state of GPVI and 2 parallel grooves on the GPVI dimer surface as collagen-binding sites with an orientation and spacing of these grooves precisely matching the dimensions of an intact collagen fiber.[64] These findings provide a structural basis for GPVI signaling mechanisms in which collagen-induced GPVI clustering triggers a signaling cascade via the FcRγ-chain. In 2007, GPVI–FcRγ-chain oligomerization on the surface of unstimulated platelets has been directly demonstrated,[71] suggesting that, like dimerization of NKG2D, oligomerization of GPVI is necessary but not sufficient to trigger the receptor.

Other MIRRs

Human TREM-1 receptor has been shown to exist as a "head-to-tail" dimer in crystal, suggesting that the dimeric TREM-1 most likely contains two distinct ligand-binding sites.[70] High-avidity ligands are thought to trigger TREM-1 and TREM-2, suggesting that formation of multivalent ligand-receptor complexes is a necessary step in TREM-1-mediated cell activation.[18,125] Murine paired immunoglobulin-like receptor (PIR)-A and human leucocyte immunoglobulin-like receptor (LILR)-A2 (ILT/LIR7) complexed with the FcRγ signaling chain through their transmembrane domains are also required to be clustered by a multivalent ligand in order to initiate TM signaling.[19,182] Recently, it has been shown that integrin signaling in neutrophils and macrophages requires ITAM-containing adaptors, DAP-12 and FcRγ, suggesting that integrin signaling-mediated activation of cellular responses in these cells proceeds by an MIRR-like mechanism.[183] Homomeric associations involving transmembrane domains have been reported to represent a driving force for integrin activation, thus providing a structural basis for the coincidence of ligand-induced integrin clustering and activation.[184,185]

Orientation

A rapidly growing body of experimental evidence strongly supports the importance of inter-receptor orientation within ligand-specific MIRR dimers/oligomers for receptor triggering and generation of an activation signal. These findings are in good agreement with the orientational restraints imposed by the SCHOOL model on the initiation of interreceptor homointeractions between signaling subunits in order to trigger MIRRs (Fig. 2).[2,47,48] Suggesting the importance of relative orientation,[2,47,48] the model explains for the first time why random encounters of MIRRs by lateral diffusion or oligomeric forms of MIRRs existing in unstimulated cells[44,69,186-189] and platelets[71] do not result in MIRR triggering and cell activation.

T-Cell Receptor

Despite direct biophysical measurements of the interreceptor relative orientation in ligand-specific TCR dimers/oligomers have not yet been performed, several lines of evidence indicate that relative orientation plays an important role in TCR-mediated cell activation. Using monoclonal antibodies (mAbs) specific for the TCR, it has been shown that T-cell activation does not correlate with the affinity of the mAbs but rather with the recognized epitope.[160] In other studies, triggering of different epitopes of the TCR-CD3-ζ_2 receptor complex has been also reported to induce different modes of T-cell activation,[156-159] suggesting that TCR signaling is not a simple on-off switch through cross-linking/clustering. In addition, high concentrations of anti-TCR, but not anti-CD3, induce a proliferative response without antibody cross-linking.[158] Also, anti-TCR and anti-CD3 have been demonstrated to be different in their capacity to induce responsiveness to interleukin-4 (IL-4)[159] and in their requirement for costimulatory signals.[156] Yang and Parkhouse have reported that stimulation of T-cells with a panel of anti-CD3 mAb recognizing different epitopes has differential functional consequences, demonstrating for the first time that differences in activation mechanisms not only exist between TCR and CD3, but also between epitopes within CD3 and postulating that occupancy of different CD3 epitopes may result in different degrees of conformational change in the receptor complex.[107] In thymocytes, only anti-TCRβ Ab but not anti-TCRα reagents cause long-term TCR downmodulation.[108] Using three-dimensional fluorescence quantitation methods, signaling-induced reorientation of T-cell receptors that cannot be mediated by simple passive diffusion has been shown to take place during immunological synapse formation.[190] In 2007, a change in the orientation of the TCR with respect to the membrane induced by binding to pMHC has been proposed to play an important role in TCR signaling.[90] Conclusions about the importance of interreceptor orientation in the ligand-specific TCR dimers/oligomers have been also made in 2007 by Minguet et al[191] who suggested the so-called permissive geometry model of TCR signaling (see also Chapter 11). In contrast to these studies, Cochran et al[168] have reported that intermolecular orientation is not critical for triggering T-cell activation. However, to address this issue, the authors have used in their studies pMHC dimers coupled via flexible chemical cross-linkers that do not prevent rotation of pMHC molecules around their long axis. This assumption is further supported by the authors' findings that estimated distances for the used cross-linkers in fully extended conformations (50, 70 and 90 Å) did not correlate with the apparent hydrodynamic diameter values experimentally determined for the corresponding crosslinked pMHC dimers in the surprisingly narrow range of 70 to 75 Å.[168] Thus, these dimers cannot be considered as conformationally constrained suggesting a lack of control over the interreceptor orientation in these experiments.[168]

The three-dimensional structures of the three A6-TCR/peptide/HLA-A2 complexes that generate very different T-cell signals have been found to be remarkably similar to each other and to the wild-type agonist complex, suggesting that different signals are not generated by different ligand-induced conformational changes in the αβTCR.[192] This is in agreement with the SCHOOL model proposing that different signaling oligomers can be formed and therefore different T-cell signals can be generated depending on the intermolecular relative orientation in the ligand-specific TCR dimers/oligomers rather than ligand-induced extracellular conformational changes.[2,47,48]

In summary, a vast majority of the experimental findings reported so far strongly support an importance of interreceptor relative orientation in ligand-specific TCR clusters for TCR triggering and cell activation.

B-Cell Receptor

BCRs have been proposed and confirmed to organize into oligomeric clusters on the B-cell surface.[44,187-189] The observed basal BCR clustering does not result in receptor triggering and subsequent cell activation suggesting that, like with TCR and EpoR, a member of cytokine receptor superfamily,[135] the oligomerization of the BCR is necessary but not sufficient for receptor activation[189] and that interreceptor relative orientation in the BCR dimers/oligomers plays an important role in receptor triggering. The differential effects of the point mutations in various parts of the TM sequence of BCR membrane Ig (mIg) have been reported to differentially affect B-cell activation induced by mono- or polyvalent anti-mIg antibodies, thus providing more evidence for importance of correct intermolecular orientation in BCR signaling.[35]

Fc Receptors

As shown for FcεRI, it is not only the number of crosslinked FcεRIs that determines the magnitude of mediator secretion-causing signal induced by different mAbs, but also the relative orientation of receptors within the produced dimers, thus suggesting the importance of the orientational restraint in ligand-specific FceRI dimers/oligomers for generating competent activation signal.[111,112,119,126,127,193] Further, in the IgA receptor, FcαRI, a positively charged arginine residue within the TM domain of ligand recognition α chain promotes association with the signaling FcRγ chain.[99] Studies on signaling through mutants of the FcαRI have shown that a vertical relocation of this TM positive charge does not have any significant effect on proximal and distal receptor functions, whereas a lateral transfer of the positive charge completely abrogates these functions.[32] A possible explanation for these findings is that a vertical relocation of the noncovalent electrostatic bond does not change interreceptor relative orientation within the receptor dimers/oligomers formed upon multivalent ligand stimulation while lateral transfer does.

NK Receptors

Existence of dimeric NKG2D receptor complexes in both NKG2D crystals and at the surface of unstimulated NK cells[69,176,177] suggests that not only dimerization but also relative orientation of receptors within ligand-specific NKG2D dimers/oligomers plays an important role in receptor triggering.

Glycoprotein VI

Similar to NKG2D receptor complexes, GPVI has been found to form a back-to-back dimer in the GPVI crystal[64] and to exist in an oligomeric state on the surface of unstimulated platelets,[71] suggesting an important role of interreceptor relative orientation within these oligomers in GPVI signaling.

Other Receptors

The type I TM glycoprotein gp130 is the commonly used signaling receptor chain of all IL-6-type cytokines (i.e., IL-6).[194] Intriguingly, signal transduction via IL-6 requires not only gp130 homodimerization but also the correct relative orientation of the gp130 cytoplasmic regions in ligand-specific receptor dimer, suggesting that subtle changes in the orientation of the receptor chains relative toward each other might result in very different responses.[138] Enforcement of gp130 dimerization is not sufficient for receptor activation but additional conformational requirements have to be fulfilled.[195] Thus, like dimerization of the MIRRs, dimerization of the cytokine receptors by monoclonal antibodies is in most cases not enough to induce signal transduction.[196]

Interestingly, many members of the tumor necrosis factor receptor superfamily were once thought to signal through ligand-induced receptor trimerization. However, recently, these receptors have been shown to exist as pre-assembled oligomers on the cell surface.[197,198] This suggests that, upon the binding of the trimeric ligand, not only oligomerization (trimerization) of these single-chain receptors but also the correct intermolecular relative orientation within trimers plays a crucial role in signaling.

Oligomerization of Signaling Subunits

According to the SCHOOL model, homooligomerization of the cytoplasmic domains of MIRR signaling subunits drives formation of competent signaling oligomers, thus leading to triggering of the receptor and initiation of the signaling cascade (Fig. 2).[2,47,48] Importantly, this homooligomerization also plays a crucial role in amplification and lateral propagation of the activation signal(s) (Fig. 2). The model also suggests that depending on the nature of stimuli, different signaling subunits can be oligomerized and become phosphorylated, thus triggering distinct signaling pathways and resulting in different functional outcomes.[2,47-49] The experimental data obtained to date for different MIRRs strongly support the main concept of the SCHOOL model.

The ability of TCR ζ cytoplasmic domain to oligomerize was first reported in 2004[78] and later confirmed in cell studies on the activity of membrane-anchored chimeric β₂m/peptide molecules fused with the cytoplasmic domain of ζ chain.[79] Similarly, the propensity of the BCR Igα and Igβ signaling subunits to oligomerize[78] has been recently confirmed and demonstrated to result in the ability of the BCR Igα/Igβ heterodimer to assemble into oligomers.[199]

Both in vitro and in vivo studies have shown that dimerization of CD3ε is critical and sufficient to substitute for a preTCR signal and drive double-positive transition, suggesting that the property of the preTCR responsible for β-selection is the autonomous formation of oligomers, which brings CD3 signaling subunits in close proximity to each other.[170,171] These findings confirm the ability of CD3ε to dimerize, first reported in 2004 for the CD3ε cytoplasmic domain[78] and proves the physiological importance of this dimerization suggested by the SCHOOL model.[2,47-49]

As reported,[200] FcεRI signaling β and γ subunits independently dissociate from a ligand-binding α chain immediately after crosslinking with multivalent ligand. Moreover, these signaling subunits dissociate in the oligomerized form. Interestingly, only γ chains are oligomerized on surfaces of cells stimulated with a suboptimal concentration of antigen, while β chains remain dispersed.[200] In contrast, stimulation of cells with an optimal concentration of antigen results in the distinct oligomerization of both signaling subunits.[200]

In cytokine receptor signaling, dimerization of not just extracellular but rather cytoplasmic domains of the gp130 signaling subunit is critically required to trigger the receptor and initiate the signaling cascade.[138,195,196] Recently, ligand-induced formation of surface receptor oligomers has been reported for the Fas receptor.[201] This single-chain receptor has a cytoplasmic death domain (DD) that, upon receptor stimulation with a trivalent ligand, binds to the homologous DD of the adaptor protein Fas-associated death domain protein (FADD) and homotrimerizes, thus initiating the caspase signaling cascade. Interestingly, a mutation in Fas cytoplasmic domain (T225K) linked to autoimmune lymphoproliferative syndrome impairs receptor oligomerization and inhibits Fas-mediated signaling but retains the ability to interact with FADD.[201] This indicates that homointeractions between Fas cytoplasmic tails have an important role in the receptor triggering. Similarly, cytoplasmic domain-mediated dimerization of toll-like receptor 4 (TLR4) has been recently reported to play an important role in the TLR4 triggering and signal transduction.[202,203]

Dissociation

Within the SCHOOL model, dissociation of competent signaling oligomers from both engaged and nonengaged ligand-recognition subunits upon multivalent ligand stimulation, plays an important role in MIRR triggering, signal amplification and propagation and initiation of the signaling cascade (Fig. 2).[2,47,48] Experimental data accumulated to date strongly support this suggestion.

In activated T-cells, the CD3 and ζ signaling chains has been shown to independently dissociate from the remaining receptor subunits.[204-207] Further, TCRs lacking ζ are endocytosed more rapidly than completely assembled receptors,[208] in line with the SCHOOL model. For BCR, it has been reported that, upon binding of moderate- to low-affinity antigen, the Igα/Igβ subunits physically dissociate from mIg resulting in BCR desensitization.[148] Interestingly, although desensitized cells fail to respond to receptor ligation by a high dose of antigen or by anti-Igλ antibodies, the dissociated Igα/Igβ signaling complex retains signaling function if aggregated by anti-Igβ antibodies.[148] In this context, similar mechanisms are proposed by the SCHOOL model to be involved in the BCR desensitization,[148,149,151] T-cell clonal anergy[141,209,210] and in the inhibition of T-cell activation by the so-called TCR core peptide (CP).[211] The ligand-mediated physical dissociation of the activated BCR complex has been later confirmed in other studies.[212] In 2005,[213] using primary murine B-cells, it has been found that while >95% of the mIg is internalized following anti-Ig-induced aggregation, 20-30% of Igβ remains on the surface, suggesting that mIg and Igβ may function independently following the initial stages of signal transduction. As mentioned, upon crosslinking of the FcεRI with multivalent ligand, oligomerized signaling β and γ chains immediately dissociate from a ligand-recognition α chain.[200]

Duration of the Ligand-Receptor Contact

The SCHOOL model suggests that the multivalent ligand-receptor contact should last long enough to bring two or more MIRRs in sufficient proximity and correct relative orientation toward each other and hold them together to promote the interreceptor homointeractions between signaling subunits, thus initiating the downstream signaling cascade (Fig. 2).[2,47,48] It should be noted that duration of the MIRR-ligand interaction generally correlates with the strength (affinity/avidity) of the ligand. Clearly, the strength of the ligand determines not only duration of the ligand-MIRR contact but also lifetime of an individual receptor in the engaged MIRR dimer/oligomer. These important aspects of the model are also consistent with the experimental data accumulated so far.

In T-cells, the results of multiple reports show a broad correlation between the duration of TCR-ligand interaction and ligand potency.[214-216] A similar interpretation is possible for the data on a revised model of kinetic proofreading in which the duration of TCR engagement regulates the efficiency with which signals trickle through the rapidly reversible early activation pathways to induce later responses[217] (see also Chapter 8). It is also known that the off-rate of ligand binding plays a role in determining the specificity of the TCR-generated signal in a population of T-cells that can discriminate between self and nonself in the thymus.[218] Also, the number of TCR ITAMs required for efficient positive or negative selection has been reported to vary depending upon the affinity of the TCR/ligand interaction.[219] In studies on T-cell activation by bacterial superantigens, a simple relationship between the affinity of the *Staphylococcus enterotoxin* C3 (SEC3)-TCR interaction and the functional responses has been proposed; stronger binding results in stronger T-cell responses.[220] As recently shown, short-lived pMHC ligands induce anergy in T-cell clones in vitro and specific memory T-cells in vivo.[221] Total signal strength has been demonstrated to determine the capacity of primed T-cells to respond to homeostatic cytokines, to survive cytokine withdrawal and to accumulate in vivo.[222] The strength of antigen stimulation is also known to regulate T-cell progression through thresholds of proliferation, differentiation and death.[223]

Similar to T-cells, the B-cell response to antigen varies as a function of antigen/BCR interaction affinity.[224] As demonstrated, above the threshold, concentration of antigen required to trigger a response decreases as the affinity increases.[224] BCR signal strength has been shown to determine B-cell fate.[225] Importantly, continuous receptor signaling of a defined amplitude appears to be critical for development and survival of mature B-cells.[226] It is also known that, upon binding of moderate- to low- but not high-affinity antigen, the Igα/Igβ subunits physically dissociate from mIg resulting in BCR desensitization.[148] A critical role of receptor affinity in antigen-driven selection of B-cell clones in vivo has been also suggested based on studies of stable B-cell transfectants.[227] Recently, the strength of the initial BCR-triggered activation signal has been proposed to finally determine the eventual duration of BCR signaling and the rate of its transmission through downstream pathways.[228]

A great body of evidence shows that the capacity of downstream signaling by an individual FcεRI depends on its capacity to remain in a cluster and is therefore influenced by the ligand affinity/avidity.[62,121,128-131,229] The ability of a similar signaling mechanism to trigger distinct FcεRI-mediated mast cell responses like mediator release and survival has been reported to be determined by the FcRγ signal strength or duration.[129,230] Interestingly, recent findings redefine FcαRI as a bifunctional inhibitory/activating receptor of the immune system that mediates both anti- and proinflammatory functions of IgA, depending on ligand multimericity and duration of multivalent ligand-induced receptor signaling.[132] In platelets, affinity/avidity of interaction of GPVI with collagen or convulxin has been suggested to play an important role in receptor signaling and GPVI-mediated platelet activation.[133,181]

For more information on the important role of the ligand-MIRR complex lifetime in MIRR triggering I refer the reader to Chapter 8 of this book and recent reviews.[216,229,231-233]

SCHOOL Model: Trinity of Description, Explanation and Prediction

Based on well-defined biochemical processes such as specific protein-protein interactions, the SCHOOL model represents the first general mechanistic model of MIRR signaling and can be also defined as a dynamic, continuous, spatially homogeneous, descriptive and explanatory model.[234] This model describes and explains molecular mechanisms and the main driving forces of TM signal transduction for functionally unrelated receptors that share a common organizing principle—extracellular recognition module(s) and intracellular signaling module(s) are found on separate subunits and are noncovalently associated through their TM domains. Thus, the basic principles of TM signaling learned from the model can be used in different fields of immunology and cell biology to describe processes that are mediated by structurally related but functionally different membrane receptors.[2,47-49] Besides the ability to describe general principles of MIRR-mediated signal transduction, the SCHOOL model provides a mechanistic explanation for specific processes behind "outside-in" MIRR signaling that remain unclear. Since it was first published in 2004,[48] the model has also predicted several experimental observations that have been later reported for different immune cells.

By definition, the utility of scientific models is evaluated on their abilities to explain past observations, predict future observations and control events as well as on their simplicity, or even aesthetic appeal. The distinct features of the SCHOOL model demonstrating its utility are described in detail below for specific MIRRs (see also Chapters 20 and 22).

SCHOOL Model of TCR Signaling

Description

The TCR is a multisubunit complex composed of the ligand-binding clonotypic $\alpha\beta$ heterodimer, as well as the heterodimeric CD3$\delta\epsilon$ and CD3$\gamma\epsilon$ signaling components and the disulfide-linked ζ homodimer that contain one (ϵ, γ and δ) or three (ζ) ITAMs, respectively (Figs. 1 and 3; Chapter 1). This receptor complex provides an intriguing ability of T-cells to discern and differentially respond to MHC-bound peptides that can differ by only a single amino acid. The mechanism by which the precise ligand-binding specificities of the TCR are converted into the distinct intracellular signaling processes and diverse functional outcomes has been one of the most controversial topics in T-cell immunology. The SCHOOL model suggests not only the mechanism of TCR triggering and cell activation that can explain the majority of immunological phenomena observed experimentally but also proposes distinct ways to control and modulate the T-cell-mediated immune response.

The overall rigid geometry and topology of the TCR is defined by electrostatic interactions between TCR$\alpha\beta$ TM domains and TM domains of different signaling dimers: CD3$\gamma\epsilon$, CD3$\delta\epsilon$ and ζ_2.[26,27,35] Interestingly, the TCR ζ subunit seems to have a unique and dynamic relationship with the TCR-CD3 complex since only this signaling homodimer appears to turn over independently from the rest of the TCR complex on the cell surface.[235] Assuming that different TCR signaling modules provide distinct signaling and T-cell functional outcomes,[47,48,83] the SCHOOL model of T-cell activation suggests[47,48] that depending on the nature of activating stimuli, two or more TCRs can be clustered to dimer/oligomer in different relative orientations that promote homointeractions between different signaling subunits. This results in formation of distinct CD3 and/or ζ signaling oligomers and their activation through the phosphorylation of the corresponding ITAM tyrosines (Fig. 3), thus initiating distinct signaling cascades and leading to distinct functional outcomes.

Within the model (Fig. 3), two or more TCRs are clustered to dimer/oligomer with sufficient interreceptor proximity upon binding with multivalent ligand and simultaneously rotate around the receptor axis perpendicular to the membrane to adopt a correct relative orientation toward each other, permissive of initiating the trans-homointeractions between ζ molecules. Until the ζ ITAM tyrosines are phosphorylated by protein tyrosine kinase (PTK), this process is reversible and its reversibility can depend on duration of the TCR-ligand contact that generally correlates with the strength (affinity/avidity) of the ligand and sufficient lifetime of a receptor in TCR dimers/oligomers. At this point of bifurcation, two alternative pathways (Fig. 3, stages IV and III) leading to partial or full T-cell activation, respectively, can take place depending on the nature of activating stimuli. As a result, either ζ or both ζ and CD3 signaling oligomers are formed with subsequent phosphorylation of ITAM tyrosines by PTKs and dissociation from remaining TCR-CD3 complexes or TCR$\alpha\beta$ chains. At this irreversible stage, downstream signaling events are triggered. Later, the remaining TCR-CD3 complexes or TCR$\alpha\beta$ chains are internalized. According to the proposed model, at least two different activation signals (shown in the Fig. 3

as signals A and B) can be provided from the ζ and CD3 signaling oligomers and both signals are required for full activation of T-cells. Thus, distinct signaling is achieved through ζ and CD3 signaling oligomers and/or through various combinations of signaling chains in oligomeric CD3 structures (Fig. 3). Then, the signaling oligomers formed from the initially engaged TCR dimer/oligomer can sequentially homointeract with the relevant signaling subunits of nonengaged TCRs resulting in formation of higher-order signaling oligomers with their subsequent phosphorylation and dissociation from ligand-binding subunits. This process leads to amplification and lateral propagation of the activation signal(s). Later, the remaining nonengaged TCR-CD3 complexes or TCRαβ chains are internalized.

Thus, in the context of the model, TCR clustering by the MHCs bound to agonist, partial agonist or antagonist peptides results in formation of receptor dimers/oligomers with similar inter-receptor proximity but different intermolecular orientation. This leads (or does not) to initiation of homointeractions between different signaling subunits with their subsequent oligomerization and activation, providing distinct signaling and T-cell functional outcomes. This mechanism is also proposed for T-cell activation mediated by other stimuli such as anti-TCRα, anti-TCRβ, anti-CD3ε, etc.

Comparison to Other Models

There exist numerous models of TCR triggering and their modifications, including but not limiting to a kinetic proofreading model,[217,231,233,236-240] serial triggering model,[110,241-244] serial encounter model,[245] conformational models,[44,113,145-147,191,246-253] permissive geometry model[191] and clustering[52,59,60,75,161,191] and segregation[254-256] models. Most of these models are discussed in detail in Chapters 6-11 of this book. However, despite the rapidly growing number of models and their modifications, no current model explains at the molecular level: (1) how ligand-induced TCR TM signaling commences and (2) how this process occurs differentially for altered ligands or in altered cellular contexts. Some of the models suggested so far were rejected in further studies, such as a conformational model that suggests a lipid-dependent folding transition of the TCR ζ cytoplasmic domain to be a molecular switch linking ligand-induced TCR clustering and phosphorylation of the ζ ITAM tyrosines.[249] Later studies have shown that binding of the ζ cytoplasmic domain to stable lipid bilayers is not accompanied by a structural transition to a folded form and that phosphorylated ζ is still able to bind to lipid, thus contradicting this model.[82] In addition, most of the current models have been developed by investigators to describe their own experimental data. As a consequence, these models are mostly descriptive and often fail by trying to explain most of the immunological data accumulated to date. Many of the models suggested to date simply describe a phenomenon but not the mechanisms underlying the phenomenon. Examples include clustering models[52,59,60,75,161,191] that describe a requirement for multivalent ligand to trigger TCR but do not explain the specific molecular mechanisms underlying those observations. Importantly, the lack of these mechanisms in a vast majority of the existing models does not permit to identify clinically important points of therapeutic intervention. Table 1 illustrates comparative features of the currently existing models and demonstrates how these distinctive models for the first time can be readily combined into one model, the SCHOOL model of TCR triggering and TM signaling.

Utility

The powerful ability of the SCHOOL model to describe, explain and predict TCR-related immunological phenomena, providing a mechanistic explanation at the molecular level, is illustrated in Table 2. Selected examples are also described below in more detail.

Clinically relevant TCR CP, or TCR mimic peptide, represents a synthetic peptide corresponding to the sequence of the TM region of the ligand-binding TCRα chain critical for TCR assembly and function (Chapter 16). This and similar TM peptides capable of inhibiting antigen-stimulated TCR-mediated T-cell activation were first reported in 1997.[257] Since that time, despite extensive basic and clinical studies of these peptides,[211,258-267] the molecular mechanisms of action of these clinically relevant peptides have not been elucidated until 2004 when the SCHOOL model was first introduced.[48] Within this model,[2,47-49] the TCR CP competes with the TCRα chain for

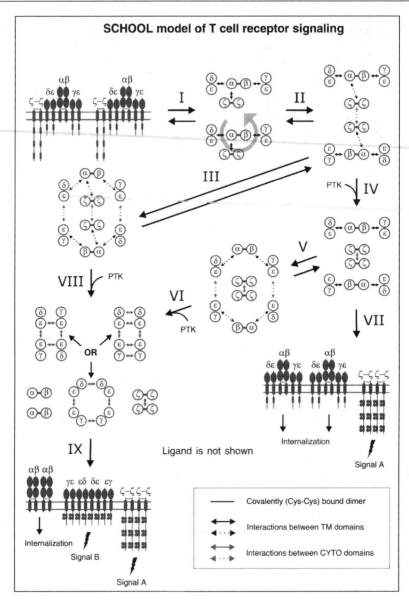

Figure 3. SCHOOL model of the T-cell receptor (TCR) activation. Immunoreceptor tyrosine-based activation motifs (ITAMs) are shown as gray rectangles. TCR-CD3-ζ components are represented as whole polypeptides and as a simplified axial view. All interchain interactions in intermediate complexes are shown by dotted arrows reflecting their transition state. Circular arrow indicates ligand-induced receptor reorientation. Interaction with multivalent ligand (not shown) clusters the receptors and pushes them to reorientate (I) and bring signaling subunits into a correct relative orientation and in sufficient proximity in the formed receptor oligomer (for illustrative purposes, receptor dimer is shown), thus starting the trans-homointeractions between ζ molecules (II). Then, two alternative pathways can take a place depending on the nature of activating stimuli. First is going through a stage IV resulting in formation of ζ₂ dimer (dimer of dimers) and phosphorylation of the ζ ITAM tyrosines, thus triggering downstream signaling events. Continued on next page.

Figure 3, continued from previous page. Then, the signaling ζ oligomers formed subsequently dissociate from the TCR-CD3 complex, resulting in internalization of the remaining engaged TCR-CD3 complexes (VII). This pathway leads to partial (or incomplete) T-cell activation. Alternatively, the intermediate complex formed at the stage II can undergo further rearrangements, starting trans-homointeractions between CD3 proteins (III) and resulting in formation of an oligomeric intermediate. Again, the stages I, II and III can be reversible or irreversible depending on interreceptor proximity and relative orientation of the receptors in TCR dimers/oligomers as well as on time duration of the TCR-ligand contact and lifetime of the receptor in TCR dimers/ oligomers that generally correlate with the nature of the stimulus and its specificity and affinity/ avidity. Next, in the signaling oligomers formed (III), the ITAM tyrosines undergo phosphorylation by PTKs that leads to generation of the activation signal, dissociation of signaling oligomers and internalization of the remaining engaged TCRαβ chains (VIII, XI). This pathway provides at least two different activation signals from the ζ and CD3 signaling oligomers (signals A and B), respectively and results in full T-cell activation. The distinct signaling through ζ and CD3 oligomers (or through various combinations of signaling chains in CD3 oligomeric structures) might be also responsible for distinct functions such as T-cell proliferation, effector functions, T-cell survival, pathogen clearance, TCR anergy, etc. In addition, the signaling oligomers formed can sequentially interact with the signaling subunits of nonengaged TCRs resulting in formation of higher-order signaling oligomers, thus amplifying and propagating the activation signal (not shown). Also, this leads to the release and subsequent internalization of the remaining nonengaged TCR complexes and/or TCRαβ chains (not shown). Abbreviations: PTK, protein tyrosine kinase. Phosphate groups are shown as filled gray circles. Reprinted from Trends Immunol, 25, Sigalov AB, Multichain immune recognition receptor signaling: Different players, same game? 583-589, copyright 2004 with permission from Elsevier.

binding to CD3δε and ζ hetero- and homodimers, respectively, thus resulting in disconnection and predissociation of the signaling subunits from the remaining receptor complex. The proposed mechanism is the only mechanism consistent with all experimental and clinical data reported up to date for TCR TM peptides and their lipid and/or sugar conjugates. The model also predicts that the same mechanisms of inhibitory action can be applied to MIRR TM peptides corresponding to the TM regions of not only the MIRR recognition subunits but the corresponding signaling subunits as well (see also Chapter 20).[2,47-49] This was recently confirmed experimentally[265,268] by showing that the synthetic peptides corresponding to the sequences of the TM regions of the signaling CD3 (δ, ε, or γ) and ζ subunits are able to inhibit the immune response in vivo. Importantly, the SCHOOL model is the first model that not only clearly explains the molecular mechanisms of action of TCR TM peptides[2,47-49] (see also Chapter 20) but also extends the concept of their action through these mechanisms to any other TM peptides of MIRRs and to the MIRR-mediated processes involved in viral pathogenesis[2,49,269] (Chapters 20 and 22).

Interestingly, the model suggests a molecular explanation for the apparent discrepancies in in vitro and in vivo activities of cell-permeable chemical inducers of dimerization.[270-272] In 1993, it has been reported that in vitro chemically induced dimerization/oligomerization of the TCR ζ cytoplasmic domain results in T-cell activation, as measured with a reporter gene assay.[270] Later, activation of a chimeric receptor, containing binding domains for chemical inducers of dimerization fused to the cytoplasmic tail of TCR ζ chain, after stimulation with chemical dimerizers in Jurkat cells has been confirmed to show tyrosine phosphorylation of the TCR ζ chain chimera, recruitment of phosphorylated Zap70 and generation of NFAT.[271] However, in vivo studies demonstrated that signaling did not lead to increased expression of activation markers, T-cell proliferation, or apoptosis.[271] The authors concluded that signaling through ζ alone is not sufficient to generate downstream events leading to full T-cell activation or thymocyte selection; instead, additional CD3 components must be required to induce a functional response in primary thymocytes and peripheral T-cells.[271] Within the model, formation of both CD3γε/δε and ζ signaling oligomers is needed to provide competent activation signal(s) resulting in full cell activation (signals A and B, Fig. 3). Formation of only ζ signaling oligomers leads to partial T-cell activation (signal A, Fig. 3).

Table 1. *Comparative features of different models of TCR triggering*

Model	Requirements/Restraints Imposed (+) or not (–) by a Model			
	Ligand Multivalency	Relative Interreceptor Orientation in TCR oligomers	Duration of Ligand-TCR Contact/Lifetime of TCR Oligomers	Ligand Affinity/Avidity
Kinetic proofreading	–	–	+	+
Serial triggering	–	–	+	+
Serial encounter	–	–	+	+
Conformational	–	–	+	+
Permissive geometry	+	+	–	–
Clustering	+	–	–	+
Segregation	–	–	+	+
SCHOOL	+	+	+	+

The model also explains the apparent discrepancy in CD3 TM peptide activity between in vitro and in vivo T-cell inhibition.[265] It has been shown that the CD3δ and CD3γ TM peptides do not impact T-cell function in vitro (the CD3ε TM peptide has not been used in the reported in vitro experiments because of solubility issues) but that all three CD3 (ε, δ and γ) TM peptides decrease signs of inflammation in an adjuvant-induced arthritis rat model in vivo and inhibit the immune response.[265] Within the SCHOOL model, the CD3δ and CD3γ TM peptides disconnect the corresponding signaling subunits (CD3δ and CD3γ, respectively) from the remaining receptor complex. Thus, these subunits do not participate in further processes upon antigen stimulation. On the other hand, the previously reported in vitro activation studies with T-cells lacking CD3γ and/or CD3δ cytoplasmic domains indicate that antigen-stimulated induction of cytokine secretion and T-cell proliferation are intact,[273-275] thus explaining the absence of inhibitory effect of the CD3δ and CD3γ TM peptides in the in vitro activation assays used.[265] However, in vivo deficiency either of CD3δ or CD3γ results in severe immunodeficiency disorders.[144,276-278] This could explain the inhibitory effect observed in the in vivo studies for all three CD3 TM peptides.[265] Thus, these experimental data confirm that our ability to selectively "disconnect" specific signaling subunits using the MIRR TM peptides in line with the SCHOOL model can provide a powerful tool to study MIRR functions and immune cell signaling.[2,49]

Interestingly, studies of T-lymphocytes expressing a TCR with a mutant TCR β TM domain have shown that upon antigen stimulation, these cells are similar to wild-type cells in terms of IL-2 secretion, IL-2 receptor expression and early activation and signaling events such as CD69 expression, Ca^{2+} flux and CD3ε and ζ phosphorylation, but are specifically defective in undergoing activation-induced cell death.[279] Considering that in the TCR-CD3-$ζ_2$ complex, the TCR β TM domain is critical for interaction with the CD3γε signaling heterodimer,[27] one can suggest the impaired association of the CD3γε with the TCR β chain in a mutant TCR. Upon antigen stimulation, this impaired (weakened) association prevents formation of CD3γε signaling oligomers and thus excludes CD3γ (but not CD3ε, because in the CD3εδ heterodimer of the TCR-CD3-$ζ_2$ complex, there is another CD3ε chain capable of signaling independently of CD3ε in the CD3γε) from further participation in signaling. Thus, within the model, only those signaling events that involve CD3γ (i.e., apoptotic response but not early activation and signaling events[274,275,280,281]) should be affected by a mutation of the TCR β TM domain. This is in a good agreement with the data reported.[279] Also, in this context, functional effect of this mutation should be and is very similar to the one observed by Collier et al for CD3γ TM peptide,[265] therefore providing more evidence for importance and utility of the proposed model and the MIRR TM peptides in studies on immune signaling.

The remarkable feature of the SCHOOL model is that it has a high predictive quality (Table 2 and Chapter 20) by generalizing molecular mechanisms of action and therefore potential therapeutic targets for all MIRRs.[2,47-49]

SCHOOL Model of FceRI Receptor Signaling: Description and Utility

Structurally, all Fc receptors can be divided into two major categories: single- (i.e., FcγRIIA and FcγRIIA) and multichain (i.e., FceRI, FcαRI, FcγRI and FcγRIIIA) receptors. Multichain Fc receptors, in turn, can be divided into two subcategories: receptors that contain one (FcαRI, FcγRI, FcγRIIIA) or two (FceRI) signaling subunits. To date, no general model has been suggested to explain at the molecular level how Fc receptor-mediated signaling commences.

As a general model of MIRR signaling, the SCHOOL model describes the molecular mechanisms underlying the receptor triggering for all multichain Fc receptors.[2,47,48] The model also suggests that the FceRI receptor that contains two different signaling subunits, β and γ (or FcRγ), has more capabilities to induce distinct signaling pathways and, therefore, lead to different functional outcomes as compared to the Fc receptors that contain only one signaling subunit (FcRγ) (Fig. 1). Below I consider the SCHOOL model of FceRI signaling in detail.

The FceRI receptor consists of a ligand-binding α subunit and two kinds of signaling subunits, a β chain and disulfide-linked homodimeric γ chains (Figs. 1 and 4). It plays a pivotal role in the initiation of allergic reactions when antigen crosslinks IgE antibodies bound to FceRI on tissue mast cells or blood basophils (Chapter 3).[40,282-284]

In resting cells, like with TCR, intrareceptor TM interactions between FceRI α, β and γ chains define the overall rigid geometry and topology of the FceRI.[29-31,33,37] Within the proposed model, upon stimulation with multivalent ligand, two or more FceRIs are brought into close proximity and adopt a correct relative orientation, initiating the interreceptor trans-homointeractions between signaling subunits and weaking the intrareceptor TM interactions (stages I and II, Fig. 4). Then, depending on the duration of the FceRI-ligand interaction (affinity/avidity of the ligand), the receptors can either go back to a resting state or forward to an active state, in which β and/or γ signaling oligomers are formed (stages III and IV, Fig. 4), thus promoting ITAM Tyr phosphorylation and generation of activation signal. Assuming that two different FceRI signaling subunits, γ and β, provide distinct signaling,[87,88,200,285-291] the model suggests[47,48] that depending on the nature of ligand, the FceRIs can be clustered to dimer/oligomer in different relative orientations that, in turn, promote homotypic interactions between different signaling subunits (Fig. 4). This leads to formation of distinct, γ and/or β, signaling oligomers, phosphorylation of the corresponding ITAM tyrosines and generation of different activation signals (signals A and B, Fig. 4), resulting in diverse functional outcomes. The formed β and/or γ oligomers can sequentially interact with β and/or γ subunits of nonengaged FceRIs, thus amplifying and propagating the activation signal.

Interestingly, several mathematical models have been recently developed for the early signaling events mediated by FceRI.[292-295] Through model simulations, it has been shown how changing the ligand concentration and consequently the concentration of receptor aggregates, can change the nature of a cellular response as well as its amplitude. These models are largely based on the recently suggested sequence of early events in FceRI signaling.[121,296] Combining the basic organizing principles of the SCHOOL model with the existing mathematical models might significantly improve our understanding the spatiotemporal organization of FceRI-mediated signal transduction as well as our ability to predict how this system will behave under a variety of experimental conditions.

Selected examples illustrating the ability of the SCHOOL model to provide a mechanistic explanation for FceRI-related immunological phenomena are shown in Table 3. These and other findings mentioned above strongly support the validity and utility of the proposed activation model for the FceRI.

SCHOOL Model of BCR Signaling: Description and Utility

The BCR is a multimeric complex composed of mIg noncovalently associated with a disulfide-linked Igα/Igβ heterodimer that is responsible for signal transduction. In the resting state, like with other MIRRs, intrareceptor TM interactions between mIg and Igα/Igβ subunits define the overall

Table 2. *Molecular mechanisms suggested or predicted by the SCHOOL model to underlie selected T-cell-mediated immunological phenomena and observations. Abbreviations: Ag, antigen; CP, core peptide; FP, fusion peptide; IFN, interferon; IS, immunological synapse; HIV, human immunodeficiency virus; mAb, monoclonal antibody; pMHC, major histocompatibility complex (MHC)-bound peptide; TCR, T-cell antigen receptor; TM, transmembrane; ζ_{cyt}, TCR ζ cytoplasmic domain*

Phenomenon	Observation	Mechanism
Inhibitory effect of TCR CP	TCR CP inhibits Ag-stimulated TM signal transduction and efficiently abrogates T-cell-mediated immune responses in mice and man in vitro and in vivo.[211,257,261,265,266]	TCR CP disrupts TCRα–CD3δε and TCRα–ζ TM interactions resulting in predissociation of these signaling subunits from the remaining complex and thus preventing the formation of signaling oligomers upon Ag stimulation and, consequently, inhibiting T-cell activation (Chapter 20).[2,47-49]
Diversity of TCR-mediated cell response	Precise ligand-binding specificities of the TCR are converted into diverse functional outcomes.[309-311] Different TCR signaling subunits engage partially distinct signaling pathways.[91-94] CD3 signaling subunits play differential biological role as revealed by human immunodeficiencies.[144]	Slightly different ligands bring two or more TCRs in different relative orientations that favor homointeractions between different signaling subunits and result in formation of different signaling oligomers or their combinations, thus initiating distinct signaling pathways and leading to diverse T-cell functional outcomes.[2,47,48] Thus, the signaling pathway and the direction of the response depends on the type of TCR signaling subunit(s) that is (are) oligomerized and ITAM-phosphorylated upon ligand stimulation.
T-cell clonal anergy	Ag-unresponsive anergic T-cells fail to produce IL-2 but produce comparable amounts of IFN-γ and proliferate to similar extents in response to immobilized anti-CD3/CD28 mAbs.[141] Ag-induced tolerance in vivo is accompanied by altered early TCR-mediated signaling events.[209] T-cell anergy is induced by activating but not by non-activating anti-CD3.[210]	Ag stimulation induces dissociation of TCR CD3 and/or ζ signaling subunits from the remaining TCRαβ subunits and/or TCRαβ-CD3 complexes, thus preventing Ag[141] or anti-TCR[143]- but not anti-CD3 mAbs[141]-mediated formation of signaling oligomers and generation of activation signal (Fig. 2B). Depending on epitope location, anti-CD3 stimulation can induce formation of CD3 but not ζ signaling oligomers, thus leading to partial cell activation and preventing Ag-mediated T-cell response. Depending on dissociated subunit(s), TCR-mediated signaling events in anergic cells and therefore the functional outcomes can be altered differently.

Continued on next page

Table 2. *Continued*

Phenomenon	Observation	Mechanism
Comodulation of nonengaged TCRs	Activation of T-cells with pMHC, bacterial superantigens, or anti-Vβ antibodies downmodulates not only directly stimulated (engaged) TCR complexes but also unstimulated (nonengaged) ones.[207,241,312-315] In the IS, only a small fraction of the TCR is bound to specific pMHCs.[255]	Upon ligand stimulation, signaling oligomers dissociate from the remaining engaged TCRs that undergo internalization. Then, the dissociated signaling oligomers sequentially interact with the signaling subunits of nonengaged TCRs resulting in the release and subsequent internalization of the remaining nonengaged TCRαβ-CD3 complexes or TCRαβ chains. Internalization and intracellular fate may be different for TCR-CD3 complexes lacking ζ chain or for TCRαβ chains remaining on the cell surface after dissociation of either ζ or both ζ and CD3 signaling oligomers, respectively.[47,48]
TCR signaling initiation and following lateral signal propagation and amplification	TCR signaling is initiated and sustained in microclusters and is terminated in the TCR-rich central supramolecular activation cluster (cSMAC), a structure from which TCR are sorted for degradation.[134]	The initially formed signaling oligomers initiate TCR signaling, dissociate from the remaining engaged TCRs and interact with the signaling subunits of nonengaged TCRs, thus propagating the activation signal to nonengaged receptors and resulting in signal amplification and lateral propagation.
Exposure of the CD3ε$_{cyt}$ epitope	Ligand engagement of TCR-results in exposure of a cryptic proline-rich CD3ε$_{cyt}$ epitope that is a binding site for the adaptor protein, Nck.[113,145] The CD3ε$_{cyt}$ epitope is recognized by antibody APA1/1 and is only detected when the TCR is fully activated.[146,147] In the IS, distribution of APA1/1 epitope is more restricted than ζ, CD3ε and tyrosine-phosphorylated proteins.[146]	During full T-cell activation, dissociation of CD3εγ and/or CD3εδ signaling oligomers from TCRαβ chains (Fig. 3) induces the release/unmasking of the CD3ε$_{cyt}$ epitope. Thus, within the SCHOOL model, the ligand-induced exposure of the epitope is effect not cause of TCR triggering. During partial T-cell activation, formation of only ζ signaling oligomers and their dissociation from the remaining TCR-CD3 complexes (Fig. 3) do not release/unmask the CD3ε$_{cyt}$ epitope.

Continued on next page

Table 2. *Continued*

Phenomenon	Observation	Mechanism
Action of HIV-1 gp41 FP	HIV-1 FP colocalizes with CD4 and TCR molecules, coprecipitates with the TCR and inhibits Ag-specific T-cell proliferation and proinflammatory cytokine secretion in vitro.[316] The peptide blocks the TCR/CD3 TM interactions needed for antigen-triggered T-cell activation.[317]	Similarly to the TCR-CP, the HIV-1 gp41 FP disrupts TCRα-CD3δε and TCRα-ζ_2 TM interactions resulting in dissociation of these signaling subunits from the remaining complex and thus preventing the formation of signaling oligomers upon antigen stimulation and, consequently, inhibiting T-cell activation.[2,269] This effect is specific: anti-CD3ε antibody-stimulated T-cell activation is not affected by the peptide.
Pre-TCR signaling	Spontaneous preTCR oligomerization mediated by the preTCR α chain results in ligand-independent receptor triggering and TM signaling crucial for early T-cell development.[170,171] Forced dimerization of CD3ε is sufficient to simulate preTCR function and promote β-selection.[170]	Oligomerization of the preTCR through the preTCR α chain brings CD3 and ζ signaling subunits in close proximity and proper relative orientation, thus promoting formation of signaling oligomers and generating the activation signal. Remarkably, as predicted by the SCHOOL model, formation of CD3ε dimers/oligomers is necessary and sufficient to induce the CD3ε ITAM Tyr phosphorylation and lead to cell response.
Epitope-dependent mAb stimulation	T-cell activation induced by mAbs specific for the TCR does not correlate with the affinity of the mAbs but rather with the recognized epitope.[160] Triggering of different epitopes of the TCR-CD3-ζ_2 receptor complex depends on the mAb specificity and induces different modes of T-cell activation.[107,156-159] In thymocytes, only anti-TCRβ but not anti-TCRα Ab reagents cause long-term TCR downmodulation.[108]	Clustering/oligomerization of TCRs by different antibodies results in different intermolecular relative orientations within receptor cluster/oligomer that promote (or do not) homointeractions between different signaling subunits, leading to the formation of different CD3 and/or ζ signaling subunit oligomers and therefore to different functional outcomes. If intermolecular relative orientation in the antibody-crosslinked TCR cluster/oligomer does not promote homointeractions between CD3 and/or ζ signaling subunits, this antibody will not stimulate T-cell response.

Continued on next page

Table 2. Continued

Phenomenon	Observation	Mechanism
Coexistence of mono- and multivalent (oligomeric) TCRs in resting cells	Monovalent TCRs coexist in intact resting cells with multivalent complexes with two or more ligand-binding TCRαβ subunits,[186,318] raising a question: why does this basal TCR clustering not lead to receptor triggering whereas ligand-induced clustering does?	In resting cells, receptors within multivalent TCR complex have the relative orientation that does not promote homointeractions between CD3 and/or ζ signaling chains. Upon stimulation with multivalent ligand, these receptors adopt proper orientation relative to each other, starting homotypic interactions between signaling subunits and resulting in generation of the activation signal. A similar mechanistic explanation can also account for the existence of dimeric or tetrameric TCR-CD3-coreceptor complexes in naïve CD4+ or CD8+ T-cells.[167]

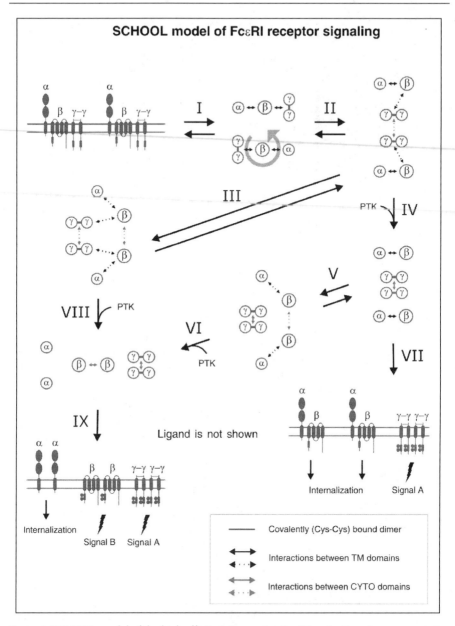

Figure 4. SCHOOL model of the high affinity IgE receptor (FcεRI) activation. Immunoreceptor tyrosine-based activation motifs (ITAMs) are shown as gray rectangles. FceRI α, β and γ components are represented as whole polypeptides and as a simplified axial view. All interchain interactions in intermediate complexes are shown by dotted arrows reflecting their transition state. Circular arrow indicates ligand-induced receptor reorientation. Interaction with multivalent ligand (not shown) clusters the receptors and pushes them to reorientate (I) and bring β and γ signaling subunits into a correct relative orientation and in sufficient proximity in the formed receptor oligomer (for illustrative purposes, receptor dimer is shown), thus starting the trans-homointeractions between γ homodimers (II). Then, two alternative pathways can take a place depending on the nature of activating stimuli. Continued on next page.

Figure 4, continued from previous page. First is going through a stage IV resulting in formation of γ_2 dimer (dimer of dimers) and phosphorylation of the γ ITAM tyrosines, thus triggering downstream signaling events. Then, the signaling γ oligomers formed subsequently dissociate from the α/β complex, resulting in internalization of the remaining engaged complexes (VII). This pathway leads to generation of the activation signal A. Alternatively, the intermediate complex formed at the stage II can undergo further rearrangements, starting trans-homointeractions between β chains (III) and resulting in formation of an oligomeric intermediate. Stages I, II and III can be reversible or irreversible depending on interreceptor proximity and relative orientation of the receptors in FcεRI dimers/oligomers as well as on time duration of the receptor-ligand contact and lifetime of the receptor in FcεRI dimers/oligomers that generally correlate with the nature of the stimulus and its specificity and affinity/avidity. Next, in the signaling oligomers formed (III), the β ITAM tyrosines undergo phosphorylation by protein tyrosine kinases (PTKs) that leads to generation of the activation signal, dissociation of signaling oligomers and internalization of the engaged α chains (VIII, XI). This pathway provides two different activation signals from the γ and β signaling oligomers (signals A and B), respectively, and results in full cell activation. In addition, the signaling oligomers formed can sequentially interact with the signaling subunits of nonengaged FcεRIs resulting in formation of higher-order signaling oligomers, thus amplifying and propagating the activation signal (not shown). Also, this leads to the release and subsequent internalization of the nonengaged α and/or $\alpha\beta$ chains (not shown). Abbreviations: PTK, protein tyrosine kinase. Phosphate groups are shown as filled gray circles.

rigid geometry and topology of the BCR.[28,34,35,297] In cells, this receptor transduces signals leading to a variety of biologic responses minimally including antigen receptor editing, apoptotic death, developmental progression, cell activation, proliferation and survival. Despite several BCR triggering and cell activation models that have been suggested,[44,116,172,298-300] no model fully explains the molecular mechanisms underlying spatiotemporal organization of BCR-triggered TM signal transduction.

Within the SCHOOL model, two or more BCRs are brought into close proximity and adopt a correct relative orientation upon receptor engagement with multivalent ligand (Fig. 5A). At this point, the trans-homointeractions between Igα and Igβ molecules are initiated, weaking the TM interactions within the BCR (Fig. 5A, stages I and II). Then, depending on the duration of ligand-BCR interaction and therefore on the affinity/avidity of ligand, the receptors can go either back to resting state or forward to active state, in which signaling Igα/Igβ oligomers are formed, thus promoting ITAM Tyr phosphorylation and generation of activation signal (Fig. 5A, stage III). Considering that Igα and Igβ chains can play different physiological roles,[84-86,301] the model suggests that depending on the nature of stimuli, different Igα/Igβ signaling oligomers can form, thus resulting in phosphorylation of Igα and/or Igβ ITAM tyrosines and induction of distinct signaling pathways. Further, once formed, these oligomers can sequentially interact with Igα/Igβ subunits of nonengaged BCRs, thus propagating and amplifying the activation signal and favoring the formation and stabilization of supramolecular complexes that can promote sustained signaling. In this context, it can also be suggested that the more BCRs are initially engaged and/or the higher is the affinity/avidity of antigen, the faster is signaling cluster formation.

In contrast to Igβ and other ITAM-containing proteins, the dynamic equilibrium between monomeric and oligomeric species of Igα is slow and this protein forms stable homooligomers (mostly, dimers and tetramers) even at very low protein concentrations.[78] Formation of the stable Igα/Igβ clusters/oligomers may be particularly important for sustained signaling during the synapse formation between B-cell and antigen-displaying target cell and subsequent antigen acquisition.[302] Also, as shown recently,[303] plasma membrane association of Igα/Igβ complexes results in generation of biologically relevant basal signaling while the ability of the BCR to interact with both conventional as well as nonconventional extracellular ligands is eliminated.

As illustrated in Table 4 by several selected examples, the proposed model is capable of providing a mechanistic explanation for BCR-related immunological phenomena. Thus, a vast majority of the experimental findings reported so far strongly support the validity and utility of this activation model for the BCR.

Table 3. Molecular mechanisms suggested or predicted by the SCHOOL model to underlie selected FcεRI-mediated immunological phenomena and observations

Phenomenon	Observation	Mechanism
FcεRI-mediated signaling and cellular responses	Signaling capacity of FcεRIs depends on the driving forces leading to their clustering and also on fine tuning provided by both lifetime (or receptor capacity to remain in a cluster that is influenced by the ligand affinity) and interreceptor relative orientation in the FcεRI dimers/oligomers.[127,229]	Triggering FcεRI requires close proximity of the receptors and a correct relative orientation in the FcεRIs clustered (or altered in preexisting clusters) by multivalent ligand binding (Fig. 4).[2,47,48] It also requires the ligand-receptor contact to last long enough to initiate the trans-homointeractions between signaling subunits and weaken the intrareceptor TM interactions, thus resulting in formation of signaling FcεRIβ and/or FcεRIγ oligomers and generation of activation signal (Fig. 4).[2,47,48]
	There is no simple correlation between multivalent ligand-promoted FcεRI clustering and FcεRI-mediated cellular responses, such as cell degranulation.[126]	Receptor clustering induced upon binding to multivalent ligand is necessary but not sufficient for the initiation of FcεRI signaling. To commence signal transduction, two or more clustered receptors should adopt a correct relative orientation toward each other, permissive of initiating the trans-homointeractions between β and/or γ subunits and remain in a cluster long enough to promote these interactions and therefore formation of signaling oligomers (Fig. 4).[2,47,48]
	The ratio of late to early FcεRI-stimulated events correlates with the affinity of a ligand for the receptor-bound IgE.[131] Orientational restraint in ligand-specific FcεRI dimers/oligomers determines the magnitude of mediator secretion–causing signal induced by different mAbs.[111,112,119,126,127,193]	
Different roles of β and γ subunits in signaling	The FcεRI β and γ subunits play different roles in signaling.[285-291] The γ chain aggregation alone can evoke cellular responses,[288,291] while the β chain acts as an amplifier for signaling.[87] Also, β chains can elicit a signal in a γ chain-independent manner.[200]	Depending on the nature of ligand (i.e., its specificity, affinity and avidity), the FcεRIs are clustered to dimer/oligomer in different relative orientations that promote homotypic interactions between different signaling subunits (Fig. 4). As a result, different, β and/or γ, signaling oligomers are formed, generating distinct activation signals and therefore distinct signaling pathways.[2,47,48]
Lateral propagation of activation signal	FcεRI-activated mast cells propagate signals from small signaling domains around dimerized/oligomerized receptors; formation of large FcεRI aggregates promotes both strong receptor triggering and rapid termination of the signaling responses.[63]	The initially formed β and/or γ signaling oligomers initiate FcεRI signaling, dissociate from the remaining engaged receptors and interact with the signaling subunits of nonengaged FcεRIs, thus propagating the activation signal to nonengaged receptors and resulting in signal amplification and lateral propagation.

Abbreviations: FcεRI, high affinity IgE receptor; mAb, monoclonal antibody; TM, transmembrane

SCHOOL Model of GPVI Signaling: Description and Utility

Studies of patients deficient in GPVI identified this platelet membrane protein as a physiological collagen receptor. This receptor is noncovalently associated with FcRγ, the ITAM-containing homodimeric signaling module. The GPVI-FcRγ receptor complex induces platelet activation when it binds to collagen or other agonists and GPVI-deficient platelets lack specifically collagen-induced aggregation and the ability to form thrombi on a collagen surface under flow conditions.[10,304] The selective inhibition of GPVI and/or its signaling is thought by most experts in the field to inhibit thrombosis without affecting hemostatic plug formation. Thus, future therapeutic strategies targeting platelet-mediated disease will depend on our detailed understanding of the molecular mechanisms underlying GPVI triggering and subsequent TM signal transduction. In addition, knowing these mechanisms would give us a new handle in dissecting the basic structural and functional aspects of thrombus formation.

In 2006, GPVI has been reported to form a back-to-back dimer in the GPVI crystal.[64] Based on these findings, a model for GPVI signaling has been suggested, in which GPVI clustering triggers a signaling cascade via the FcRγ chain coreceptor.[64] Despite its apparent similarity to the SCHOOL model,[48] it does not explain the existence of oligomeric GPVI-FcRγ complexes at the surface of unstimulated platelets[71] and does not suggest specific protein-protein interactions involved in the molecular mechanisms underlying the GPVI-triggered signaling. These findings[64,71] raise an important and intriguing question: why does the observed basal receptor dimerization not lead to receptor triggering and subsequent platelet activation whereas agonist-induced receptor crosslinking/clustering does?

Despite extensive studies of the GPVI-FcRγ receptor complex and its mechanism of action,[10,178,305,306] the only model that can answer this question and even more important, mechanistically explain how GPVI-mediated TM signaling begins, is the SCHOOL model.[2,47-49,307] Within this model, GPVI-mediated platelet activation is a result of the interplay between GPVI-FcRγ TM interactions, the association of two TM Asp residues in the FcRγ homodimer with the TM Arg residue of GPVI,[30] that maintain receptor integrity in platelets under basal conditions and homointeractions between FcRγ subunits that lead to formation of signaling oligomers and initiation of a signaling response (Fig. 5B). Binding of the multivalent ligand (collagen) to two or more GPVI-FcRγ receptor complexes pushes the receptors to cluster, rotate and adopt an appropriate orientation relative to each other (Fig. 5B, stages I and II), at which point the trans-homointeractions between FcRγ molecules are initiated. Upon formation of FcRγ signaling oligomers, the Src-family kinases Fyn or Lyn phosphorylate the tyrosine residues in the FcRγ ITAM that leads to generation of the activation signal (Fig. 5B, stage III) and subsequent dissociation of FcRγ signaling oligomers and downmodulation of the remaining engaged GPVI subunits (Fig. 5B, stage IV). Later, the dissociated oligomeric FcRγ chains can interact with FcRγ subunits of the nonengaged GPVI-FcRγ complexes, resulting in formation of higher-order signaling oligomers and their subsequent phosphorylation, thus providing lateral signal propagation and amplification (not shown).

Thus, for the preformed oligomeric receptor complexes existing in unstimulated platelets as found by Berlanga et al[71] the proposed model suggests that under basal conditions, the overall geometry of the receptor dimer keeps FcRγ chains apart, whereas stimulation by collagen results in breakage of GPVI-GPVI extracellular interactions and reorientation of signaling FcRγ homodimers, thus bringing them into close proximity and appropriate relative orientation permissive of initiating the FcRγ homointeractions and receptor triggering.

Intriguingly, suggesting how binding to collagen triggers the GPVI-mediated signal cascade at the molecular level, the SCHOOL model reveals GPVI-FcRγ TM interactions as a novel therapeutic target for the prevention and treatment of platelet-mediated thrombotic events (Chapter 20).[2,49,307,308] Preliminary experimental results provided support for this novel concept of platelet inhibition and resulted in the development of novel class of promising platelet inhibitors.[307,308]

Thus, the experimental evidence accumulated to date on the GPVI-mediated TM signal transduction and platelet activation strongly support the validity and utility of the proposed activation model for this receptor.

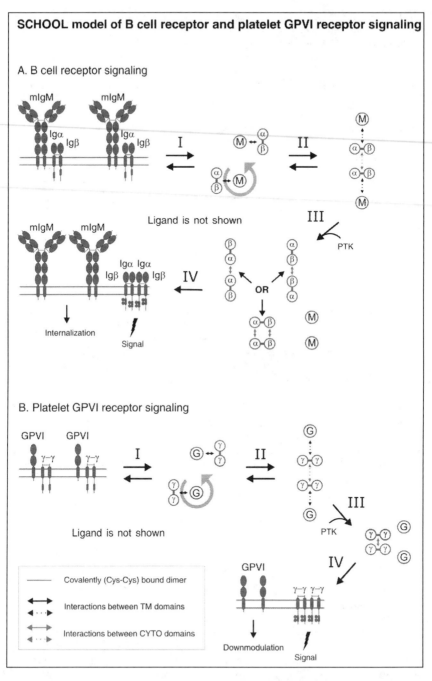

Figure 5. SCHOOL model of the B-cell receptor (BCR, panel A) and platelet collagen receptor glycoprotein VI (GPVI, panel B) activation. Immunoreceptor tyrosine-based activation motifs (ITAMs) are shown as gray rectangles. Receptor components are represented as whole polypeptides and as a simplified axial view. All interchain interactions in intermediate complexes are shown by dotted arrows reflecting their transition state. Continued on next page.

Figure 5, continued from previous page. Circular arrows indicate ligand-induced receptor reorientation. Interaction with multivalent ligand (not shown) clusters the receptors and pushes them to reorientate (I) and bring signaling subunits into a correct relative orientation and in sufficient proximity in the receptor oligomer (for illustrative purposes, receptor dimers are shown), thus starting the trans-homointeractions between Igα/Igβ heterodimers (panel A, II) or FcRγ homodimers (panel B, II). On a stage III, formation of signaling oligomers results in phosphorylation of the ITAM tyrosines, thus triggering downstream signaling events. Then, the signaling oligomers formed subsequently dissociate from the mIg or GPVI (panels A and B, respectively), resulting in generation of the activation signal and internalization of the remaining engaged receptor chains (IV). Stages I and II can be reversible or irreversible depending on interreceptor proximity and relative orientation of the receptors in ligand-specific dimers/ oligomers as well as on time duration of the receptor-ligand contact and lifetime of the receptor in these dimers/oligomers that generally correlate with the nature of the stimulus and its specificity and affinity/avidity. In contrast to homodimeric FcRγ signaling subunit in GPVI-FcRγ receptor complex, the BCR signaling module contains two different signaling chains, Igα and Igβ, providing possibility of the signal and cell response diversity depending on the particular set of the Igα and/or Igβ ITAM tyrosines that become phosphorylated. Further, the signaling oligomers formed can sequentially interact with the signaling subunits of nonengaged receptors resulting in formation of higher-order signaling oligomers, thus amplifying and propagating the activation signal (not shown). Also, this leads to the release and subsequent internalization/ downmodulation of the nonengaged mIg or GPVI chains (not shown). Abbreviations: PTK, protein tyrosine kinase. Phosphate groups are shown as filled gray circles.

SCHOOL Model of Other MIRR Signaling

As illustrated in Figure 1, a structural assembly of many MIRRs, such as FcαRI, FcγRI, FcγRIIIA, ILT/LIR receptors, DCAR, NK and TREM receptors, etc., is very similar to that of the GPVI receptor; all these receptors have a ligand-recognition subunit and one homodimeric signaling subunit. Thus, the basic principles of GPVI triggering and TM signaling suggested by the SCHOOL model can be easily applied to these and other, structurally related, MIRRs. Selected examples illustrating the capability of the SCHOOL model to provide a mechanistic explanation for immunological phenomena mediated by these receptors are shown in Table 5.

Conclusions

Despite growing interest in targeting MIRR signaling as a potential treatment strategy for different immune-mediated diseases (see also Chapters 20 and 22), the molecular mechanisms underlying MIRR triggering and subsequent TM signal transduction are unknown, impeding our fundamental understanding of MIRR-mediated immunological phenomena and thus preventing the development of novel pharmacological approaches.

Considering MIRR triggering as a result of ligand-induced interplay between well-defined protein-protein interactions, the proposed SCHOOL model is the first general model that provides a set of basic principles of MIRR signaling and mechanistically explains how MIRR-mediated signaling commences and what the main driving forces and restraints of MIRR triggering/signaling are. Furthermore, this model is the first model that can describe, explain and predict numerous MIRR-mediated immunological phenomena. Thus, this model represents a powerful tool that can be used in dissecting the basic structural and functional aspects of the immune response and using this knowledge in both fundamental and clinical fields. In addition, revealing the main driving forces and fundamental stages of MIRR triggering and TM signal transduction, the model identifies effective ways of modulating the immune response.

Importantly, by generalizing mechanistic features of MIRR signaling, the SCHOOL model shows how the similar structural architecture of the MIRRs dictates similar mechanisms of MIRR triggering and subsequent TM signal transduction and furthermore, reveals similar therapeutic targets in seemingly unrelated diseases (see also Chapter 20). This permits the transfer of accumulated knowledge and pharmacological approaches between seemingly disparate immune disorders and builds the molecular basis for existing and future therapeutic strategies. Impressively,

Table 4. *Molecular mechanisms suggested or predicted by the SCHOOL model to underlie selected BCR-mediated immunological phenomena and observations*

Phenomenon	Observation	Mechanism
BCR-mediated signaling and cellular responses	B-cell response is induced by multivalent but not monovalent ligand stimulation[56] and Ag valency influences B-cell responses by modulating the stability of BCR-signaling microdomains and BCR trafficking.[54]	Triggering BCR requires multivalent ligand-induced clustering of the BCRs in a close proximity and a correct relative orientation in the formed clusters (or reorientation of the receptors in preexisting oligomers/clusters) (Fig. 5A).[2,47,48] Also, the more BCRs are initially engaged, the faster is Igα/Igβ signaling oligomer formation and the stronger is the amplified activation signal.
Formation of Igα/Igβ oligomers	Signaling Igα/Igβ heterodimer assembles into oligomers upon ligand stimulation.[199]	Upon multivalent ligand stimulation, BCRs are clustered in close proximity and correct relative orientation, thus promoting homotypic interactions between Igα/Igβ signaling subunits. This leads to formation of signaling oligomers and phosphorylation of the ITAM tyrosines, thus initiating the signaling cascade.[2,47,48]
Comodulation of nonengaged BCRs	Unligated BCRs cluster with BCRs engaged by multivalent ligands.[56] The extent of BCR internalization is not correlated with Ag valency, suggesting that BCR signaling and internalization are distinct processes.[56] Upon anti-Ig-induced BCR clustering, >95% of the mIg is internalized, whereas 20-30% of Igβ remains on the surface.[213]	Similar to TCR (Table 2), upon multivalent ligand stimulation, signaling Igα/Igβ oligomers dissociate from the remaining engaged mIgs that undergo internalization. Then, the dissociated oligomers sequentially interact with the signaling subunits of nonengaged receptors resulting in their activation and therefore the signal amplification and propagation. This also leads to the release and subsequent internalization of the remaining nonengaged mIg chains.[47,48]
B-cell tolerance/BCR desensitization	Monomeric hen egg lysozyme (HEL) efficiently engages the specific BCR, however, presentation of HEL-derived epitopes is impaired compared to multivalent antigens.[319] Soluble monovalent antigen, administered intravenously, induces B-cell tolerance.[320,321] Upon binding of moderate- to low-affinity Ag, physical dissociation of the Igα/Igβ subunits from mIg results in BCR desensitization.[148] However, these desensitized cells can be still activated by anti-Igβ antibodies.[148]	Monovalent or moderate to low-affinity Ag stimulation induces dissociation of BCR Igα/Igβ signaling subunits from the remaining mIg, thus preventing Ag- or anti-Ig- but not anti-Igβ mAbs-mediated formation of signaling oligomers and generation of activation signal (Fig. 2B).[47,48] The remaining mIg chains are internalized. Within the SCHOOL model, the ligand-induced dissociation of signaling subunits from ligand recognition subunits is suggested to be a general molecular mechanism underlying T- and B-cell tolerance, BCR desensitization and TM peptide-modulated T and other immune cell response.[2,47,48,269,307]

Abbreviations: Ag, antigen; BCR, B-cell antigen receptor; mAb, monoclonal antibody; mIg, membrane immunoglobulin; TM, transmembrane

Table 5. *Molecular mechanisms suggested or predicted by the SCHOOL model to underlie selected MIRR-mediated immunological phenomena and observations*

Phenomenon	Observation	Mechanism
FcαRI-mediated signaling and cellular responses	Vertical relocation of the TM positive charge responsible for FcαRI-FcRγ association does not effect on calcium flux, MAPK phosphorylation and IL-2 release, whereas its lateral transfer completely abrogates these functions.[32]	Vertical relocation of the noncovalent electrostatic bond does not affect interreceptor relative orientation within the FcαRI dimers/oligomers formed upon multivalent ligand stimulation, whereas lateral transfer does, thus preventing formation of FcRγ signaling oligomers and initiation of signaling cascade.
NKR-mediated signaling and cellular response	Short CPs derived from the TM sequence of NKRs inhibit NK cell cytolytic activity.[268]	NKR CPs disrupt the TM interactions between NKR ligand-binding subunits and associated homodimeric signaling subunits, such as ζ–ζ, γ–γ or DAP-12 (Chapter 20).[2]
Immune escape in hCMV pathogenesis	hCMV tegument protein pp65 interacts directly with NKp30, leading to dissociation of the linked ζ signaling subunit and, consequently, to reduced killing.[152]	Binding to pp65 protein affects the NKp30-ζ_2 TM interactions resulting in dissociation of the ζ signaling subunit from the remaining complex and thus preventing the formation of ζ signaling oligomers upon ligand stimulation and, consequently, inhibiting NK cell cytolytic activity (Chapter 20).[2]
TREM-mediated signaling	Structurally similar receptors, TREM-1 and TREM-2 (Fig. 1) that contain the same signaling subunit, DAP-12, show activating and inhibitory functions.[18,125]	Depending on the affinity/avidity of the ligand, ligand stimulation can result in: 1) receptor clustering, formation of oligomeric signaling subunits and generation of the activation signal (TREM-1, activating function), or 2) dissociation of signaling subunit from the engaged receptor and unmasking a specific "inhibitory" epitope(s) in the cytoplasmic tail of ligand recognition subunit (TREM-2, inhibitory function).

Abbreviations: Ag, antigen; CPs, core peptides; hCMV, human cytomegalovirus; DAP-12, DNAX activation protein 12; mAb, monoclonal antibody; MAPK, mitogen-activated protein kinase; NKRs, natural killer cell receptors; TM, transmembrane; TREM, triggering receptors expressed on myeloid cells

applications of this model have already illustrated how do the similar molecular mechanisms of MIRR signaling revealed by the model work in seemingly unrelated fields, such as the treatment of T-cell-mediated skin diseases, HIV entry into target cells and the development of a novel concept of platelet inhibition (see also Chapters 20 and 22).

In conclusion, I sincerely hope that the model and issues presented in this Chapter will stimulate debate and new research to further test and apply the proposed model, thus opening new horizons in our knowledge about the immune system and generating new perspectives for the effective prevention and/or treatment of numerous immune disorders.

Acknowledgements
I would like to thank Walter M. Kim for critical reading of this manuscript.

References

1. Keegan AD, Paul WE. Multichain immune recognition receptors: Similarities in structure and signaling pathways. Immunol Today 1992; 13:63-68.
2. Sigalov AB. Immune cell signaling: A novel mechanistic model reveals new therapeutic targets. Trends Pharmacol Sci 2006; 27:518-524.
3. Krogsgaard M, Davis MM. How T-cells 'see' antigen. Nat Immunol 2005; 6:239-245.
4. DeFranco AL. B-cell activation 2000. Immunol Rev 2000; 176:50-9.
5. Dal Porto JM, Gauld SB, Merrell KT et al. B-cell antigen receptor signaling 101. Mol Immunol 2004; 41:599-613.
6. Takai T. Fc receptors and their role in immune regulation and autoimmunity. J Clin Immunol 2005; 25:1-18.
7. Takai T. Fc receptors: Their diverse functions in immunity and immune disorders. Springer Semin Immunopathol 2006; 28:303-304.
8. Colonna M, Nakajima H, Navarro F et al. A novel family of Ig-like receptors for HLA class I molecules that modulate function of lymphoid and myeloid cells. J Leukoc Biol 1999; 66:375-381.
9. Borrego F, Kabat J, Kim DK et al. Structure and function of major histocompatibility complex (MHC) class I specific receptors expressed on human natural killer (NK) cells. Mol Immunol 2002; 38:637-660.
10. Moroi M, Jung SM. Platelet glycoprotein VI: Its structure and function. Thromb Res 2004; 114:221-233.
11. Barclay AN, Brown MH. The SIRP family of receptors and immune regulation. Nat Rev Immunol 2006; 6:457-464.
12. Kanazawa N, Tashiro K, Miyachi Y. Signaling and immune regulatory role of the dendritic cell immunoreceptor (DCIR) family lectins: DCIR, DCAR, dectin-2 and BDCA-2. Immunobiology 2004; 209:179-190.
13. Biassoni R, Cantoni C, Falco M et al. Human natural killer cell activating receptors. Mol Immunol 2000; 37:1015-1024.
14. Biassoni R, Cantoni C, Marras D et al. Human natural killer cell receptors: Insights into their molecular function and structure. J Cell Mol Med 2003; 7:376-387.
15. Aoki N, Kimura S, Xing Z. Role of DAP12 in innate and adaptive immune responses. Curr Pharm Des 2003; 9:7-10.
16. Bakker AB, Baker E, Sutherland GR et al. Myeloid DAP12-associating lectin (MDL)-1 is a cell surface receptor involved in the activation of myeloid cells. Proc Natl Acad Sci USA 1999; 96:9792-9796.
17. van den Berg TK, Yoder JA, Litman GW. On the origins of adaptive immunity: Innate immune receptors join the tale. Trends Immunol 2004; 25:11-16.
18. Klesney-Tait J, Turnbull IR, Colonna M. The TREM receptor family and signal integration. Nat Immunol 2006; 7:1266-1273.
19. Takai T. Paired immunoglobulin-like receptors and their MHC class I recognition. Immunology 2005; 115:433-440.
20. Nakahashi C, Tahara-Hanaoka S, Totsuka N et al. Dual assemblies of an activating immune receptor, MAIR-II, with ITAM-bearing adapters DAP12 and FcRgamma chain on peritoneal macrophages. J Immunol 2007; 178:765-770.
21. Fujimoto M, Takatsu H, Ohno H. CMRF-35-like molecule-5 constitutes novel paired receptors, with CMRF-35-like molecule-1, to transduce activation signal upon association with FcRgamma. Int Immunol 2006; 18:1499-1508.
22. Stewart CA, Vivier E, Colonna M. Strategies of natural killer cell recognition and signaling. Curr Top Microbiol Immunol 2006; 298:1-21.
23. Reth M. Antigen receptor tail clue. Nature 1989; 338:383-384.
24. Songyang Z, Shoelson SE, Chaudhuri M et al. SH2 domains recognize specific phosphopeptide sequences. Cell 1993; 72:767-778.
25. Wu J, Cherwinski H, Spies T et al. DAP10 and DAP12 form distinct, but functionally cooperative, receptor complexes in natural killer cells. J Exp Med 2000; 192:1059-1068.
26. Manolios N, Bonifacino JS, Klausner RD. Transmembrane helical interactions and the assembly of the T-cell receptor complex. Science 1990; 249:274-277.
27. Call ME, Pyrdol J, Wiedmann M et al. The organizing principle in the formation of the T-cell receptor-CD3 complex. Cell 2002; 111:967-979.
28. Michnoff CH, Parikh VS, Lelsz DL et al. Mutations within the NH2-terminal transmembrane domain of membrane immunoglobulin (Ig) M alters Ig alpha and Ig beta association and signal transduction. J Biol Chem 1994; 269:24237-24244.
29. Daeron M. Fc receptor biology. Annu Rev Immunol 1997; 15:203-234.
30. Feng J, Garrity D, Call ME et al. Convergence on a distinctive assembly mechanism by unrelated families of activating immune receptors. Immunity 2005; 22:427-438.

31. Feng J, Call ME, Wucherpfennig KW. The assembly of diverse immune receptors is focused on a polar membrane-embedded interaction site. PLoS Biol 2006; 4:e142.
32. Bakema JE, de Haij S, den Hartog-Jager CF et al. Signaling through mutants of the IgA receptor CD89 and consequences for Fc receptor gamma-chain interaction. J Immunol 2006; 176:3603-3610.
33. Varin-Blank N, Metzger H. Surface expression of mutated subunits of the high affinity mast cell receptor for IgE. J Biol Chem 1990; 265:15685-15694.
34. Stevens TL, Blum JH, Foy SP et al. A mutation of the mu transmembrane that disrupts endoplasmic reticulum retention. Effects on association with accessory proteins and signal transduction. J Immunol 1994; 152:4397-4406.
35. Zidovetzki R, Rost B, Pecht I. Role of transmembrane domains in the functions of B- and T-cell receptors. Immunol Lett 1998; 64:97-107.
36. Blum JH, Stevens TL, DeFranco AL. Role of the mu immunoglobulin heavy chain transmembrane and cytoplasmic domains in B-cell antigen receptor expression and signal transduction. J Biol Chem 1993; 268:27236-27245.
37. Ra C, Jouvin MH, Kinet JP. Complete structure of the mouse mast cell receptor for IgE (Fc epsilon RI) and surface expression of chimeric receptors (rat-mouse-human) on transfected cells. J Biol Chem 1989; 264:15323-15327.
38. Rudd CE. Disabled receptor signaling and new primary immunodeficiency disorders. N Engl J Med 2006; 354:1874-1877.
39. Gomes MM, Herr AB. IgA and IgA-specific receptors in human disease: structural and functional insights into pathogenesis and therapeutic potential. Springer Semin Immunopathol 2006; 28:383-395.
40. Honda Z. Fcepsilon- and Fcgamma-receptor signaling in diseases. Springer Semin Immunopathol 2006; 28:365-375.
41. Moretta A, Bottino C, Vitale M et al. Activating receptors and coreceptors involved in human natural killer cell-mediated cytolysis. Annu Rev Immunol 2001; 19:197-223.
42. Clemetson KJ. Platelet receptors and their role in diseases. Clin Chem Lab Med 2003; 41:253-260.
43. Ortega E. How do multichain immune recognition receptors signal? A structural hypothesis. Mol Immunol 1995; 32:941-945.
44. Reth M. Oligomeric antigen receptors: A new view on signaling for the selection of lymphocytes. Trends Immunol 2001; 22:356-360.
45. Langlet C, Bernard AM, Drevot P et al. Membrane rafts and signaling by the multichain immune recognition receptors. Curr Opin Immunol 2000; 12:250-255.
46. Dykstra M, Cherukuri A, Pierce SK. Rafts and synapses in the spatial organization of immune cell signaling receptors. J Leukoc Biol 2001; 70:699-707.
47. Sigalov A. Multi-chain immune recognition receptors: Spatial organization and signal transduction. Semin Immunol 2005; 17:51-64.
48. Sigalov AB. Multichain immune recognition receptor signaling: Different players, same game? Trends Immunol 2004; 25:583-589.
49. Sigalov AB. Transmembrane interactions as immunotherapeutic targets: Lessons from viral pathogenesis. Adv Exp Med Biol 2007; 601:335-344.
50. Heldin CH. Dimerization of cell surface receptors in signal transduction. Cell 1995; 80:213-223.
51. Metzger H. Transmembrane signaling: The joy of aggregation. J Immunol 1992; 149:1477-1487.
52. Boniface JJ, Rabinowitz JD, Wulfing C et al. Initiation of signal transduction through the T-cell receptor requires the multivalent engagement of peptide/MHC ligands [corrected]. Immunity 1998; 9:459-466.
53. Holowka D, Baird B. Antigen-mediated IGE receptor aggregation and signaling: A window on cell surface structure and dynamics. Annu Rev Biophys Biomol Struct 1996; 25:79-112.
54. Thyagarajan R, Arunkumar N, Song W. Polyvalent antigens stabilize B-cell antigen receptor surface signaling microdomains. J Immunol 2003; 170:6099-6106.
55. Bankovich AJ, Raunser S, Juo ZS et al. Structural insight into preB cell receptor function. Science 2007; 316:291-294.
56. Puffer EB, Pontrello JK, Hollenbeck JJ et al. Activating B-cell signaling with defined multivalent ligands. ACS Chem Biol 2007; 2:252-262.
57. Deng L, Langley RJ, Brown PH et al. Structural basis for the recognition of mutant self by a tumor-specific, MHC class II-restricted T-cell receptor. Nat Immunol 2007; 8:398-408.
58. Adams EJ, Chien YH, Garcia KC. Structure of a gammadelta T-cell receptor in complex with the nonclassical MHC T22. Science 2005; 308:227-231.
59. Bachmann MF, Ohashi PS. The role of T-cell receptor dimerization in T-cell activation. Immunol Today 1999; 20:568-576.
60. Bachmann MF, Salzmann M, Oxenius A et al. Formation of TCR dimers/trimers as a crucial step for T-cell activation. Eur J Immunol 1998; 28:2571-2579.

61. DeFranco AL, Gold MR, Jakway JP. B-lymphocyte signal transduction in response to anti-immunoglobulin and bacterial lipopolysaccharide. Immunol Rev 1987; 95:161-176.

62. Posner RG, Savage PB, Peters AS et al. A quantitative approach for studying IgE-FcepsilonRI aggregation. Mol Immunol 2002; 38:1221-1228.

63. Draberova L, Lebduska P, Halova I et al. Signaling assemblies formed in mast cells activated via Fcepsilon receptor I dimers. Eur J Immunol 2004; 34:2209-2219.

64. Horii K, Kahn ML, Herr AB. Structural basis for platelet collagen responses by the immune-type receptor glycoprotein VI. Blood 2006; 108:936-942.

65. Cochran JR, Cameron TO, Stern LJ. The relationship of MHC-peptide binding and T-cell activation probed using chemically defined MHC class II oligomers. Immunity 2000; 12:241-250.

66. Fahmy TM, Bieler JG, Schneck JP. Probing T-cell membrane organization using dimeric MHC-Ig complexes. J Immunol Methods 2002; 268:93-106.

67. Kiessling LL, Gestwicki JE, Strong LE. Synthetic multivalent ligands as probes of signal transduction. Angew Chem Int Ed Engl 2006; 45:2348-2368.

68. Klemm JD, Schreiber SL, Crabtree GR. Dimerization as a regulatory mechanism in signal transduction. Annu Rev Immunol 1998; 16:569-592.

69. Garrity D, Call ME, Feng J et al. The activating NKG2D receptor assembles in the membrane with two signaling dimers into a hexameric structure. Proc Natl Acad Sci USA 2005; 102:7641-7646.

70. Radaev S, Kattah M, Rostro B et al. Crystal structure of the human myeloid cell activating receptor TREM-1. Structure 2003; 11:1527-1535.

71. Berlanga O, Bori-Sanz T, James JR et al. Glycoprotein VI oligomerization in cell lines and platelets. J Thromb Haemost 2007; 5:1026-1033.

72. Symer DE, Dintzis RZ, Diamond DJ et al. Inhibition or activation of human T-cell receptor transfectants is controlled by defined, soluble antigen arrays. J Exp Med 1992; 176:1421-1430.

73. Schweitzer-Stenner R, Tamir I, Pecht I. Analysis of Fc(epsilon)RI-mediated mast cell stimulation by surface-carried antigens. Biophys J 1997; 72:2470-2478.

74. Patrick SM, Kim S, Braunstein NS et al. Dependence of T-cell activation on area of contact and density of a ligand-coated surface. J Immunol Methods 2000; 241:97-108.

75. Germain RN. T-cell signaling: the importance of receptor clustering. Curr Biol 1997; 7:R640-644.

76. Alam SM, Davies GM, Lin CM et al. Qualitative and quantitative differences in T-cell receptor binding of agonist and antagonist ligands. Immunity 1999; 10:227-237.

77. Pribluda VS, Pribluda C, Metzger H. Transphosphorylation as the mechanism by which the high-affinity receptor for IgE is phosphorylated upon aggregation. Proc Natl Acad Sci USA 1994; 91:11246-11250.

78. Sigalov A, Aivazian D, Stern L. Homooligomerization of the cytoplasmic domain of the T-cell receptor zeta chain and of other proteins containing the immunoreceptor tyrosine-based activation motif. Biochemistry 2004; 43:2049-2061.

79. Berko D, Carmi Y, Cafri G et al. Membrane-anchored beta 2-microglobulin stabilizes a highly receptive state of MHC class I molecules. J Immunol 2005; 174:2116-2123.

80. Talbott M, Hare M, Nyarko A et al. Folding is coupled to dimerization of Tctex-1 dynein light chain. Biochemistry 2006; 45:6793-6800.

81. Sigalov AB, Zhuravleva AV, Orekhov VY. Binding of intrinsically disordered proteins is not necessarily accompanied by a structural transition to a folded form. Biochimie 2007; 89:419-421.

82. Sigalov AB, Aivazian DA, Uversky VN et al. Lipid-binding activity of intrinsically unstructured cytoplasmic domains of multichain immune recognition receptor signaling subunits. Biochemistry 2006; 45:15731-15739.

83. Pitcher LA, van Oers NS. T-cell receptor signal transmission: Who gives an ITAM? Trends Immunol 2003; 24:554-560.

84. Pike KA, Baig E, Ratcliffe MJ. The avian B-cell receptor complex: Distinct roles of Igalpha and Igbeta in B-cell development. Immunol Rev 2004; 197:10-25.

85. Storch B, Meixlsperger S, Jumaa H. The Ig-alpha ITAM is required for efficient differentiation but not proliferation of preB cells. Eur J Immunol 2007; 37:252-260.

86. Gazumyan A, Reichlin A, Nussenzweig MC. Ig beta tyrosine residues contribute to the control of B-cell receptor signaling by regulating receptor internalization. J Exp Med 2006; 203:1785-1794.

87. Lin S, Cicala C, Scharenberg AM et al. The Fc(epsilon)RIbeta subunit functions as an amplifier of Fc(epsilon)RIgamma-mediated cell activation signals. Cell 1996; 85:985-995.

88. Sanchez-Mejorada G, Rosales C. Signal transduction by immunoglobulin Fc receptors. J Leukoc Biol 1998; 63:521-533.

89. Lysechko TL, Ostergaard HL. Differential Src family kinase activity requirements for CD3 zeta phosphorylation/ZAP70 recruitment and CD3 epsilon phosphorylation. J Immunol 2005; 174:7807-7814.

90. Kuhns MS, Davis MM. Disruption of extracellular interactions impairs T-cell receptor-CD3 complex stability and signaling. Immunity 2007; 26:357-369.
91. Chau LA, Bluestone JA, Madrenas J. Dissociation of intracellular signaling pathways in response to partial agonist ligands of the T-cell receptor. J Exp Med 1998; 187:1699-1709.
92. Jensen WA, Pleiman CM, Beaufils P et al. Qualitatively distinct signaling through T-cell antigen receptor subunits. Eur J Immunol 1997; 27:707-716.
93. Pitcher LA, Mathis MA, Young JA et al. The CD3 gammaepsilon/deltaepsilon signaling module provides normal T-cell functions in the absence of the TCR zeta immunoreceptor tyrosine-based activation motifs. Eur J Immunol 2005; 35:3643-3654.
94. Kesti T, Ruppelt A, Wang JH et al. Reciprocal Regulation of SH3 and SH2 Domain Binding via Tyrosine Phosphorylation of a Common Site in CD3{epsilon}. J Immunol 2007; 179:878-885.
95. Chae WJ, Lee HK, Han JH et al. Qualitatively differential regulation of T-cell activation and apoptosis by T-cell receptor zeta chain ITAMs and their tyrosine residues. Int Immunol 2004; 16:1225-1236.
96. Wines BD, Trist HM, Monteiro RC et al. Fc receptor gamma chain residues at the interface of the cytoplasmic and transmembrane domains affect association with FcalphaRI, surface expression and function. J Biol Chem 2004; 279:26339-26345.
97. Kikuchi-Maki A, Catina TL, Campbell KS. Cutting edge: KIR2DL4 transduces signals into human NK cells through association with the Fc receptor gamma protein. J Immunol 2005; 174:3859-3863.
98. DeFranco AL, Mittelstadt PR, Blum JH et al. Mechanism of B-cell antigen receptor function: trans-membrane signaling and triggering of apoptosis. Adv Exp Med Biol 1994; 365:9-22.
99. Morton HC, van den Herik-Oudijk IE, Vossebeld P et al. Functional association between the human myeloid immunoglobulin A Fc receptor (CD89) and FcR gamma chain. Molecular basis for CD89/FcR gamma chain association. J Biol Chem 1995; 270:29781-29787.
100. Kim MK, Huang ZY, Hwang PH et al. Fcgamma receptor transmembrane domains: role in cell surface expression, gamma chain interaction and phagocytosis. Blood 2003; 101:4479-4484.
101. Arkin M. Protein-protein interactions and cancer: Small molecules going in for the kill. Curr Opin Chem Biol 2005; 9:317-324.
102. Fry DC. Protein-protein interactions as targets for small molecule drug discovery. Biopolymers 2006; 84:535-552.
103. Loregian A, Palu G. Disruption of protein-protein interactions: Towards new targets for chemotherapy. J Cell Physiol 2005; 204:750-762.
104. Angelov GS, Guillaume P, Cebecauer M et al. Soluble MHC-peptide complexes containing long rigid linkers abolish CTL-mediated cytotoxicity. J Immunol 2006; 176:3356-3365.
105. Rotzschke O, Falk K, Strominger JL. Superactivation of an immune response triggered by oligomerized T-cell epitopes. Proc Natl Acad Sci USA 1997; 94:14642-14647.
106. Tamir I, Schweitzer-Stenner R, Pecht I. Immobilization of the type I receptor for IgE initiates signal transduction in mast cells. Biochemistry 1996; 35:6872-6883.
107. Yang H, Parkhouse RM. Differential activation requirements associated with stimulation of T-cells via different epitopes of CD3. Immunology 1998; 93:26-32.
108. Niederberger N, Buehler LK, Ampudia J et al. Thymocyte stimulation by anti-TCR-beta, but not by anti-TCR-alpha, leads to induction of developmental transcription program. J Leukoc Biol 2005; 77:830-841.
109. Carreno LJ, Gonzalez PA, Kalergis AM. Modulation of T-cell function by TCR/pMHC binding kinetics. Immunobiology 2006; 211:47-64.
110. Holler PD, Lim AR, Cho BK et al. CD8(-) T-cell transfectants that express a high affinity T-cell receptor exhibit enhanced peptide-dependent activation. J Exp Med 2001; 194:1043-1052.
111. Ortega E, Schweitzer-Stenner R, Pecht I. Possible orientational constraints determine secretory signals induced by aggregation of IgE receptors on mast cells. EMBO J 1988; 7:4101-4109.
112. Pecht I, Ortega E, Jovin TM. Rotational dynamics of the Fc epsilon receptor on mast cells monitored by specific monoclonal antibodies and IgE. Biochemistry 1991; 30:3450-3458.
113. Gil D, Schamel WW, Montoya M et al. Recruitment of Nck by CD3 epsilon reveals a ligand-induced conformational change essential for T-cell receptor signaling and synapse formation. Cell 2002; 109:901-912.
114. Tamir I, Cambier JC. Antigen receptor signaling: Integration of protein tyrosine kinase functions. Oncogene 1998; 17:1353-1364.
115. Sulzer B, Perelson AS. Immunons revisited: Binding of multivalent antigens to B-cells. Mol Immunol 1997; 34:63-74.
116. Reth M, Wienands J. Initiation and processing of signals from the B-cell antigen receptor. Annu Rev Immunol 1997; 15:453-479.
117. Nimmerjahn F. Activating and inhibitory FcgammaRs in autoimmune disorders. Springer Semin Immunopathol 2006; 28:305-319.

118. Maurer D, Fiebiger E, Reininger B et al. Fc epsilon receptor I on dendritic cells delivers IgE-bound multivalent antigens into a cathepsin S-dependent pathway of MHC class II presentation. J Immunol 1998; 161:2731-2739.
119. Lara M, Ortega E, Pecht I et al. Overcoming the signaling defect of Lyn-sequestering, signal-curtailing FcepsilonRI dimers: Aggregated dimers can dissociate from Lyn and form signaling complexes with Syk. J Immunol 2001; 167:4329-4337.
120. Metzger H. The high affinity receptor for IgE, FcepsilonRI. Novartis Found Symp 2004; 257:51-59; discussion 59-64, 98-100, 276-185.
121. Metzger H, Eglite S, Haleem-Smith H et al. Quantitative aspects of signal transduction by the receptor with high affinity for IgE. Mol Immunol 2002; 38:1207-1211.
122. Holowka D, Sil D, Torigoe C et al. Insights into immunoglobulin E receptor signaling from structurally defined ligands. Immunol Rev 2007; 217:269-279.
123. Radaev S, Sun PD. Structure and function of natural killer cell surface receptors. Annu Rev Biophys Biomol Struct 2003; 32:93-114.
124. Natarajan K, Dimasi N, Wang J et al. Structure and function of natural killer cell receptors: multiple molecular solutions to self, nonself discrimination. Annu Rev Immunol 2002; 20:853-885.
125. Turnbull IR, Colonna M. Activating and inhibitory functions of DAP12. Nat Rev Immunol 2007; 7:155-161.
126. Posner RG, Paar JM, Licht A et al. Interaction of a monoclonal IgE-specific antibody with cell-surface IgE-Fc epsilon RI: characterization of equilibrium binding and secretory response. Biochemistry 2004; 43:11352-11360.
127. Schweitzer-Stenner R, Ortega E, Pecht I. Kinetics of Fc epsilon RI dimer formation by specific monoclonal antibodies on mast cells. Biochemistry 1994; 33:8813-8825.
128. Kent UM, Mao SY, Wofsy C et al. Dynamics of signal transduction after aggregation of cell-surface receptors: studies on the type I receptor for IgE. Proc Natl Acad Sci USA 1994; 91:3087-3091.
129. Yamasaki S, Ishikawa E, Kohno M et al. The quantity and duration of FcRgamma signals determine mast cell degranulation and survival. Blood 2004; 103:3093-3101.
130. Gibbs BF, Rathling A, Zillikens D et al. Initial Fc epsilon RI-mediated signal strength plays a key role in regulating basophil signaling and deactivation. J Allergy Clin Immunol 2006; 118:1060-1067.
131. Torigoe C, Inman JK, Metzger H. An unusual mechanism for ligand antagonism. Science 1998; 281:568-572.
132. Pasquier B, Launay P, Kanamaru Y et al. Identification of F calphaRI as an inhibitory receptor that controls inflammation: dual role of FcRgamma ITAM. Immunity 2005; 22:31-42.
133. Miura Y, Takahashi T, Jung SM et al. Analysis of the interaction of platelet collagen receptor glycoprotein VI (GPVI) with collagen. A dimeric form of GPVI, but not the monomeric form, shows affinity to fibrous collagen. J Biol Chem 2002; 277:46197-46204.
134. Varma R, Campi G, Yokosuka T et al. T-cell receptor-proximal signals are sustained in peripheral microclusters and terminated in the central supramolecular activation cluster. Immunity 2006; 25:117-127.
135. Jiang G, Hunter T. Receptor signaling: when dimerization is not enough. Curr Biol 1999; 9:R568-571.
136. Livnah O, Johnson DL, Stura EA et al. An antagonist peptide-EPO receptor complex suggests that receptor dimerization is not sufficient for activation. Nat Struct Biol 1998; 5:993-1004.
137. Syed RS, Reid SW, Li C et al. Efficiency of signalling through cytokine receptors depends critically on receptor orientation. Nature 1998; 395:511-516.
138. Greiser JS, Stross C, Heinrich PC et al. Orientational constraints of the gp130 intracellular juxtamembrane domain for signaling. J Biol Chem 2002; 277:26959-26965.
139. Gay NJ, Gangloff M, Weber AN. Toll-like receptors as molecular switches. Nat Rev Immunol 2006; 6:693-698.
140. Mescher MF, Agarwal P, Casey KA et al. Molecular basis for checkpoints in the CD8 T-cell response: Tolerance versus activation. Semin Immunol 2007; 19:153-161.
141. Colombetti S, Benigni F, Basso V et al. Clonal anergy is maintained independently of T-cell proliferation. J Immunol 2002; 169:6178-6186.
142. Choi S, Schwartz RH. Molecular mechanisms for adaptive tolerance and other T-cell anergy models. Semin Immunol 2007; 19:140-152.
143. Chiodetti L, Choi S, Barber DL et al. Adaptive tolerance and clonal anergy are distinct biochemical states. J Immunol 2006; 176:2279-2291.
144. Recio MJ, Moreno-Pelayo MA, Kilic SS et al. Differential biological role of CD3 chains revealed by human immunodeficiencies. J Immunol 2007; 178:2556-2564.
145. Gil D, Schrum AG, Alarcon B et al. T-cell receptor engagement by peptide-MHC ligands induces a conformational change in the CD3 complex of thymocytes. J Exp Med 2005; 201:517-522.

146. Risueno RM, Gil D, Fernandez E et al. Ligand-induced conformational change in the T-cell receptor associated with productive immune synapses. Blood 2005; 106:601-608.
147. Risueno RM, van Santen HM, Alarcon B. A conformational change senses the strength of T-cell receptor-ligand interaction during thymic selection. Proc Natl Acad Sci USA 2006; 103:9625-9630.
148. Vilen BJ, Nakamura T, Cambier JC. Antigen-stimulated dissociation of BCR mIg from Ig-alpha/Ig-beta: implications for receptor desensitization. Immunity 1999; 10:239-248.
149. Vilen BJ, Famiglietti SJ, Carbone AM et al. B-cell antigen receptor desensitization: Disruption of receptor coupling to tyrosine kinase activation. J Immunol 1997; 159:231-243.
150. Cambier JC, Fisher CL, Pickles H et al. Dual molecular mechanisms mediate ligand-induced membrane Ig desensitization. J Immunol 1990; 145:13-19.
151. Vilen BJ, Burke KM, Sleater M et al. Transmodulation of BCR signaling by transduction-incompetent antigen receptors: Implications for impaired signaling in anergic B-cells. J Immunol 2002; 168:4344-4351.
152. Arnon TI, Achdout H, Levi O et al. Inhibition of the NKp30 activating receptor by pp65 of human cytomegalovirus. Nat Immunol 2005; 6:515-523.
153. Levit MN, Liu Y, Stock JB. Stimulus response coupling in bacterial chemotaxis: Receptor dimers in signalling arrays. Mol Microbiol 1998; 30:459-466.
154. Sourjik V, Berg HC. Functional interactions between receptors in bacterial chemotaxis. Nature 2004; 428:437-441.
155. Sourjik V. Receptor clustering and signal processing in E. coli chemotaxis. Trends Microbiol 2004; 12:569-576.
156. Kawaguchi M, Eckels DD. Differential activation through the TCR-CD3 complex affects the requirement for costimulation of human T-cells. Hum Immunol 1995; 43:136-148.
157. Lanier LL, Ruitenberg JJ, Allison JP et al. Distinct epitopes on the T-cell antigen receptor of HPB-ALL tumor cells identified by monoclonal antibodies. J Immunol 1986; 137:2286-2292.
158. Schlitt HJ, Kurrle R, Wonigeit K. T-cell activation by monoclonal antibodies directed to different epitopes on the human T-cell receptor/CD3 complex: Evidence for two different modes of activation. Eur J Immunol 1989; 19:1649-1655.
159. Schwinzer R, Franklin RA, Domenico J et al. Monoclonal antibodies directed to different epitopes in the CD3-TCR complex induce different states of competence in resting human T-cells. J Immunol 1992; 148:1322-1328.
160. Yoon ST, Dianzani U, Bottomly K et al. Both high and low avidity antibodies to the T-cell receptor can have agonist or antagonist activity. Immunity 1994; 1:563-569.
161. Reich Z, Boniface JJ, Lyons DS et al. Ligand-specific oligomerization of T-cell receptor molecules. Nature 1997; 387:617-620.
162. Garboczi DN, Ghosh P, Utz U et al. Structure of the complex between human T-cell receptor, viral peptide and HLA-A2. Nature 1996; 384:134-141.
163. Abastado JP, Lone YC, Casrouge A et al. Dimerization of soluble major histocompatibility complex-peptide complexes is sufficient for activation of T-cell hybridoma and induction of unresponsiveness. J Exp Med 1995; 182:439-447.
164. Lindstedt R, Monk N, Lombardi G et al. Amino acid substitutions in the putative MHC class II "dimer of dimers" interface inhibit CD4+ T-cell activation. J Immunol 2001; 166:800-808.
165. Hayball JD, Lake RA. The immune function of MHC class II molecules mutated in the putative superdimer interface. Mol Cell Biochem 2005; 273:1-9.
166. Watts TH. T-cell activation by preformed, long-lived Ia-peptide complexes. Quantitative aspects. J Immunol 1988; 141:3708-3714.
167. Rubin B, Knibiehler M, Gairin JE. Allosteric Changes in the TCR/CD3 Structure Upon Interaction With Extra- or Intra-cellular Ligands. Scand J Immunol 2007; 66:228-237.
168. Cochran JR, Cameron TO, Stone JD et al. Receptor proximity, not intermolecular orientation, is critical for triggering T-cell activation. J Biol Chem 2001; 276:28068-28074.
169. Cebecauer M, Guillaume P, Hozak P et al. Soluble MHC-peptide complexes induce rapid death of CD8+ CTL. J Immunol 2005; 174:6809-6819.
170. Yamasaki S, Ishikawa E, Sakuma M et al. Mechanistic basis of pre-T-cell receptor-mediated autonomous signaling critical for thymocyte development. Nat Immunol 2006; 7:67-75.
171. Yamasaki S, Saito T. Molecular basis for preTCR-mediated autonomous signaling. Trends Immunol 2007; 28:39-43.
172. Geisberger R, Crameri R, Achatz G. Models of signal transduction through the B-cell antigen receptor. Immunology 2003; 110:401-410.
173. Yamashita T, Mao SY, Metzger H. Aggregation of the high-affinity IgE receptor and enhanced activity of p53/56lyn protein-tyrosine kinase. Proc Natl Acad Sci USA 1994; 91:11251-11255.

174. Kane P, Erickson J, Fewtrell C et al. Cross-linking of IgE-receptor complexes at the cell surface: Synthesis and characterization of a long bivalent hapten that is capable of triggering mast cells and rat basophilic leukemia cells. Mol Immunol 1986; 23:783-790.
175. Kanamaru Y, Blank U, Monteiro RC. IgA Fc receptor I is a molecular switch that determines IgA activating or inhibitory functions. Contrib Nephrol 2007; 157:148-152.
176. Li P, Morris DL, Willcox BE et al. Complex structure of the activating immunoreceptor NKG2D and its MHC class I-like ligand MICA. Nat Immunol 2001; 2:443-451.
177. Strong RK. Asymmetric ligand recognition by the activating natural killer cell receptor NKG2D, a symmetric homodimer. Mol Immunol 2002; 38:1029-1037.
178. Farndale RW. Collagen-induced platelet activation. Blood Cells Mol Dis 2006; 36:162-165.
179. Smethurst PA, Onley DJ, Jarvis GE et al. Structural basis for the platelet-collagen interaction: The smallest motif within collagen that recognizes and activates platelet Glycoprotein VI contains two glycine-proline-hydroxyproline triplets. J Biol Chem 2007; 282:1296-1304.
180. Lu Q, Navdaev A, Clemetson JM et al. Snake venom C-type lectins interacting with platelet receptors. Structure-function relationships and effects on haemostasis. Toxicon 2005; 45:1089-1098.
181. Kato K, Furihata K, Cheli Y et al. Effect of multimer size and a natural dimorphism on the binding of convulxin to platelet glycoprotein (GP)VI. J Thromb Haemost 2006; 4:1107-1113.
182. Singer KL, Mostov KE. Dimerization of the polymeric immunoglobulin receptor controls its transcytotic trafficking. Mol Biol Cell 1998; 9:901-915.
183. Mocsai A, Abram CL, Jakus Z et al. Integrin signaling in neutrophils and macrophages uses adaptors containing immunoreceptor tyrosine-based activation motifs. Nat Immunol 2006; 7:1326-1333.
184. Li R, Mitra N, Gratkowski H et al. Activation of integrin alphaIIbbeta3 by modulation of transmembrane helix associations. Science 2003; 300:795-798.
185. Hantgan RR, Lyles DS, Mallett TC et al. Ligand binding promotes the entropy-driven oligomerization of integrin alpha IIb beta 3. J Biol Chem 2003; 278:3417-3426.
186. Schamel WW, Arechaga I, Risueno RM et al. Coexistence of multivalent and monovalent TCRs explains high sensitivity and wide range of response. J Exp Med 2005; 202:493-503.
187. Iber D, Gruhn T. Organisation of B-cell receptors on the cell membrane. Syst Biol (Stevenage) 2006; 153:401-404.
188. Matsuuchi L, Gold MR. New views of BCR structure and organization. Curr Opin Immunol 2001; 13:270-277.
189. Schamel WW, Reth M. Monomeric and oligomeric complexes of the B-cell antigen receptor. Immunity 2000; 13:5-14.
190. Moss WC, Irvine DJ, Davis MM et al. Quantifying signaling-induced reorientation of T-cell receptors during immunological synapse formation. Proc Natl Acad Sci USA 2002; 99:15024-15029.
191. Minguet S, Swamy M, Alarcon B et al. Full activation of the T-cell receptor requires both clustering and conformational changes at CD3. Immunity 2007; 26:43-54.
192. Ding YH, Baker BM, Garboczi DN et al. Four A6-TCR/peptide/HLA-A2 structures that generate very different T-cell signals are nearly identical. Immunity 1999; 11:45-56.
193. Ortega E, Lara M, Lee I et al. Lyn dissociation from phosphorylated Fc epsilon RI subunits: A new regulatory step in the Fc epsilon RI signaling cascade revealed by studies of Fc epsilon RI dimer signaling activity. J Immunol 1999; 162:176-185.
194. Simpson RJ, Hammacher A, Smith DK et al. Interleukin-6: Structure-function relationships. Protein Sci 1997; 6:929-955.
195. Muller-Newen G, Kuster A, Wijdenes J et al. Studies on the interleukin-6-type cytokine signal transducer gp130 reveal a novel mechanism of receptor activation by monoclonal antibodies. J Biol Chem 2000; 275:4579-4586.
196. Autissier P, De Vos J, Liautard J et al. Dimerization and activation of the common transducing chain (gp130) of the cytokines of the IL-6 family by mAb. Int Immunol 1998; 10:1881-1889.
197. Chan KF, Siegel MR, Lenardo JM. Signaling by the TNF receptor superfamily and T-cell homeostasis. Immunity 2000; 13:419-422.
198. Chan FK. Three is better than one: Preligand receptor assembly in the regulation of TNF receptor signaling. Cytokine 2007; 37:101-107.
199. Siegers GM, Yang J, Duerr CU et al. Identification of disulfide bonds in the Ig-alpha/Ig-beta component of the B-cell antigen receptor using the Drosophila S2 cell reconstitution system. Int Immunol 2006; 18:1385-1396.
200. Asai K, Fujimoto K, Harazaki M et al. Distinct aggregation of beta- and gamma-chains of the high-affinity IgE receptor on cross-linking. J Histochem Cytochem 2000; 48:1705-1716.
201. Siegel RM, Muppidi JR, Sarker M et al. SPOTS: signaling protein oligomeric transduction structures are early mediators of death receptor-induced apoptosis at the plasma membrane. J Cell Biol 2004; 167:735-744.

202. Lee HK, Dunzendorfer S, Tobias PS. Cytoplasmic domain-mediated dimerizations of toll-like receptor 4 observed by beta-lactamase enzyme fragment complementation. J Biol Chem 2004; 279:10564-10574.
203. Weber AN, Moncrieffe MC, Gangloff M et al. Ligand-receptor and receptor-receptor interactions act in concert to activate signaling in the Drosophila toll pathway. J Biol Chem 2005; 280:22793-22799.
204. Kishimoto H, Kubo RT, Yorifuji H et al. Physical dissociation of the TCR-CD3 complex accompanies receptor ligation. J Exp Med 1995; 182:1997-2006.
205. Kosugi A, Saitoh S, Noda S et al. Translocation of tyrosine-phosphorylated TCRzeta chain to glycolipid-enriched membrane domains upon T-cell activation. Int Immunol 1999; 11:1395-1401.
206. La Gruta NL, Liu H, Dilioglou S et al. Architectural changes in the TCR: CD3 complex induced by MHC: Peptide ligation. J Immunol 2004; 172:3662-3669.
207. Liu H, Rhodes M, Wiest DL et al. On the dynamics of TCR: CD3 complex cell surface expression and downmodulation. Immunity 2000; 13:665-675.
208. D'Oro U, Munitic I, Chacko G et al. Regulation of constitutive TCR internalization by the zeta-chain. J Immunol 2002; 169:6269-6278.
209. McKay DB, Irie HY, Hollander G et al. Antigen-induced unresponsiveness results in altered T-cell signaling. J Immunol 1999; 163:6455-6461.
210. Willems F, Andris F, Xu D et al. The induction of human T-cell unresponsiveness by soluble anti-CD3 mAb requires T-cell activation. Int Immunol 1995; 7:1593-1598.
211. Wang XM, Djordjevic JT, Kurosaka N et al. T-cell antigen receptor peptides inhibit signal transduction within the membrane bilayer. Clin Immunol 2002; 105:199-207.
212. Kim JH, Cramer L, Mueller H et al. Independent trafficking of Ig-alpha/Ig-beta and mu-heavy chain is facilitated by dissociation of the B-cell antigen receptor complex. J Immunol 2005; 175:147-154.
213. Kremyanskaya M, Monroe JG. Ig-independent Ig beta expression on the surface of B lymphocytes after B-cell receptor aggregation. J Immunol 2005; 174:1501-1506.
214. Davis MM, Boniface JJ, Reich Z et al. Ligand recognition by alpha beta T-cell receptors. Annu Rev Immunol 1998; 16:523-544.
215. Gascoigne NR, Zal T, Alam SM. T-cell receptor binding kinetics in T-cell development and activation. Expert Rev Mol Med 2001; 2001:1-17.
216. Germain RN, Stefanova I. The dynamics of T-cell receptor signaling: Complex orchestration and the key roles of tempo and cooperation. Annu Rev Immunol 1999; 17:467-522.
217. Rosette C, Werlen G, Daniels MA et al. The impact of duration versus extent of TCR occupancy on T-cell activation: a revision of the kinetic proofreading model. Immunity 2001; 15:59-70.
218. Werlen G, Hausmann B, Naeher D et al. Signaling life and death in the thymus: Timing is everything. Science 2003; 299:1859-1863.
219. Love PE, Lee J, Shores EW. Critical relationship between TCR signaling potential and TCR affinity during thymocyte selection. J Immunol 2000; 165:3080-3087.
220. Andersen PS, Geisler C, Buus S et al. Role of the T-cell receptor ligand affinity in T-cell activation by bacterial superantigens. J Biol Chem 2001; 276:33452-33457.
221. Mirshahidi S, Ferris LC, Sadegh-Nasseri S. The magnitude of TCR engagement is a critical predictor of T-cell anergy or activation. J Immunol 2004; 172:5346-5355.
222. Gett AV, Sallusto F, Lanzavecchia A et al. T-cell fitness determined by signal strength. Nat Immunol 2003; 4:355-360.
223. Langenkamp A, Casorati G, Garavaglia C et al. T-cell priming by dendritic cells: Thresholds for proliferation, differentiation and death and intraclonal functional diversification. Eur J Immunol 2002; 32:2046-2054.
224. Batista FD, Neuberger MS. Affinity dependence of the B-cell response to antigen: A threshold, a ceiling and the importance of off-rate. Immunity 1998; 8:751-759.
225. Casola S, Otipoby KL, Alimzhanov M et al. B-cell receptor signal strength determines B-cell fate. Nat Immunol 2004; 5:317-327.
226. Benschop RJ, Cambier JC. B-cell development: signal transduction by antigen receptors and their surrogates. Curr Opin Immunol 1999; 11:143-151.
227. George J, Penner SJ, Weber J et al. Influence of membrane Ig receptor density and affinity on B-cell signaling by antigen. Implications for affinity maturation. J Immunol 1993; 151:5955-5965.
228. Singh DK, Kumar D, Siddiqui Z et al. The strength of receptor signaling is centrally controlled through a cooperative loop between Ca2+ and an oxidant signal. Cell 2005; 121:281-293.
229. Kinet JP. The high-affinity IgE receptor (Fc epsilon RI): From physiology to pathology. Annu Rev Immunol 1999; 17:931-972.
230. Yamasaki S, Saito T. Regulation of mast cell activation through FcepsilonRI. Chem Immunol Allergy 2005; 87:22-31.

231. Hlavacek WS, Redondo A, Wofsy C et al. Kinetic proofreading in receptor-mediated transduction of cellular signals: Receptor aggregation, partially activated receptors and cytosolic messengers. Bull Math Biol 2002; 64:887-911.
232. Iwashima M. Kinetic perspectives of T-cell antigen receptor signaling. A two-tier model for T-cell full activation. Immunol Rev 2003; 191:196-210.
233. Hlavacek WS, Redondo A, Metzger H et al. Kinetic proofreading models for cell signaling predict ways to escape kinetic proofreading. Proc Natl Acad Sci USA 2001; 98:7295-7300.
234. Haefner J. Modeling Biological Systems: Principles and Applications, 2nd ed. New York: Springer; 2005:480.
235. Ono S, Ohno H, Saito T. Rapid turnover of the CD3 zeta chain independent of the TCR-CD3 complex in normal T-cells. Immunity 1995; 2:639-644.
236. McKeithan TW. Kinetic proofreading in T-cell receptor signal transduction. Proc Natl Acad Sci USA 1995; 92:5042-5046.
237. Rabinowitz JD, Beeson C, Lyons DS et al. Kinetic discrimination in T-cell activation. Proc Natl Acad Sci USA 1996; 93:1401-1405.
238. Rabinowitz JD, Beeson C, Wulfing C et al. Altered T-cell receptor ligands trigger a subset of early T-cell signals. Immunity 1996; 5:125-135.
239. Kersh GJ, Kersh EN, Fremont DH et al. High- and low-potency ligands with similar affinities for the TCR: the importance of kinetics in TCR signaling. Immunity 1998; 9:817-826.
240. Schamel WW, Risueno RM, Minguet S et al. A conformation- and avidity-based proofreading mechanism for the TCR-CD3 complex. Trends Immunol 2006; 27:176-182.
241. Valitutti S, Muller S, Cella M et al. Serial triggering of many T-cell receptors by a few peptide-MHC complexes. Nature 1995; 375:148-151.
242. Itoh Y, Hemmer B, Martin R et al. Serial TCR engagement and down-modulation by peptide:MHC molecule ligands: relationship to the quality of individual TCR signaling events. J Immunol 1999; 162:2073-2080.
243. Borovsky Z, Mishan-Eisenberg G, Yaniv E et al. Serial triggering of T-cell receptors results in incremental accumulation of signaling intermediates. J Biol Chem 2002; 277:21529-21536.
244. Sousa J, Carneiro J. A mathematical analysis of TCR serial triggering and down-regulation. Eur J Immunol 2000; 30:3219-3227.
245. Friedl P, Gunzer M. Interaction of T-cells with APCs: The serial encounter model. Trends Immunol 2001; 22:187-191.
246. Davis MM. A new trigger for T-cells. Cell 2002; 110:285-287.
247. Alarcon B, Gil D, Delgado P et al. Initiation of TCR signaling: Regulation within CD3 dimers. Immunol Rev 2003; 191:38-46.
248. Reth M, Wienands J, Schamel WW. An unsolved problem of the clonal selection theory and the model of an oligomeric B-cell antigen receptor. Immunol Rev 2000; 176:10-18.
249. Aivazian D, Stern LJ. Phosphorylation of T-cell receptor zeta is regulated by a lipid dependent folding transition. Nat Struct Biol 2000; 7:1023-1026.
250. Rojo JM, Janeway CA Jr. The biologic activity of anti-T-cell receptor V region monoclonal antibodies is determined by the epitope recognized. J Immunol 1988; 140:1081-1088.
251. Krogsgaard M, Prado N, Adams EJ et al. Evidence that structural rearrangements and/or flexibility during TCR binding can contribute to T-cell activation. Mol Cell 2003; 12:1367-1378.
252. Kjer-Nielsen L, Dunstone MA, Kostenko L et al. Crystal structure of the human T-cell receptor CD3 epsilon gamma heterodimer complexed to the therapeutic mAb OKT3. Proc Natl Acad Sci USA 2004; 101:7675-7680.
253. Levin SE, Weiss A. Twisting tails exposed: The evidence for TCR conformational change. J Exp Med 2005; 201:489-492.
254. Anton van der Merwe P, Davis SJ, Shaw AS et al. Cytoskeletal polarization and redistribution of cell-surface molecules during T-cell antigen recognition. Semin Immunol 2000; 12:5-21.
255. Monks CR, Freiberg BA, Kupfer H et al. Three-dimensional segregation of supramolecular activation clusters in T-cells. Nature 1998; 395:82-86.
256. Davis SJ, van der Merwe PA. The kinetic-segregation model: TCR triggering and beyond. Nat Immunol 2006; 7:803-809.
257. Manolios N, Collier S, Taylor J et al. T-cell antigen receptor transmembrane peptides modulate T-cell function and T-cell-mediated disease. Nat Med 1997; 3:84-88.
258. Ali M, De Planque MRR, Huynh NT et al. Biophysical studies of a transmembrane peptide derived from the T-cell antigen receptor. Lett Pept Sci 2002; 8:227-233.
259. Bender V, Ali M, Amon M et al. T-cell antigen receptor peptide-lipid membrane interactions using surface plasmon resonance. J Biol Chem 2004; 279:54002-54007.

260. Gerber D, Quintana FJ, Bloch I et al. D-enantiomer peptide of the TCRalpha transmembrane domain inhibits T-cell activation in vitro and in vivo. FASEB J 2005; 19:1190-1192.
261. Enk AH, Knop J. T-cell receptor mimic peptides and their potential application in T-cell-mediated disease. Int Arch Allergy Immunol 2000; 123:275-281.
262. Gollner GP, Muller G, Alt R et al. Therapeutic application of T-cell receptor mimic peptides or cDNA in the treatment of T-cell-mediated skin diseases. Gene Ther 2000; 7:1000-1004.
263. Manolios N, Huynh NT, Collier S. Peptides in the treatment of inflammatory skin disease. Australas J Dermatol 2002; 43:226-227.
264. Kurosaka N, Bolte A, Ali M et al. T-cell antigen receptor assembly and cell surface expression is not affected by treatment with T-cell antigen receptor-alpha chain transmembrane peptide. Protein Pept Lett 2007; 14:299-303.
265. Collier S, Bolte A, Manolios N. Discrepancy in CD3-transmembrane peptide activity between in vitro and in vivo T-cell inhibition. Scand J Immunol 2006; 64:388-391.
266. Amon MA, Ali M, Bender V et al. Lipidation and glycosylation of a T-cell antigen receptor (TCR) transmembrane hydrophobic peptide dramatically enhances in vitro and in vivo function. Biochim Biophys Acta 2006.
267. Ali M, Salam NK, Amon M et al. T-cell antigen receptor-alpha chain transmembrane peptides: Correlation between structure and function. Int J Pept Res Ther 2006; 12:261-267.
268. Vandebona H, Ali M, Amon M et al. Immunoreceptor transmembrane peptides and their effect on natural killer (NK) cell cytotoxicity. Protein Pept Lett 2006; 13:1017-1024.
269. Sigalov AB. Interaction between HIV gp41 fusion peptide and T-cell receptor: Putting the puzzle pieces back together. FASEB J 2007; 21:1633-1634; author reply 1635.
270. Spencer DM, Wandless TJ, Schreiber SL et al. Controlling signal transduction with synthetic ligands. Science 1993; 262:1019-1024.
271. Soldevila G, Castellanos C, Malissen M et al. Analysis of the individual role of the TCRzeta chain in transgenic mice after conditional activation with chemical inducers of dimerization. Cell Immunol 2001; 214:123-138.
272. Crabtree GR, Schreiber SL. Three-part inventions: Intracellular signaling and induced proximity. Trends Biochem Sci 1996; 21:418-422.
273. Buferne M, Luton F, Letourneur F et al. Role of CD3 delta in surface expression of the TCR/CD3 complex and in activation for killing analyzed with a CD3 delta-negative cytotoxic T-lymphocyte variant. J Immunol 1992; 148:657-664.
274. Luton F, Buferne M, Legendre V et al. Role of CD3gamma and CD3delta cytoplasmic domains in cytolytic T-lymphocyte functions and TCR/CD3 down-modulation. J Immunol 1997; 158:4162-4170.
275. Haks MC, Cordaro TA, van den Brakel JH et al. A redundant role of the CD3 gamma-immunoreceptor tyrosine-based activation motif in mature T-cell function. J Immunol 2001; 166:2576-2588.
276. Haks MC, Pepin E, van den Brakel JH et al. Contributions of the T-cell receptor-associated CD3gamma-ITAM to thymocyte selection. J Exp Med 2002; 196:1-13.
277. de Saint Basile G, Geissmann F, Flori E et al. Severe combined immunodeficiency caused by deficiency in either the delta or the epsilon subunit of CD3. J Clin Invest 2004; 114:1512-1517.
278. Roifman CM. CD3 delta immunodeficiency. Curr Opin Allergy Clin Immunol 2004; 4:479-484.
279. Teixeiro E, Daniels MA, Hausmann B et al. T-cell division and death are segregated by mutation of TCRbeta chain constant domains. Immunity 2004; 21:515-526.
280. Torres PS, Zapata DA, PachecoCastro A et al. Contribution of CD3 gamma to TCR regulation and signaling in human mature T-lymphocytes. Int Immunol 2002; 14:1357-1367.
281. Rodriguez-Tarduchy G, Sahuquillo AG, Alarcon B et al. Apoptosis but not other activation events is inhibited by a mutation in the transmembrane domain of T-cell receptor beta that impairs CD3zeta association. J Biol Chem 1996; 271:30417-30425.
282. Sutton BJ, Gould HJ. The human IgE network. Nature 1993; 366:421-428.
283. Gould HJ, Sutton BJ, Beavil AJ et al. The biology of IGE and the basis of allergic disease. Annu Rev Immunol 2003; 21:579-628.
284. Kraft S, Novak N. Fc receptors as determinants of allergic reactions. Trends Immunol 2006; 27:88-95.
285. Adamczewski M, Paolini R, Kinet JP. Evidence for two distinct phosphorylation pathways activated by high affinity immunoglobulin E receptors. J Biol Chem 1992; 267:18126-18132.
286. Dombrowicz D, Lin S, Flamand V et al. Allergy-associated FcRbeta is a molecular amplifier of IgE- and IgG-mediated in vivo responses. Immunity 1998; 8:517-529.
287. Eiseman E, Bolen JB. Signal transduction by the cytoplasmic domains of Fc epsilon RI-gamma and TCR-zeta in rat basophilic leukemia cells. J Biol Chem 1992; 267:21027-21032.

288. Jouvin MH, Adamczewski M, Numerof R et al. Differential control of the tyrosine kinases Lyn and Syk by the two signaling chains of the high affinity immunoglobulin E receptor. J Biol Chem 1994; 269:5918-5925.
289. Scharenberg AM, Lin S, Cuenod B et al. Reconstitution of interactions between tyrosine kinases and the high affinity IgE receptor which are controlled by receptor clustering. EMBO J 1995; 14:3385-3394.
290. Shiue L, Green J, Green OM et al. Interaction of p72syk with the gamma and beta subunits of the high-affinity receptor for immunoglobulin E, Fc epsilon RI. Mol Cell Biol 1995; 15:272-281.
291. Wilson BS, Kapp N, Lee RJ et al. Distinct functions of the Fc epsilon R1 gamma and beta subunits in the control of Fc epsilon R1-mediated tyrosine kinase activation and signaling responses in RBL-2H3 mast cells. J Biol Chem 1995; 270:4013-4022.
292. Goldstein B, Faeder JR, Hlavacek WS. Mathematical and computational models of immune-receptor signalling. Nat Rev Immunol 2004; 4:445-456.
293. Goldstein B, Faeder JR, Hlavacek WS et al. Modeling the early signaling events mediated by FcepsilonRI. Mol Immunol 2002; 38:1213-1219.
294. Faeder JR, Hlavacek WS, Reischl I et al. Investigation of early events in Fc epsilon RI-mediated signaling using a detailed mathematical model. J Immunol 2003; 170:3769-3781.
295. Hlavacek WS, Faeder JR, Blinov ML et al. The complexity of complexes in signal transduction. Biotechnol Bioeng 2003; 84:783-794.
296. Nadler MJ, Matthews SA, Turner H et al. Signal transduction by the high-affinity immunoglobulin E receptor Fc epsilon RI: coupling form to function. Adv Immunol 2000; 76:325-355.
297. Liu KJ, Parikh VS, Tucker PW et al. Role of the B-cell antigen receptor in antigen processing and presentation. Involvement of the transmembrane region in intracellular trafficking of receptor/ligand complexes. J Immunol 1993; 151:6143-6154.
298. Pleiman CM, D'Ambrosio D, Cambier JC. The B-cell antigen receptor complex: Structure and signal transduction. Immunol Today 1994; 15:393-399.
299. Cambier JC. Signal transduction by T- and B-cell antigen receptors: Converging: Structures and concepts. Curr Opin Immunol 1992; 4:257-264.
300. Cambier JC, Pleiman CM, Clark MR. Signal transduction by the B-cell antigen receptor and its coreceptors. Annu Rev Immunol 1994; 12:457-486.
301. Kraus M, Pao LI, Reichlin A et al. Interference with immunoglobulin (Ig)alpha immunoreceptor tyrosine-based activation motif (ITAM) phosphorylation modulates or blocks B-cell development, depending on the availability of an Igbeta cytoplasmic tail. J Exp Med 2001; 194:455-469.
302. Batista FD, Iber D, Neuberger MS. B-cells acquire antigen from target cells after synapse formation. Nature 2001; 411:489-494.
303. Bannish G, Fuentes-Panana EM, Cambier JC et al. Ligand-independent signaling functions for the B- lymphocyte antigen receptor and their role in positive selection during B-lymphopoiesis. J Exp Med 2001; 194:1583-1596.
304. Kato K, Kanaji T, Russell S et al. The contribution of glycoprotein VI to stable platelet adhesion and thrombus formation illustrated by targeted gene deletion. Blood 2003; 102:1701-1707.
305. Gawaz M. Role of platelets in coronary thrombosis and reperfusion of ischemic myocardium. Cardiovasc Res 2004; 61:498-511.
306. Gibbins JM, Okuma M, Farndale R et al. Glycoprotein VI is the collagen receptor in platelets which underlies tyrosine phosphorylation of the Fc receptor gamma-chain. FEBS Lett 1997; 413:255-259.
307. Sigalov AB. More on: glycoprotein VI oligomerization: A novel concept of platelet inhibition. J Thromb Haemost 2007; 5:2310-2312.
308. Sigalov AB. Inhibiting Collagen-induced Platelet Aggregation and Activation with Peptide Variants. US 12/001,258 and PCT PCT/US2007/025389 patent applications were filed on 12/11/2007 and 12/12/2007, respectively, claiming a priority to US provisional patent application 60/874,694 filed on 12/13/2006.
309. Leitenberg D, Balamuth F, Bottomly K. Changes in the T-cell receptor macromolecular signaling complex and membrane microdomains during T-cell development and activation. Semin Immunol 2001; 13:129-138.
310. Janeway CA Jr, Bottomly K. Responses of T-cells to ligands for the T-cell receptor. Semin Immunol 1996; 8:108-115.
311. Janeway CA Jr, Bottomly K. Signals and signs for lymphocyte responses. Cell 1994; 76:275-285.
312. Niedergang F, Dautry-Varsat A, Alcover A. Peptide antigen or superantigen-induced down-regulation of TCRs involves both stimulated and unstimulated receptors. J Immunol 1997; 159:1703-1710.
313. Bonefeld CM, Rasmussen AB, Lauritsen JP et al. TCR comodulation of nonengaged TCR takes place by a protein kinase C and CD3 gamma di-leucine-based motif-dependent mechanism. J Immunol 2003; 171:3003-3009.

314. San Jose E, Borroto A, Niedergang F et al. Triggering the TCR complex causes the downregulation of nonengaged receptors by a signal transduction-dependent mechanism. Immunity 2000; 12:161-170.
315. von Essen M, Nielsen MW, Bonefeld CM et al. Protein kinase C (PKC) alpha and PKC theta are the major PKC isotypes involved in TCR down-regulation. J Immunol 2006; 176:7502-7510.
316. Quintana FJ, Gerber D, Kent SC et al. HIV-1 fusion peptide targets the TCR and inhibits antigen-specific T-cell activation. J Clin Invest 2005; 115:2149-2158.
317. Bloch I, Quintana FJ, Gerber D et al. T-Cell inactivation and immunosuppressive activity induced by HIV gp41 via novel interacting motif. FASEB J 2007; 21:393-401.
318. Alarcon B, Swamy M, van Santen HM et al. T-cell antigen-receptor stoichiometry: Preclustering for sensitivity. EMBO Rep 2006; 7:490-495.
319. Kim YM, Pan JY, Korbel GA et al. Monovalent ligation of the B-cell receptor induces receptor activation but fails to promote antigen presentation. Proc Natl Acad Sci USA 2006; 103:3327-3332.
320. Eynon EE, Parker DC. Small B-cells as antigen-presenting cells in the induction of tolerance to soluble protein antigens. J Exp Med 1992; 175:131-138.
321. Gahring LC, Weigle WO. The induction of peripheral T-cell unresponsiveness in adult mice by monomeric human gamma-globulin. J Immunol 1989; 143:2094-2100.

CHAPTER 13

Visualization of Cell-Cell Interaction Contacts—Synapses and Kinapses

Michael L. Dustin*

Abstract

T-cell activation requires interactions of T-cell antigen receptors (TCR) and peptides presented by major histocompatibility complex molecules (MHCp) in an adhesive junction between the T-cell and antigen-presenting cell (APC). Stable junctions with bull's eye supramolecular activation clusters (SMACs) have been defined as immunological synapses. The term synapse works in this case because it joins roots for "same" and "fasten," which could be translated as "fasten in the same place." These structures maintain T-cell-APC interaction and allow directed secretion. We have proposed that SMACs are not really clusters, but are analogous to higher order membrane-cytoskeleton zones involved in amoeboid locomotion including a substrate testing lamellipodium, an adhesive lamella and anti-adhesive uropod. Since T-cells can also integrate signaling during locomotion over antigen presenting cells, it is important to consider adhesive junctions maintained as cells move past each other. This combination of movement (kine-) and fastening (-apse) can be described as a kinapse or moving junction. Synapses and kinapses operate in different stages of T-cell priming. Optimal effector functions may also depend upon cyclical use of synapses and kinapses. Visualization of these structures in vitro and in vivo presents many distinct challenges that will be discussed in this chapter.

Introduction

The partnership between dendritic cells (DC) and T-lymphocytes (T-cells) defends the body against microbes, parasites, abnormal cells and environmental toxins that breach the barrier function of skin and epithelial surfaces.[1,2] Diverse tools including those of biochemistry, cell biology, genetics and imaging have been employed to understand the mechanistic basis of this partnership. In recent years, imaging approaches have become increasingly useful as molecular technologies for labeling cells and proteins and imaging hardware and software have improved. In vitro imaging led to the initial definition of the immunological synapse (IS, or synapse) based on the organization of polarity and adhesion molecules to fasten (-apse) the T-cells to the same (syn-) antigen presenting cells (APCs) or place.[3-8] Advances in near-field in vitro imaging have led to the description of TCR microclusters that sustain signaling in the periphery of synapses.[9-11] Introduction of two-photon laser-scanning microscopy and methods for long-term in vivo observation have led to a basic understanding of the dynamics of T-cell-APC interactions in the living lymph node and the affects of antigen, which leads to signal integration via both short and long-lived T-cell-APC contacts.[12-14] The long-lived interactions can be defined as synapses since they fasten the T-cell to

*Michael L. Dustin—Program in Molecular Pathogenesis, Skirball Institute of Biomolecular Medicine and Department of Pathology, New York University School of Medicine, 540 1st Ave, New York, NY 10016, USA. Email: dustin@saturn.med.nyu.edu

Multichain Immune Recognition Receptor Signaling: From Spatiotemporal Organization to Human Disease, edited by Alexander B. Sigalov. ©2008 Landes Bioscience and Springer Science+Business Media.

the same APC. The short-lived interactions both early and late in the process appear to be the product of combining motility and cell-cell communication as a continuous kinetic process.[15] The T-cells move while maintaining extensive contact with the APC. This dynamic interaction, for which there is no convenient descriptor, could be described as a kinapse—combining roots indicating movement (kine-) and fastening (-apse) at the same time. A challenge for in vitro molecular imaging is to provide insight into how T-cells integrate signals from synapses and kinapses. It is likely that TCR microclusters will be common structures in this process.

One of the basic biological questions in immunology is what distinguishes T-cell responses to DC that lead to tolerance or priming. One concept is that the outcome of antigen presentation depends upon the activation status of the DC.[2] Immature DCs patrol the tissue spaces and boundaries of the body and gather antigenic structures, both self and foreign. Induced or spontaneous maturation of DC triggers their migration to the lymph node and concurrent processing of antigens to generate peptides that bind to major histocompatibility complex molecules (MHCp) that are then presented at the cell surface. DC migrate to the lymph node via the lymphatics and then migrate in the parenchyma and join DC networks in the T-cell zones where they encounter many T-cells.[16,17] The level of costimulatory molecules expressed by the DC is determined by the level of cytokines-like TNF produced in response to various endogenous or exogenous activators of innate immunity.[18] This level of innate stimulation appears to control whether the antigen-dependent T-DC interactions lead to tolerance or priming of an immune response over a period of 5-7 days.[19] While some have argued that tolerance induction does not involve synapses in vitro or in vivo,[20,21] we have found that the TCR-MHCp interactions alone control in vitro synapse formation[22] and that T-cells do synapse with DC during tolerance induction in vivo.[14] T-cells also synapse with DC during induction of oral tolerance.[23]

Once T-cells are primed they may take on a number of fates. They may become memory cells that continue to recirculate,[24] exit the secondary lymphoid tissues altogether to sites of inflammation,[25-27] remain in T-cells zones to help CD8+ T-cell responses,[28] or move to follicles within the lymph node to help B-cells.[29] It has been demonstrated that effector CD8+ T-cells are active in killing targets within lymph nodes.[30] The manner in which these fates are established is poorly understood, but may involve processes such as asymmetric cell division set up by synapses or different cytokine milieus encountered by daughter cells as the migrate.[31] Memory T-cells have been shown to accumulate in the bone marrow and to interact with bone marrow DC during secondary stimulation.[32]

Peripheral tissue scanning by DC is only one mode of innate immune surveillance of tissues. Two striking examples are the surveillance of the brain by the dynamic processes of microglial cells[33] and the active patrolling of liver sinusoides by natural killer T-cells, an innate-like T-cell.[34]

In this review we will summarize a new view of sustained T-cells activation through the synapses and kinapses. Then, how the synapse and kinapse work together in T-cell tolerance and immune surveillance will be discussed. Throughout the chapter the various visualization methods that are employed will be described and critiqued with respect to potential and limitations.

New Model for Sustained Signaling through the Synapse

Studies on the synapse bring together three parallel lines of experimentation in immunology through high-resolution fluorescence microscopy: TCR signal transduction, T-cell adhesion and polarity mechanisms. TCR signaling is based on a tyrosine kinase cascades that leads to rapid activation of phospholipase C γ.[35] The key tyrosine kinases are Lck, which initiates phosphorylation of immunoreceptor tyrosine-based activation motifs (ITAMs) in the cytoplasmic domain of the TCR, ZAP-70, which is recruited to phosphorylated ITAMs and phosphorylates LAT and ITK, which phosphorylates phospholipase C γ that is recruited to phosphorylated LAT. Phospholipase C γ activation leads to generation of inositol-1,4,5-triphosphate, leading to Ca^{2+} mobilization and diacylglycerol leading to activation of protein kinase C and Ras exchange factors.[31] The triggering of the cascade is based on recruitment of Lck-associated coreceptors to the TCR and on TCR oligomer formation (see also Chapters 6 and 11).

Members of the integrin and immunoglobulin families mediate T-cell adhesion to APCs. These interactions greatly extend the sensitivity of TCR to small numbers of MHCp-bearing agonist peptides.[36] Costimulatory molecules also are configured as adhesion molecules and the line between adhesion and costimulation molecules if often blurry.[22,37] By definition, adhesion enhances the physical interaction of T-cells with APC and the interaction of TCR and MHCp, while costimulation enhances TCR signaling or produces independent signals that integrate with the TCR signal to influence T-cell activation. However, the major T-cell adhesion molecules have some costimulatory activity. For example, LFA-1 contributes to the adhesion of T-cells in many contexts, contributes to TCR-MHCp interactions and provides signals that enhance Ca^{2+}, phosphatidylinositol-3-kinase and MAPK pathway activation.[38] There are also negative costimulators. For example, CTLA-4 and PD-1 negatively regulate T-cell expansion at intermediate and late periods of activation.[39,40]

Polarity of secretion is a hallmark of the neural synapses and is one of the most compelling parallels between the IS and neural synapse.[7] Early studies on the mechanism of T cell-mediated killing suggested that killing worked by exocytosis of preformed granules containing lytic molecules with activity-like complement.[41] It was later found that the primary role of perforin was to induce the target cell to take a "poison pill" by introducing granzyme A or B into the cytoplasm, which initiates a pro-apoptotic caspase cascade.[42,43] Evidence that cell T-cell polarity was related to directed secretion was provided by seminal studies of Geiger and Kupfer showing that the microtubule organizing center and Golgi apparatus reorients toward the target cell for killing.[44,45] Kupfer published a series of studies on molecular makeup of the T-cell-B-cell interface with the first demonstration of CD4, LFA-1, IL-4, Talin and protein kinase C-θ polarization to the interface.[46-50] All of these studies were performed with fixed cells so temporal information was deduced from populations of images for cells fixed at different times.

In 1998, Kupfer published a paper on the organization of LFA-1, Talin, TCR and protein kinase C-θ in the interface between antigen-specific T-cells and antigen-presenting B-cells.[4] LFA-1 and Talin were shown to form a ring in the interface and TCR and protein kinase C-θ were shown to cluster in the middle. These structures were defined as supramolecular activation clusters (SMAC). The TCR cluster marked the central SMAC (cSMAC), while the LFA-1 ring marked the peripheral SMAC (pSMAC). It was implied that TCR signaling was initiated and sustained by the cSMAC. My collaborators and I published a paper in parallel in which live T-cells interacting with supported planar bilayers were imaged in real time to visualize segregated adhesive domains composed of LFA-1-ICAM-1 and CD2-CD58 interactions.[3] It was posited that the segregation of the adhesion molecules was driven by the different topology of the LFA-1-ICAM-1 (40 nm domain) and CD2-CD58 (15 nm domain) interactions.[3,51] The antigen-dependent organization of these domains into a bull's eye pattern, similar to that reported in several international meetings by Kupfer, was an active process.[3] We proposed the definition of "immunological synapse" for the bull's eye pattern described by Kupfer and colleagues and our studies with adhesion molecules, linking a specific molecular pattern to the widely discussed concept.[52,53] Taking these two studies together, the synapse was defined as a specialized cell-cell junction composed of a cSMAC and a pSMAC. The term immunological synapse has subsequently been applied to a more diverse array of structures, but we will focus on the positional stability and polarity and how these are generated by the cell, which are the functionally important aspects.

The formation of the synapse was first evaluated in live T-cell-supported planar bilayer models.[5] Supported planar bilayers are formed from phospholipid liposomes on clean glass coverslips and can contain physiological densities of purified adhesion molecules that are anchored to the upper leaflet of the bilayer such that they are laterally mobile.[54,55] When ICAM-1 and MHC-peptide complexes are included, T-cells are fully activated by the substrates and organize SMACs similarly to T-cell-B-cell synapses. These systems can be imaged by wide-field or confocal fluorescence microscopy and near-field methods as will be discussed below. It was shown that TCR are engaged first in the periphery within 30 seconds and then these TCR clusters translocate to the center of the synapse to form the cSMAC by five minutes. Examination of cell-cell systems showed a similar pattern with peripheral TCR clusters merging in the center to form the cSMAC.[56,57] This process could take up to 30 minutes with naive T-cells. IS formation is enhanced by CD28-CD80 mediated costimulation,[58] but CD28-CD80 interactions are dependent upon TCR-MHCp interactions,[22,59,60] perhaps due to local rearrangements of actin that are compatible with enhanced CD28-CD80 interaction. Thus, costimulation mediated by CD28 interaction with CD80/86 is a positive feedback loop in IS formation and function.

The IS pattern was highly correlated with full T-cell activation in vitro in multiple studies using both T-cell-B-cell, T-cell-MHCp and ICAM-1 bearing planar bilayer and NK cell-targeT-cell IS.[4,5,61] While it was recognized early on that TCR signaling was initiated well before the cSMAC was formed, it was still posited that the cSMAC might be involved in sustained signaling. While kinases can be localized to the cSMAC at 1–5 minutes, there is a consensus that the cSMAC has relatively low levels of phosphotyrosine, activated phospho-Lck, or activated phospho-ZAP-70 at later times.[57,62,63] Phosphotyrosine staining was retained in the cSMAC at one hour in CD2AP deficient T-cells. Since CD2AP regulates TCR degradation, it was argued that the cSMAC is engaged in continuous signaling, which is made occult by TCR degradation processes.[63]

A striking property of the cSMAC is that TCR-MHCp interactions in the cSMAC are stable as measured by fluorescence photobleaching recovery.[5] Since each TCR-agonist MHCp interaction has a half-life of 5–30 seconds it would be expected that half of the TCR-MHCp interactions in the interface would exchange with free MHCp over a period of several minutes. In fact, the TCR-MHCp interactions in the cSMAC do not exchange with MHCp in the bilayer over a period of one hour. The basis of this stabilization is not clear, but may be as simple as the very high local density of TCR and MHCp creating diffusion barriers that exclude free MHCp or favor rebinding of the same MHCp to the TCR following spontaneous dissociation.[63] In contrast, sustained signaling by T-cells appears to be maintained by new TCR-MHCp interactions since it can be acutely inhibited by antibodies to MHCp that compete for TCR binding.[64] This acute inhibition by anti-MHCp appears to be inconsistent with a central role of stable TCR-MHCp interactions in the cSMAC with TCR-MHCp interactions; however, in the absence of other TCR-containing structures after 5–30 minutes, attention has continued to focus on the cSMAC as a signaling structure, even though the formation of a cSMAC is not required for T-cell activation.[63,65]

To search for other TCR-containing structures in the IS, my lab has employed total internal reflection fluorescence microscopy (TIRFM) with the live T-cell-supported planar bilayer system. This is a uniquely advantageous combination for increasing sensitivity to small, low-contrast structures. Through the lens, total internal reflection fluorescence microscopy is based on using very high-resolution oil immersion objectives with a laser focused at the outer edge of the back aperture to steer the shaft of illumination to the sample at an angle that exceeds the "critical angle."[66] Under these conditions, all the light is reflected off of the interface between the coverglass and the cell, but an evanescent wave is generated that can excite fluorescence in the "near-field" (within 200 nm of the interface) (Fig. 1A,B). This method is generally not useful in cell-cell interfaces because these structures are many microns away from the interface, but it can be very effective for examination of the interface with cells and the supported planar bilayer, which is only 2 nm off the surface of the coverglass. Thus, the entire IS can be illuminated with lateral and axial resolution of ~ 200 nm, a uniquely optimal situation in light microscopy. TIRFM is used for single-fluorophore imaging, so as a long as contrast exists, a small fluorescent structure can be detected.[67,68]

Application of TIRFM to the IS led to a striking discovery. While the field had seen the IS formation process as a single wave of TCR-MHCp movement from the periphery to the cSMAC, TIRFM revealed continued formation of TCR-MHCp microclusters in the periphery of the IS.[9-11] TCR clustering had been recognized as an important concept on theoretical grounds based on work with growth factor receptors[69,70] and the FcεRI receptor[71] (see also Chapter 6), but the minimal clusters sufficient to sustain TCR signaling had been assumed to be too small for direct visualization.[72,73] Initially, TCR microcluster can contain up to ~150 TCR each at which point they are readily detectable by conventional methods, but by 60 minutes of sustained signaling the TCR clusters contain only ~10 TCR per cluster and could only be detected by TIRFM.[9]

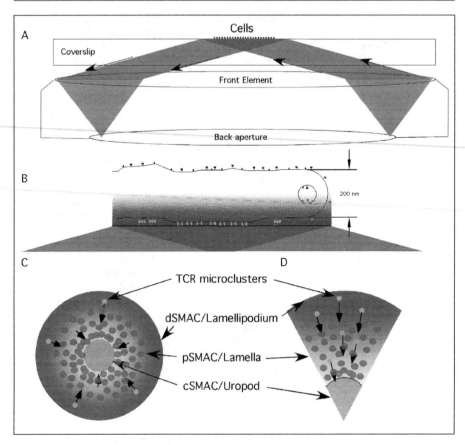

Figure 1. Total internal reflection fluorescence microscopy and schematics of synapse and kinapse. A) Schematic of TIRFM objective (NA > 1.45) with laser light path indicated and incidence on supported planar bilayer with attached cells. B) Expanded view of evanescent wave (gradient region above the coverslip) in which two sides of a thin cellular structure can be selectively imaged. Molecules in contact area with the bilayer are visualized whereas objected on the upper side or weakly or not visualized. This greatly increases contrast. However, it should be pointed out that some cytoplasmic membrane systems such as exocytic vesicles can be detected in the TIRFM field. C) En face view of the synapse with cSMAC, pSMAC and dSMAC as defined by Kupfer.[4,62] The grayscale gradient approximates the pattern of f-actin, which is highest in dSMAC, lower in the pSMAC and lowest in the cSMAC. TCR (fine speckling) and LFA-1 (coarse speckling) microclusters are indicated. The SMACs are not homogeneous clusters, but at high resolution are collections of different types of micron scale molecular clusters that function in microcluster initiation (dSMAC), microcluster translocation (pSMAC) and degradation (cSMAC) with respect to the TCR. Arrows on schematic indicate direction of TCR microcluster movement. D) En face view of a kinapse. In migrating cells the leading edge is defined as a lamellipodium, followed by the lamella and trailed by the uropod. The kinapse has not been extensively studied, so this model is based on the hypothesis that the structures in the synapse and kinapse can be paired as indicated.

TCR microclusters are detected over the entire range of MHCp densities leading to cSMAC formation.[9] Each microcluster lasts about two minutes prior to capture by the cSMAC. TCR microclusters are stained with anti-phosphotyrosine antibodies; they recruit ZAP-70-GFP and anti-MHCp Ab that block Ca^{2+} signaling within two minutes by blocking formation of new TCR microclusters.[9-11] Interestingly, TCR-MHCp interactions in the cSMAC persisted for many minutes after anti-MHCp addition, demonstrating that stable interactions in the cSMAC are not sufficient to sustain TCR proximal signaling. Therefore, contrary to initial perceptions we could demonstrate with near-field imaging methods that TCR microclusters are newly formed in the periphery of the IS to sustain signaling. This finding redirects our attention from the cSMAC to the more dynamic T TCR structures in the periphery of the IS. It also emphasizes the essential heterogeneity of SMACs. These are not homogeneous clusters but complex membrane-cytoskeletal zones composed of distinct microclusters (Fig. 1C).

The TCR microclusters are formed in the periphery of the IS–a structure that Kupfer described as a distal SMAC (dSMAC).[62] The dynamics of the dSMAC are very similar to the lamellipodium of a spreading or migrating fibroblasts in that it displays cycles of extension and retraction referred to as contractile oscillations.[74,75] Consistent with this, the dSMAC is also highly enriched in dynamic f-actin. Based on studies with Jurkat T-cells on anti-CD3 coated surfaces, the dynamic f-actin structures are dependent upon the Rac1 effector WAVE2 and the lymphocyte cortactin homology HS-1[76,77] (Fig. 1C). The dSMAC is also rich in CD45, the transmembrane tyrosine phosphatase that primes activation of Src family kinases including Lck and Fyn.[62,78] The near-field imaging clearly demonstrated that CD45 is excluded from peripheral TCR microclusters (see also Chapter 7), but is included in the cSMAC.[9] A model for CD45 function could be stated as Lck is activated by CD45 outside the TCR microclusters and then active Lck diffuses a short distance to capture in the TCR microcluster where it operates until rephosphorylated on the inhibitory site by Csk.[67]

The TCR microclusters traverse the pSMAC, a domain rich in the larger LFA-1-ICAM-1 interaction. The pSMAC is not a solid ring, but a meshwork of micron-scale, LFA-1 rich clusters with interspersed spaces lacking LFA-1-ICAM-1 interaction.[5] The TCR microclusters appear to navigate these holes in a tortuous fashion.[9,10] Thus, the TCR microclusters do not take a straight-line path to the cSMAC, but zigzag through the pSMAC, perhaps because of the obstacles formed by the dynamic LFA-1 clusters (Fig. 1C). It is notable that ZAP-70 recruitment and activation appears to be maximal as the TCR microclusters traverse the pSMAC.[9-11] Thus, the close juxtaposition of TCR microclusters and surrounding integrin microclusters appears to be an optimal condition for TCR signaling, but signaling turns off as, or shortly after, the TCR clusters join the cSMAC, which is typically free of LFA-1-ICAM-1 interactions.

It has been observed in many studies that when T-cells migrate on ICAM-1 containing surfaces they produce "focal zones" of LFA-1-ICAM-1 interaction that accumulate in the lamella, the force generating structure in amoeboid cell locomotion.[5,37,79-83] These focal zones can vary in shape from crescents to wedges that are analogous to a half or quarter synapses. We would posit that the focal zone is similar or identical to the "lamella" zone of a migrating tissue cell.[82,83] In a synapse, the inward-directed forces generated in the pSMAC/lamella are balanced and the cell moves slowly or not at all.[84] If the symmetry of the synapse is broken, then the asymmetric lamella mediates rapid cell movement, but this does not necessarily terminate signaling. Thus, we propose that the contact structure used by a migrating T-cell to integrate signal from an APC is not a synapse, but can be defined as a "kinapse" translated as a moving junction. Kinapses signal by forming TCR microclusters in the leading lamellipodia that signal as they translocate through the lamella and are inactivated in the uropod, the trailing structure that is analgous to the cSMAC (Fig. 1D). This model suggests a way to resolve the controversy regarding synapse formation and signaling between groups that typically observe synapse formation during T-cell activation and those who observe migration only.[85] Since TCR signaling is initiated and sustained in leading lamellipodium by TCR microclusters, the formation of these structures by migrating T-cells is fully compatible with integration of signals. The uropod and cSMAC may differ in that a cSMAC receives and can preserve, to some degree, TCR-MHCp complexes and associated molecules, although there is no evidence that the molecules accumulated in the cSMAC continue to signal in normal T-cells, while the uropod may only maintain long-term connections using membrane nanotubes, which may prolong signaling connections between immune cells, but are likely to break beyond a few 10s of micrometers.[86,87] The major differences between synapse and kinapse include that a synapse maintains contact with the same APC over hours, whereas a kinapse may maintain communication with one APC for only a few minutes. Synapses polarize some secretion toward the APC, whereas kinapses would not appear to have the same possibility.

The current challenge in the field is to image TCR microclusters in T-DC or other T-APC interfaces. This is a challenge because the near-field imaging technologies that are available cannot access the cell-cell junction. Closer examination of a study from Kupfer using wide-field microscopy and computational deconvolution suggests that these methods can detect early microclusters, but its not clear if the later-sustained microclusters, which are smaller, would be visible.[62] T-DC synapses appear to have multiple foci of TCR-MHCp interaction,[88] but other studies observe classical synapses with rings of pSMACs and cSMACs formed with mature DC.[20] The observation that the classical synapses have multiple signaling TCR microclusters may reconcile these observations, but dynamic imaging of T-DC microclusters would be required to validate this position.[89] T-cell synapses with CHO cells as model APC appeared multifocal and demonstrated a surprising segregation of TCR clusters from CD28 clusters,[90] which have been observed to colocalize in other model systems.[22,59,60] Davis and colleagues have visualized single MHCp complexes in the interface between T-cells and APC by labeling peptides with phycoerythrin, but the limitation of this approach is that its not clear how many TCR are involved in recognition of these peptides[61,91,92] (see also Chapter 6). In fact, the evidence that these peptides are recognized at all is circumstantial rather than direct. When a single peptide-agonist-peptide complex is seen in the interface, the T-cell fluxes Ca^{2+} transiently, but the limitation that the single peptide can only be imaged once due to fluorophore bleaching makes it impossible to follow dynamic processes that could provide more direct evidence for interaction. Technologies with the potential to collect such dynamic interaction include spinning disc confocal and line-scan technologies that have recently been commercialized.[82] The problem is not single molecule sensitivity, but contrast and the ability to distinguish small clusters from random fluctuations.

In Vivo Functions of Synapse and Kinapse

TCR signal integration through microclusters formed in synapses or kinapses provides a framework for thinking about results from recent in vivo studies. Since we can now understand the ability of T-cells to signal while migrating, we can consider regulation of T-cell migration not in terms of signaling changes, but in terms of controlling the network of spatiotemporal cell-cell interactions that recruit different T-cells into the immune response or allow them to execute effector functions. Another consideration is directed secretion, which is likely best delivered through a synapse, whereas a kinapse may be much less efficient at delivering secreted molecules to specific APC.

Naive T-cell priming is a central process in initiation of immune responses. Priming of naive T-cells requires interactions with DCs in vivo.[93] However, not all antigen-specific T-DC interactions will lead to priming. When an antigen is presented by DCs in the steady state (absence of inflammation), the result is induction of peripheral tolerance.[94,95] Presentation under steady-state conditions leads to T-cell proliferation followed by induction of antigen-specific nonresponsiveness in the effector cells or their deletion.[96] A second mechanism of peripheral tolerance is the induction of antigen-specific regulatory T-cells.[97] The steady-state process of inducing tolerance preconditions the active peripheral T-cell repertoire to react selectively with pathogen-associated foreign antigens under conditions of infection or tissue damage since T-cells specific for self antigens and benign foreign antigens are deleted or anergic.[2] The dynamics of priming or priming versus tolerance has been the subject of several papers.

Stand-alone studies on priming of naive T-cells have been carried out using three different TCR transgenic models, two different adjuvant systems and both explanted intact lymph node and intravital imaging approaches. Consistent themes are emerging and hypotheses can be developed regarding the basis for differences between studies.

In Vivo Analysis of CD4⁺ T-Cell Priming and Tolerance Induction

Stoll et al[98] examined dynamics of T-cell priming by transferred mature DC. CFSE labeled 5C.C7 TCR Tg T-cells were introduced into the lymph nodes by intravenous adoptive transfer and DiI- or DiD-labeled DC were introduced to specific draining lymph nodes by sub-cutaneous injection. 5C.C7 is an I-E^k restricted TCR that binds moth cytochrome C (MCC) peptide 91-103 as an agonist, so DC were either pulsed with this peptide or a control peptide. The lymph nodes were then explanted, embedded and imaged with a conventional confocal microscope. This is the only study of lymph node T-cell dynamics by conventional confocal microscopy, which limits depth of penetration to ~50 μm. T-cells displayed low basal motility in the absence of antigen, which may be attributed to low perfusion (media movement over the surface of the lymph node). Nonetheless, many more 5C.C7 T-cells interacted with antigen-pulsed DC compared to unpulsed DC. The antigen-specific interactions appeared to be stable in that T-cells and DC formed extensive interfaces that excluded CD43, a characteristic of in vitro synapse. Naive 5C.C7 T-cells expressing CD43-GFP were prepared by retroviral transduction of bone marrow stem cells followed by reconstitution of irradiate mice. A remarkable transformation took place at 36 hours when the activated antigen-specific blasts initiated rapid migration. This 36-hour period took place in vivo, not in the organ culture; nonetheless, the imaging conditions were the same so it is apparent that recently activated blasts "break" synapse-like interactions after 36 hours when they initiate rapid migration under conditions where naive T-cells were incapable of migration. Subsequent encounters with antigen-positive DC did not lead to synapses. This study suggested that CD4 T-cells remain in synapse until they start to proliferate such that all communication prior to the first division is through synapses and all communication after the first division takes place through kinapses.

A series of studies by Miller et al using two-photon laser-scanning microscopy (TPLSM) from 2002 to 2004[13,99-101] shed light on repertoire scanning and priming in a similar hybrid in vivo/explant system with some modifications compared to Stoll et al.[98] For information on TPLSM, which allows imaging of cells at depths of up to 500 μm in tissues, I refer the reader to recent reviews.[102-104] Briefly, TPLSM uses pulsed infrared light to excite fluorophores with two photons of red light, a nonlinear process that can be tuned to provide excitation in a tightly defined focal point of a laser within the tissues followed by wide-field detection of all photons emitted by the tissue. The advantage of infrared light is that it penetrates tissues better than visible light due to lower scattering, except in the wavelength range around 950 nm were infrared light is absorbed strongly by water. Since fluorescence is excited only in the focal volume bleaching above and below the focal plane in which the laser beam is scanned, photobleaching is minimized and no pin-hole is needed for detection. In fact, the photomultiplier should be very close to the objective to collect scattered photons in the appropriate wavelength range. Two-photon excitation spectra tend to be broad, so a single laser wavelength can be used to excite three fluorescent proteins and all quantum dots colors simultaneously. The limitation then becomes the number of photomultiplier tubes (PMTs), which is limited to two or three on most commercial systems. The ideal geometry of having the PMTs very close to the objective means that tradeoffs in depth are required to gain more colors. Figure 2 shows a spectral unmixing strategy to use two PMTs to identify cells expressing CFP, GFP and YFP, with a second channel for a quantum dot or red organic dye (e.g., Texas red™). Despite these advantages, great care in optimizing excitation power and collection is essential since deep-tissue imaging with brightly labeled cells is typically performed near the damage threshold for the tissue and the line between optimal imaging and tissue damage is easily crossed in a 2–3 fold range.[104] Adoptively transferred DO11.10 TCR Tg T-cells were activated with ovalbumin in alum.[13,99] Alum is an adjuvant that primes Th2 responses perhaps due to the activation of IL-4 producing APC.[105] Unlike the Stoll et al system where the node was immobilized by implanting it in a drop of agarose and imaging it from the immobilized side,[98] Miller et al attached the lymph node to the bottom surface of a dish via the hillus and then imaged from above with a continuous perfusion of highly oxygenated media.[99] These conditions and perfusion may be the key difference, resulted in the first observations of dramatic motility of naive T-cells in the lymph node, which is similar to that observed by intravital microscopy,[100] and established the

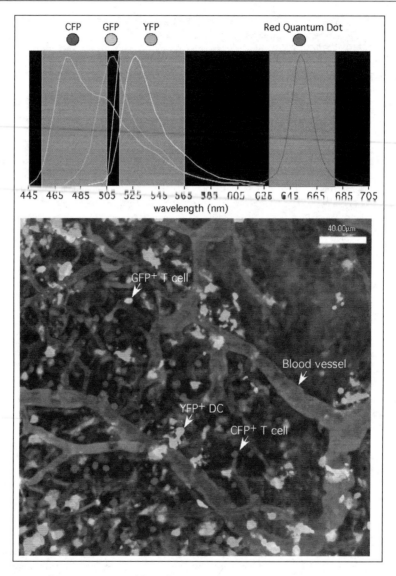

Figure 2. Two-photon excitation of three fluorescent proteins and intravascular quantum dots in the inguinal lymph node. The emission spectra of enhanced cyan fluorescent protein (ECFP), enhanced green fluorescent protein (EGFP), enhanced yellow fluorescent protein (EYFP) and red quantum dots, and the band-pass limits of the filters used for emitted light collection (grey). Images utilizing this approach can be found in Shakhar et al.[14]

current paradigm for stochastic repertoire scanning,[101] In studies on priming purified DO11.10, T-cells were labeled with CMTMR and adoptively transferred i.v., while DC were labeled in situ by subcutaneous injection of CFSE with the alum/antigen mixture. Therefore, all DC that were present at the injection site and them migrated to the lymph node would be labeled. Resident DC in the lymph node that were exposed to antigen draining via the lymph or immigrant DC that entered the tissue hours after the injection would not be labeled because the reactive dye would be hydrolyzed in a few minutes. It is not clear how effective the alum is at preventing drainage of soluble antigen to the lymph node ahead of immigrant DC, which may have an impact on early priming.[106] Nonetheless, this was a highly effective method for examining T-cell interactions with DC that migrated from the injection site.

The fundamental findings from this series of experiments were: (1) T-cells move rapidly and, to a first approximation, randomly in the T-cell zones,[99] (2) similar movement of T-cells is observed in vivo using intravital microscopy of the inguinal lymph node,[100] (3) DC move slowly, but contact many T-cells by probing with many veil-like processes such that each DC can contact up to 5000 T-cell per hour,[101] and (4) antigen-specific interactions can be more dynamic that those observed by Stoll et al[98] with stable interaction between 3-16 hours, with resumption of antigen-specific swarming and rapid migration by 16-24 hours and beyond.[13] Even at the most stable phases of interaction, the apparent contact area size and the changes over time suggested a more dynamic situation than in vitro synapse; however, no molecular imaging was performed. The same model was also used to study oral tolerance in mesenteric and peripheral lymph nodes.[23] In conditions of priming and tolerance, T-cell clusters were formed that were similar to those in Miller et al. 2004.[13] Under conditions of priming, the clusters were larger and longer-lived, but synapses were observed in both priming and tolerance induction. The prevalence of a spectrum of interaction types from rapid migration to swarming in addition to stable interactions suggest that kinapses also have a role in in vivo priming and tolerance.

Steady-state DC have been visualized in vivo using CD11c promoter YFP transgenic (CD11c-YFP) mice.[17] Lymph node DCs include immigrants from the tissues and resident cells that may enter directly from the blood. While tissue inflammation stimulates synchronous migration of many DC to the draining lymph node, there is a poorly understood steady-state migration of DC from the tissues to the lymph nodes that is likely to be critical for peripheral tolerance to tissue antigens.[107] These immigrant DC are mature in that they express high levels of MHC class II on their surfaces, but they are not activated since they only express low to intermediate levels of CD80 and CD86. The CD11c promoter was used to generate a number of transgenic strains that were tested for bright enough fluorescence to be useful in two-photon intravital microscopy in up to 300 μm depth. One founder of several tested had high expression in CD11c^{hi} myeloid DC. In the T-cell zones of inguinal lymph nodes of live mice, steady-state DC form extensive sessile networks.[17] Distinct DC behaviors were described in the subcapular sinus (migrating DC), superficial surface of B-cell follicles (layer of dim stationary DC), interfollicular zones (clusters of DC trapping T and B-cells) and T-cell zones (DC networks). Immigrant DC migrated from the subcapular sinus to the T-cell zones rapidly and join the T-cell zone networks.[17] Thus, rapid T-cell migration through the DC networks and DC outreach to passing T-cells through formation of long membrane processes are the mechanisms that drive repertoire scanning.

CD4$^+$ T-cell priming and tolerance was studied in vivo using the CD11c-YFP Tg mice and a strategy for targeting antigen to DC by attaching antigenic peptides to mAb specific for the scavenger receptor DEC-205.[14,94] Under these conditions all of the DEC-205 positive DC, which include CD11c high and low DC types, will present antigen. In the absence of innate immune stimulation, presentation of antigen to three different MHC class II-restricted TCR tested thus far (3A9/HEL, 2D2/MOG, OTII/OVA) leads to tolerance.[14,94,95] Tolerance involves initial activation and expansion of T-cells followed by death of the expanded cells between days three and seven after initial activation. Induction of tolerance requires both the early exposure to antigen, which induces activation and proliferation and late exposure to antigen after day three to induce deletion (M. Nussenzweig, personal communication).

To perform these experiments, OTII TCR Tg/chicken β actin promoter-GFP (GFP) Tg T-cells (specific) and non-TCR Tg/CFP Tg T-cells (nonspecific) were transferred into CD11c-YFP Tg hosts that had been injected four hours prior with DEC-205-Ova peptide such that the specific and nonspecific T-cells enter lymph nodes that already contain antigen-presenting DC. Blood flow in the intravital inguinal lymph node preparation was followed using red quantum dots, which excite well at the same TPLSM wavelength, 900-920 nm. Tolerance was induced with the DEC-205-Ova alone and priming was induced by addition of anti-CD40 mAb to induce activation of DC. This study is distinguished from other studies on CD4[+] T-cells and other studies on tolerance by using intravital microscopy in which blood and lymph flow are intact. Imaging was initiated in three time frames from 0-6 hours, 6-12 hours and 12-18 hours after T-cell transfer. Specific T-cells showed rapid arrest near high endothelial venules (HEV) within an hour of injection, regardless of whether the conditions favored tolerance or priming. Over the next 18 hours, the specific T-cell regained rapid motility in the DC networks and did not systemically form stable interactions at later times. Some statistically significant differences were detected between tolerance and priming in that under conditions of priming, the rate of return to control migration velocity was slower than for tolerizing conditions. Since activation and proliferation are induced under both tolerizing and priming conditions, there was a positive correlation between synapses and proliferation. It appeared that critical signals for tolerance were integrated during later, T-DC kinapse formation. In fact, continued interaction of T-cells with antigen-positive DC appears to be critical for full expansion of CD4[+] T-cells.[108] Since all three studies on CD4[+] T-cells, Stoll et al,[98,] Miller et al[13] and Shakhar et al[14] concur that T-DC interactions are dynamic after proliferation is initiated, it seems very likely that both synapses and kinapses play a key role in CD4[+] T-cells priming and tolerance.

In Vivo Analysis of CD8[+] T-Cell Priming and Tolerance Induction

The priming of CD8 T-cells has been shown to require a short time of interaction with antigen-presenting cells followed by a long antigen-independent expansion process.[109-111] As mentioned in the previous section, this is very different from the current understanding of antigen requirements for CD4[+] T-cell expansion.[108,112] However, induction of antigen-dependent tolerance of CD8[+] T-cells appears to require sustained contact with antigen-positive DC after 72 hours.[19] Thus, while priming of CD8[+] T-cell responses may be imprinted by early interactions with mature, activated DC, the peripheral deletion of auto-reactive CD8[+] T-cells appears not to be imprinted, but to require sustained interactions with DC for longer than three days. These biological issues need to be considered in interpretations of the imaging data.

Bousso and Robey studied the priming of LCMV gp33 specific P14 TCR Tg T-cells using transferred DCs[113] and a similar explanted lymph node imaging approach as Miller et al.[99] They focused on one time point that was 24 hours after transfer. It is assumed that DC migrated rapidly to the lymph node since they were injected into very proximal subcutaneous sites. Bousso et al[113] provided the first estimate of DC repertoire scanning rate at 500 T-cells per DC per hour. This is 10-fold lower than the subsequent estimate by Miller et al of 5000 T-cells per DC per hour and the major difference may be in the visualization of fine dendritic processes that greatly increase the effective target size of the DC in the Miller/Cahalan study.[101] This may be an imaging issue or an issue related to handling of the DC altering their morphology. Like Stoll et al,[98] Bousso et al[113] emphasize the formation of synapses, but only at one (24 hour) time point. A unique aspect of this study is that it is the only intact lymph node study where antigen dose was varied. They found that synapses were formed over the entire range that would lead to priming.

The first kinetic study of T-cell-DC interaction in vivo was Mempel et al[114] performed extensive studies with the P14 line also used before by Bousso et al.[113] Unlike Bousso, Mempel et al imaged the popliteal lymph node of live anesthetized mice with intact blood and lymph flow. While studies were performed with the DO11.10 CD4[+] T-cells, this data set was very limited compared to the more complete study by Miller et al.[13] Mempel et al used LPS, treated, mature CMTMR-labeled DC that were injected into the foot pad, which then drained uniquely to the ipsilateral popliteal lymph node. They injected CFSE labeled P14 T-cells, allowed the cells an hour to enter the lymph node via the HEV and then injected mAb to L-selectin throughout the rest of the experiment to block further entry. This ensured that the one temporally defined cohort of T-cells is followed. Mempel et al[114] is the first study to clearly demonstrate three phases of interaction during priming. Antigen-specific T-cells began to encounter DC soon after entering the lymph node, but for the first eight hours the encounters were short-lived kinapses. Early kinapse formation activated the T-cells since CD69 was upregulated in this time frame, referred to as phase 1. In the 8-12 hour time period synapses formed. After this synapse period, referred to as phase 2, cytokine production was initiated. At 24-26 hours there was a mixed picture of synapses and kinases and by 44-48 hours all interactions were kinapses. This period, referred to as phase 3, was characterized by proliferation. Thus, although it was found that CD8+ T-cells only require a few hours of stimulation to fully commit to many rounds of cells division, the in vivo profile of T-DC interaction is very similar to what was reported subsequently for CD4+ T-cells, which integrate signals over longer periods. It is possible that CD8+ T-cells in vitro immediately form synapses with antigen-rich DC and integrate signals quickly to commit fully to an effector program. In contrast, the sparse antigen-positive DC in vivo may require a longer period of signal integration requiring phase 1 and 2 before becoming antigen independent in phase 3. More work would be required to determine if CD8+ T-cells do or do not integrate meaningful signals through kinapses in phase 3.

Hugues et al[21] studied tolerance versus priming of CD8+ OTI TCR Tg T-cells using an explant system very similar to that of Miller et al.[99] They used the DEC-205 antigen delivery approach after CFSE labeled OTI T-cells were transferred and labeled all DC in the explanted lymph nodes by injecting fluorescently labeled anti-CD11c IgG into the parenchyma prior to imaging. Thus, unlike Shakhar et al[14] Hugues et al[21] delivered antigen to DC after T-cells were equilibrated in the lymph nodes, rather than injecting T-cells into mice that had been equilibrated with the antigen and performed the imaging with explanted lymph nodes in which DC had CD11c and perhaps FcR engaged by anti-CD11c mAb, rather than performing intravital microscopy in a mouse in which DC express YFP. In this study, OTI T-cells were found to engage in a spectrum of interactions with DC that was biased toward synapses in priming conditions and kinapses in conditions of tolerance. It is not clear if the differences between Hugues et al[21] and Shakhar et al[14] are due to differences between CD4+ and CD8+ systems or due to technical differences in the way the experiments were performed. In Shakhar et al[14] the area around the HEV played a key role in early synapses and it might be argued this region may not function in the same way in the absence of blood flow. Zinselmeyer et al[23] used a totally different antigen-delivery route and also found that both priming and tolerance involved stable, IS-like T-cell migration patterns. It will be important to revisit the issue of T-DC dynamics during CD8+ T-cell tolerance induction in light of these discrepancies.

Priming vs. Tolerance

Priming and tolerance require T-cell activation. Tolerance induction by deletion of effector cells appears to require prolonged TCR signaling beyond three days. Thus far, activation is associated with synapses in 9 of 10 data sets. Only one data set on tolerance induction is associated with activation without stable, IS-like interactions and this is contradicted by two other studies that readily detected these interactions under conditions of tolerance induction following primary activation. In most cases the formation of synapses is highly correlated to T-cell activation, but not to the generation of effector cells or memory, which is only correlated with more subtle changes in the dynamics of interaction. CD4+ T-cells and CD8+ T-cells undergoing tolerance induction all integrate signals at a later time (>24 hours) and all show rapid migration with short duration interactions with DC in this time frame. Thus, kinapse signal integration is likely to be critical for full activation of CD4+ T-cells and tolerance induction and CD4+ and CD8+ T-cells.

Dynamics of CD4+ T-Cell Help for CD8+ T-Cell Responses

Germain and colleagues[28] have examined the dynamics of CD4 T-cell help under conditions where the delivery of help and direct stimulation via MHC class I peptide complexes was independently controlled. A key result from this study was that early CD8+ T-cell stimulation by antigen or innate signals upregulates CCR5 and appears to render the baseline movement of CD8+ T-cell dependent upon expression of CCR5. DC that have interacted with CD4+ T-cells produce CCR5 ligands and attract these partially activated CD8+ T-cells to make additional contacts with these licensed DCs. This increases the probability that CD8+ T-cells will interact with the licensed DC and improves production of memory CD8+ T-cells. Whether other inflammatory chemokine receptors are used by CD8+ T-cells is not known, but is likely. This study suggests an explanation for earlier observations of "swarming" in which particular DC appeared to act as local attractors for subsets of T-cells.

Effector Sites

At the time of writing of this chapter, there are only a few papers on intravital microscopy of T-cell-APC interactions at effector sites. While there are a number of other studies that report on various lymphoid cells in the intestine and in tumors, I have not included these because there was no concept of antigen-specific interaction; rather, the movement of cell types was documented often with only a limited understanding of the cell type involved. The primary studies that I will discuss have focused on natural killer T (NKT) cells in the liver,[34] helper T-cells in the lymph node,[29] memory T-cells in the bone marrow,[32] and activated effector cells in the central nervous system (CNS).[115] These studies all suggest that effector cells form synapses with antigen-positive APC.

Geissmann et al[34] took advantage of the expression pattern of the chemokine receptor CXCR6 in the liver to follow NKT-cells in vivo. CXCR6 is highly expressed on activated T-cells and NKT-cells. In the liver of healthy mice, NKT-cells represent 30% of the mononuclear cells and 70% of the CXCR6+ cells. Unutmaz et al[116] replaced the major coding exon of CXCR6 with GFP by homologous recombination. The CXCR6$^{gfp/+}$ mice have a normal number of NKT-cells in the liver, which are GFP^h, whereas the CXCR6$^{gfp/gfp}$ mice have 3-5 fold reduced numbers of NKT-cells in the liver, which remain GFP^{hi}.[34] The reason for the reduced numbers of NKT-cells in the CXCR6$^{gfp/gfp}$ mice appears to be reduced NKT-cell survival in the absence of CXCR6. The high levels of GFP expressed in the NKT-cells afforded the opportunity to track these cells in the liver of mice by intravital microscopy and to determine the effect of antigen. It had been reported that a population of leukocytes identified as Kupffer cells migrated within the sinusoids of the liver.[117] Geissmann et al[34] discovered that Kupffer cells stained by high molecular weight rhodamine dextran are stationary cells, while NKT-cells migrate rapidly within the sinusoids with or against blood flow. This rapid migration within the fenestrated sinusoids allowed WT NKT-cells to visit each hepatocyte in the liver on average every 15 minutes. Geissmann et al[34] never observed GFP+ cells extravasating. NKT-cells have a dominant Vα14 rearrangement that produces a high affinity TCR for CD1d with α-galactosylceramide (AGC), which corresponds structurally to a class of bacterial lipids.[118,119] When AGC was injected i.v., most of the NKT-cells become activated within two hours based on cytokine production and 60% of the GFP+ cells stopped migrating and formed synapses within 20 minutes,[34] apparently with Kupffer cells. NKT-cells have a previously activated memory/effector phenotype in vivo, such that these observations provide the first evidence of in vivo synapses in the effector phase of an immune response. We have also found that conventional effector T-cells patrol liver sinusoids. This suggests that patrolling of sinusoids is typical of immune surveillance by activated T-cells in the liver. By reducing the dose of AGC we have recently found that NKT-cells can be induced to produce cytokines without reduction in patrolling migration (P. Velazquez and MLD, unpublished observations). This suggests that NKT-cells are flexible in using synapses or kinapses for signal integration, depending upon the strength of signal.

Okada et al[29] studied the T-B interactions during recognition of hen egg white lyozyme (HEL) or I-A^b-HEL 74-88 complexes by MD4 IgG Tg B-cells and TCR7 TCR Tg T-cells, respectively. The mice were immunized with HEL in alum and explanted lymph nodes were imaged exactly as in Miller et al 2004.[13] T-cell proliferation was extensive by day two, whereas B-cells did not proliferate extensively until day three. The B-cells slowed down in the first 1-3 hours and then again regained speed and migrated toward the junction between the T zone-B zone interface. This migration was guided by CCL21 gradients in the follicle and required CCR7 expression on the B-cells. This was the first demonstration of chemokine directed migration in an intact lymph node. By 30 hours the B and T-cells are activated and long-lived antigen specific T-B conjugates begin to form at the T zone-B zone interface. The interactions are strikingly different than priming T-DC interactions in that the B-cells continue to migrate rapidly and drag the T-cells behind them. This could be explained by the T-cell forming a synapse with the B-cell via the B-cell uropod, but the B-cell continuing to migrate under the influence of chemokinetic and chemotactic factors in the environment. This is also different from an earlier study in which activated T-cells interacted with naive B-cells as viewed by conventional confocal microscopy in the inguinal lymph node.[120] In this case, the activated T-cells pushed the B-cells, which rounded up and became sessile after T-cell contact. While Gunzer et al[120] opens an interesting physical possibility that the lateral surfaces of a T-cell can function in migration while the front part forms an IS, it is less physiological than Okada et al[29] which studied a classical T-cell dependent antibody response. Again, effector cells form synapses rapidly after encountering APC.

von Andrian and colleagues[32] studied memory T-cells in the bone marrow and their rapid antigen-specific interactions with DC. They found that DC travel to the bone marrow and interact with central memory T-cells in an unexpected migration route for both of these cell types. This helps explain the well-known challenges of mature T-cells in bone marrow transplantation. Central memory T-cells and DC migrate randomly in the bone marrow cavity in the steady state. Introduction of antigenic peptide pulsed mature DCs and P14 central memory T-cells followed by imaging in the bone marrow cavity revealed rapid formation of synapses between the central memory T-cells and DC. This mechanism may help protect the bone marrow from infection and neoplastic events in the highly transformation sensitive hematopoetic system.

Kawakami et al[115] studied the dynamics of T-cells in acute spinal cord sliced from rats with early experimental allergic encephalitis (EAE) lesions. The EAE lesions were induced by injecting 5 million GFP+ myelin basic protein (MBP) specific cloned T-cells i.v. Acute spinal cord slices were prepared four days after T-cell transfer, at which time antigen-specific and non-antigen-specific ovalbumin-specific cloned T-cells were also found in the lesions. This is characteristic of inflamed sites, which are equally attractive for extravasation of antigen-specific and nonspecific cells. Tracking of MBP-specific T-cells revealed that 40% were immobile over long periods, whereas only 5% of ovalbumin specific cells were similarly stationary. The remaining cells were migrating in the living brain tissue. Staining of the live sections in real time with antibodies to LFA-1, TCR and MHC class II revealed polarization of these molecules toward the shared interface between MBP specific T-cells and class II positive APC, but rarely with Ova-specific T-cells. These studies provide evidence from positional stability and molecular composition for in vivo synapses. Prior histological analysis in lymph nodes,[121] the meninges during the CD8+ T-cell response to lymphocytic chriomeningitis virus,[122] and recent studies in adenovirus infection with thick section confocal imaging[123] have detected molecular signatures of synapses in fixed tissue with increasing acuity, but Kawakami et al[115] was the first study to be able to reference this staining to the dynamics of interaction in ex vivo live tissue. While the use of intact antibodies could be criticized—for example, Fab fragments with nonblocking Ab would be a better choice—this study provided a strong step toward imaging of SMACs in vivo.

Innate immune surveillance of the CNS was studied by Davalos et al[33] and Nimmerjahn et al[124] Microglial cells are the innate immune cells of the central nervous system in the steady state. These cells can be visualized in live mice in which the CX3CR1 chemokine receptor's major coding exon is replaced by eGFP. Microglial cells are the only cells in the CNS that express GFP in CX3CR1+/gfp mice. Both of these studies used the thinned skull technology developed by Gan and colleagues[125] to image microglial cells in the intact brain of anesthetized mice. Parenchymal microglial cells are referred to as "ramified" because each cell projects many long processes in three dimensions to cover a territory of 65,000 μm.[3] The processes extend and retract at a velocity of 1-2 μm/min. Focal injury in the CNS results in a rapid response by the ramified microglial cells in which the cell bodies remain in place, but all the processes within a radius of 75 μm from the site of injury, dozens of individual processes, converge on the site while maintaining tethers to the cell bodies. While the biological function of this response is not known with certainty, the release ATP is necessary and sufficient to trigger the response and a reasonable biological hypothesis is that the cells are sealing off local injuries within 30 minutes. This does not appear to be a classical phagocytic response and in fact, classical phagocytes like neutrophils and monocytes appear to be excluded from small injuries that are completely surrounded by microglial foot processes.[126]

Whether microglial cells that respond to focal injuries are competent to present antigen to CD4⁺ or CD8⁺ T-cells is not known.

Summary

Recent evidence suggests that the basic unit of sustained T-cell receptor signaling are TCR microclusters dynamically generated near the leading edge of migrating T-cells forming kinapses or in the dSMAC of T-cells forming synapses. Retention of many TCR in the cSMAC is only possible when the T-cell stays with one APC, which requires a stop signal, but even then the continuous formation of peripheral TCR microclusters is required to sustain signaling. The specific function of the cSMAC is unknown, but it is not sufficient to sustain Ca^{2+} signals in T-cells. In vivo, activation of T-cells associated with activation and tolerance requires both synapses and kinapses. Synapses are a common feature of effector phase and these interactions are initiated rapidly after antigen recognition.

Acknowledgements

I thank R. Varma, G. Campi, T. Sims, T. Cameron, P. Velazquez J. Kim, G. Shakhar and M. Nussenzweig for valuable discussions and permission to discuss unpublished data.

References

1. Banchereau J and Steinman RM. Dendritic cells and the control of immunity. Nature 1998; 392:245-252.
2. Steinman RM, Hawiger D and Nussenzweig MC. Tolerogenic dendritic cells. Annu Rev Immunol 2003; 21:685-711.
3. Dustin ML, Olszowy MW, Holdorf AD et al. A novel adapter protein orchestrates receptor patterning and cytoskeletal polarity in T-cell contacts. Cell 1998; 94:667-677.
4. Monks CR, Freiberg BA, Kupfer H et al. Three-dimensional segregation of supramolecular activation clusters in T-cells. Nature 1998; 395:82-86.
5. Grakoui A, Bromley SK, Sumen C et al. The immunological synapse: A molecular machine controlling T-cell activation. Science 1999; 285:221-227.
6. Stinchcombe JC, Bossi G, Booth S et al. The immunological synapse of CTL contains a secretory domain and membrane bridges. Immunity 2001; 15:751-761.
7. Dustin ML and Colman DR. Neural and immunological synaptic relations. Science 2002; 298:785-789.
8. Ludford-Menting MJ, Oliaro J, Sacirbegovic F et al. A network of PDZ-containing proteins regulates T-cell polarity and morphology during migration and immunological synapse formation. Immunity 2005; 22:737-748.
9. Varma R, Campi G, Yokosuka T et al. T-cell receptor-proximal signals are sustained in peripheral microclusters and terminated in the central supramolecular activation cluster. Immunity 2006; 25:117-127.
10. Yokosuka T, Sakata-Sogawa K, Kobayashi W et al. Newly generated T-cell receptor microclusters initiate and sustain T-cell activation by recruitment of Zap70 and SLP-76. Nat Immunol 2005; 6:1253-1262.
11. Campi G, Varma R and Dustin ML. Actin and agonist MHC-peptide complex-dependent T-cell receptor microclusters as scaffolds for signaling. J Exp Med 2005; 202:1031-1036.
12. Mempel TR, Scimone ML, Mora JR et al. In vivo imaging of leukocyte trafficking in blood vessels and tissues. Curr Opin Immunol 2004; 16:406-417.
13. Miller MJ, Safrina O, Parker I et al. Imaging the single cell dynamics of CD4⁺ T-cell activation by dendritic cells in lymph nodes. J Exp Med 2004; 200:847-856.
14. Shakhar G, Lindquist RL, Skokos D et al. Stable T-cell-dendritic cell interactions precede the development of both tolerance and immunity in vivo. Nat Immunol 2005; 6:707-714.
15. Friedl P and Brocker EB. TCR triggering on the move: Diversity of T-cell interactions with antigen-presenting cells. Immunol Rev 2002; 186:83-89.
16. Randolph GJ. Dendritic cell migration to lymph nodes: Cytokines, chemokines and lipid mediators. Semin Immunol 2001; 13:267-274.
17. Lindquist RL, Shakhar G, Dudziak D et al. Visualizing dendritic cell networks in vivo. Nat Immunol 2004; 5:1243-1250.
18. Fujii S, Liu K, Smith C et al. The linkage of innate to adaptive immunity via maturing dendritic cells in vivo requires CD40 ligation in addition to antigen presentation and CD80/86 costimulation. J Exp Med 2004; 199:1607-1618.

19. Redmond WL, Hernandez J and Sherman LA. Deletion of naive CD8 T-cells requires persistent antigen and is not programmed by an initial signal from the tolerogenic APC. J Immunol 2003; 171:6349-6354.
20. Benvenuti F, Lagaudriere-Gesbert C, Grandjean I et al. Dendritic cell maturation controls adhesion, synapse formation and the duration of the interactions with naive T-lymphocytes. J Immunol 2004; 172:292-301.
21. Hugues S, Fetler L, Bonifaz L et al. Distinct T-cell dynamics in lymph nodes during the induction of tolerance and immunity. Nat Immunol 2004; 5:1235-1242.
22. Bromley SK, Iaboni A, Davis SJ et al. The immunological synapse and CD28-CD80 interactions. Nat Immunol 2001; 2:1159-1166.
23. Zinselmeyer BH, Dempster J, Gurney AM et al. In situ characterization of CD4⁺ T-cell behavior in mucosal and systemic lymphoid tissues during the induction of oral priming and tolerance. J Exp Med 2005; 201:1815-1823.
24. Lanzavecchia A and Sallusto F. Antigen decoding by T-lymphocytes: From synapses to fate determination. Nat Immunol 2001; 2:487-492.
25. Matloubian M, Lo CG, Cinamon G et al. Lymphocyte egress from thymus and peripheral lymphoid organs is dependent on S1P receptor 1. Nature 2004; 427:355-360.
26. Schwab SR, Pereira JP, Matloubian M et al. Lymphocyte sequestration through S1P lyase inhibition and disruption of S1P gradients. Science 2005; 309:1735-1739.
27. Lo CG, Xu Y, Proia RL et al. Cyclical modulation of sphingosine-1-phosphate receptor 1 surface expression during lymphocyte recirculation and relationship to lymphoid organ transit. J Exp Med 2005; 201:291-301.
28. Castellino F, Huang AY, Altan-Bonnet G et al. Chemokines enhance immunity by guiding naive CD8⁺ T-cells to sites of CD4⁺ T-cell-dendritic cell interaction. Nature 2006; 440:890-895.
29. Okada T, Miller MJ, Parker I et al. Antigen-engaged B-cells undergo chemotaxis toward the T zone and form motile conjugates with helper T-cells. PLoS Biol 2005; 3:e.150
30. Mempel TR, Pittet MJ, Khazaie K et al. Regulatory T-cells reversibly suppress cytotoxic T-cell function independent of effector differentiation. Immunity 2006; 25:129-141.
31. Dustin ML and Chan AC. Signaling takes shape in the immune system. Cell 2000; 103:283-294.
32. Cavanagh LL, Bonasio R, Mazo IB et al. Activation of bone marrow-resident memory T-cells by circulating, antigen-bearing dendritic cells. Nat Immunol 2005; 6:1029-1037.
33. Davalos D, Grutzendler J, Yang G et al. ATP mediates rapid microglial response to local brain injury in vivo. Nat Neurosci 2005; 8:752-758.
34. Geissmann F, Cameron TO, Sidobre S et al. Intravascular immune surveillance by CXCR6+ NKT-cells patrolling liver sinusoids. PLoS Biol 2005; 3:e.113
35. Weiss A and Littman DR. Signal transduction by lymphocyte antigen receptors. Cell 1994; 76:263-274.
36. Bachmann MF, McKall-Faienza K, Schmits R et al. Distinct roles for LFA-1 and CD28 during activation of naive T-cells: adhesion versus costimulation. Immunity 1997; 7:549-557.
37. Somersalo K, Anikeeva N, Sims TN et al. Cytotoxic T-lymphocytes form an antigen-independent ring junction. J Clin Invest 2004; 113:49-57.
38. Dustin ML, Bivona TG and Philips MR. Membranes as messengers in T-cell adhesion signaling. Nat Immunol 2004; 5:363-372.
39. Egen JG, Kuhns MS and Allison JP. CTLA-4: new insights into its biological function and use in tumor immunotherapy. Nat Immunol 2002; 3:611-618.
40. Okazaki T, Iwai Y and Honjo T. New regulatory coreceptors: Inducible costimulator and PD-1. Curr Opin Immunol 2002; 14:779-782.
41. Young JD, Cohn ZA and Podack ER. The ninth component of complement and the pore-forming protein (perforin 1) from cytotoxic T-cells: Structural, immunological and functional similarities. Science 1986; 233:184-190.
42. Darmon AJ, Nicholson DW and Bleackley RC. Activation of the apoptotic protease CPP32 by cytotoxic T-cell-derived granzyme B. Nature 1995; 377:446-448.
43. Keefe D, Shi L, Feske S et al. Perforin triggers a plasma membrane-repair response that facilitates CTL induction of apoptosis. Immunity 2005; 23:249-262.
44. Geiger B, Rosen D and Berke G. Spatial relationships of microtubule-organizing centers and the contact area of cytotoxic T-lymphocytes and targeT-cells. J Cell Biol 1982; 95:137-143.
45. Kupfer A, Dennert G and Singer SJ. Polarization of the Golgi apparatus and the microtubule-organizing center within cloned natural killer cells bound to their targets. Proc Natl Acad Sci USA 1983; 80:7224-7228.

46. Kupfer A and Singer SJ. The specific interaction of helper T-cells and antigen-presenting B-cells. IV. Membrane and cytoskeletal reorganizations in the bound T-cell as a function of antigen dose. J Exp Med 1989; 170:1697-1713.
47. Kupfer A, Swain SL, Janeway Jr CA et al. The specific direct interaction of helper T-cells and antigen-presenting B-cells. Proc Natl Acad Sci USA 1986; 83:6080-6083.
48. Kupfer A, Singer SJ, Janeway Jr CA et al. Coclustering of CD4 (L3T4) molecule with the T-cell receptor is induced by specific direct interaction of helper T-cells and antigen-presenting cells. Proc Natl Acad Sci USA 1987; 84:5888-5892.
49. Kupfer A, Mosmann TR and Kupfer H. Polarized expression of cytokines in cell conjugates of helper T-cells and splenic B-cells. Proc Natl Acad Sci USA 1991; 88:775-779.
50. Monks CR, Kupfer H, Tamir I et al. Selective modulation of protein kinase C-theta during T-cell activation. Nature 1997; 385:83-86.
51. Springer TA. Adhesion receptors of the immune system. Nature 1990; 346:425-434.
52. Norcross MA. A synaptic basis for T-lymphocyte activation. Ann Immunol (Paris) 1984; 135D:113-134.
53. Paul WE and Seder RA. Lymphocyte responses and cytokines. Cell 1994; 76:241-251.
54. McConnell HM, Watts TH, Weis RM et al. Supported planar membranes in studies of cell-cell recognition in the immune system. Biochim Biophys Acta 1986; 864:95-106.
55. Groves JT and Dustin ML. Supported planar bilayers in studies on immune cell adhesion and communication. J Immunol Methods 2003; 278:19-32.
56. Krummel MF, Sjaastad MD, Wulfing C et al. Differential clustering of CD4 and CD3z during T-cell recognition. Science 2000; 289:1349-1352.
57. Lee KH, Holdorf AD, Dustin ML et al. T-cell receptor signaling precedes immunological synapse formation. Science 2002; 295:1539-1542.
58. Wülfing C, Sumen C, Sjaastad MD et al. Costimulation and endogenous MHC ligands contribute to T-cell recognition. Nat Immunol 2002; 3:42-47.
59. Egen JG and Allison JP. Cytotoxic T-lymphocyte antigen-4 accumulation in the immunological synapse is regulated by TCR signal strength. Immunity 2002; 16:23-35.
60. Andres PG, Howland KC, Dresnek D et al. CD28 signals in the immature immunological synapse. J Immunol 2004; 172:5880-5886.
61. Irvine DJ, Purbhoo MA, Krogsgaard M et al. Direct observation of ligand recognition by T-cells. Nature 2002; 419:845-849.
62. Freiberg BA, Kupfer H, Maslanik et al. Staging and resetting T-cell activation in SMACs. Nat Immunol 2002; 3:911-917.
63. Lee KH, Dinner AR, Tu C et al. The immunological synapse balances T-cell receptor signaling and degradation. Science 2003; 302:1218-1222.
64. Huppa JB, Gleimer M, Sumen C et al. Continuous T-cell receptor signaling required for synapse maintenance and full effector potential. Nat Immunol 2003; 4:749-755.
65. Purtic B, Pitcher LA, Van Oers NS et al. T-cell receptor (TCR) clustering in the immunological synapse integrates TCR and costimulatory signaling in selected T-cells. Proc Natl Acad Sci USA 2005; 102:2904-2909.
66. Axelrod D. Total internal reflection fluorescence microscopy. Methods Cell Biol 1989; 30:245-270.
67. Douglass AD and Vale RD. Single-Molecule Microscopy Reveals Plasma Membrane Microdomains Created by Protein-Protein Networks that Exclude or Trap Signaling Molecules in T-cells. Cell 2005; 121:937-950.
68. Klopfenstein DR, Tomishige M, Stuurman N et al. Role of phosphatidylinositol(4,5)bisphosphate organization in membrane transport by the Unc104 kinesin motor. Cell 2002; 109:347-358.
69. Schreiber AB, Libermann TA, Lax I et al. Biological role of epidermal growth factor-receptor clustering: investigation with monoclonal anti-receptor antibodies. J Biol Chem 1983; 258:846-853.
70. Fanger BO, Austin KS, Earp HS et al. Cross-linking of epidermal growth factor receptors in intacT-cells: detection of initial stages of receptor clustering and determination of molecular weight of high-affinity receptors. Biochemistry 1986; 25:6414-6420.
71. Erickson J, Kane P, Goldstein B et al. Cross-linking of IgE-receptor complexes at the cell surface: a fluorescence method for studying the binding of monovalent and bivalent haptens to IgE. Mol Immunol 1986; 23:769-781.
72. Germain RN. T-cell signaling: the importance of receptor clustering. Curr Biol 1997; 7:R640-644.
73. Van der Merwe PA, Davis SJ, Shaw AS et al. Cytoskeletal polarization and redistribution of cell-surface molecules during T-cell antigen recognition. Semin Immunol 2000; 12:5-21.
74. Giannone G, Dubin-Thaler BJ, Dobereiner HG et al. Periodic lamellipodial contractions correlate with rearward actin waves. Cell 2004; 116:431-443.

75. Dobereiner HG, Dubin-Thaler BJ, Hofman JM et al. Lateral membrane waves constitute a universal dynamic pattern of motile cells. Phys Rev Lett 2006; 97:038102.
76. Nolz JC, Gomez TS, Zhu P et al. The WAVE2 complex regulates actin cytoskeletal reorganization and CRAC-mediated calcium entry during T-cell activation. Curr Biol 2006; 16:24-34.
77. Gomez TS, McCarney SD, Carrizosa E et al. HS1 Functions as an Essential Actin-Regulatory Adaptor Protein at the Immune Synapse. Immunity 2006; 24:741-752.
78. Johnson KG, Bromley SK, Dustin ML et al. A supramolecular basis for CD45 regulation during T-cell activation. Proc Natl Acad Sci USA 2000; 97:10138-10143.
79. Heissmeyer V, Macian F, Im SH et al. Calcineurin imposes T-cell unresponsiveness through targeted proteolysis of signaling proteins. Nat Immunol 2004; 5:255-265.
80. Sumen C, Dustin ML and Davis MM. T-cell receptor antagonism interferes with MHC clustering and integrin patterning during immunological synapse formation. J Cell Biol 2004; 166:579-590.
81. Smith A, Carrasco YR, Stanley P et al. A talin-dependent LFA-1 focal zone is formed by rapidly migrating T-lymphocytes. J Cell Biol 2005; 170:141-151.
82. Ponti A, Machacek M, Gupton SL et al. Two distinct actin networks drive the protrusion of migrating cells. Science 2004; 305:1782-1786.
83. Gupton SL anderson KL, Kole TP et al. Cell migration without a lamellipodium: Translation of actin dynamics into cell movement mediated by tropomyosin. J Cell Biol 2005; 168:619-631.
84. Dustin ML. Stop and go traffic to tune T-cell responses. Immunity 2004; 21:305-314.
85. Gunzer M, Schafer A, Borgmann S et al. Antigen presentation in extracellular matrix: Interactions of T-cells with dendritic cells are dynamic, short lived and sequential. Immunity 2000; 13:323-332.
86. Onfelt B and Davis DM. Can membrane nanotubes facilitate communication between immune cells? Biochem Soc Trans 2004; 32:676-678.
87. Onfelt B, Nedvetzki S, Yanagi K et al. Cutting edge: Membrane nanotubes connect immune cells. J Immunol 2004; 173:1511-1513.
88. Brossard C, Feuillet V, Schmitt A et al. Multifocal structure of the T-cell—Dendritic cell synapse. Eur J Immunol 2005; 35:1741-1753.
89. Trautmann A. Microclusters initiate and sustain T-cell signaling. Nat Immunol 2005; 6:1213-1214.
90. Tseng SY, Liu M and Dustin ML. CD80 Cytoplasmic Domain Controls Localization of CD28, CTLA-4 and Protein Kinase C{theta} in the Immunological Synapse. J Immunol 2005; 175:7829-7836.
91. Purbhoo MA, Irvine DJ, Huppa JB et al. T-cell killing does not require the formation of a stable mature immunological synapse. Nat Immunol 2004; 5:524-530.
92. Krogsgaard M, Li QJ, Sumen C et al. Agonist/endogenous peptide-MHC heterodimers drive T-cell activation and sensitivity. Nature 2005; 434:238-243.
93. Jung S, Unutmaz D, Wong P et al. In vivo depletion of CD11c(+) dendritic cells abrogates priming of CD8(+) T-cells by exogenous cell-associated antigens. Immunity 2002; 17:211-220.
94. Hawiger D, Inaba K, Dorsett Y et al. Dendritic cells induce peripheral T-cell unresponsiveness under steady state conditions in vivo. J Exp Med 2001; 194:769-779.
95. Hawiger D, Masilamani RF, Bettelli E et al. Immunological unresponsiveness characterized by increased expression of CD5 on peripheral T-cells induced by dendritic cells in vivo. Immunity 2004; 20:695-705.
96. Vidard L, Colarusso LJ and Benacerraf B. Specific T-cell tolerance may be preceded by a primary response. Proc Natl Acad Sci USA 1994; 91:5627-5631.
97. Mucida D, Kutchukhidze N, Erazo A et al. Oral tolerance in the absence of naturally occurring Tregs. J Clin Invest 2005; 115:1923-1933.
98. Stoll S, Delon J, Brotz TM et al. Dynamic imaging of T-cell-dendritic cell interactions in lymph nodes. Science 2002; 296:1873-1876.
99. Miller MJ, Wei SH, Parker I et al. Two-photon imaging of lymphocyte motility and antigen response in intact lymph node. Science 2002; 296:1869-1873.
100. Miller MJ, Wei SH, Cahalan MD et al. Autonomous T-cell trafficking examined in vivo with intravital two-photon microscopy. Proc Natl Acad Sci USA 2003; 100:2604-2609.
101. Miller MJ, Hejazi AS, Wei SH et al. T-cell repertoire scanning is promoted by dynamic dendritic cell behavior and random T-cell motility in the lymph node. Proc Natl Acad Sci USA 2004; 101:998-1003.
102. Zipfel WR, Williams RM and Webb WW. Nonlinear magic: Multiphoton microscopy in the biosciences. Nat Biotechnol 2003; 21:1369-1377.
103. Cahalan MD, Parker I, Wei SH et al. Two-photon tissue imaging: Seeing the immune system in a fresh light. Nat Rev Immunol 2002; 2:872-880.
104. Germain RN, Miller MJ, Dustin ML et al. Dynamic imaging of the immune system: Progress, pitfalls and promise. Nat Rev Immunol 2006; 6:497-507.
105. Jordan MB, Mills DM, Kappler J et al. Promotion of B-cell immune responses via an alum-induced myeloid cell population. Science 2004; 304:1808-1810.

106. Itano AA, McSorley SJ, Reinhardt RL et al. Distinct dendritic cell populations sequentially present antigen to CD4 T-cells and stimulate different aspects of cell-mediated immunity. Immunity 2003; 19:47-57.
107. Ohl L, Mohaupt M, Czeloth N et al. CCR7 governs skin dendritic cell migration under inflammatory and steady-state conditions. Immunity 2004; 21:279-288.
108. Obst R, Van Santen HM, Mathis D et al. Antigen persistence is required throughout the expansion phase of a CD4(+) T-cell response. J Exp Med 2005; 201:1555-1565.
109. Kaech SM and Ahmed R. Memory CD8+ T-cell differentiation: Initial antigen encounter triggers a developmental program in naive cells. Nat Immunol 2001; 2:415-422.
110. Wong P and Pamer EG. Cutting edge: antigen-independent CD8 T-cell proliferation. J Immunol 2001; 166:5864-5868.
111. Van Stipdonk MJ, Hardenberg G, Bijker MS et al. Dynamic programming of CD8$^+$ T-lymphocyte responses. Nat Immunol 2003; 4:361-365.
112. Iezzi G, Karjalainen K and Lanzavecchia A. The duration of antigenic stimulation determines the fate of naive and effector T-cells. Immunity 1998; 8:89-95.
113. Bousso P and Robey E. Dynamics of CD8+ T-cell priming by dendritic cells in intact lymph nodes. Nat Immunol 2003; 4:579-585.
114. Mempel TR, Henrickson SE and Von Andrian UH. T-cell priming by dendritic cells in lymph nodes occurs in three distinct phases. Nature 2004; 427:154-159.
115. Kawakami N, Nagerl UV, Odoardi F et al. Live imaging of effector cell trafficking and autoantigen recognition within the unfolding autoimmune encephalomyelitis lesion. J Exp Med 2005; 201:1805-1814.
116. Unutmaz D, Xiang W, Sunshine MJ et al. The primate lentiviral receptor Bonzo/STRL33 is coordinately regulated with CCR5 and its expression pattern is conserved between human and mouse. J Immunol 2000; 165:3284-3292.
117. MacPhee PJ, Schmidt EE and Groom AC. Evidence for Kupffer cell migration along liver sinusoids, from high-resolution in vivo microscopy. Am J Physiol 1992; 263:G17-23.
118. Mattner J, Debord KL, Ismail N et al. Exogenous and endogenous glycolipid antigens activate NKT-cells during microbial infections. Nature 2005; 434:525-529.
119. Kinjo Y, Wu D, Kim G et al. Recognition of bacterial glycosphingolipids by natural killer T-cells. Nature 2005; 434:520-525.
120. Gunzer M, Weishaupt C, Hillmer A et al. A spectrum of biophysical interaction modes between T-cells and different antigen presenting cells during priming in 3-D collagen and in vivo. Blood 2004; 104:2801-2809.
121. Reichert P, Reinhardt RL, Ingulli E et al. Cutting edge: In vivo identification of TCR redistribution and polarized IL-2 production by naive CD4 T-cells. J Immunol 2001; 166:4278-4281.
122. McGavern DB, Christen U and Oldstone MB. Molecular anatomy of antigen-specific CD8(+) T-cell engagement and synapse formation in vivo. Nat Immunol 2002; 3:918-925.
123. Barcia C, Thomas CE, Curtin JF et al. In vivo mature immunological synapses forming SMACs mediate clearance of virally infected astrocytes from the brain. J Exp Med 2006; 203:2095-2107.
124. Nimmerjahn A, Kirchhoff F and Helmchen F. Resting microglial cells are highly dynamic surveillants of brain parenchyma in vivo. Science 2005; 308:1314-1318.
125. Grutzendler J, Kasthuri N and Gan WB. Long-term dendritic spine stability in the adult cortex. Nature 2002; 420:812-816.
126. Kim JV and Dustin ML. Innate response to focal necrotic injury inside the blood-brain barrier. J Immunol 2006; 177:5269-5277.

CHAPTER 14

Visualization of Protein Interactions in Living Cells

Tomasz Zal*

Abstract

Ligand binding to cell membrane receptors sets off a series of protein interactions that convey the nuances of ligand identity to the cell interior. The information may be encoded in conformational changes, the interaction kinetics and, in the case of multichain immunoreceptors, by chain rearrangements. The signals may be modulated by dynamic compartmentalization of the cell membrane, cellular architecture, motility, and activation—all of which are difficult to reconstitute for studies of receptor signaling in vitro. In this chapter, we will discuss how protein interactions in general and receptor signaling in particular can be studied in living cells by different fluorescence imaging techniques. Particularly versatile are methods that exploit Förster resonance energy transfer (FRET), which is exquisitely sensitive to the nanometer-range proximity and orientation between fluorophores. Fluorescence correlation microscopy (FCM) can provide complementary information about the stoichiometry and diffusion kinetics of large complexes, while bimolecular fluorescence complementation (BiFC) and other complementation techniques can capture transient interactions. A continuing challenge is extracting from the imaging data the quantitative information that is necessary to verify different models of signal transduction.

Introduction

Recognition of extracellular ligands by cell surface receptors depends on membrane compartmentalization, subcellular organization, and whole cell dynamics. Ligand-engaged or free receptors can interact with numerous proteins that co-inhabit the cell membrane, partition in different membrane domains, as well as being targets for intracellular adaptors, effector enzymes, cytoskeleton terminals, and the recycling machinery. Any of these interactions may modulate the activity of receptor components and all are themselves subject to continuous change according to subcellular localization, cell motility, polarity, state of activation, and the extracellular environment. Not surprisingly, the mechanisms of ligand recognition by different receptors can be difficult to understand based on in vitro studies alone and have to be verified in the milieu of the living cell.

Particularly puzzling is the signal transduction by the multichain immunoreceptors that use dedicated chains for ligand binding and a number of noncovalently associated signaling chains to interface with the intracellular effector enzymes. How the information about the quality of binding between the ligand and the extracellular domain is projected by multichain receptors to the cell interior is of great general interest; especially for understanding antigen recognition, cytokine communication, and homeostasis in the immune system and beyond. According to the structural models, binding of a ligand to the extracellular domain induces a range of structural changes that

*Tomasz Zal—Department of Immunology, University of Texas, MD Anderson Cancer Center, Unit 902, 7455 Fannin, Houston TX, USA. Email: tzal@mdanderson.org

Multichain Immune Recognition Receptor Signaling: From Spatiotemporal Organization to Human Disease, edited by Alexander B. Sigalov. ©2008 Landes Bioscience and Springer Science+Business Media.

propagate to the intracellular domains to expose sites for docking of various adaptors and signaling enzymes. The structural changes could be conformational, chain rearrangements, ligand-driven dimerization, or multimerization.[1] In contrast, the kinetic models favor the view that the information is conveyed by the net balance of otherwise unstructured interactions between the receptors and the membrane-resident kinases and phosphatases, which have opposing effects on signaling.[2] To distinguish between the alternative mechanisms requires quantitative characterization of various parameters of protein interactions in living cells. Verifying the structural mechanisms of ligand recognition requires determining distances and orientations between protein domains within and between receptors, while the kinetic models call for determination of affinities, lifetimes, diffusion coefficients, and frequencies of random collisions—all with subcellular resolution in living cells.

The last two decades witnessed significant refinement of fluorescence microscopy techniques that allowed looking non-invasively inside cells and visualizing receptor dynamics in situ. The most powerful approaches harness fluorescence to provide information about protein interactions (Fig. 1). The general strategy is to hyperlink the structural data on a pixel-by-pixel basis to additional parameters of fluorescence that are sensitive to the local environment. The most direct and versatile are imaging modalities based on Förster (fluorescence) resonance energy transfer (FRET), which is sensitive to the proximity and orientation between fluorophores and is amenable to the structural and the kinetic analysis. Complementary information about diffusion and stoichiometry of large protein complexes can be obtained by fluorescence correlation microscopy (FCM), while bimolecular fluorescence complementation (BiFC) and other complementation techniques can be used to determine protein interactions.

FRET

FRET microscopy is the most powerful and popular approach to study protein interactions in living cells. Occurring through dipole-dipole resonance between the excited donor fluorophore and a nearby acceptor, FRET allows direct detection of nanometer-range proximity between appropriately labeled proteins as well as conformational changes. FRET can be imaged based on several parameters that are detectable by wide field, confocal, multiphoton, as well as total internal reflection fluorescence microscopy. Being a proximity effect, FRET can be used to detect both the specific complex formation as well as random collisions—both of which may be important for signaling by multichain immunoreceptors. We will focus later on how quantitative FRET imaging can be leveraged to study the underlying mechanisms of protein interactions.

Bimolecular Fluorescence Complementation

The BiFC technique is based on nonfluorescent, complementary fragments of fluorescent proteins (FPs) that can refold into a fluorescing product.[3] By genetically attaching the fragments to different proteins, their interactions can be detected based on de novo fluorescence.[4] Due to irreversible refolding, BiFC is not a general approach to monitor the dynamics of protein interactions but it excels as an end-point kinetic assay.[5] Quantitative application of BiFC is possible by multiplexing fragments from different color FPs. That way, the relative efficiency of competing interactions can be evaluated ratiometrically.[6,7] Recent improvements include new fragments of the Cerulean and Venus FPs that offer faster refolding kinetics and better sensitivity.[8] The BiFC assay can complement FRET to determine and screen for protein interactions.[9]

Fluorescence Correlation Techniques

The formation of large protein complexes that exceed the nanometer range of FRET can be studied at the single-molecule level in living cells by FCM.[10] This method uses highly sensitive detectors to detect bursts of fluorescence due to diffusion of single fluorophores through a small observation volume, which can come from a confocal or multiphoton excitation. The diffusion coefficient, which depends on the mass of freely diffusing complexes, can be discerned by applying the autocorrelation function to the fluctuations of fluorescence. Classic FCM is performed under free diffusion conditions to quantify the absolute molecular mass

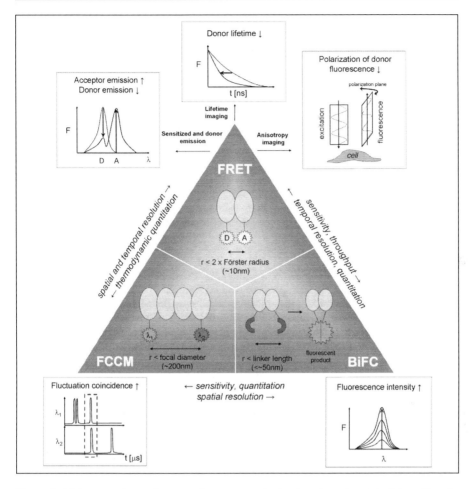

Figure 1. Major experimental approaches to quantitative imaging of protein interactions in living cells.

and the relative representation of the different weight species—in cells, it is best suited to follow interactions in the cytosol. Nevertheless, importantly to study membrane receptors, FCM is applicable to cell membranes as well.[11]

FCM is robust only when detecting interactions of a small labeled ligand with a large partner. This limitation is avoided by labeling two proteins with different color fluorophores and enumerating the coincidence of diffusion, hence complex formation, by fluorescence cross-correlation microscopy (FCCM).[12-16] FCCM is a powerful approach to directly measure the concentrations of the free and complexed species and from these, to calculate the affinity constant of complex formation in solution. Furthermore, the stoichiometry of the complex can be determined from the relative intensities during the coordinate bursts of fluorescence. By cross-correlating fluorescence in three or more colors, complex formation between more than two components can be studied.[17] Recent advances in FCCM improved the sensitivity and cross-talk separation by using time- and space-correlated single photon counting as well as interleaved excitation.[18,19] F(C)CM techniques are not suitable for full frame imaging, however, and are typically used for spot measurements in predetermined sites of the cell body. An imaging variant of FCM is achieved by cross-correlating fluorescence fluctuations across space instead of time; the technique is termed image cross-correlation microscopy (ICCM). ICCM allows imaging of the degree of aggregation and colocalization in living cells by confocal or multiphoton laser scanning.[20] As all fluorescence correlation techniques, ICCM performs well at low (physiological) concentrations of fluorophores but may require prolonged acquisition times.

Fluorescent Labeling of Proteins in Living Cells

Common to all techniques, the critical first step to visualizing protein interactions in living cells is fluorescent tagging of proteins with fluorophores that have suitable spectral properties and minimal impact on the biological functions. Foremost, proteins can be genetically fused with different color FPs that may be attached to the C or the N terminus or, if the structure of the carrier protein permits, spliced into the sequence of the protein.[21] The growth of FRET imaging studies in living cells is particularly indebted to the development of FPs that are monomeric and have favorable spectral overlap, low cross-detection, high quantum yield, and low sensitivity of fluorescence to environmental changes.[22-24] Currently, the recommended pairs of fluorescent proteins for FRET include mCerulean, CyPet, or SCFP3A as the donors and the yellow mVenus, mCitrine, or SYFP2 as the acceptors.[23,25,26] Less often used but advantageous due to lesser photobleaching are the green-red pairs: EGFP as the donor and mRFP1, mKO, mOrange, or mCherry as acceptors.[27,28] The nonfluorescent yellow chromoprotein REACh can be used as an acceptor-quencher with EGFP for lifetime and anisotropy-based FRET imaging.[29]

The biggest drawback of FPs is their bulkness, which may alter the cellular distribution or interfere with ligand binding. An alternative approach relies on small biarsenical fluorophores green FlAsH or red ReAsH that react specifically with short tetracysteine motifs, which can be incorporated at almost any place by genetic modification. A recent optimization of the tetracysteine motif improved the selectivity and lessened the conditions required for labeling in living cell.[30] For FRET, FlAsH and ReAsH can be acceptors for cyan FP and GFP, respectively.[30-32] Additional possibilities for multiplexed labeling and pulse-chase studies are provided by attaching to the protein a binding domain that is specific for a small fluorophore. The O6-alkylguanine-DNA alkyltransferase (AGT) domain can be labeled with fluorescent O6-benzylguanine (O6-BG) derivatives,[33,34] oligohistidine sequences on cell surfaces can be labeled with nitroloacetate fluorophores,[35,36] and acyl carrier protein (ACP) can be labeled with acyl-fluorophores.[37] Highly fluorescent nanocrystals (quantum dots) are advantageous for imaging of low abundance cell surface proteins and for FRET thanks to their brightness and good spectral separation of emission from excitation (Stokes shift).[38,39] These and many other fluorophores can be attached to cell surface proteins in living cells using antibodies that do not interfere with biological functions. In permeabilized cells, imaging FRET between GFP-tagged receptors and fluorescent-labeled anti-phosphotyrosine antibodies allowed specific detection of receptor phosphorylation.[40-42] A more extensive review of different classes of fluorophores and dyes suitable for FRET experiments is provided by Sapsford et al.[43]

Quantitative FRET Imaging

Determining the efficiency of FRET is the key to analyze the structures and kinetics of protein interactions. FRET efficiency depends on the distance between donor and acceptor, as well as the relative orientation of the electrical dipoles, the spectral overlap of donor emission with acceptor absorption, and the refractive index of the medium. In principle, one could triangulate the topology of multiprotein complexes by attaching donors and acceptors at various positions, measuring efficiencies of FRET, and calculating the distances according to the Förster equation. (The orientation factor (κ^2) is equal 2/3 if the donor or the acceptor has freedom of rotation.) However, when imaging FRET in a heterogeneous population of donors and acceptors, a typical situation in living cells, an important distinction has to be made between the intrinsic FRET efficiency, which characterizes individual donor-acceptor pairs, and the apparent efficiency (E_{app}), which is actually measured by most techniques. E_{app} is a weighted average of intrinsic efficiencies for all donors in the measurement volume; therefore, a particular E_{app} value can be due to a combination of distances, orientations, and degrees of donor occupancy by acceptor, as well as it could be due to random collisions. While E_{app} of a heterogeneous population cannot be used to calculate the donor-acceptor distance, a systematic analysis of the dependence of E_{app} on the local concentrations of donors and acceptors can shed light on the mechanism of protein interactions, which will be discussed later.

FRET efficiency can be quantified based on the donor fluorescence intensity, lifetime, and polarization, as well as from sensitized emission of acceptors (if these are fluorescent)—each modality offers a different balance of sensitivity, speed, and quantitation.

Donor Dequenching

FRET quenches donor fluorescence; therefore, the rebound of donor fluorescence after photo-destruction of acceptors provides a straightforward means to measure E_{app}. The raw data consist of two images: donor fluorescence taken before (D_{before}) and after (D_{after}) acceptor photobleaching:

$$E_{app} = 1 - D_{before} / D_{after}$$

An alternative approach is to monitor the rate of donor photobleaching, which is decreased in presence of FRET.[44,45] Since the range of apparent FRET efficiencies may be in the order of only a few percent, low noise, precise detectors, and strong signals are essential to obtain reliable data—the accuracy can be improved by gradual acceptor photobleaching.[46] Due to a finite time required to substantially eliminate acceptors, photobleaching methods are limited by cell motility and instrument drift. Accordingly, donor dequenching tends to be used primarily for fixed or immobile specimens. Despite limitations, donor dequenching is a robust method to image E_{app} and is often used to corroborate FRET imaged by other methods. Imaging of FRET by photobleaching has been applied extensively to study the subcellular regulation of immunoreceptor interactions.[47-52]

An elegant extension of the donor quenching approach takes advantage of photoactivatable GFP (PA-GFP),[53] which can be instantaneously activated by illumination with a 405 nm laser and accept FRET from cyan FP donors.[54] When photo-activated locally in a subcellular compartment, the spreading of FRET (detected by the drop of donor fluorescence) provides invaluable information about diffusion, stability of protein complexes, and the rates of dissociation and association.[54]

Sensitized Fluorescence Imaging

Unlike donor dequenching, sensitized emission-based FRET imaging does not require harsh irradiation and can be performed repeatedly on live cells with a high temporal and three-dimensional resolution.[55-58] Numerous methods evolved over the years that take advantage of sensitized emission to detect FRET. In general, the sample is illuminated at the donor excitation wavelength and the measurement (imaging) is done at the donor as well as the acceptor emission wavelengths or, the full emission spectra are collected. The latter approach is often termed spectral FRET. For heterologous FRET experiments (as opposed to the internal FRET in covalently linked donor-acceptor sensors), the acceptor concentration has to be accounted for, hence an additional exposure is taken to acquire the acceptor-only fluorescence. E_{app} is calculated on a pixel-by-pixel basis based on the three intensities that are linearly unmixed from any spectral overlap:[57-59]

$$E_{app} = S/(S + GD)$$

where S is sensitized fluorescence, D, donor fluorescence, and the G parameter [60] can be calibrated by acceptor photobleaching,[57,59] lifetime measurements,[58] or by using pairs of donor-acceptor constructs having different FRET efficiencies.[61] Additional calculations allow determination of local stoichiometry of donors, acceptors, and FRET complexes.[58]

In the non-imaging mode, sensitized emission FRET was applied to study receptor aggregation in platelets,[62] interactions between antibody-labeled IL-1 receptors,[63] ligand-dependent rearrangements of IL-2 receptor subunits,[64] the multivalent structure of T-cell receptor (TCR),[65] as well as MHC-I-dependent[66,67] and -independent[68] interactions between TCR and CD8. Quantitative FRET imaging based on sensitized emission allowed time-lapse, three-dimensional visualization of interactions between the TCRζ chain and CD4[57] or CD8[69] in immunological synapses. MHC-II interactions were tracked in subcellular compartments by confocal sensitized emission FRET.[70] In B-cells, quantitative sensitized emission allowed dissecting chain interactions and the lyn kinase recruitment during B-cell receptor (BCR) activation.[71,72] Sensitized emission FRET is perhaps easiest to implement using wide field microscopy but it is applicable to confocal detection as well as two-photon, near-field scanning, or atomic force microscopy.[70,73-75]

Fluorescence Lifetime Imaging

FRET shortens the time donors spend in the excited state, which can be imaged by fluorescence lifetime imaging microscopy (FLIM).[76] Two FLIM modalities are available, using pulsed excitation with time-gated detection (time-domain) or modulated excitation with phase-shifted detection (frequency-domain). Time-domain FLIM, especially when using time and space-correlated single photon counting mode (TSCSPC), allows recording entire fluorescence decay profiles for each voxel in three dimensions.[77-79] This is a distinct advantage of time-domain FLIM over all other FRET modalities because the fluorescence decay curves can be deconvolved into individual lifetime exponents that are proportional to E_{int}, which is the basis for distance determinations. The average lifetime is proportional to E_{app}:[80,81]

$$E_{int} = 1 - \tau_i/\tau_0 \qquad E_{app} = 1 - \bar{\tau}/\tau_0$$

where τ_0 is the lifetime of free donors.

Important for the development of FRET imaging in vivo, time-domain lifetime imaging is an excellent match for multiphoton, femtosecond-pulsed excitation.[82-85] Frequency-domain FLIM offers improved temporal resolution but is limited to average lifetimes,[42,76,86] except when using more advanced, nonsinusoid modulation[87] or two-component analysis.[88] Acquisition times may be shortened by using a streak camera-based detection that can be combined with multiphoton excitation.[83,89] In practice, however, a compromise is necessary between the temporal, spatial, and lifetime resolution. For this reason, application of FLIM to image fast interaction dynamics has been scarce. A general advantage of FLIM is the ability to detect FRET between spectrally similar donors and acceptors, in which case the combined donor and acceptor lifetime is increased.[90] Like all FRET imaging methods, FLIM has its caveats. One is the dependence of lifetimes on the refractive index around the fluorophore—the property that can be used to monitor the local environment in cells but can also interfere with the quantitation of FRET.[91-93] Another difficulty is the sensitivity to photobleaching, which may be lessened by using GFP and mCherry instead of the more common CFP and YFP pair.[27]

FRET Imaging by Polarization Anisotropy

Yet another technique to image FRET is based on changes in fluorescence polarization anisotropy. The degree of fluorescence depolarization (in relation to the excitation light) depends on the rotational dynamics of the fluorophore and FRET.[94] Therefore, FPs are particularly suitable for polarization-based FRET imaging because, due to their large size, FPs are only minimally depolarized by the rotational mechanism and all depolarization can be attributed to FRET. Fluorescence polarization anisotropy is calculated from the intensities of fluorescence that is detected through polarizing filters placed parallel ($D_=$) and perpendicular ($D_\perp$) to excitation:

$$r = (D_= - D_\perp)(D_= + 2D_\perp)$$

Through additional calculations, polarization measurements can be converted to apparent FRET efficiencies.[95]

Unlike other FRET imaging modalities, anisotropy imaging can detect FRET between fluorophores of the same type, obviating the need for double labeling with different donors and acceptors.[96,97] This way, anisotropy imaging is particularly convenient to visualize homotypic aggregation and multimerization of FP-labeled proteins. Polarization FRET can be combined with lifetime microscopy for comprehensive characterization of rotational coefficients[98] or with sensitized emission to allow imaging of heterologous FRET in a single-exposure, which reduces motility errors.[99,100] A minor drawback of polarization-based FRET imaging is somewhat lower spatial resolution and sensitivity due to using low numerical aperture objectives to maintain polarization.

Multiphoton Imaging of Heterologous FRET

Multiphoton microscopy has been used to image FRET based on donor lifetimes, polarization anisotropy[101] and sensitized emission of acceptors.[102] Nevertheless, due to single wavelength excitation, multiphoton-excited FRET could be imaged only when using internally linked biosensors but not in heterologous protein-protein interaction experiments, whereby donors and acceptors are attached to independently expressed proteins.

In heterologous experiments, FRET can result from molecular crowding and/or complex formation while no FRET can be due to a lack of interaction or insufficient acceptors. Therefore, regardless of the imaging modality employed to detect FRET, it is critical to account for the local acceptor concentrations that requires selective excitation at donor and acceptor specific wavelengths. By using dual, interline-switched femtosecond laser excitation and dual channel detection, we recently realized truly heterologous FRET imaging by multiphoton microscopy.[103] The system was tested using dimer-forming TCRζ chains tagged with cyan and yellow FPs (Fig. 2).[103] Future application of sensitized emission multiphoton FRET imaging will be to study the motions of the intracellular signaling domains of TCR and other receptors at the sites of antigen exposure in vivo.

Using FRET to Analyze Receptor (Re)arrangements

Dimerization and multimerization of membrane receptors can be studied by co-expressing the receptors labeled with donor or acceptor and imaging FRET using any imaging modality.[104] For example, FRET images revealed differences in the oligomerization of B7-1 and B7-2 family members.[105] By fitting alternative mathematical models of molecule distribution to E_{app}, FRET can give insight into the spatial arrangement of clustered proteins.[106] Such analysis was applied to study IgA-ligand-receptor complexes in the endocytic membranes in MDCK cells and indicated that single receptors in microclusters are surrounded by 2.5-3 neighbors. Similar quantitative FRET imaging confirmed multimerization of otherwise dimeric receptors of transferrin upon ligand binding,[107] dimerization of galanin-1 receptor,[108] and the dimer-tetramer transition of epidermal growth factor.[109]

Internal rearrangements of multichain receptors were studied most extensively using cytokine receptors, whose signaling does not necessarily coincide with large scale clustering and is therefore more likely explained by conformational changes or chain rearrangements. Changes in the distance between the common signaling chain and ligand-specific chains were detected by FRET in IL-2 family receptors.[64] Likewise, in the case of the leukemia inhibitory factor receptor (LIFR), FRET between the LIF-specific chain and gp130 increased upon binding of LIF indicating ligand-induced heterodimerization or an intra-complex rearrangement.[9] The latter possibility was supported by another study, which found that FRET between gp130 and LIFR was constitutive at the steady state and increased above the basal level upon LIF binding.[110] An opposite effect was observed for the IL-17 receptor: FRET between intracellular domains of IL-17R (labeled with CFP and YFP) was constitutive at the steady state and decreased upon binding of IL-17.[111] The decrease indicated a scissor-like opening of the intracellular domains,[111] which was reminiscent of earlier observations in the interferon-γ receptor.[112] In the case of the IL-10 receptor, binding of IL-10 did not cause a change in FRET between the cytoplasmic domains.[113] The homotypic lateral interaction

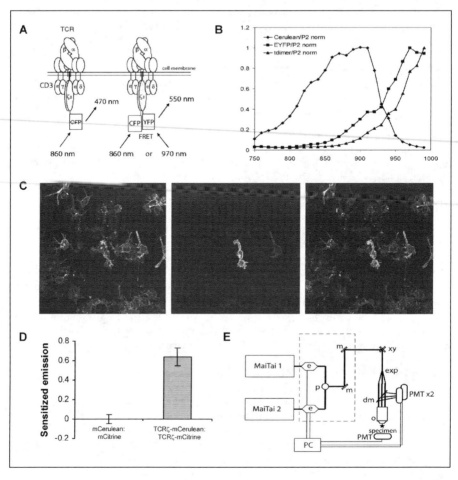

Figure 2. Imaging of heterologous FRET by dual laser, two-photon, interline excitation. A) Experimental system to detect the proximity between the intracellular domains of T-cell Receptor (TCR)ζ chains. CFP denotes the mCerulean fluorescent protein. YFP denotes the enhanced yellow FP. B) Normalized two-photon excitation spectra of CFP and YFP or CFP and the red fluorescing tdimer(12) have poor overlap, which precludes efficient excitation at a single wavelength but allows selective excitation at 860 nm, 970 nm, and 990 nm, respectively. C) Raw two-photon excited images of mCerulean, EYFP, and the raw FRET at the indicated excitation/emission wavelengths, respectively, in a mixture of T-cells expressing different ratios of TCRζ-mCerulean and TCRζ-EYFP. D) Quantitation of donor-normalized sensitized emission in cells co-expressing donor and acceptor. Donor normalized sensitized emission (F_c/D) was calculated after subtracting the directly excited donor and acceptor signals from the raw FRET channel, as generally described for single photon FRET.[59] $F_c/D = (\ I_{860/550} - dI_{860/470} - aI_{970/550})/ I_{860/470}$. The cross-talk coefficients d and a were calibrated based on donor or acceptor only cells imaged under identical conditions. Non-interacting: co-expressed cytoplasmic (free) fluorescent proteins. E) The optical path of the dual laser multiphoton microscope setup. *MaiTai 1, 2*: femtosecond lasers, e: electrooptical modulators, *p*: polarization merge optics, *m*: mirrors. xy: resonant scanner (Leica SP2 RS), *exp*: beam expander, *dm*: dichroic mirror, *PMT*: photomultipliers, *o*: water dipping objective (Olympus 20 × NA = 0.95 or Leica 63 × NA = 0.9). The figure is adapted from Zal et al, Proceedings of SPIE, with permission from the Society of Photo-Optical Instrumentation Engineers 2007.[103]

between gp130 chains was studied in IL-6R using CFP and YFP fusions of gp130. FRET was constitutive and did not increase upon IL-6 ligation, which indicated that the gp130 chains in IL-6R are pre-associated and are not further cross-linked by IL-6.[9] Ligand-induced internal rearrangements in BCR were also demonstrated by FRET.[71] A largely undeveloped is the issue of the orientation of receptor domains with respect to the cell membrane, which may be tackled by introducing FRET donors or acceptors to the lipid environment.[114]

Imaging TCR-Coreceptor Interactions in the Immunological Synapse

According to the kinetic proofreading model of peptide-MHC recognition by TCR, generation of activation signals depends on a transient complex that needs to be stable enough to allow phosphorylation of CD3 chains by the lck kinase. Lck is brought to the immunological synapse by CD8 or CD4 glycoproteins. Like TCR, these molecules can bind to MHC-I or MHC-II, respectively, and are brought in the vicinity of TCR coincident with TCR ligation.[66,115] It was therefore possible that the dynamics of the interaction between TCR and coreceptors is regulated by the ligand quality. This hypothesis was tested by imaging FRET between the intracellular domains of TCRζ-CFP (donor) and CD4-YFP (acceptor) using the sensitized emission method, which allowed three-dimensional time-lapse imaging of the immunological synapse and quantitation in terms of E_{app}. Indeed, TCRζ and CD4 are brought together as early as 30 s after T-cell encounter of agonist peptide-loaded antigen-presenting cells, i.e., before a cSMAC is formed, indicating that cSMAC formation is not prerequisite for the TCR-CD4 association.[57] Moreover, E_{app} was decreased in a dominant fashion by presentation of antagonist peptides that inhibit T-cell activation. A similar although not identical effect was observed between TCRζ-CFP and CD8-YFP.[116] The TCRζ-CD8 associations were transient and had lower peak FRET efficiencies than the association of CD4 with TCRζ, indicating a kinetic and/or structural difference between the ways CD8 and CD4 associate with TCR. Nevertheless, the kinetics of association between TCRζ and CD8 in immune synapses correlated with the biological activities of presented peptides: agonists drove a fast raise of FRET and antagonists caused delayed FRET.[116] Overall, FRET imaging supports the kinetic proofreading role of CD4 and CD8. However, it remains unclear to what extent exactly is FRET due to the formation of relatively stable complexes, i.e., affinity-driven interaction, or due to the regulation of diffusion-driven collisions in the immunological synapse.

Affinity versus Random Collisions: Acceptor Titration FRET

Irrespective which imaging modality is used to obtain quantitative FRET images, further analysis of the data, preferably in terms of FRET efficiency, is the key to study the mechanisms of protein interactions. Receptor signaling is often coincident with the clustering of receptors in a small area of the cell membrane; for example TCR in the immunological synapse, BCR cross-linking, or receptor clustering in lipid rafts. When using FRET to image receptor interactions, the question comes up on how to distinguish FRET due to the formation of specific complexes from FRET due to random collisions in the areas of receptor clustering.

In general, the strategy is to discriminate specific complexes from random collisions by titrating donors and acceptors and measuring local changes of FRET efficiency. For high affinity interactions, E_{app} is relatively independent of concentration (beyond stoichiometric acceptor concentration), while FRET due to random collisions will be evident only at a high local concentration of acceptor.[117,118] The acceptor titration approach found extensive use to study how proteins are arranged in lipid rafts, which are submicrometer-sized assemblies of lipids and membrane proteins that are often involved in signaling. Titration analysis of FRET between raft-resident glycosylphosphatdylinositol (GPI)-linked fluorescent proteins showed that FRET depends on concentration, which is indicative of random interactions.[119] More detailed analysis involved fitting alternative theoretical raft models to the observed relationship between apparent FRET efficiencies and concentrations of donors and acceptors.[120] The best fit was with the model where lipid rafts are small and harbor only several GPI-linked proteins in equilibrium with dispersed monomers. A poor fit was observed with models that assumed larger and more populous rafts. Concentration dependent FRET was also noted in other systems such as MHC-I, MHC-II, CD48, the IL-2R/IL-15R subunits, or ErbB transmembrane receptor tyrosine kinases.[121-123] Some of these concentration effects could be attributed to partitioning in lipid rafts. Overall, these studies demonstrated that fitting the data to mathematical models is a powerful tool to distinguish complex formation from random interactions as well as to evaluate the distribution of receptors in membrane microdomains.[124,125] Moreover, acceptor titration is also applicable to high throughput FRET screening to distinguish high affinity ligands in living (*E. coli*) cells.[126]

Mathematical Model of FRET for Simultaneous Complex Formation and Random Collisions

It would be valuable to have a mathematical model to quantify FRET due to donors and acceptors forming specific complexes concurrent with interacting randomly due to diffusion-driven collisions. Both of these processes have been proposed to occur in immunological synapses. The frequency of random collisions can be characterized by the bimolecular interaction constant that is related to the diffusion coefficient in the Stern-Volmer equation of diffusion quenching of donor lifetimes.[127,128] We modeled E in response to titration of donors by acceptors in solution by combining the Stern-Volmer equations with affinity and taking into account the relationship

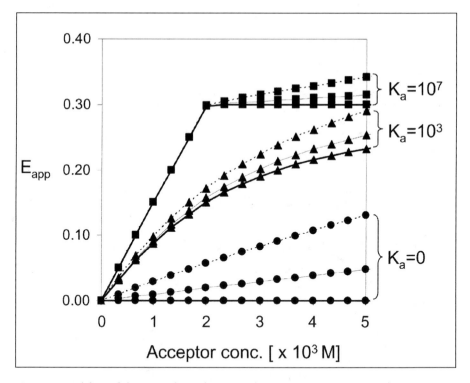

Figure 3. Modeling of the FRET dependence on the acceptor concentration for concurrent processes, formation of specific complexes and diffusion-driven random collisions. The model was derived by combining the Stern-Volmer equation for collision quenching with the affinity constant and the fluorescence lifetime—FRET efficiency relation. Assuming a 1:1 stoichiometry and donor concentration 2×10^{-3} M. Squares: a high affinity interaction, $K_a = 10^7$, triangles: a low affinity interaction $K_a = 10^3$, dots: no affinity, $K_a = 0$. Solid lines: no collisions, dashed lines: an intermediate rate of random collisions, dotted lines: a high rate of random collisions.

between donor lifetimes and FRET efficiency. Figure 3 shows a family of analytical solutions for different affinities and diffusion kinetics. The model exhibits the characteristic biphased response of E_{app}. The initial raise in E_{app} is determined by the affinity constant, while the slope of E_{app} above the stoichiometric concentration reflects the bimolecular interaction constant. We envisage that by curve fitting the two-dimensional version of this model to the experimental data, it will be possible to evaluate the relative contribution of affinity and random collisions in membrane compartments as well. In general, continued development of acceptor (and donor) titration analysis will help in discriminating which mechanisms of protein interactions are modulated by ligand engagement.

Conclusion

Rapid development of quantitative imaging techniques allows non-invasive study of protein biochemistry in living cells. Arguably, a particularly promising approach is a combination of FRET imaging with computational modeling of FRET efficiency at different concentrations of donor and acceptor. Through such analysis, FRET imaging can be used to evaluate the affinity of complex formation and the frequency of diffusion-driven random collisions, both of which may contribute to signal transduction by multichain receptor complexes. Future developments will improve the quantitative analysis of FRET as well as applying other, complementary imaging approaches to uncover the structure and internal dynamics of cell surface receptors in living cells.

References

1. Sigalov A. Multi-chain immune recognition receptors: Spatial organization and signal transduction. Semin Immunol 2005; 17:51-64.
2. Davis SJ, Van der Merwe PA. The kinetic-segregation model: TCR triggering and beyond. Nat Immunol 2006; 7(8):803-809.
3. Ghosh I, Hamilton AD, Regan L. Antiparallel leucine zipper-directed protein reassembly: Application to the green fluorescent protein. J Am Chem Soc 2000; (122):5658-5659.
4. Hu CD, Chinenov Y, Kerppola TK. Visualization of interactions among bZIP and Rel family proteins in living cells using bimolecular fluorescence complementation. Mol Cell 2002; 9(4):789-798.
5. Kerppola TK. Visualization of molecular interactions by fluorescence complementation. Nat Rev Mol Cell Biol 2006; 7(6):449-456.
6. Hu CD, Kerppola TK. Simultaneous visualization of multiple protein interactions in living cells using multicolor fluorescence complementation analysis. Nat Biotechnol 2003; 21(5):539-545.
7. Grinberg AV, Hu CD, Kerppola TK. Visualization of Myc/Max/Mad family dimers and the competition for dimerization in living cells. Mol Cell Biol 2004; 24(10):4294-4308.
8. Shyu YJ, Liu H, Deng X et al. Identification of new fluorescent protein fragments for bimolecular fluorescence complementation analysis under physiological conditions. Biotechniques 2006; 40(1):61-66.
9. Giese B, Roderburg C, Sommerauer M et al. Dimerization of the cytokine receptors gp130 and LIFR analysed in single cells. J Cell Sci 2005; 118(Pt 21):5129-5140.
10. Berland KM. Fluorescence correlation spectroscopy: A new tool for quantification of molecular interactions. Methods Mol Biol 2004; 261:383-398.
11. Schwille P, Korlach J, Webb WW. Fluorescence correlation spectroscopy with single molecule sensitivity on cell and model membranes. Cytometry 1999; 36:176-182.
12. Bacia K, Majoul IV, Schwille P. Probing the endocytic pathway in live cells using dual-color fluorescence cross-correlation analysis. Biophys J 2002; 83:1184-1193.
13. Kohl T, Heinze KG, Kuhlemann R et al. A protease assay for two-photon crosscorrelation and FRET analysis based solely on fluorescent proteins. Proc Natl Acad Sci USA 2002; 99(19):12161-12166.
14. Kuroyama H, Ikeda T, Kasai M et al. Identification of a novel isoform of ZAP-70, truncated ZAP kinase. Biochem Biophys Res Commun 2004; 315(4):935-941.
15. Haustein E, Schwille P. Ultrasensitive investigations of biological systems by fluorescence correlation spectroscopy. Methods 2003; 29:153-166.
16. Bacia K, Kim SA, Schwille P. Fluorescence cross-correlation spectroscopy in living cells. Nat Methods 2006; 3(2):83-89.
17. Heinze KG, Jahnz M, Schwille P. Triple-color coincidence analysis: One step further in following higher order molecular complex formation. Biophys J 2004; 86:506-516.
18. Muller BK, Zaychikov E, Brauchle C et al. Pulsed interleaved excitation. Biophys J 2005; 89(5): 3508-3522.
19. Lamb DC, Muller BK, Brauchle C. Enhancing the sensitivity of fluorescence correlation spectroscopy by using time-correlated single photon counting. Curr Pharm Biotechnol 2005; 6(5):405-414.

20. Wiseman PW, Squier JA, Ellisman MH et al. Two-photon image correlation spectroscopy and image cross-correlation spectroscopy. J Microsc 2000; 200:14-25.
21. Rocheleau JV, Edidin M, Piston DW. Intrasequence GFP in class I MHC molecules, a rigid probe for fluorescence anisotropy measurements of the membrane environment. Biophys J 2003; 84(6): 4078-4086.
22. Shaner NC, Steinbach PA, Tsien RY. A guide to choosing fluorescent proteins. Nat Methods 2005; 2(12):905-909.
23. Giepmans BN, Adams SR, Ellisman MH et al. The fluorescent toolbox for assessing protein location and function. Science 2006; 312(5771):217-224.
24. Patterson GH, Piston DW, Barisas BG. Forster distances between green fluorescent protein pairs. Anal Biochem 2000; 284(2):438-440.
25. Rizzo MA, Springer GH, Granada B et al. An improved cyan fluorescent protein variant useful for FRET. Nat Biotechnol 2004; 22(4):445-449.
26. Kremers GJ, Goedhart J, van Munster EB et al. Cyan and yellow super fluorescent proteins with improved brightness, protein folding, and FRET Forster radius. Biochemistry 2006; 45(21):6570-6580.
27. Tramier M, Zahid M, Mevel JC et al. Sensitivity of CFP/YFP and GFP/mCherry pairs to donor photobleaching on FRET determination by fluorescence lifetime imaging microscopy in living cells. Microsc Res Tech 2006; 69(11):933-939.
28. Peter M, Ameer-Beg SM, Hughes MK et al. Multiphoton-FLIM quantification of the EGFP-mRFP1 FRET pair for localization of membrane receptor-kinase interactions. Biophys J 2005; 88(2):1224-1237.
29. Ganesan S, Ameer-Beg SM, Ng TT et al. A dark yellow fluorescent protein (YFP)-based Resonance Energy-Accepting Chromoprotein (REACh) for Forster resonance energy transfer with GFP. Proc Natl Acad Sci USA 2006; 103(11):4089-4094.
30. Martin RM, Leonhardt H, Cardoso MC. DNA labeling in living cells. Cytometry A 2005; 67(1):45-52.
31. Hoffmann C, Gaietta G, Bunemann M et al. A FlAsH-based FRET approach to determine G protein-coupled receptor activation in living cells. Nat Methods 2005; 2(3):171-176.
32. Nakanishi J, Takarada T, Yunoki S et al. FRET-based monitoring of conformational change of the beta2 adrenergic receptor in living cells. Biochem Biophys Res Commun 2006; 343(4):1191-1196.
33. Keppler A, Gendreizig S, Gronemeyer T et al. A general method for the covalent labeling of fusion proteins with small molecules in vivo. Nat Biotechnol 2003; 21(1):86-89.
34. Keppler A, Arrivoli C, Sironi L et al. Fluorophores for live cell imaging of AGT fusion proteins across the visible spectrum. BioTechniques. 2006; 41(2):167-170,172,174-175.
35. Guignet EG, Hovius R, Vogel H. Reversible site-selective labeling of membrane proteins in live cells. Nat Biotechnol 2004; 22(4):440-444.
36. Lata S, Gavutis M, Tampe R et al. Specific and stable fluorescence labeling of histidine-tagged proteins for dissecting multi-protein complex formation. J Am Chem Soc 2006; 128(7):2365-2372.
37. Meyer BH, Segura JM, Martinez KL et al. FRET imaging reveals that functional neurokinin-1 receptors are monomeric and reside in membrane microdomains of live cells. Proc Natl Acad Sci USA 2006; 103(7):2138-2143.
38. Grecco HE, Lidke KA, Heintzmann R et al. Ensemble and single particle photophysical properties (two-photon excitation, anisotropy, FRET, lifetime, spectral conversion) of commercial quantum dots in solution and in live cells. Microsc Res Tech 2004; 65(4-5):169-179.
39. Clapp AR, Medintz IL, Mattoussi H. Forster resonance energy transfer investigations using quantum-dot fluorophores. Chemphyschem 2006; 7(1):47-57.
40. Treanor B, Lanigan PM, Kumar S et al. Microclusters of inhibitory killer immunoglobulin-like receptor signaling at natural killer cell immunological synapses. J Cell Biol 2006; 174(1):153-161.
41. Ng T, Squire A, Hansra G et al. Imaging protein kinase Calpha activation in cells. Science 1999; 283(5410):2085-2089.
42. Haj FG, Verveer PJ, Squire A et al. Imaging sites of receptor dephosphorylation by PTP1B on the surface of the endoplasmic reticulum. Science 2002; 295(5560):1708-1711.
43. Sapsford KE, Berti L, Medintz IL. Materials for fluorescence resonance energy transfer analysis: Beyond traditional donor-acceptor combinations. Angew Chem Int Ed 2006; 45:4562-4588.
44. Kubitscheck U, Kircheis M, Schweitzer-Stenner R et al. Fluorescence resonance energy transfer on single living cells. Application to binding of monovalent haptens to cell-bound immunoglobulin E. Biophys J 1991; 60(2):307-318.
45. Young RM, Arnette JK, Roess DA et al. Quantitation of fluorescence energy transfer between cell surface proteins via fluorescence donor photobleaching kinetics. Biophys J 1994; 67(2):881-888.
46. van Munster EB, Gadella TW. Fluorescence lifetime imaging microscopy (FLIM). Adv Biochem Eng Biotechnol 2005; 95:143-175.
47. Damjanovich S, Vereb G, Schaper A et al. Structural hierarchy in the clustering of HLA class I molecules in the plasma membrane of human lymphoblastoid cells. Proc Natl Acad Sci USA 1995; 92(4):1122-1126.

48. Szaba GJ, Pine PS, Weaver JL et al. Epitope mapping by photobleaching fluorescence resonance energy transfer measurements using a laser scanning microscope system. Biophys J 1992; 61(3):661-670.
49. Szabo GJ, Weaver JL, Pine PS et al. Cross-linking of CD4 in a TCR/CD3-juxtaposed inhibitory state: a pFRET study. Biophys J 1995; 68(3):1170-1176.
50. Jurgens L, Arndt-Jovin D, Pecht I et al. Proximity relationships between the type I receptor for Fc epsilon (Fc epsilon RI) and the mast cell function-associated antigen (MAFA) studied by donor photobleaching fluorescence resonance energy transfer microscopy. Eur J Immunol 1996; 26(1):84-91.
51. Bacso Z, Bene L, Bodnar A et al. A photobleaching energy transfer analysis of CD8/MHC-I and LFA-1/ICAM-1 interactions in CTL-target cell conjugates. Immunol Lett 1996; 54(2-3):151-156.
52. Kim M, Carman CV, Springer TA. Bidirectional transmembrane signaling by cytoplasmic domain separation in integrins. Science 2003; 301(5640):1720-1725.
53. Patterson GH, Lippincott-Schwartz J. A photoactivatable GFP for selective photolabeling of proteins and cells. Science 2002; 297:1873-1877.
54. Demarco IA, Periasamy A, Booker CF et al. Monitoring dynamic protein interactions with photoquenching FRET. Nat Methods 2006; 3(7):519-524.
55. Youvan DC, Silva CM, Bylina EJ et al. Calibration of fluorescence resonance energy transfer in microscopy using genetically engineered GFP derivatives on nickel chelating beads. Biotechnology et alia 1997; 3:1-18.
56. Erickson MG, Alseikhan BA, Peterson BZ et al. Preassociation of calmodulin with voltage-gated Ca(2+) channels revealed by FRET in single living cells. Neuron 2001; 31(6):973-985.
57. Zal T, Zal MA, Gascoigne NR. Inhibition of T-cell receptor-coreceptor interactions by antagonist ligands visualized by live FRET imaging of the T-hybridoma immunological synapse. Immunity 2002; 16(4):521-534.
58. Hoppe A, Christensen K, Swanson JA. Fluorescence resonance energy transfer-based stoichiometry in living cells. Biophys J 2002; 83(6):3652-3664.
59. Zal T, Gascoigne NR. Photobleaching-corrected FRET efficiency imaging of live cells. Biophys J 2004; 86(6):3923-3939.
60. Gordon GW, Berry G, Liang XH et al. Quantitative fluorescence resonance energy transfer measurements using fluorescence microscopy. Biophys J 1998; 74:2702-2713.
61. Chen H, Puhl HL, Koushik SV et al. Measurement of FRET efficiency and ratio of donor to acceptor concentration in living cells. Biophys J 2006; 91(5):L39-41.
62. Steiner M. Changes in the distribution of platelet membrane proteins revealed by energy transfer. Biochim Biophys Acta 1984; 805(1):53-58.
63. Guo C, Dower SK, Holowka D et al. Fluorescence resonance energy transfer reveals interleukin (IL)-1-dependent aggregation of IL-1 type I receptors that correlates with receptor activation. J Biol Chem 1995; 270(46):27562-27568.
64. Damjanovich S, Bene L, Matko J et al. Preassembly of interleukin 2 (IL-2) receptor subunits on resting Kit 225 K6 T-cells and their modulation by IL-2, IL-7, and IL-15: A fluorescence resonance energy transfer study. Proc Natl Acad Sci USA 1997; 94(24):13134-13139.
65. Fernandez-Miguel G, Alarcon B, Iglesias A et al. Multivalent structure of an alphabetaT-cell receptor. Proc Natl Acad Sci USA 1999; 96(4):1547-1552.
66. Block MS, Johnson AJ, Mendez-Fernandez Y et al. Monomeric class I molecules mediate TCR/CD3 epsilon/CD8 interaction on the surface of T-cells. J Immunol 2001; 167(2):821-826.
67. Lee PU, Kranz DM. Allogeneic and syngeneic class I MHC complexes drive the association of CD8 and TCR on 2C T-cells. Mol Immunol 2003; 39(12):687-695.
68. Buslepp J, Kerry SE, Loftus D et al. High affinity xenoreactive TCR: MHC interaction recruits CD8 in absence of binding to MHC. J Immunol 2003; 170(1):373-383.
69. Yachi PP, Ampudia J, Gascoigne NR et al. Nonstimulatory peptides contribute to antigen-induced CD8-T-cell receptor interaction at the immunological synapse. Nat Immunol 2005; 6(8):785-792.
70. Zwart W, Griekspoor A, Kuijl C et al. Spatial separation of HLA-DM/HLA-DR interactions within MIIC and phagosome-induced immune escape. Immunity 2005; 22(2):221-233.
71. Tolar P, Sohn HW, Pierce SK. The initiation of antigen-induced B-cell antigen receptor signaling viewed in living cells by fluorescence resonance energy transfer. Nat Immunol 2005; 6(11):1168-1176.
72. Sohn HW, Tolar P, Jin T et al. Fluorescence resonance energy transfer in living cells reveals dynamic membrane changes in the initiation of B-cell signaling. Proc Natl Acad Sci USA 2006; 103(21):8143-8148.
73. van Rheenen J, Langeslag M, Jalink K. Correcting confocal acquisition to optimize imaging of fluorescence resonance energy transfer by sensitized emission. Biophys J 2004; 86(4):2517-2529.
74. Vickery SA, Dunn RC. Scanning near-field fluorescence resonance energy transfer microscopy. Biophys J 1999; 76(4):1812-1818.
75. Vickery SA, Dunn RC. Combining AFM and FRET for high resolution fluorescence microscopy. J Microsc 2001; 202(Pt 2):408-412.

76. Gadella TW Jr, Jovin TM. Oligomerization of epidermal growth factor receptors on A431 cells studied by time-resolved fluorescence imaging microscopy. A stereochemical model for tyrosine kinase receptor activation. J Cell Biol 1995; 129(6):1543-1558.

77. Verveer PJ, Wouters FS, Reynolds AR et al. Quantitative imaging of lateral ErbB1 receptor signal propagation in the plasma membrane. Science 2000; 290(5496):1567-1570.

78. Verveer PJ, Squire A, Bastiaens PI. Improved spatial discrimination of protein reaction states in cells by global analysis and deconvolution of fluorescence lifetime imaging microscopy data. J Microsc 2001; 202(Pt 3):451-456.

79. Duncan RR, Bergmann A, Cousin MA et al. Multi-dimensional time-correlated single photon counting (TCSPC) fluorescence lifetime imaging microscopy (FLIM) to detect FRET in cells. J Microsc 2004; 215(Pt 1):1-12.

80. Elangovan M, Day RN, Periasamy A. Nanosecond fluorescence resonance energy transfer-fluorescence lifetime imaging microscopy to localize the protein interactions in a single living cell. J Microsc 2002; 205(Pt 1):3-14.

81. Clegg RM, Murchie AI, Zechel A et al. Fluorescence resonance energy transfer analysis of the structure of the four-way DNA junction. Biochemistry 1992; 31(20):4846-4856.

82. Chen Y, Periasamy A. Characterization of two-photon excitation fluorescence lifetime imaging microscopy for protein localization. Microsc Res Tech 2004; 63(1):72-80.

83. Krishnan RV, Masuda A, Centonze VE et al. Quantitative imaging of protein-protein interactions by multiphoton fluorescence lifetime imaging microscopy using a streak camera. J Biomed Opt 2003; 8(3):362-367.

84. Peter M, Ameer-Beg SM. Imaging molecular interactions by multiphoton FLIM. Biol Cell 2004; 96(3):231-236.

85. Liu Y, Walter S, Stagi M et al. LPS receptor (CD14): A receptor for phagocytosis of Alzheimer's amyloid peptide. Brain 2005; 128(Pt 8):1778-1789.

86. Clegg RM. FRET tells us about proximities, distances, orientations and dynamic properties. J Biotechnol 2002; 82(3):177-179.

87. Van Munster EB, Gadella TWJ. phiFLIM: a new method to avoid aliasing in frequency-domain fluorescence lifetime imaging microscopy. J Microsc 2004; 213(Pt 1):29-38.

88. Clayton AH, Hanley QS, Verveer PJ. Graphical representation and multicomponent analysis of single-frequency fluorescence lifetime imaging microscopy data. J Microsc 2004; 213(Pt 1):1-5.

89. Gertler A, Biener E, Ramanujan KV et al. Fluorescence resonance energy transfer (FRET) microscopy in living cells as a novel tool for the study of cytokine action. J Dairy Res 2005; 72 Spec No:14-19.

90. Calleja V, Ameer-Beg SM, Vojnovic B et al. Monitoring conformational changes of proteins in cells by fluorescence lifetime imaging microscopy. Biochem J 2003; 372(Pt 1):33-40.

91. Suhling K, Siegel J, Phillips D et al. Imaging the environment of green fluorescent protein. Biophys J 2002; 83(6):3589-3595.

92. McCann FE, Suhling K, Carlin LM et al. Imaging immune surveillance by T-cells and NK cells. Immunol Rev 2002 2; 189:179-192.

93. Treanor B, Lanigan PM, Suhling K et al. Imaging fluorescence lifetime heterogeneity applied to GFP-tagged MHC protein at an immunological synapse. J Microsc 2005; 217(Pt 1):36-43.

94. Lidke DS, Nagy P, Barisas BG et al. Imaging molecular interactions in cells by dynamic and static fluorescence anisotropy (rFLIM and emFRET). Biochem Soc Trans 2003; 31(Pt 5):1020-1027.

95. Cohen-Kashi M, Moshkov S, Zurgil N et al. Fluorescence resonance energy transfers measurements on cell surfaces via fluorescence polarization. Biophys J 2002; 83(3):1395-1402.

96. Harpur AG, Wouters FS, Bastiaens PI. Imaging FRET between spectrally similar GFP molecules in single cells. Nat Biotechnol 2001; 19(2):167-169.

97. Squire A, Verveer PJ, Rocks O et al. Red-edge anisotropy microscopy enables dynamic imaging of homo-FRET between green fluorescent proteins in cells. J Struct Biol 2004; 147(1):62-69.

98. Clayton AH, Hanley QS, Arndt-Jovin DJ et al. Dynamic fluorescence anisotropy imaging microscopy in the frequency domain (rFLIM). Biophys J 2002; 83(3):1631-1649.

99. Mattheyses AL, Hoppe AD, Axelrod D. Polarized fluorescence resonance energy transfer microscopy. Biophys J 2004; 87(4):2787-2797.

100. Rizzo MA, Piston DW. High-contrast imaging of fluorescent protein FRET by fluorescence polarization microscopy. Biophys J 2005; 88(2):L14-16.

101. Wallrabe H, Periasamy A. Imaging protein molecules using FRET and FLIM microscopy. Curr Opin Biotechnol 2005; 16(1):19-27.

102. Stockholm D, Bartoli M, Sillon G et al. Imaging calpain protease activity by multiphoton FRET in living mice. J Mol Biol 2005; 346(1):215-222.

103. Zal MA, Nelson M, Zal T. Interleaved dual-wavelength multiphoton imaging system for heterologous FRET and versatile fluorescent protein excitation. Proceedings of SPIE 2007; 6442.

104. Tertoolen LG, Blanchetot C, Jiang G et al. Dimerization of receptor protein-tyrosine phosphatase alpha in living cells. BMC Cell Biol 2001; 2:8.
105. Bhatia S, Edidin M, Almo SC et al. Different cell surface oligomeric states of B7-1 and B7-2: Implications for signaling. Proc Natl Acad Sci USA 2005; 102(43):15569-15574.
106. Wallrabe H, Stanley M, Periasamy A et al. One- and two-photon fluorescence resonance energy transfer microscopy to establish a clustered distribution of receptor-ligand complexes in endocytic membranes. J Biomed Opt 2003; 8(3):339-346.
107. Wallrabe H, Chen Y, Periasamy A et al. Issues in confocal microscopy for quantitative FRET analysis. Microsc Res Tech 2006; 69(3):196-206.
108. Wirz SA, Davis CN, Lu X et al. Homodimerization and internalization of galanin type 1 receptor in living CHO cells. Neuropeptides 2005; 39(6):535-546.
109. Clayton AH, Walker F, Orchard SG et al. Ligand-induced dimer-tetramer transition during the activation of the cell surface epidermal growth factor receptor-A multidimensional microscopy analysis. J Biol Chem 2005; 280(34):30392-30399.
110. Tenhumberg S, Schuster B, Zhu L et al. gp130 dimerization in the absence of ligand: preformed cytokine receptor complexes. Biochem Biophys Res Commun 2006; 346(3):649-657.
111. Kramer JM, Yi L, Shen F et al. Evidence for ligand-independent multimerization of the IL-17 receptor. J Immunol 2006; 176(2):711-715.
112. Krause CD, Mei E, Xie J et al. Seeing the light: Preassembly and ligand-induced changes of the interferon gamma receptor complex in cells. Mol Cell Proteomics 2002; 1(10):805-815.
113. Krause CD, Mei E, Mirochnitchenko O et al. Interactions among the components of the interleukin-10 receptor complex. Biochem Biophys Res Commun 2006; 340(2):377-385.
114. Nazarov PV, Koehorst RB, Vos WL et al. FRET study of membrane proteins: determination of the tilt and orientation of the N-terminal domain of M13 major coat protein. Biophys J 2007; 92(4):1296-1305.
115. Mittler RS, Goldman SJ, Spitalny GL et al. T-cell receptor-CD4 physical association in a murine T-cell hybridoma: induction by antigen receptor ligation. Proc Natl Acad Sci USA 1989; 86(21):8531-8535.
116. Yachi PP, Ampudia J, Zal T et al. Altered peptide ligands induce delayed CD8-T-Cell receptor Interaction-a role for CD8 in distinguishing antigen quality. Immunity 2006; 25(2):203-211.
117. Matko J, Edidin M. Energy transfer methods for detecting molecular clusters on cell surfaces. Methods Enzymol 1997; 278:444-462.
118. Grailhe R, Merola F, Ridard J et al. Monitoring protein interactions in the living cell through the fluorescence decays of the cyan fluorescent protein. Chemphyschem 2006; 7(7):1442-1454.
119. Glebov OO, Nichols BJ. Lipid raft proteins have a random distribution during localized activation of the T-cell receptor. Nat Cell Biol 2004; 6(3):238-243.
120. Sharma P, Varma R, Sarasij RC et al. Nanoscale organization of multiple GPI-anchored proteins in living cell membranes. Cell 2004; 116(4):577-589.
121. Matko J, Bodnar A, Vereb G et al. GPI-microdomains (membrane rafts) and signaling of the multi-chain interleukin-2 receptor in human lymphoma/leukemia T-cell lines. Eur J Biochem 2002; 269(4):1199-1208.
122. Vamosi G, Bodnar A, Vereb G et al. IL-2 and IL-15 receptor alpha-subunits are coexpressed in a supramolecular receptor cluster in lipid rafts of T-cells. Proc Natl Acad Sci USA 2004; 101(30):11082-11087.
123. Nagy P, Vereb G, Sebestyen Z et al. Lipid rafts and the local density of ErbB proteins influence the biological role of homo- and heteroassociations of ErbB2. J Cell Sci 2002; 115(Pt 22):4251-4262.
124. Kiskowski MA, Kenworthy AK. In silico characterization of resonance energy transfer for disk-shaped membrane domains. Biophys J 2007; 92(9):3040-3051.
125. Nazarov PV, Koehorst RB, Vos WL et al. FRET study of membrane proteins: Simulation-based fitting for analysis of membrane protein embedment and association. Biophys J 2006; 91(2):454-466.
126. You X, Nguyen AW, Jabaiah A et al. Intracellular protein interaction mapping with FRET hybrids. Proc Natl Acad Sci USA 2006; 103(49):18458-18463.
127. Lakowicz JR. Principles of Fluorescence Spectroscopy. New York: Plenum Publishing Corporation; 1999.
128. Matyus L, Szollosi J, Jenei A. Steady-state fluorescence quenching applications for studying protein structure and dynamics. J Photochem Photobiol B 2006; 83(3):223-236.

CHAPTER 15

Immunogenicity in Peptide-Immunotherapy:
From Self/Nonself to Similar/Dissimilar Sequences

Darja Kanduc*

Abstract

The nature of the relationship between an antigenic amino acid sequence and its capability to evoke an immune response is still an unsolved problem. Although experiments indicate that specific (dis)continuous amino acid sequences may determine specific immune responses, how immunogenic properties and recognition informations are mapped onto a non-linear sequence is not understood.

Immunology has invoked the concept of self/nonself discrimination in order to explain the capability of the organism to selectively immunoreact. However, no clear, logical and rational pathway has emerged to relate a structure and its immuno-nonreactivity. It cannot yet be dismissed what Koshland wrote in 1990: "Of all the mysteries of modern science, the mechanism of self versus nonself recognition in the immune system ranks at or near the top."[1]

This chapter reviews the concept of self/nonself discrimination in the immune system starting from the historical perspective and the conceptual framework that underlie immune reaction pattern. It also introduces future research directions based on a proteomic dissection of the immune unit, qualitatively defined as a low-similarity sequence and quantitatively delimitated by the minimum amino acid requisite able to evoke an immune response, independently of any, microbial or viral, "foreignness".

Introduction

Peptides and anti-peptide antibodies are widely used in biochemistry and molecular biology mostly for purification and characterization of specific oligopeptides and proteins, characterization of protein-protein, enzyme-substrate, or enzyme-inhibitor interactions, as well as for identification and mapping of the binding sites of antibodies. In addition, the last two decades have seen the exploitation of peptide antigens and anti-peptide antibodies in disease diagnosis and synthetic vaccine development. Previous and current clinical trials test a number of peptide-based vaccines against cancer[1-4] and both autoimmune and infectious diseases.[5-8] These vaccines suppose that short amino acid fragments derived from the parent protein antigen may induce or augment an immune response in cancer or, viceversa, alternatively the vaccines may neutralize autoreactive autoantibodies in autoimmune pathologies. As a matter of fact, peptide-immunotherapy appears able to obtain antibodies of predetermined specificity and without the complications associated with whole cells or entire protein vaccines.

*Darja Kanduc—Department of Biochemistry and Molecular Biology, University of Bari, Via Orabona 4, 70126 Bari, Italy. Email: d.kanduc@biologia.uniba.it

Multichain Immune Recognition Receptor Signaling: From Spatiotemporal Organization to Human Disease, edited by Alexander B. Sigalov. ©2008 Landes Bioscience and Springer Science+Business Media.

Whatever the purpose of using peptide antigens and anti-peptide antibodies or whether trying to evoke or neutralize an immune response, success depends on the precise and exact identification of the antigenic peptide sequences at the root of an immune response. In this regard, a central concern is the understanding of the molecular basis of immunogenicity.

The Question: What Renders a Peptide Immunogenic?

We currently do not know why a peptide sequence is non-immunogenic, how the changing of only one amino acid residue can dramatically alter the peptide non-immunogenicity,[9] and when and where it arises the premises of the tight physico-chemical interaction between paratope and epitope. Empirically, epitope mapping shows the amino acid sequence interacting with the antibody under analysis. In abstract, we talk about the fine discrimination of the immune system's ability to sense and understand that specific single amino acid residue which is changed in the sequence. Experimentally, we are able to use harsh pH conditions and high salt reagents to break the strong bonds between the epitope and the paratope. But we do not know how the immunogenic epitope potency and the high paratope specificity originate.

Our ignorance is mainly due to (and partly justified by) the complexity of the system. As a general definition, the epitope is a set of atom groups in three-dimensional space that form the "target" of the immunoglobulin antigen binding site (the paratope). A typical epitope is roughly 5-6 residues long, but both trimer and octamer epitopes have been described.[10] The variability of the epitope length itself adds several orders of complexity to the analysis of epitope specificity. Indeed, peptide diversity is enormous and fits well with the enormous potential of antibody diversity. Using the 20 naturally occurring amino acids, one can generate about 3×10^6 different 5-mer peptides and about 2.5×10^{10} different 8-mer peptides.

Protein epitopes (or antigenic determinants) are classified into linear and nonlinear determinants. The latter are composed of noncontiguous residues that are not adjacent in the parent protein primary sequence but become so by three-dimensional folding. That makes practically infinite the possible determinant configurations, especially when considering a medium to high molecular weight protein. Moreover, numerically limited linear determinants might exist in a few configurations as components of linear determinants endowed with some mobility. Nonetheless, antibodies are exquisitely specific by hitting only a few of the numerous possible epitopic sequences.

The factors by which specific peptides are able to induce a B-cell response remain elusive.[11,12] A parallel question is present in our understanding in T-cell recognition: although the structural characteristics of the trimeric complex is clarifying[13,14] and we have learned that the TCRs recognize peptide-MHC with diagonal orientation and the CDR3 domains interact with the peptide bound to the MHC,[15,16] we remain ignorant of the functional process by which specific peptides are able to induce a T-cell response.

So, the inescapable question is: what characteristics render a peptide capable to evoke an immune response?

From a Historical Point of View—The Immune Response and Self/Nonself Sequences

The question of "what characteristics enable a structure to evoke an immune response" was easily answered in the nascent immunology of the late 19th-century concerned with understanding harmful infectious diseases.[17] In that context, immunology started as the study of defence mechanisms against the foreign pathogens. The patient as the attacked host became the "self" whose integrity had been threatened by external, foreign, nonself enemies. Slowly and tacitly the basis of the self/nonself dichotomy were dogmatically established in immunology. Potential immunogens were catalogued according to this self/nonself discrimination principle[18] with "nonself" strictly defined as belonging to a foreign organism, as opposed to "self", which were tolerated elements eliciting no response by being part of the organism itself.[19] The language of self and nonself had its foundations in a metchnikovian image of competitive struggle between organisms and infectious agents (e.g., bacteria and viruses)[20] and reflects the antinomy between benign and toxic, protection and damage, internal and external.[21-23] In this perspective, intentionality and teleology became the molecular biochemistry of the immune response. The self became (and still is) a human category with ethical, political, psychological and existential meanings. The same immune system is viewed as 'recognizing', 'remembering', 'learning' and 'acting'—terms borrowed from the cognitive sciences.[18]

The immunological self/nonself antinomy became an example of "coincidentia oppositorum" by which everything could be intuitively explained, from cancer (tumor escape from immunosurveillance) to autoimmunity (self-defense excess). With the antibody molecule chemistry and pathology overlapped and the clear-cut physico-chemical coordinates that marked the three distinct domains of antigenicity, immunogenicity and pathogenicity fused together, surrounded by the emotional involvement of good against bad in a war-peace scenario.

This dominant self/nonself perspective remained unaltered during the century from Metchnikoff through Burnet[24,25] and still lingers.[26-28] Upon that metaphor, a theory of immunological tolerance was constructed that still dominates the field. Changes amount to little more than new terminology such as, Matzinger's danger model.[22,29] "Standing on the shoulders of the Self/Nonself",[29] the danger model proposes that antigen-presenting-cells are activated by danger/alarm signals from injured cells, such as those exposed to pathogens, toxins, or mechanical damage.

The danger model pari passu reproposes the Metchnikoff's overall representation, where the phagocyte is an agent[30] able to "sense" and "understand" the danger and, consequently, mount a response with a sense of independent arbitration.[17]

In this regard, Oldstone's molecular mimicry hypothesis, which defined molecular mimicry as similar structures—either linear amino acid sequences or their conformational foldings—shared by the host and virus, made significant scientific progress. The hypothesis suggested cross-reactivity between similar microbial determinants and host 'self' antigens as a pathogenic mechanism for autoimmune disease. In the hypothesis, the immune response against the determinant evokes a destructive tissue-specific immune response and the induction of cross-reactivity does not require a replicating agent, since the immune-mediated injury could occur after the immunogen has been removed—a hit-and-run event. The Oldstone's hypothesis marks a breaking point with the perspective of bad attacking spirits and good defensive intentions and introduces the immune response in molecular terms. For the first time in the immunology history, foreign entities have been reductionistically defined as bacterial or viral molecular sequences that mimic host molecular sequences.

The hypothesis has had an enormous impact on the science of the time and has greatly contributed to developing the sequence bioinformatic tools all of us utilize routinely. An intensive effort was undertaken in the attempt to validate the association of infectious agents with autoimmunity using molecular mimicry models to dissect the parameters required for the activation and association of virus-induced autoimmune disease. For decades the attention focused (and still focuses) on possible associations between infectious agents and autoimmunity. A list of examples includes, but are not limited to: Mycobacterium tuberculosis[33] and adjuvant arthritis; beta haemolytic streptococci and rheumatic fever;[34-36] herpes and autoimmune reactions against corneal tissues;[37] B3 coxsackieviruses and myocarditis;[38] Trypanosoma cruzi and Chagas' disease;[39] diverse viruses and multiple sclerosis;[40-44] Borrelia burgdorfii and Lyme arthritis;[45,46] and B4 Coxsackievirus, cytomegalovirus or rubella and type 1 diabetes.[47-52] However, many of the postulated associations remain unproven.[50,51,53]

Exempli gratia, the fact that the nitrogenase enzyme of K. pneumoniae, a bacterium present in the bowels of many individuals including ankylosing spondylitis patients, contains a 6-mer amino acid motif in common with HLA B27 protein sequence has repeatedly been reported as a model of molecular mimicry that might have a role in ankylosing spondylitis autoimmune disease. However, it is not a new observation that ankylosing spondylitis is mainly limited to the synovial joints of the spine, whereas HLA B27 molecules are expressed on almost all somatic cells.[54]

The weakness in the molecular mimicry hypothesis appears to be that molecular sequences are not analysed by themselves as a function of their own intrinsic qualities such as hydrophobicity/hydrophylicity, function, reactivity, 3-D conformation, masking (by glycosylation, polymerization, pairing to other molecules, etc.), spatiotemporal expression, quantitative level of expression, stability/

Table 1. From literature: examples of epitopic peptides characterized by being (or containing) sequences with low similarity to the host proteome

Protein	Amino Acid Position	Sequence*	Matches**	Proteome
Large tumour antigen (tag) of simian virus 40[67]	91-95	WEQWW	0	Murine
Duffy glycoprotein[70]	22-26	FEDVW	0	Murine
Receptor of vascular endothelial growth factor[71]	262-266	YPSSK	3	Murine
	256/257/261/ 313/315***	IDELT	2	Murine
p185HER2[80]	235-243	cCHEQCAag	0	Murine
Bovine leukemia virus transactivator protein tax[81]	261-280	HVWSSpqalqrflhdptltw	2	Murine
HIV gp41[82]	683-689	NWFDlt	0	Murine
Bordetella pertussis FIM2[83]	74-80	gRTPFli	1	Human
Bordetella pertussis FIM3[83]	53-69	kvvqlpklSKNAlrndg	2	Human
	91-97	IkLYFEP	2	Human
Gluthathione-s-transferase from *Schistosoma bovis*[84]	58-67	iTDNHGHvkw	0	Murine
Leishmania infantum GRP94[85]	281-300	tqgvvkerrwtlvneNRPIW	0	Human
λ Repressor CII[86]	12-26	ledarrLKAIYekkk	1	Murine
Ovalbumin[87]	325-336	isqavhaaHAEINe	1	Murine
Toxic shock syndrome toxin-1[88]	47-56	fpSPYYSpaf	2	Murine
Staphylococcal enterotoxin B[88]	83-92	dvfgaNYYYQ	0	Murine
HOXD4 protein[89]		VYPWMK	0	Murine
α-subunit CK2[90]	319-324	MEHPYf	0	Murine
HLA class I H chain[91]	55-64	egpEYWDR(n/e)t	1	Murine
Cytochrome P4502D6[92]	193-212	RRFEYddprflrlldlaqeg	2	Human
Acetylcholinesterase[93]	112-119	tpvLVWIY	0	Murine
	143-151	rtvlVSMNY	2	Murine
	294-302	VFRFSfvpv	2	Murine
	332-341	kdegsYFLVY	2	Murine
	496-503	kapQWPPY	1	Murine
	523-532	glraqACAFW	0	Murine
Acetyl-Choline Receptor[94]	111-126	qytGHITWTppaifks****	0	Human
	122-138	aifkSYCEIlvthfpfd****	0	Human
	182-198	gwkhsvTYSCCpdtpy****	1	Human

Continued on next page

Table 1. Continued

Protein	Amino Acid Position	Sequence*	Matches**	Proteome
Myelin Basic Protein[95-97]	1-11	asqkrPSQRHg	1	Murine
	83-99	adpgsRPHLlrlfsrda	1	Murine
	70-89	tadPKNAWQD ahpadpgsrp	0	Human
Proteo-Lipid Protein [97]	139-151	chCLGKWlghpdk****	0	Murine
	178-191	FNTWTtcqsiafps****	1	Murine
Myelin Oligodendrocyte Glycoprotein[97]	1-22	gqfrVIGPRhpiralvgdevel	1	Murine
	92-106	deggFTCFFrDHSYQ****	0	Murine
Thyroglobulin[98]	2339-2358	qvaaltWVQTHirgfggdpr	0	Human
	2471-2490	pparalkRSLWVevdlligs	3	Human
	2651-2670	yefsrkvptfaTPWPDfvp	1	Human

*Low similarity 5-mers given in capital letters. **Matches: refer to the 5-mer in capital letters; correspond to the number of times a 5-mer occurs in the set of proteins that comprehensively constitutes the host proteome; calculated as already described in detail.[72-79] Low-similarity numerically defined as ≤ 3. ***Conformational epitope. ****All 5-mers forming the determinant have low similarity to the host proteome.

half-life time, proteolytic susceptibility, etc. The oldstonian analysis of the molecular sequences in the immunological context is still based mostly upon their derivance from bacterial or viral organisms.

From a Logical Point of View—The Immune Response and Similar/Dissimilar Sequences

Recently, it has been proposed that sequence similarity to the host proteome may modulate peptide immunogenicity.[55-58] The rationale is the following. If it is true that normal autoantigens are tolerated through the elimination of the antigen-reactive cells[59,60] and that the receptor repertoire must be purged of all antigen receptors that could possibly recognize self-antigens,[61-64] then it is logical to postulate that the sequences/patterns never or uniquely expressed in a proteome have more chances to escape the deletion process and, consequently, have more chances to induce an immune response.

But How to Define Sequence Similarity in the Immunological Context ?

Similarity between biological sequences is represented as sequence identity: the number of aligned positions where the corresponding characters (e.g., amino acids in proteins) are identical.[65] This protein similarity characterization by amino acid sequence comparison utilizes (multi)alignment programs and represents a very accurate method for predicting an evolutionary relationship among sequences.[66] High sequence identity (i.e., a high number of identical aligned amino acid residues) between two biological sequences indicates they belong to the same family. In other words, amino acid sequence similarity is a property that describes evolutionary history and whether biological sequences have a common ancestor.

In the immunological context, similarity analysis among biological sequences identifies amino acid groupings that represent rare or common sequences and, consequently, might or might not be considered as possible epitopes. To this aim, similarity search for immunogenic amino acid groupings (that we named Immunogenic Peptide Blocks, IPBs) utilizes perfect peptide match programs and, by so doing, transforms sequence similarity from an evolutionary quality (when two sequences are compared point by point, i.e., amino acid by amino acid) into a mathematical quantity that describes "IPB percent identity" and can be numerically measured by the match number, i.e., the number of times that an IPB is present in the set of proteins analysed.

In such a context, the immunological significance of the IPB percent similarity to the host proteome primarily depends on the length definition of the shorter sequence that can constitute a linear determinant. Since literature data indicate five to six amino acids are sufficient minimal antigenic determinants,[67-71] IPB was defined as delimited by a minimal epitopic length of five amino acids. Therefore, immunologically the similarity between a pair of aligned biological sequences may be represented by the number of aligned IPBs (e.g., 5-mers in proteins) with perfect identity matching. Using this definition, the similarity level of a peptide sequence to a proteome is calculated as the number of times the peptide pentamers occur in the analysed proteome. More precisely, the similarity level of a peptide is zero when the 5-mers forming the peptide are absent in the proteome under analysis, whereas the similarity level of a peptide is high when its 5-mers are repeatedly represented in the protein set that comprehensively forms the proteome. As an important collateral notation, the relationship between peptide and proteome introduces the difference between similarity and redundancy, where similarity applies to peptide sequences from heterologous proteins and redundancy refers to autologous peptide sequences.

Browsing Through Literature: Similarity Level of Identified Epitopes

IPB similarity analysis has been successfully applied to define epitopic sequences in different experimental models.[72-79] In addition, the data obtained by analysing the scientific literature on identified epitopes are even more eloquent. Table 1 illustrates the concrete application of this IBP similarity rationale in analyzing the literature data. It shows how a first screening produced dozens of well-defined epitopic sequences that are or harbor IPB(s) with no or low similarity to the host proteome.

Concluding Remarks

In 1859 Darwin demonstrated that complex, gradual adaptation processes arise over time without outside agency and, in so doing, he demolished teleology in science. Nonetheless, today we still have an immunology science dominated by the teleology of intentionality: explaining immune reactions in terms of self entities against nonself enemies and interpreting immune processes as meditated actions against enemies and protective conduct towards self entities.

In this context, the development of high-throughput technologies and the nascent peptidomics research offer exciting new opportunities to comprehensively analyse peptides in the immune subsystem, that is to define the immuno-peptidome. The time for a more precise answer to the logical question, "what are the molecular features that make a peptide immunogenic?" appears closer. The time for a geometrical definition of the limits and intersections among the three distinct domains of peptide antigenicity, immunogenicity and pathogenicity is getting closer as well. The challenges for these goals lie in archiving and functionally relating the vast majority of data derived from immunoassay experiments and bioinformatic predictions into a coherent informational mass relevant to physio- and pathological processes. To this end, it will be necessary to establish universally accepted criteria for positive identification of immunoreactive peptides to design effective peptide-immunotherapies.

References

1. Koshland DJ Jr. Recognizing self from nonself. Science 1990; 248:4961.
2. Gjertsen MK, Bakka A, Breivik J et al. Vaccination with mutant ras peptides and induction of T-cell responsiveness in pancreatic carcinoma patients carrying the corresponding RAS mutation. Lancet 1995; 346:1399-400.
3. Noguchi M, Kobayashi K, Suetsugu N et al. Induction of cellular and humoral immune responses to tumor cells and peptides in HLA-A24 positive hormone-refractory prostate cancer patients by peptide vaccination. Prostate 2003; 57:80-92.

4. Moulton HM, Yoshihara PH, Mason DH et al. Active specific immunotherapy with a beta-human chorionic gonadotropin peptide vaccine in patients with metastatic colorectal cancer: Antibody response is associated with improved survival. Clin Cancer Res 2002; 8:2044-2051.

5. Offner H, Vandenbark AA. Congruent effects of estrogen and T-cell receptor peptide therapy on regulatory T-cells in EAE and MS. Int Rev Immunol 2005; 24:447-477.

6. Herrington DA, Clyde DF, Losonsky G et al. Safety and immunogenicity in man of a synthetic peptide malaria vaccine against Plasmodium falciparum sporozoites. Nature 1987; 328:257-259.

7. El Kasmi KC, Muller CP. New strategies for closing the gap of measles susceptibility in infants: Towards vaccines compatible with current vaccination schedules. Vaccine 2001; 19:2238-2244.

8. Sabhanini L, Manocha M, Sridevi K et al. Developing subunit immunogens using B- and T-cell epitopes and their constructs derived from F1 antigen of Yersinia pestis using novel delivery vehicles. FEMS Immunol Med Microbiol 2003; 1579:1-15.

9. Wiens GD, Pascho R, Winton JR. A single Ala139-to-Glu substitution in the Renibacterium salmoninarum virulence-associated protein p57 results in antigenic variation and is associated with enhanced p57 binding to chinook salmon leukocytes. Appl Environ Microbiol 2002; 68:3969-3977.

10. Stephen CW, Helminen P, Lane DP. Characterisation of epitopes on human p53 using phage-displayed peptide libraries: Insights into antibody-peptide interactions. J Mol Biol 1995; 248:58-78.

11. van Regenmortel MHV. The recognition of proteins and peptides by antibodies. J Immunoassay 2000; 21:85-108.

12. van Regenmortel MHV. Antigenicity and immunogenicity of synthetic peptides. Biologicals 2001; 29:209-213.

13. Kellenberger C, Porciero S, Roussel A. Expression, refolding, crystallization and preliminary crystallographic study of MHC H-2K(k) complexed with octapeptides and non-apeptides. Acta Crystallogr D Biol Crystallogr 2004; 60:1278-1280.

14. Webb AI, Borg NA, Dunstone MA et al. The structure of H-2K(b) and K(bm8) complexed to a Herpes Simplex virus determinant: evidence for a conformational switch that governs T-cell repertoire selection and viral resistance. J Immunol 2004; 173:402-409.

15. Bankovich AJ, Garcia KC. Not just any T-cell receptor will do. Immunity 2003; 18:7-11.

16. Garcia KC, Adams EJ. How the T-cell receptor sees antigen—A structural view. Cell 2005; 122:333-336.

17. Tauber AI. The biological notion of self and nonself. In: Zalta EN, ed. The Stanford Encyclopedia of Philosophy. Spring 2006 Edition. (http:// plato.stanford.edu/ archives/spr2006/entries/biology-self/).

18. Tauber AI. Moving beyond the immune self? Semin Immunol 2000; 12:241-248.

19. Silverstein AM, Rose NR. On the mystique of the immunological self. Immunol Rev 1997; 159:197-206.

20. Tauber AI, Chernyak L. Metchnikoff and the origins of immunology: From metaphor to theory. Oxford: Oxford University Press, 1991.

21. Mitchison NA, Katz DR, Chain B. Self/nonself discrimination among immunoregulatory (CD4) T-cells. Semin Immunol 2000; 12:179-183.

22. Matzinger P. Tolerance, danger and the extended family. Annu Rev Immunol 1994; 12:991-1045.

23. Cohn M, Langman RE. The protecton: The evolutionarily selected unit of humoral immunity. Immunol Rev 1990; 115:1-131.

24. Burnet FM, Fenner F. The Production of Antibodies, 2nd edition. Melbourne: Macmillan and Co., 1949; 1-142.

25. Burnet FM. The concept of immunological surveillance. Prog Exp Tumor Res 1970; 13:1-27

26. Medzhitov R, Janeway CA Jr. Decoding the patterns of self and nonself by the innate immune system. Science 2002; 296:298-300.

27. Natarajan K, Dimasi N, Wang J et al. Structure and function of natural killer cell receptors: multiple molecular solutions to self, nonself discrimination. Annu Rev Immunol 2002; 20:853-885.

28. Hickman HD, Luis AD, Buchli R et al. Toward a definition of self: Proteomic evaluation of the class I peptide repertoire. J Immunol 2004; 172:2944-2952.

29. Matzinger P. The Danger model: A renewed sense of self. Science 296; 2002:301-305.

30. Crist E, Tauber AI. The phagocyte, the antibody and agency: Contending turn-of-the-century approaches to immunity. In: Moulin AM, Cambrosio A, eds. Singular Selves: Historical Issues and Contemporary Debates in Immunology. Amsterdam: Elsevier, 2001:115-139.

31. Fujinami RS, Oldstone MB, Wroblewska Z et al. Molecular mimicry in virus infection: Crossreaction of measles virus phosphoprotein or of herpes simplex virus protein with human intermediate filaments. Proc Natl Acad Sci USA 1983; 80:2346-2350.

32. Fourneau JM, Bach JM, van Endert PM et al. The elusive case for a role of mimicry in autoimmune diseases. Mol Immunol 2004; 40:1095-1102.

33. Van Bilsen JH, Wagenaar-Hilbers JP, Boot EP et al. Searching for the cartilage-associated mimicry epitope in adjuvant arthritis. Autoimmunity 2002; 35:201-210.
34. McDonald M, Currie BJ, Carapetis JR. Acute rheumatic fever: A chink in the chain that links the heart to the throat? Lancet Infect Dis 2004; 4:240-245.
35. Dinkla K, Rohde M, Jansen WT et al. Rheumatic fever-associated Streptococcus pyogenes isolates aggregate collagen. J Clin Invest 2003; 111:1905-1912.
36. Guilherme L, Cunha-Neto E, Coelho V et al. Human heart-infiltrating T-cell clones from rheumatic heart disease patients recognize both streptococcal and cardiac proteins. Circulation 1995; 92:415-420.
37. Panoutsakopoulou V, Sanchirico ME, Huster KM et al. Analysis of the relationship between viral infection and autoimmune disease. Immunity 2001; 15:137-147.
38. Kim KS, Hufnagel G, Chapman NM et al. The group B coxsackieviruses and myocarditis. Rev Med Virol 2001; 11:355-368.
39. Girones N, Cuervo H, Fresno M. Trypanosoma cruzi-induced molecular mimicry and Chagas' disease. Curr Top Microbiol Immunol 2005; 296:89-123.
40. Katz-Levy Y, Neville KL, Girvin AM et al. Endogenous presentation of self myelin epitopes by CNS-resident APCs in Theiler's virus-infected mice. J Clin Invest 1999; 104:599-610.
41. Kuchroo VK, Anderson AC, Waldner H et al. T-cell response in experimental autoimmune encephalomyelitis (EAE): role of self and cross-reactive antigens in shaping, tuning and regulating the autopathogenic T-cell repertoire. Annu Rev Immunol 2002; 20:101-123.
42. Carrizosa AM, Nicholson LB, Farzan M et al. Expansion by self antigen is necessary for the induction of experimental autoimmune encephalomyelitis by T-cells primed with a cross-reactive environmental antigen. J Immunol 1998; 161:3307-3314.
43. Lang HL, Jacobsen H, Ikemizu S et al. A functional and structural basis for TCR cross-reactivity in multiple sclerosis. Nat Immunol 2002; 3:940-943.
44. Miller SD, Vanderlugt CL, Begolka WS et al. Persistent infection with Theiler's virus leads to CNS autoimmunity via epitope spreading. Nat Med 1997; 3:1133-1136.
45. Salazar CA, Rothemich M, Drouin EE et al. Human Lyme arthritis and the immunoglobulin G antibody response to the 37-kilodalton arthritis-related protein of Borrelia burgdorferi. Infect Immun 2005; 73:2951-2957.
46. Raveche ES, Schutzer SE, Fernandes H et al. Evidence of Borrelia autoimmunity-induced component of Lyme carditis and arthritis. J Clin Microbiol 2005; 43:850-856.
47. Atkinson MA, Bowman MA, Campbell L et al. Cellular immunity to a determinant common to glutamate decarboxylase and Coxsackie virus in insulin-dependent diabetes. J Clin Invest 1994; 94:2125-2129.
48. Bach JM, Otto H, Jung G et al. Identification of mimicry peptides based on sequential motifs of epitopes derived from 65-kDa glutamic acid decarboxylase. Eur J Immunol 1998; 28:1902-1910.
49. Hiemstra HS, Schloot NC, van Veelen PA et al. Cytomegalovirus in autoimmunity: T-cell crossreactivity to viral antigen and autoantigen glutamic acid decarboxylase. Proc Natl Acad Sci USA 2001; 98:3988-3991.
50. Horwitz MS, Bradley LM, Harbertson J et al. Diabetes induced by Coxsackie virus: initiation by bystander damage and not molecular mimicry. Nat Med 1998; 4:781-785.
51. Schloot NC, Willemen SJ, Duinkerken G et al. Molecular mimicry in type 1 diabetes mellitus revisited: T-cell clones to GAD65 peptides with sequence homology to Coxsackie or proinsulin peptides do not crossreact with homologous counterpart. Hum Immunol 2001; 62:299-309.
52. Uemura Y, Senju S, Maenaka K et al. Systematic analysis of the combinatorial nature of epitopes recognized by TCR leads to identification of mimicry epitopes for glutamic acid decarboxylase 65-specific TCRs. J Immunol 2003; 170:947-960.
53. Benoist C, Mathis D. Autoimmunity provoked by infection: How good is the case for T-cell epitope mimicry? Nat Immunol 2001; 2:797-801.
54. Benjamin R, Parham P. Guilt by association: HLA-B27 and ankylosing spondylitis. Immunol Today 1990; 11:137-142.
55. Willers J, Lucchese A, Kanduc D et al. Molecular mimicry of phage displayed peptides mimicking GD3 ganglioside. Peptides 1999; 20:1021-1026.
56. Natale C, Giannini T, Lucchese A et al. Computer-assisted analysis of molecular mimicry between HPV16 E7 oncoprotein and human protein sequences. Immunol Cell Biol 2000; 78:580-585.
57. Kanduc D. Peptimmunology: immunogenic peptides and sequence redundancy. Curr Drug Discov Technol 2005; 2:239-244.
58. Kanduc D. Defining peptide sequences: from antigenicity to immunogenicity through redundancy. Curr Pharmacogenomics 2006; 4:33-37.
59. Burnet FM. The Clonal Selection Theory of Acquired Immunity. Cambridge University Press, Cambridge: 1959; 1-209.
60. Burnet FM. Self and Not-Self. Cambridge University Press, Cambridge: 1969; 1-318.

61. Nossal GJ. How is tolerance generated? Ciba Found Symp 1987; 129:59-72.
62. Gonzalo JA, de Alboran IM, Kroemer G. Dissociation of autoaggression and self-superantigen reactivity. Scand J Immunol 1993; 37:1-6.
63. Touma M, Mori KJ, Hosono M. Failure to remove autoreactive Vbeta6+ T-cells in Mls-1 newborn mice attributed to the delayed development of B-cells in the thymus. Immunology 2000; 100:424-431.
64. Mapara MY, Sykes M. Tolerance and cancer: Mechanisms of tumor evasion and strategies for breaking tolerance. J Clin Oncol 2004; 22:1136-1151.
65. May AC. Percent sequence identity; the need to be explicit. Structure 2004; 12:737-738.
66. Park J, Karplus K, Barrett C et al. Sequence comparisons using multiple sequences detect three times as many remote homologues as pairwise methods. J Mol Biol 1998; 284:1201-1210.
67. Lindner K, Mole SE, Lane DP et al. Epitope mapping of antibodies recognising the N-terminal domain of simian virus large tumour antigen. Intervirology 1998; 41:10-16.
68. Reddehase MJ, Rothbard JB, Koszinowski UH. A pentapeptide as minimal antigenic determinant for MHC class I-restricted T-lymphocytes. Nature 1989; 337:651-653.
69. Hemmer B, Kondo T, Gran B et al. Minimal peptide length requirements for CD4(+) T-cell clones-implications for molecular mimicry and T-cell survival. Int Immunol 2000; 12:375-383.
70. Wasniowska K, Petit-LeRoux Y, Tournamille C et al. Structural characterization of the epitope recognized by the new anti-Fy6 monoclonal antibody NaM 185-2C3. Transfus Med 2002; 12:205-211
71. Lu D, Kussie P, Pytowski B et al. Identification of the residues in the extracellular region of KDR important for interaction with vascular endothelial growth factor and neutralizing anti-KDR antibodies. J Biol Chem 2000; 275:14321-14330.
72. Kanduc D, Lucchese A, Mittelman A. Individuation of monoclonal anti-HPV16 E7 antibody linear peptide epitope by computational biology. Peptides 2001; 22:1981-1985.
73. Mittelman A, Lucchese A, Sinha AA et al. Monoclonal and polyclonal humoral immune response to EC HER-2/neu peptides with low similarity to the host's proteome. Int J Cancer 2002; 98:741-747.
74. Lucchese A, Stevanovic S, Sinha AA et al. Role of MHC II affinity and molecular mimicry in defining anti-HER-2/neu MAb-3 linear peptide epitope. Peptides 2003; 24:193-197.
75. Kanduc D, Fanizzi FP, Lucchese G et al. NMR probing of in silico identification of anti-HPV16 E7 mAb linear peptide epitope. Peptides 2004; 25:243-250.
76. Mittelman A, Tiwari R, Lucchese G et al. Identification of monoclonal anti-HMW-MAA antibody linear peptide epitope by proteomic database mining. J Invest Dermatol 2004; 123:670-675.
77. Dummer R, Mittelman A, Fanizzi FP et al. Nonself discrimination as a driving concept in the identification of an immunodominant HMW-MAA epitopic peptide sequence by autoantibodies from melanoma cancer patients. Int J Cancer 2004; 111:720-726.
78. Lucchese A, Mittelman A, Lin MS et al. Epitope definition by proteomic similarity analysis: identification of the linear determinant of the anti-Dsg3 MAb 5H10. J Transl Med 2004; 2:43.
79. Lucchese A, Willers J, Mittelman A et al. Proteomic scan for tyrosinase peptide antigenic pattern in vitiligo and melanoma. Role of sequence similarity and HLA-DR1 affinity. J Immunol 2005; 175:7009-7020.
80. Orlandi R, Formatici C, Menare S et al. X A linear region of a monoclonal antibody conformational epitope mapped on p185HER2 oncoprotein. Biol Chem 2005; 378:1387-1392.
81. Sakakibara N, Kabeya H, Ohashi K et al. Epitope mapping of bovine leukemia virus transactivator protein tax. J Vet Med Sci 1998; 60:599-605.
82. Cardoso RM, Zwick MB, Stanfield RL et al. Broadly neutralizing anti-HIV antibody 4E10 recognizes a helical conformation of a highly conserved fusion-associated motif in gp41. Immunity 2005; 22:163-173.
83. Williamson P, Matthews R. Epitope mapping the Fim2 and Fim3 proteins of Bordetella pertussis with sera from patients infected with or vaccinated against whooping cough. FEMS Immunol Med Microbiol 1996; 13:69-78.
84. da Costa AV, Lafitte S, Fontaine J et al. Definition and mapping of epitopes recognized by specific monoclonal antibodies to Schistosoma bovis 28 kDa glutathione S-ransferase: Relation with anti-egg viability immunity. Parasite Immunol 1999; 21:9-18.
85. Larreta R, Guzman F, Patarroyo ME et al. Antigenic properties of the Leishmania infantum GRP94 and mapping of linear B-cell epitopes. Immunol Lett 2002; 80:199-205.
86. Lai MZ, Huang SY, Briner TJ et al. T-cell receptor gene usage in the response to lambda repressor cI protein. An apparent bias in the usage of a V alpha gene element. J Exp Med 1988; 168:1081-1097.
87. Shimonkevitz R, Colon S, Kappler JW et al. Antigen recognition by H-2-restricted T-cells. II. A tryptic ovalbumin peptide that substitutes for processed antigen. J Immunol 1984; 133:2067-2074.
88. Pang LT, Kum WW, Chow AW. Inhibition of staphylococcal enterotoxin B-induced lymphocyte proliferation and tumor necrosis factor alpha secretion by MAb5, an anti-toxic shock syndrome toxin 1 monoclonal antibody. Infect Immun 2000; 68:3261-3268.

89. Burgess K, Han I, Zhang A et al. DiSSiMiL: Diverse Small Size Mini-Libraries applied to simple and rapid epitope mapping of a monoclonal antibody. J Pept Res 2001; 57:68-76.

90. Nastainczyk W, Issinger OG, Guerra B. Epitope analysis of the MAb 1AD9 antibody detection site in human protein kinase CK2alpha-subunit. Hybrid Hybridomics 2003; 22:87-90.

91. Perosa F, Luccarelli G, Prete M et al. Beta 2-microglobulin-free HLA class I heavy chain epitope mimicry by monoclonal antibody HC-10-specific peptide. J Immunol 2003; 171:1918-1926.

92. Kerkar N, Choudhuri K, Ma Y et al. Cytochrome P4502D6(193-212): A new immunodominant epitope and target of virus/self cross-reactivity in liver kidney microsomal autoantibody type1-positive liver disease. J Immunol 2003; 170:1481-1489.

93. Zhang XM, Liu G, Sun MJ. Epitopes of human brain acetylcholinesterase. Brain Res 2000; 868:157-164.

94. Deitiker P, Ashizawa T, Atassi MZ. Antigen mimicry in autoimmune disease. Can immune responses to microbial antigens that mimic acetylcholine receptor act as initial triggers of Myasthenia gravis? Hum Immunol 2000; 61:255-265.

95. He XL, Radu C, Sidney J et al. Structural snapshot of aberrant antigen presentation linked to autoimmunity: the immunodominant epitope of MBP complexed with I-Au. Immunity 2002; 17:83-94.

96. Crowe PD, Boehme SA, Wong T et al. Differential signaling and hierarchical response thresholds induced by an immunodominant peptide of myelin basic protein and an altered peptide ligand in human T-cells. Hum Immunol 1998; 59:679-689.

97. Dharmasaroja P. Specificity of autoantibodies to epitopes of myelin proteins in multiple sclerosis. J Neurol Sci 2003; 206:7-16.

98. Thrasyvoulides A, Sakarellos-Daitsiotis M, Philippou G et al. B-cell autoepitopes on the acetylcholinesterase-homologous region of human thyroglobulin: Association with Graves' disease and thyroid eye disease. Eur J Endocrinol 2001; 145:119-127.

CHAPTER 16

Therapeutic Application of Transmembrane T and Natural Killer Cell Receptor Peptides

Nicholas Manolios,* Marina Ali, Michael Amon and Veronika Bender

Abstract

Autoimmune diseases primarily mediated by T-cells effect a significant proportion of the population and include common and distressing conditions such as diabetes, multiple sclerosis, inflammatory bowel disease, skin diseases and arthritis. Current treatments are restrictive in terms of range of options and side-effect profiles and new drugs and new approaches are always eagerly sought. With the T-cell antigen receptor (TCR) as a model system we have identified a new approach to inhibit T-cell activation. By means of peptides derived from the transmembrane TCR-alpha chain region we have shown that T-cells, the major effector cells of disease, can be inhibited in vitro and the immune responses leading to disease ameliorated in animal models.

The exact molecular mechanism of peptide action is still uncertain and assumed to involve a disturbance in transmembrane protein-protein interactions mediated by amino acid charges that disrupt normal signaling pathways. This chapter summarizes the results to date of TCR core peptide (CP); the most effective peptide noted so far, in terms of function, behavior in membranes and future development and application as a therapeutic agent. The lessons learned from this model can be applied to other multi-subunit receptors that serve critical cellular functions and open new doors for drug design, development and application.

Introduction

Autoimmune and T-cell-mediated diseases effect a significant proportion of the general population worldwide and encompass disorders such as asthma, diabetes mellitus, multiple sclerosis, inflammatory bowel disease, psoriasis and arthritis. The spectrum of diseases is huge and yet the drug options available are limited. Corticosteroids and cyclosporine are the most widely used immuno-suppressants available to date. However, both have significant toxicity and side-effect profiles when administered over long periods of time, increasing the need for alternative treatment options. Since the identification of the TCR-MHC complex, biotechnology companies have focused intensively on finding a means to disrupt this interaction. Various methods and techniques have included the use of vaccines, monoclonal antibodies, or peptide mimetics with various levels of success.[1-4]

*Corresponding Author: Nicholas Manolios—Department of Rheumatology, Westmead Hospital, Corner of Hawkesbury and Darcy Road, Westmead, New South Wales, 2145, Australia. Email: nickm@westgate.wh.usyd.edu.au

Multichain Immune Recognition Receptor Signaling: From Spatiotemporal Organization to Human Disease, edited by Alexander B. Sigalov. ©2008 Landes Bioscience and Springer Science+Business Media.

There is no dispute surrounding the structural and functional complexity of the TCR. Yet despite this complexity, the constituents of this intricate structure are simply held together by noncovalent bonds and transmembrane electrostatic interactions. It is the recognition of the latter interactions between dimer pairs that has allowed us to develop an innovative approach to disrupting receptor function by using peptides able to influence protein-protein associations via inhibition of opposing transmembrane charges. Using the TCR as a model receptor, proof of principle has been demonstrated in that peptides with sequences derived from the transmembrane segments of the TCR can inhibit the immune response both in vitro and in animal models of T-cell-mediated inflammation.[5] This principle may be applicable to any ubiquitous cell surface multi-chain complex containing transmembrane charges, e.g., natural killer (NK) cells. The availability of a variety of critical receptors such as those found on T-cells, B-cells, NK-cells, Fc receptors and platelet glycoprotein VI and the essential cellular function they serve is large, giving rise to a considerable diversity of therapeutic application.

Therapeutic Application

Core peptide (CP; GLRILLLKV), a lead compound, has the potential of being used as an alternative therapy either as a single agent or as an adjunct to existing treatment protocols for immunosuppression. In animal studies, CP given subcutaneously significantly reduced the induction of T-cell-mediated inflammation in animal models with adjuvant-induced arthritis, allergic encephalomyelitis and delayed type contact hypersensitivity.[5] More recently the mode of presentation (subcutaneous versus intra-peritoneal), various solvents and conjugations to CP have been examined in the arthritis model. The mode and site of delivery are important and various CP conjugates have been shown to be comparable in effect to cyclosporine in reducing the acute phase of inflammation in the adjuvant-induced rat arthritis model (discussed further in this chapter).[6]

As an evolving platform technology, new approaches to therapy, innovative methods of delivery and fresh applications are being explored, providing exciting new options to current therapies. The application of topical CP peptide or CP cDNA in the treatment of T-cell-mediated skin diseases (mouse and human) has already been reported by Gollner et al.[7] In murine contact sensitivity, direct topical application of the peptide inhibited the elicitation of contact sensitivity following application of a contact allergen in sensitized animals.[7,8] When naked cDNA encoding the peptide sequence was injected into skin before application of contact allergen to sensitized animals, local immunosuppression was also observed.[7] This implies that the expressed peptide is structurally intact and does not undergo problems with folding or posttranslational modification. When the topical effects of CP were examined in humans with psoriasis, atopic eczema, lichen planus, or contact dermatitis all patients except one reported a marked improvement of their skin disease. These data support peptide effectiveness and indicate that CP, as peptide or cDNA treatment, might be a possible substitute for corticosteroids or immunosuppressive agents such as cyclosporine.

Furthermore, dendritic cells engineered to secrete CP have been shown to induce antigen-specific immunosuppression in vivo by Mahnke et al.[9] In a CD8-driven allergy model, the injection of dendritic cells transduced with CP significantly reduced inflammation. In a CD4+ T-cell-dependent model of multiple sclerosis, injection of CP-secreting dendritic cells abrogated symptoms and prolonged survival. These effects were antigen-specific as transduced dendritic cells that did not express the respective antigen failed to convey protection. This implies that dendritic cells engineered to secrete CP are able to suppress T-cell activation in an antigen specific and localized manner without affecting other immune cells or other cell types. This denotes CP as a very useful new agent for in vivo immune suppression.[9]

The large range of T-cell-mediated diseases allows CP to be used in a variety of clinical conditions, some of which are listed in Table 1. Since T-cells are effectors of inflammation in every organ, the list shown is representative rather than exhaustive. Recently, CP has been under investigation for the treatment of psoriasis as a topical agent and is currently under evaluation as an active ingredient for the treatment of asthma.

Table 1. Potential therapeutic targets

Organ System	Specific Disease Examples
Neural	Multiple sclerosis, Guillain Barre Syndrome
Endocrine	Diabetes, Hashimotos disease, pernicious anaemia
Skeletal	Rheumatoid arthritis, ankylosing spondylitis, reactive arthritis, systemic lupus erythematosus
Immune	Transplant rejection syndrome, urticaria, drug allergy
Dermal	Pemphigus, eczema, contact dermatitis, psoriasis
Gastro-intestinal	Chrons disease, Ulcerative colitis
Respiratory	Asthma

Despite the increasing evidence indicating therapeutic efficacy of CP, little is known about its basic mode of action and biophysical interactions with cell surface membranes. Recently, focus within our group has been aimed at investigating the physical interactions between TCR transmembrane peptides with their receptor and lipid-membrane in the hope of allowing new insight conceptually, methodologically and technologically for the development and application of these compounds.

Biophysical Attributes of CP

Many of the biological processes that occur at the cellular level require peptide–membrane interactions. These processes include the insertion and folding of membrane proteins, the formation of ion channels, translocation of polypeptides through membranes, the interaction of peptide hormones with membrane receptors, signal transduction and the actions of antimicrobial and cytotoxic peptides and lipolysis; this makes characterization of the molecular details of these processes very important.[10-13] A wide variety of biophysical techniques have been combined with the use of model membrane systems to study peptide-membrane interactions. These techniques such as circular dichroism (CD), nuclear magnetic resonance (NMR), fluorescence spectroscopy, fourier transform infrared spectroscopy, attenuated total reflection fourier transform infrared spectroscopy and surface plasmon resonance (SPR) have provided important information on the relationship between membrane-active peptide structures and their biological functions. In this section we focus on the use of several of these techniques to study the biophysical features of CP.

Surface Plasmon Resonance

SPR was used by Bender et al[14] to understand CP's affinity for different membranes. The dibasic CP was found to bind to both zwitterionic and anionic model membranes (Fig. 1) as well as to a T-cell-membrane preparation. By contrast, switching one or both of the basic residues to acidic residues on the peptide sequence led to a loss of binding to model membranes. In addition, the position of the charged amino acids in the sequence, the number of hydrophobic amino acids between the charged residues and substitution of one or both basic to neutral amino acids were found to effect binding. These results, when compared with in vitro T-cell stimulation assays and in vivo adjuvant-induced arthritis models, showed very close correlation and confirmed the findings that both amino acid charge and location have a critical role in CP activity. Table 2 summarizes the biological activity of CP and its analogues and the binding affinity of these peptides to model membranes. Interaction time of CP with the model membranes was found to be directly related with the level of bound peptide, indicating a step in the mechanism of binding, where the initially bound CP undergoes a change in conformation.

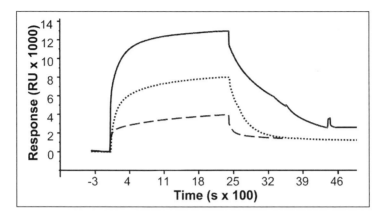

Figure 1. Sensorgram showing the binding of CP to DMPC (---), DMPG (—) and DMPC:DMPG 70:30 (...) liposomes immobilised on an L1 sensor chip. CP concentration is 50 μM. Binding is expressed as the difference in signals between points before and after sample injection. Reproduced from Bender et al.[14]

Circular Dichroism

Secondary structure analysis of CP using CD revealed that in aqueous solution the peptide contains only a small percentage of α helical and about 75% random coil structures, while CP in a hydrophobic solution (Trifluoroethanol; TFE) has about 40% α-helical structure. CD spectrum in the presence of 1,2-dimyristoyl-sn-glycero-3-phosphocholine/1,2-dimyristoyl-sn-glycero-3-phosphoglycerol (DMPC/DMPG) membrane (70:30 ratio) was found to be characteristic for peptides primarily in the β-conformation at lower concentration, whereas at higher CP concentration the spectrum was typical for an exclusively β-structure, attributed to aggregated peptides associated with the negatively charged DMPG components.[15] CD studies also revealed a dose-dependent conformational change of CP from a dominantly random coil structure to that of beta-structure as the concentration of lipid increased relative to CP. This occurred only in the presence of the anionic DMPG at a lipid:peptide molar ratio of 1.6:1, as no conformational change was observed when the neutral DMPC was tested up to a lipid:peptide ratio of 8.4:1.

Nuclear Magnetic Resonance

In the study of CP, both solution and solid-state NMR spectroscopy was applied. Initially,[1] H NMR spectroscopy was used to study the effect of CP on the mobile lipid of T-cell membranes (2B4 cells). ^{1}H NMR spectroscopy did not detect any changes in the mobile lipid in the presence of CP. This could be due to the technique not being sensitive enough or to the changes occurring in the less mobile cholesterol and/or phospholipids part of the membrane, which would not be NMR visible.[16] Extending the above studies, the effect of CP on model membranes (liposomes) was assessed by ^{31}P and ^{2}H solid state NMR spectroscopy.[15] ^{31}P and ^{2}H NMR measurements were performed to investigate whether CP is capable of perturbing the model membrane structure. A measurement temperature of 34 °C was used to ensure that the model membranes were in the biologically relevant liquid crystalline (Lα) phase. ^{31}P NMR spectra give information about the organization and dynamics of the lipid head groups of the lipid of interest. The conformation of the lipid acyl chains can be assessed by ^{2}H NMR measurements on membranes that contain lipids with perdeuterated acyl chains. These experiments found that CP did not significantly influence the structure of DMPC membranes, but the small spectral changes that were caused by CP showed that at least part of the peptide population associates with the membranes, most likely with the lipid head groups. In negatively charged model membranes, the peptide, at lower concentration (2%), seems to aggregate on the surface without disrupting the membrane structure. However,

Table 2. Percentage binding of CP and analogues to the model membrane using SPR and in vitro studies

Code	Peptide Sequence	% IL-2 Inhibition Relative To CP**	DMPC % Binding	DMPG % Binding
CP	G L R I L L L K V	100	100	100
A*	M G L R I L L L - -	0	8	5
B*	- - - I L L L K V A G	0	5	3
C*	- L G I L L L G V	0	16	0
D*	- L K I L L L R V	121	80	98
E*	G L D I L L L E V	0	0	0
G*	- L R I L L K V -	76	73	70
H*	- L R I L L L G V	73	73	40
I*	- L G I L L L K V	0	17	10
M*	I I V T D V I A T L	0	0	0
5	S S G L R I L K L L K V	0	0	0
6	G L R I L K L L K S S	0	0	0
7	G L R I L L K L V K	40	30	65
8	G L R I L L L K K V	0	0	15
9	G L K K I L L K V	0	0	0

*Published data.[14]
**Percentage of IL-2 inhibition of CP analogues relative to CP. Peptides that had no effect or stimulated IL-2 production are denoted as 0.
Abbreviations: DMPC, 1, 2-dimyristoyl-sn-glycero-3-phosphocholine; DMPG, 1, 2-dimyristoyl-sn-glycero-3-phosphoglycerol.
Reproduced with permission from Springer NL.[17]

at the higher peptide concentration (10%), most of the lipids were no longer in a planar lamellar phase and the spectra indicated that peptides and lipids were part of amorphous aggregate structures. These results indicate that CP is capable of disrupting membrane vesicles, but only at very high concentrations. Electron microscopy and flow cytometry studies also had shown that cell membranes remained intact in the presence of CP. Hence it appears that CP exerts its biological action by a more complex mechanism than simply perturbing the lipid bilayer membrane.

Biological Features of CP

CP translocates across cell membranes in a nonreceptor-mediated process, penetrating cell membranes and translocating into the cytoplasm and nucleus.[5] This feature has been used to conjugate oligodeoxynucleotides to CP and to examine CP as a potential carrier peptide for transfection (Chan et al manuscript submitted). At present, little is known regarding the extent of cellular specificity of CP and studies are in progress to determine the effects on macrophages, dendritic cells and endothelial cells. Immunocompetent cells examined to date include T-cells, B-cells and NK-cells. Further work is required to fully understand the mechanism of action and cellular effects of CP.

Cellular Specificity

T-Cells

The specificity of CP for T-cells is dependent on amino acid composition as well as the need for two critically placed positive charges within its sequence.[17] Initial experiments aimed at examining

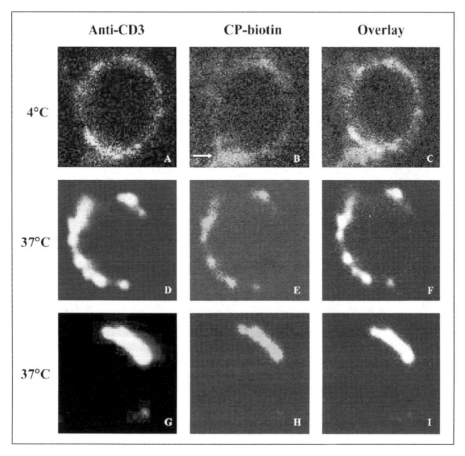

Figure 2. CP localizes with TCR in the cell membrane of human T-cells. PBMCs were incubated with FITC-conjugated antihuman immunoglobulin and STR following antiCD3 and CP-biotin incubation. After incubation, cells were paraformaldehyde-fixed (A-C) or were held at 37°C to allow TCR clustering into patches or caps before fixing (D-I). The samples were viewed by confocal microscopy. Reproduced with permission from Elsevier.[18]

CP-TCR interactions within the plasma membrane have shown that CP localizes within membranes and associates with the TCR upon T-cell activation (Fig. 2). This is a specific interaction, not noted with other transmembrane proteins such as CD45, transferrin receptor (TfR), IL-2 receptor β chain (IL-2Rβ), or the glycosylphosphatidyl inositol (GPI)-anchored protein CD14. This association upon T-cell activation prevents the TCR ζ chain phosphorylation and results in a subsequent decrease in IL-2 production.[18,19]

B-Cells

CP had no effects on B-cell proliferative responses when used at concentrations less than 50 μM known to effect T-cell function. To determine the effect of the TCR transmembrane peptides on B-cell activation, a number of different B-cell mitogens (CD40L, lipopolysaccharide (LPS) and antiIgM and antiIgD) similar to those used for T-cell activation were analyzed. These three mitogens act at different surface receptors and on different sites of the B-cell activation cascade leading to proliferation. In the first method, naïve B-cells isolated from the spleens of CBA/H mice were cultured with antiIgM in the presence or absence of peptides. The cells cultured were pulsed on the third day with a [³H]-thymidine and a thymidine incorporation assay was used to measure cellular proliferation. This method of activation is analogous to B-cells being activated by an antigen. The second mitogen used was CD40L, which acts as a ligand for the surface CD40 receptor on B-cells. This downstream method of activation involves lipid rafts and BCR co-accessory molecules. The third method used LPS to stimulate the surface LPS receptors on B-cells to cause proliferation. This is an antigen-independent method of activation. These experiments showed that TCR peptides had a differential inhibitory effect on BCR function and extra studies are required for further clarification of CP action at higher concentrations.[8,20]

NK-Cells

TCR peptides were assessed for their effect on direct NK cytotoxicity. This involved the use of the standard NK cytotoxicity assay. The target cell used was the NK sensitive K562, a human leukaemic cell line. K562 cells were labeled with radioactive chromium isotope and incubated with freshly isolated peripheral blood leukocytes (PBLs) containing fresh NK-cells. The amount of chromium released reflected the magnitude of direct NK cytotoxicity. Both CP and analogues (LP; Lipopeptide, peptide D; L-K-I-L-L-L-R-V) showed a dose-dependent inhibition of direct NK cytotoxicity. These peptides inhibited direct NK cytotoxicity by an average of 47-59%. Other peptides with negative and neutral charge substitutions had minimum or no effect on direct NK cytotoxicity.[20]

Short peptides derived from the transmembrane sequence of NK activating receptors and associated molecules (NKp46, NKp30, NKG2D and TCR ζ peptides) have been tested in vitro without any significant inhibition of NK cell cytotoxicity noted.[21] These results may have been due to technical difficulties and further investigations are warranted. The structural similarities between these immunoreceptors and in particular the need for transmembrane electrostatic interactions for receptor function, provide the basis for ongoing and future targeted therapeutic strategies.

Site of CP Action

CP appears to act at the transmembrane level to inhibit signal transduction. The exact molecular mechanism has not been defined and further studies are required. In T-cells, to determine the molecular site of action of CP, various in vitro activation assays were employed, each activating T-cells at different points throughout the signal transduction cascade. CP was not able to inhibit activation upon PMA/ionomycin activation or CD3 cross-linking, suggesting that CP works up-stream of the calcium mobilization phase of T-cell activation and further up-stream of the CD3 chains, possibly between the TCR and the CD3 chains.[6,19] To test this further, T-cells were activated using the super antigen, staphylococcal entertoxin A (SEA). In this system, which requires a complete and functional TCR complex, CP was able to inhibit T-cell activation. This was in contrast to TCRβ cross-linking in which CP was ineffective.[6] These in vitro results are summarized in Table 3.[6] Taken together, this suggests that CP interacts specifically at the level of the TCRα and its associated CD3 and ζ subunits. While this mechanism has yet to be conclusively proven, several studies have added weight to this hypothesis.

Mechanism of Action

Call et al[22,23] have recently proposed a model of TCR assembly highlighting the importance of the charged amino acid residues within the TCR, CD3, and ζ transmembrane regions. These results—taken in combination with Sigalov's recent model of T-cell activation—suggest that CP interferes with the TCRα and its associated CD3 chains, resulting in "predissociated" CD3δε and ζ dimers.[6,24] Upon antigen activation, which requires a fully assembled TCR-CD3-ζζ complex, when in the presence of CP, the CD3δε and ζζ subunits are prematurely dissociated from the TCRα, therefore, no activation proceeds. Neither PMA/ionomycin activation nor CD3 cross-linking require the CD3 chains to be associated with the TCRαβ, which may explain why CP has no effect on these stimulants. By contrast, CP successfully inhibits IL-2 production where stimulants require complete association of the CD3 and ζ chains with the TCRαβ (antigen and SEA activation).[6] However, CP is not capable of inhibiting IL-2 production when activation is

Table 3. Summary of CP activity in different activation models

Activation Method		IL-2 Production
TCR/dependent	Antigen	–
	SEA	–
	TCR cross-linking	+
CD3 dependent	CD3 cross-linking	+
non TCR-CD3 dependent	PMA/ionomycin	+

"–" represents an inhibition of IL-2 production to levels significantly below that presented by the DMSO controls ($P < 0.05$); "+" represents IL-2 production equal to or greater than levels presented by DMSO controls. Reproduced with permission from Elsevier.[6]

through cross-linking with a monoclonal antibody directed against the TCRβ chain.[6] One potential reason for this lies in the aforementioned specificity of the transmembrane charges, suggesting that CP may not affect the interactions between the TCRβ and the CD3γε, which has been shown to involve extracellular[25-27] as well as transmembrane charge interactions,[22] making this interaction possibly harder to disrupt. These results taken together, suggest that CP activity may be centered around the interactions between the TCRα and its associated CD3δε and ζζ subunits (Fig. 3).[6,24] Another possibility, that could happen concurrently, is the possibility that CP inhibits the formation of higher order oligomerization complexes in the transmembrane region between TCR and co-accessory molecules necessary for signal transduction.

Peptide Bioavailability

CP and other TCR transmembrane peptides have an intrinsic ability to insert into membranes. The low solubility makes these peptides difficult to dissolve in aqueous solutions and work with in biological assays. Modifications by our group and collaborators to improve CP's bioavailability have included the covalent attachment of lipids, lipoamino acids, carbohydrates and additional basic amino acids.[6,28]

Lipidation

Three different techniques of lipidation to improve CP's bioavailability have been attempted. These include: (i) conjugation of a myristyl group on the N-terminal of the peptide sequence; (ii) C-terminally attached mono-, di- and tripalmitates; and (iii) lipoaminoacid conjugation. The effects of these conjugates on T-cell activation were examined by IL-2 production in vitro and by using a rat model of adjuvant-induced arthritis in vivo.[6]

N-Terminal Myristoylation

N-terminal lipidation, in particular myristoylation, is a well-known posttranslational modification used by cells to anchor proteins to plasma membranes. The myristoylated alanine rich C kinase substrate (MARCKS) such an example. Partitioning of MARCKS into membranes is a consequence of hydrophobic interaction of the myristoyl moiety with membrane acyl chains and the electrostatic interaction of the cationic peptide with an anionic membrane surface.[29] A significant improvement in the biological activity was observed when CP was myristoylated on the N-terminal.[6] This resulted in an increase in CP's ability to inhibit IL2 production.

TRIS-Lipidation

Lipid conjugation using Tris as a linker was reported by Wells et al[30] to introduce mono-, di- and tripalmitate conjugates to improve the lipophilic character of therapeutic agents. The structural similarity of the Tris molecule to glycerol allows fatty acids to attach and mimic naturally occurring mono-, di- and triglycerides. Unlike glycerol, however, the central carbon atom in the Tris molecule

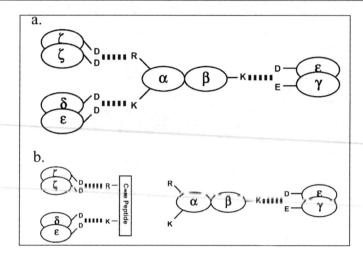

Figure 3. A) TCR α/β and the CD3εδ, CD3γε and ζζ chains essential in the formation of a functional T-cell receptor. B) Proposed transmembrane interactions between CP's arginine and lysine residues and those of the CD3δε and ζζ chains. This model predicts that CP disrupts the transmembrane charges between the TCRα and the CD3δε and ζζ chains. This results in a nonfunctioning T-cell receptor consisting of only the TCRα, TCRβ and CD3γε subunits. Reproduced with permission from Elsevier.[6]

is symmetrical and lends itself to derivatisation by avoiding structural isomerisation which can be a major difficulty in the synthesis and purification of glycerol derivatives. In addition, there are three hydroxyl groups available for acylation. Facile variation of the length, number and saturation of fatty acid chains is possible and therefore the properties of the resulting conjugates can be manipulated. The amino group attached to the central carbon atom provides a convenient attachment site for the covalent linkage to a wide variety of compounds with or without a spacer molecule.

Mono-, di- and tripalmitate derivatives of gly-Tris where conjugated to the carboxyl terminus of CP and to a control peptide analogue in which the basic amino acids (arginine and lysine) were substituted with the neutral amino acid alanine. Preliminary results[5] indicated greater inhibition of IL-2 production in vitro than CP alone. More recent investigations revealed that although the monopalmitate conjugates of CP and the control peptide exhibited the strongest inhibition of IL-2 production, these compounds were cytotoxic.[28] GTP1, the building block of the monopalmitate derivative of CP, also causes cytotoxicity and is, therefore, unsuitable as conjugates. Tripalmitate conjugation resulted in improved inhibition of IL-2 production, whilst the dipalmitate was not effective in these bioassays.[28] Lipoconjugates of the di- and tripalmitate of the control peptide did not inhibit IL-2 production. It appears, therefore, that it is not only the palmitoyl moiety but also the peptide sequence that is critical for exhibiting the biological effect.

Lipoamino Acids

Toth and his colleagues[31] have undertaken a series of studies in designing novel lipidic conjugates which combined the structural features of amino acids and fatty acids. These lipoamino acids can be reacted either by their amino or by their carboxylic acid function or both. They can be used in conventional solid-phase peptide synthesis and can be linked in any position of the peptide sequence. This class of compounds combines the structural properties of peptides with the characteristics of lipids and membranes. Conjugation with a lipoaminoacid on the N-terminal of its sequence enhanced CP's ability to lower IL-2 production considerably.

Glycosylation

In an attempt to render CP more water soluble to improve its bioavailability in aqueous environments, Toth has utilized the same technology used to create lipoamino acids to incorporate sugar groups to peptides.[31] The potential of this technique is that it enables hydrophobic molecules, such as peptides, to be rendered water-soluble and opens the possibility of orally administered peptides.[31] Conjugation of CP with a glucose succinate moiety, while rendering CP water soluble, was incapable of inhibiting T-cell activation based on IL-2 production.[6] In this case, rendering CP hydrophilic likely removed its ability to interact with the membrane bilayer (see below) and in doing so, removed its ability to inhibit T-cell activation. This suggests that membrane-active peptides require a certain degree of hydrophobicity to allow peptide-membrane interactions to occur.

Polar Residue Tagging of Peptides

In our studies on the transmembrane peptide sequence derived from NKp44, measurement of NK-cell inhibition was hindered by the insolubility of this peptide. In order to improve the solubility of this peptide, five extra lysine residues were added onto the N-terminus as outlined by Melnyk et al.[32] Significant inhibitory effect was noted (unpublished data) when the tagged peptide was assayed for biological activity. In analogues of CP (Table 1) where additional nonconsecutive additional lysine residues were incorporated in the sequence, inhibition of IL-2 production was diminished compared to the parent compound.

Effects of CP Conjugation

Biophysical Studies

The initial step taken by CP to effect TCR activity is proposed as peptide-membrane binding.[14] Using SPR, a surprising correlation between binding of the peptide and its analogues to model membranes and the in vitro biological effect (IL-2 inhibition) was observed. Lipid conjugates of CP all exhibited irreversible binding to model membranes. Even extensive washing with elution buffer was incapable of removing any of the lipid conjugates of CP, which implies irreversible binding to the lipid bilayer. Hydrophobic interaction of the myristoyl group and the electrostatic attraction of the cationic peptide moiety to the anionic lipid headgroups seems sufficient to anchor myristoyl-CP to membranes. Similar binding was found with the lipoamino acid conjugated CP.

All the gly-Tris palmitate conjugates of CP bound more strongly to model membranes than CP. The strongest binding was exhibited by the mono-palmitate derivative and likely relates to the cytotoxic nature of this conjugate. Di- and tripalmitate conjugates of the control peptide exhibited no binding to model membranes at all; this is in agreement with the result of the IL-2 inhibition assay.[28] These results demonstrate that the charged amino acids in the peptide sequence and the lipid conjugation are both important factors in CP's function.

The glucosesuccinate-conjugated CP did not bind to the model membrane system used, reflecting the results in IL-2 production.[6]

Adjuvant-Induced Arthritis

Mirroring the in vitro results, significant effects were recorded by the myristoylated- and lipoamino acid-conjugated CP on the treatment of T-cell-mediated, adjuvant-induced arthritis. By contrast, the glucose succinate derivative of CP, while presenting as inactive in the IL-2 inhibition assay and incapable of binding to the model membranes in SPR experiments, exhibited remarkable effects in treating arthritis with almost complete inhibition of disease progression; this result is similar to that obtained with cyclosporine (Fig. 4).[6] This discrepancy between in vitro and in vivo peptide activity has also been noted and reported by Collier et al.[33]

Conclusion

The ability of CP to inhibit immune reactions may provide a new and exciting alternative to existing therapies for T-cell-mediated diseases such as rheumatoid arthritis, multiple sclerosis, psoriasis, or transplantation. Since the TCR is viewed as a model for multi-subunit transmembrane protein

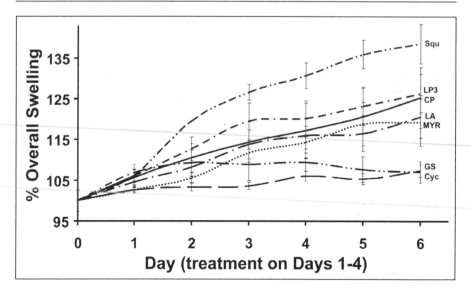

Figure 4. The influence of CP conjugates as a treatment in the rat adjuvant-induced arthritis model. Rats were treated subcutaneously with 6 mg of peptide for four consecutive days starting one day after arthritis was diagnosed. Results are presented as the overall paw swelling averaged for all measurements over time (days). 100% is set at Day 0 (the day arthritis was diagnosed). CP:—; Squalane: –– • •; LP3:– •; LA:– • –; MYR: • • •; Cyclosporin (Cyc):– • •–; GS:––. Reproduced with permission from Elsevier.[6]

interactions, a similar approach may be used to inhibit other membrane subunit receptors that rely on transmembrane electrostatic protein-protein interactions for biological function. These disorders may involve malignancy (by inhibiting protein-protein interactions involved with uncontrolled cell cycle division) or even viral infections such as HIV where the focus on glycolipid-enriched domains for attachment and entry into the cell may be a new site of focus for therapeutic intervention.

Peptides represent a new class of compounds for potential therapy. Understanding physical interactions between peptides, receptors and membranes is crucial to improving their stability, bioavailability and effectiveness. Conjugation of peptides with lipids and sugars enhances their functional ability. The capacity to develop different delivery systems with peptides instead of conventional drugs is also exciting and increases the scope of application. For instance, the potential to transduce dendritic cells to express CP in an antigen-specific manner is an exciting prospect that provides hope for the treatment of T-cell-mediated diseases without the complications of generalized, systemic effects.

References

1. Cohen-Kaminsky S, Jambou F. Prospects for a T-cell receptor vaccination against myasthenia gravis. Expert Rev Vaccines 2005; 4(4):473-492.
2. Bluestone J, Tang Q. Therapeutic vaccination using CD4+ CD25+ antigen-specific regulatory T-cells. Proc Natl Acad Sci USA 2004; 101(Suppl 2):14622-14626.
3. Koller M. Targeted therapy in rheumatoid arthritis. Wiener Medizinische Wochenschrift 2006; 156(1-2):53-60.
4. Griffin M, Holman P, Tang Q et al. Development and applications of surface-linked single chain antibodies against T-cell antigens. J Immunol Methods 2001; 248(1-2):77-90.
5. Manolios N, Collier S, Taylor J et al. T-cell antigen receptor transmembrane peptides modulate T-cell function and T-cell-mediated disease. Nat Med 1997; 3:84-88.
6. Amon MA, Ali M, Bender V et al. Lipidation and glycosylation of a T-cell antigen receptor (TCR) transmembrane hydrophobic peptide dramatically enhances in vitro and in vivo function. Biochim Biophys Acta—Mol Cell Res 2006; 1763(8):879-888.

7. Gollner GP, Muller G, Alt R et al. Therapeutic application of T-cell receptor mimic peptides or cDNA in the treatment of T-cell-mediated skin diseases. Gene Therapy 2000; 7:1000-1004.
8. Huynh NT. T-cell antigen receptor transmembrane peptides and their effects on B and Natural Killer Cells. Masters Thesis, University of Sydney, 2005.
9. Mahnke K, Qian Y, Knop J et al. Dendritic cells engineered to secrete a T-cell receptor mimic peptide, induce antigen-specific immunosuppression in vivo. Nat Biotech 2003; 21:903-908.
10. Mozsolits H, Aguilar M-I. Surface Plasmon Resonance Spectroscopy: An emerging tool for the study of peptide-membrane interactions. Biopolymers (Peptide Science) 2002; 66:3-18.
11. Shai Y. Mechanism of binding insertion and destabilization of phospholipid bilayer membranes by alpha-helical antimicrobial and cell non selective membrane lytic peptides. Biochim Biophys Acta 1999; 1462:55-70.
12. Deber CM, Li SC. Peptides in membranes: Helicity and hydrophobicity. Biopolymers 1995; 37:295-318.
13. White SH, Wimley WC. Peptides in lipid bilayers: Structural and thermodynamic basis for partitioning and folding. Curr Opin Struct Biol 1994; 4:79-86.
14. Bender V, Ali M, Amon M et al. T-cell antigen receptor peptide-lipid membrane interactions using Surface Plasmon Resonance. J Biol Chem 2004; 279(52):54002-54007.
15. Ali M, De Planque M, Huynh N et al. Biophysical studies of transmembrane peptide derived from the T-cell antigen receptor. Lett Peptide Sci 2002; 8:227-233.
16. Wang X. Biological and biophysical properties of T-cell antigen receptor (TCR) transmembrane peptides in cell membranes. Masters Thesis Sydney, University of Sydney 2001.
17. Ali M, Salam NK, Amon MA et al. T-cell antigen receptor—Alpha chain transmembrane peptides: Correlation between structure and function. Int J Pept Res Ther 2006; 12:261-267.
18. Wang XM, Djordjevic JT, Bender V et al. T-cell antigen receptor (TCR) transmembrane peptides colocalise with TCR, not lipid rafts, in surface membranes. Cell Immunol 2002; 215:12-19.
19. Wang XM, Djordjevic JT, Kurosaka N et al. T-cell antigen receptor peptides inhibit signal transduction within the membrane bilayer. Clin Immunol 2002; 105:199
20. Huynh NT, Ffrench RA, Boadle RA et al. Transmembrane TCR peptides inhibit B and NK cell function. Immunology 2003; 108(4):458-464.
21. Vandebona H, Ali M, Amon M et al. Immunoreceptor transmembrane peptides and their effect on Natural Killer (NK) cell cytotoxicity. Protein Pept Lett 2006; 13(10):1017-1024.
22. Call ME, Pyrdol J, Wiedmann M et al. The organizing principle in the formation of the T-cell receptor-CD3 complex. Cell 2002; 111:967-979.
23. Call ME, Wucherpfennig KW. The T-cell receptor: Critical role of the membrane environment in receptor assembly and function. Annu Rev Immunol 2005; 23:101-125.
24. Sigalov A. Multi-chain immune recognition receptors: Spatial organization and signal transduction. Semin Immunol 2005; 17:51-64.
25. Manolios N, Li ZG. The T-cell antigen receptor beta chain interacts with the extracellular domain of CD3-gamma. Immunol Cell Biol 1995; 73:532-536.
26. Wang J-h, Lim K, Smolyar A et al. Atomic structure of an αβ T-cell receptor (TCR) heterodimer in complex with an antiTCR Fab fragment derived from a mitogenic antibody. EMBO J 1998; 17(1):10-26.
27. Ghendler Y, Smolyar A, Chang H-C et al. One of the CD3ε Subunits within a T-cell receptor complex lies in close proximity to the Cβ FG loop. J Exp Med 1998; 187(9):1529-1536.
28. Ali M, Amon M, Bender V et al. Hydrophobic transmembrane-peptide lipid conjugations enhance membrane binding and functional activity in T-cells. Bioconjug Chem 2005; 16(6):1556-1563.
29. Kim J, Shishido T, Jiang X et al. Phosphorylation, high ionic strength and calmodulin reverse the binding of MARCKS to phospholipid vesicles. J Biol Chem 1994; 269:28214-28219.
30. Wells XE, Bender VJ, Francis CL et al. Tris and the ready production of drug-fatty acyl conjugates. Drug Dev Res 1999; 46(3-4):302-308.
31. Blanchfield JT, Toth I. Modification of peptides and other drugs using lipoamino acids and sugars. Methods Mol Biol 2005; 298:45-61.
32. Melnyk RA, Partridge AW, Yip J et al. Polar residue tagging of transmembrane peptides. Biopolymers (Peptide Science) 2003; 71:675-685.
33. Collier S, Bolte A, Manolios N. Discrepancy in CD3-transmembrane peptide activity between in vitro and in vivo T-cell inhibition. Scand J Immunol 2006; 64(4):388-391.

Fc Receptor Targeting in the Treatment of Allergy, Autoimmune Diseases and Cancer

Akira Nakamura,* Tomohiro Kubo and Toshiyuki Takai

Abstract

Fc receptors (FcRs) play an important role in the maintenance of an adequate activation threshold of various cells in antibody-mediated immune responses. Analyses of murine models show that the inhibitory FcR, FcγRIIB plays a pivotal role in the suppression of antibody-mediated allergy and autoimmunity. On the other hand, the activating-type FcRs are essential for the development of these diseases, suggesting that regulation of inhibitory or activating FcR is an ideal target for a therapeutic agent. Recent experimental or clinical studies also indicate that FcRs function as key receptors in the treatment with monoclonal antibodies (mAbs) therapy. This review summarizes FcR functions and highlights possible FcR-targeting therapies including mAb therapies for allergy, autoimmune diseases and cancer.

Introduction

FcRs are widely expressed on hemopoietic cells and distinguished by their structure, function, distribution and ligands, such as IgG, IgM, IgE and IgA. FcRs have homologous extracellular immunoglobulin domains, however, there are functionally two kinds of FcRs, the activation type and the inhibitory type FcRs. Recent analysis using FcRs-deficient mice have revealed that immune responses by antibodies depend upon regulation of activating and inhibitory FcR.[1-4] Deletion of activating FcRs does not elicit antibody-mediated responses, whereas deletion of inhibitory FcRs causes excessive antibody responses, resulting in the development of allergic or autoimmune disorders.

Immunotherapy using mAbs is a new strategy against allergy, autoimmune diseases and cancer. Genetic engineering enabled the development of humanized antibodies, leading to rapid progress of mAb therapy. In particular, activating FcRs play a pivotal role in the effects of mAb therapy. In addition to mAbs therapy, recent studies reveal that an inhibitory FcR, FcγRIIB, contributes to the effect of intravenous immunoglobulin (IVIg) therapy. This review summarizes the immunological functions of FcRs and focuses on FcR-targeting therapy.

*Corresponding Author: Akira Nakamura—Department of Experimental Immunology and CREST program of Japan Science and Technology Agency, Institute of Development, Aging and Cancer, Tohoku University, Seiryo 4-1, Sendai 980-8575, Japan. Email: aki@idac.tohoku.ac.jp

Multichain Immune Recognition Receptor Signaling: From Spatiotemporal Organization to Human Disease, edited by Alexander B. Sigalov. ©2008 Landes Bioscience and Springer Science+Business Media.

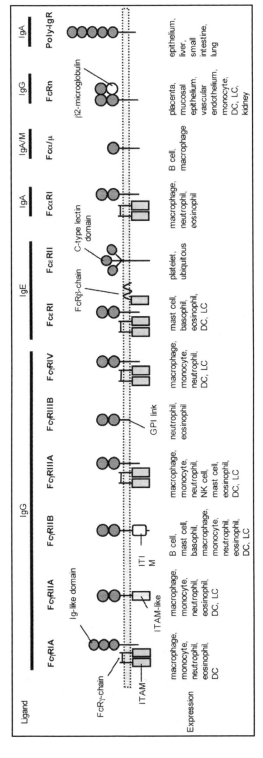

Figure 1. Schematic structure of human and murine Fc receptors. Structures of human and murine Fc receptors with their subunits, FcRγ- and FcRβ-chain, are shown. Abbreviations: DC, dendritic cell; FcRn, neonatal Fc receptor; GPI, glycosylphosphatidylinositol; Ig, immunoglobulin; ITA, immunoreceptor tyrosine-based activation motif; ITIM, immunoreceptor tyrosine-base inhibitory motif; LC, Langerhans cell; NK, natural killer; poly-IgR, polymeric immunoglobulin receptor.

FcR Function

Structure

Most FcRs belong to immunoglobulin superfamily proteins. Since the detailed biochemistry of FcRs has been reviewed in Chapter 3 of Section I, this chapter briefly introduces the FcR structure and function. Figure 1 shows the schematic structure of representative FcγRs, FcγRIA, FcγRIIA, FcγRIIB, FcγRIIIA, FcγRIIIB and a recently discovered murine FcγRIV.[5,6] While FcγRIA binds to IgG with high affinity, FcγRIB, FcγRIC, FcγRIIs (FcγRIIA, FcγRIIB and FcγRIIC), FcγRIIIs (FcγRIIIA and FcγRIIIB) and FcγRIV have low affinity to IgG. FcγRIA, FcγRIIIA and murine FcγRIV require the FcR γ-chain (FcRγ) for their surface expression and signal delivery. Human FcRs for IgE, FcεRs, are encoded by two genes, FcεRI and FcεRII. FcεRI, which associates with the FcRγ and FcεRI β-chain (FcεγRIβ), binds to IgE with high affinity (>10^{-10}M). Therefore, monomeric IgE constitutively binds to FcεRI on the surface of mast cells or basophils. Low-affinity FcεRII is a C-type lectin family protein. In humans, there are three kinds of FcRs for IgA, FcαRI, polymeric Ig receptor (poly-IgR) and Fcα/μR. IgA antibodies are enriched in serum and in mucosal tissue and have three forms, monomeric, dimeric and secretory forms. The poly-IgR can transport dimeric IgA across the epithelium. Fcα/μR can bind both IgA and IgM. The neonatal FcR for IgG, FcRn, which is an MHC class I-like molecule, is responsible for perinatal IgG transport and for IgG homeostasis in adults.[7]

Regarding FcR structural information, crystal structures of IgG-FcγRIII, IgE-FcεRI and IgA-FcαRI complexes have been reported, showing the detailed information of the binding interaction between Ig-Fc fragments and FcRs.[8-12] Atomic-level structural information on the extracellular domains of FcγRIIa, FcγRIIb, FcεRI and FcαRI was also elucidated.[13-16] These FcRs contain two Ig-like domains in their extracellular regions and bear similar structures in that the two ectodomains bend at an acute angle to each other. The structural analysis of IgG1-Fc fragment-FcγRIII complex reveals that FcγRIII binds the lower hinge region and constant region 2 (Cγ2) of the Fc fragment.[8] The IgE-Fc fragment-FcεRI complex has similar conformation to that of FcγRIII complex, in which the binding site is located in the Cε2-Cε3 linker region.[9] Furthermore, the binding site of both FcγRIII and FcεRI to the Ig-Fc fragment is situated in the D2 domain, which is the carboxyl-terminal of ectodomains.[8,9] The structure of IgA-Fc fragment-FcαRI complex is different from that of other FcRs.[10] The Ig domains of FcαRI are rotated by ~180° from the positions seen in other FcRs.[17,18] FcαRI binds to the interface of Cα2 and Cα3 domains.

Activating and Inhibitory FcRs—ITAM and ITIM Signaling

Antibody-mediated cellular activation is regulated by activating and inhibitory FcRs. Activating-type FcRs, FcγRI, FcγRIII, FcγRIV and FcεRI trigger cellular activation through FcRγ that contains an intracellular immunoreceptor tyrosine-based activation motif (ITAM). FcεRI associates with both FcRγ and FcεRIβ that also contains ITAM in the cytoplasm. On the contrary, FcγRIIB bears an immunoreceptor tyrosine-based inhibitory motif (ITIM) and inhibits ITAM-mediated cellular activation triggered through receptors upon cross-linking through immune complexes (ICs) with activating-type receptors such as FcγRI, FcγRIII, FcγRIV and FcεRI or B cell antigen receptor (BCR). Figure 2A shows the activating and inhibitory signaling in B cell responses. Upon clustering activating FcRs, intracellular Src family protein tyrosine kinases such as Lyn, Fyn, Fgr and Hck phosphorylate the tyrosine residues of the ITAM,[19,20] which then become the target for cytosolic protein kinase Syk. Simultaneously, the coengagement of activating-type receptor and FcγRIIB by ICs results in the tyrosine phosphorylation of the ITIM and the recruitment of Src-homology-2 (SH2)-domain-containing inositol 5'-phosphatase (SHIP). SHIP hydrolyzes phosphatidylinositol-3,4,5-triphosphate, PtdIns(3,4,5)P$_3$, to PtdIns(3,4)P$_2$, leading to the dissociation of Bruton's tyrosine kinase (Btk) and phospholipase-γ (PLC-γ) from the membrane and the inhibition of Ca influx into the cell.[21,22] PtdIns(3,4,5)P$_3$ also serves as the docking site of the anti-apoptotic kinase, Akt. Thus, Akt cannot be recruited to the membrane when SHIP hydrolyzes PtdIns(3,4,5)P$_3$ to PtdIns(3,4)P$_2$. Furthermore, phosphorylated SHIP provides binding sites for the phosphotyrosine-binding (PTB) domain of the adaptor proteins, Shc and Dok, leading to the

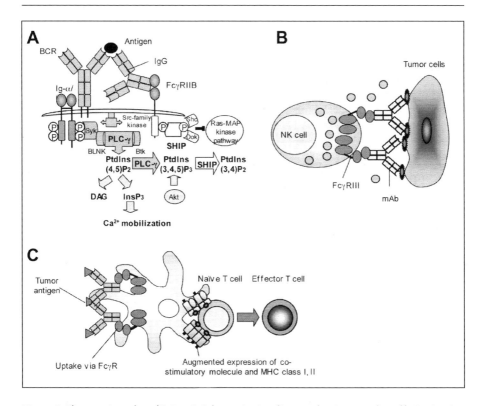

Figure 2. Three major roles of FcRs. A) Schematic signaling mechanisms mediated by activating and inhibitory receptors on B-cells. Upon cross-linking of BCR by immune complexes (ICs) on B-cells, Src family protein kinases such as Lyn, Blk and Fyn phosphorylate tyrosine residues in ITAM of Igα/β. Syk tyrosine kinase binds to the phosphorylated ITAMs and becomes activated by Src family protein kinases. The activated kinases activates BLNK, phospholipase C-γ (PLC-γ) and Tec kinases such as Btk. PLC-γ cleaves phosphatidylinositol-4,5-bisphosphate phosphatidylinositol, PtdIns(4,5)P$_2$, to inositol-1,4,5-triphosphate (InsP$_3$) and DAG, then initiating Ca^{2+} mobilization pathways. In contrast, SHIP hydrolyses PtdIns(3,4,5)P$_3$ to PtdIns(3,4) P$_2$. PtdIns(3,4,5)P$_3$ is the docking site of pleckstrin homology (PH) domain-containing proteins, including Btk and PLC-γ. Therefore, Btk and PLC-γ cannot be recruited to membrane PtdIns(3,4,5)P$_3$, resulting in the inhibition of Ca^{2+} mobilization pathways. PtdIns(3,4,5)P$_3$ also serves as the binding site of PH domain of Akt. Thus, Akt cannot be recruited to the membrane after the recruitment of SHIP to the ITIM. Phosphorylated SHIP provides binding sites for phosphotyrosine-binding (PTB) domain of adaptor proteins, Shc and Dok, leading to the blocking the downstream Ras-MAP kinase pathways. B) Antibody-dependent cell-mediated cytotoxicity (ADCC). ADCC is the killing attack of antibody-coated target cells, such as tumor cells, by NK cells or macrophages with FcγRIII. ADCC has a major role in a mAb therapy for cancer. C) FcγR-mediated effective antigen presentation on dendritic cells (DCs). FcγR efficiently uptakes antigen-antibody complexes and upregulates the expression of both MHC class I and II complexes bearing epitopes and costimulatory molecules, such as CD40, CD80 and CD86. DCs stimulated by FcγR can effectively present antigens to T-cells and initiate strong cellular and humoral responses. Abbreviations: BCR, B-cell receptor; BLNK, B-cell linker; Btk, Bruton's tyrosine kinase; DAG, diacylglycerol; Grb 2, growth-factor-receptor-bound protein 2; IC, immune complex; Ig: immunoglobulin; MHC, major histocompatibility complex; SHIP, Src-homology-2 (SH2)-domain-containing inositol 5'-phosphatase; SHP, SH2-domain-containing protein tyrosine phosphatase.

inhibition of the Ras-MAP kinase pathway.[23] This inhibitory signaling through FcγRIIB was also observed in other immune cells including mast cells and macrophages.[3]

Antibody-Dependent Cell-Mediated Cytotoxicity (ADCC)

The destruction of antibody-coated target cells by NK cells or macrophages is called ADCC (Fig. 2B). Currently, ADCC seems to be the main mechanism of cytotoxicity against tumors mediated by tumor antigen-specific antibodies. In a model of B16F10 melanoma lung metastasis, the monoclonal antibody TA77 against tumor antigen gp75 is effective in the protection of metastatic expansion.[24] Since the efficiency against tumor by TA77 is significantly impaired in FcRγ-deficient mice, ADCC is mediated by an FcRγ-dependent mechanism.[25] In contrast, deletion of FcγRIIB augments cytotoxicity induced by tumor specific antibody in vivo.[26] This increased cytotoxicity in FcγRIIB deficiency is attributed to macrophage-mediated ADCC.

FcγR-Mediated Antigen Presentation on Dendritic Cells

Antigen uptake through FcγRs induces strong antigen-specific T-cell responses (Fig. 2C).[27-29] When compared to antigen alone, ICs are efficiently internalized into dendritic cells (DCs) through FcγRs. The FcγR-mediated efficient antigen-internalization initiates the DC-specific antigen transport pathway in the cytosol. In addition, FcγR-mediated antigen loading leads DCs to enhance the expression of costimulatory molecules, such as CD86 and MHC class II molecules.[27,28,30] Although targeting ICs to FcγRs on DCs significantly enhances the efficiency of antigen presentation,[27,28,30,31] it remains controversial whether inhibitory FcγRIIB contribute to antigen presentation. Our previous paper demonstrated the positive contribution of FcγRIIB to the class I and II-restricted antigen presentation because FcγRIIB as well as FcγRI and FcγRIII efficiently internalize ICs,[27,28] however, another group has shown that deletion of FcγRIIB enhances the efficiency of OVA-specific cytotoxic T-lymphocyte (CTL) activity, indicating that FcγRIIB negatively regulates the FcγR-mediated augmenting pathway for antigen presentation.[32] Moreover, a recent paper by Bergtold A, et al shows a unique role of FcγRIIB in ICs-mediated antigen presentation by DCs.[33] They showed that ICs taken up through FcγRIIB are inefficiently degraded and do not reach a vesicular compartment but are instead detected intracellularly in recycling endosomes as well as on the cell surface. The recycling effect of ICs by FcγRIIB on DCs activates B-cells well, resulting in efficient T-independent humoral responses. Since this unique effect of FcγRIIB is observed in FcRγ-deficient DCs but not in wild-type DCs, further study will be required for its physiological role.

FcR-Targeting Therapy

Antibody Therapy

Recently developed immunotherapy using mAbs is expected to provide new therapeutic agents against various disorders, especially allergy, autoimmune diseases and cancer. This is mainly due to advanced biotechnology by which genetic engineering has developed recombinant humanized antibodies. The humanize antibody is about 95% human antibody and hardly has immunogenicity, leading to the current rapid progress of mAb therapy. In addition to mAb therapy, recently, FcR-directed bispecific antibodies that target both tumor antigens and FcRs on immune effector cells are generated. In this chapter, we discuss current possible FcR-targeting mAb therapy in allergy, autoimmunity and malignant disease.

General Mechanism of Antibody Therapy

Although a number of potential mechanisms have been pointed out in mAb therapy, the main mechanism is as follows:[34,35]

1. Blocking effect. Blocking effect by mAb mainly targets the interruption of pro-inflammatory cytokines and/or cell-cell crosstalk such as receptor-ligand interaction.
2. Antigen targeting. In treatment of malignancy, the target of mAb therapy is a tumor-specific antigen. MAb bound to antigen is recognized by FcγR on immune cells, such as NK

cells and macrophages, leading to the induction of ADCC and also activates classical pathway of complement such as C1q.

3. Induction of signaling. In the mAb therapy of B-cell lymphoma, mAb crosslinks the target molecules on target cell surface and induces the target cells to signaling that impairs cell function.

Antibody Against Allergy

IgE-FcεRI interaction on mast cells or basophils triggers allergic reaction. In almost all patients who suffer allergic disease, the titers of serum IgE is elevated and thus, several passive anti-IgE therapies that aim at a neutralizing effect against IgE have been developed. One of these, a humanized anti-IgE monoclonal IgG1 antibody, Omalizumab (anti-IgE monoclonal antibody E25, E25, humanised anti-IgE MAb, IGE 025, monoclonal antibody E25, olizumab, rhuMAb-E25, Xolair), which is a nonanaphylactogenic murine anti-human IgE antibody directly against the FcεRI-binding domain of human IgE, is used for the treatment of allergic rhinitis and bronchial asthma.[36-40] Omalizumab has three separate mechanistic components based on the results of the clinical trails. Firstly, Omalizumab down-regulates IgE production. Secondly, it down-regulates FcεRI expression in receptor-expressing cells such as basophils, mast cells and DCs. Thirdly, it reduces both early allergic responses such as hypersensitivity reaction through IgE-FcεRI in mast cells and late allergic responses such as the infiltration of inflammatory cells after the release of chemical mediators from mast cells. In particular, it is important that Omalizumab binds not only to free IgE but also to the surface IgE on B-cells, which produce IgE, leading to the down-regulation of IgE production. The possible mechanism is that the Fc region of Omalizumab crosslinks the FcγRIIB on IgE-producing B-cells, resulting in the reduction of their IgE production.

Antibody Therapy Against Autoimmune Diseases

Although the exact mechanism of loss of self-tolerance remains unclear, the pathogenesis of these autoimmune diseases is mainly due to the excessive inflammation mediated by pro-inflammatory cytokines or autoantibodies that are initiated by autoreactive T-or B-cells.[41,42] Therefore, current mAb therapies against autoimmune diseases aim at blocking cytokines and depletion of B-cells. Particularly, passive anti-tumor necrosis factor (TNF)-α (infliximab, adalimumab) or anti-IL-6 receptor mAb (MRA) therapies show a remarkable effect in the suppression of disease progression and severity in rheumatoid arthritis (RA) patients.[43-45] However, anti-cytokine antibody treatments were considered to have no relation to FcR-mediated mechanisms. On the other hand, another strategy of mAb treatment in autoimmune diseases that aims at the depletion of autoreactive B-cells is considered as one of the FcR-targeting therapy.

Currently, in patients with RA, systemic lupus erythematosus (SLE), myasthenia gravis, idiopathic thrombocytopenic purpura (ITP) and lymphomas, a humanized anti-CD20 IgG1 mAb (Rituximab) is used for B-lymphocyte depletion therapy.[46-49] CD20 is a B-cell-specific antigen and is expressed on almost all B-cell stages from preB-cells to plasma cells but not on hematopoietic stem cell or other normal cells, indicating that CD20 is one of the ideal target molecules in B-cell depletion treatment.[50] Several studies have shown that CD20 mAb therapy is effective in improving the symptoms and disease progression. The Rituximab has three putative mechanisms that contribute to deplete B-cells as follows:[51]

1. ADCC mediated by mainly monocytes, macrophages and NK cells. The therapeutic effect is absence in FcRγ-deficient mice. In addition, a recent study shows that FcγRIIIa polymorphism, which affects affinity to IgG1, may be associated with the response rate in the treatment of SLE by affecting ADCC.[52,53]
2. Complement-dependent cytotoxicity induced by B-cell surface CD20-CD20mAb complexes and subsequent binding C1q.
3. Induction of B-cell apoptosis by CD20mAb crosslinking.

Antibody Therapy Against Malignant Disease

Recently, a targeted mAb therapy has made rapid progress in the treatment against cancers including lymphoma and solid tumor.[49,54-56] Particularly, mAb therapies for malignant lymphoma, rituximab, alemtuzumab and trastuzumab, are partially mediated by FcR-mediated immune response, ADCC.

Rituximab, which is also used in the treatment of autoimmune diseases, is the first anti-tumor mAb drug admitted by the US Food and Drug Administration (FDA) in 1997. As mentioned previously, Rituximab targets CD20 antigen, which is expressed on normal B-cells. CD20 is also expressed on 95% of B-cell lymphomas.[57] A combination of rituximab plus chemotherapy showed a remarkable effect in patients with diffuse large B-cell lymphoma and follicular lymphoma.[49] As observed in patients with SLE, the clinical response of nonHodgkin lymphoma patients is also associated with FcγRIIIa polymorphism because the patients having FcγRIIIa-V158 genotype show slightly higher response to rituximab than the patients having FcγRIIIa-F158 genotype.[58]

Alemtuzumab targets CD52, which is a glycoprotein that is expressed on malignant lymphocytes as well as peripheral lymphocytes, but not on haemopoietic stem cells. The main tumor killing mechanism with alemtuzumab involves FcγRIII on macrophages that mediate ADCC and cross-linking-induced apoptosis.[59]

Trastuzumab is a humanized anti-HER2 antibody that targets a tyrosine-kinase receptor. HER2 is overexpressed on 30% of breast cancer cells and other solid tumors, such as nonsmall cell lung cancer, ovarian and prostate cancer.[60,61] Trastuzumab binds to HER2 with high affinity and is internalized in the cytosol with its receptor through FcγR, resulting in the blockade of downstream signaling pathways. A recent report shows that trastuzumab evokes ADCC,[62] and also augments CTL against HER2-expressed cancer cells, which may be due to the FcγR-mediated effective cross-presentation by DCs.[63,64]

FcγRIIB-Targeting Therapy

FcγRIIB-Targeting Chimeric Recombinant Protein and Bispecific Antibody

As mentioned above, the mAb therapy targets activating FcRs, mainly FcγRIII on NK cells and macrophages. On the other hand, many murine studies demonstrate the inhibitory role of FcγRIIB in allergy and autoimmunity. For example, FcγRIIB-deficient mice are sensitive to IgE or IgG-mediated passive systemic anaphylaxis. FcγRIIB-deficient B6 or B6.Fas*lpr/lpr* mice also show spontaneous autoimmune glomerulonephritis,[65,66] providing supporting evidence that FcγRIIB functions as a critical suppressor gene in the development of murine autoimmunity. In human disorders, several reports indicate that FcγRIIB polymorphism may associate with SLE.[67,68] The FcγRIIB polymorphism, threonine 232-FcγRIIB, have a single amino acid substitution (from isoleucine to threonine) at position 232 within the transmembrane domain. Recently, Floto R.A. et al demonstrated that human monocytes transfected with 232Th-FcγRIIB fail to inhibit FcγRI-mediated cellular activation due to the impaired recruitment to sphingolipid rafts.[69,70] Although these findings show the involvement of FcγRIIB in the development of autoimmunity in humans and strongly suggest that FcγRIIB could be a potential therapeutic target, FcγRIIB-targeting therapy has not been generated because there is an other type of human FcγRII, activating-type FcγRIIA, that is very homologous to FcγRIIB. However, recently, FcγRIIB-targeting chimeric recombinant protein and bispecific antibody that target both FcγRI-expressing cells and FcγRIIB have been generated. The strategy of these chimeric molecules aims at the induction of inhibitory signaling by crosslinking of FcγRI with FcγRIIB on human mast cells or basophils (Fig. 3). One of these, a human IgG1 Fc fragment (γHinge-CHγ2-CHγ3) linked with IgE Fc fragment (CHε2-CHε3-CHε4), compulsorily links FcεRI to FcγRIIB (Fig. 3A).[71] This Fcγ-Fcε fusion protein decreases IgE-mediated basophil histamine release and also inhibits IgE-mediated passive cutaneous anaphylaxis in human FcεRIα transgenic mice. Another fusion protein is composed of a human IgG1 Fc fragment (Fcγ1) and a cat allergen (Fel d1) (Fig. 3B).[72] The Fcγ1-Fel d1 protein aims at the crosslink between FcγRIIB and Fel d1-specific IgE antibody bound to FcεRI. As observed in the effect of the Fcγ-Fcε protein, this Fcγ1-Fel d1 protein inhibits Fel d1-induced activation

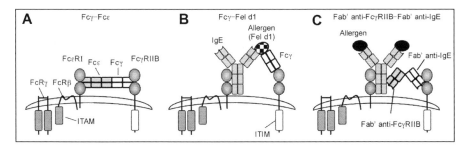

Figure 3. FcγRIIB-targeting chimeric recombinant protein and bispecific antibody. A) The Fcγ-Fcε fusion protein binds to both FcεRI and FcγRIIB, resulting in competing with IgE-FcεRI binding and the possible induction of inhibitory signaling through FcγRIIB. B) The Fcγ1-Fel d1 protein aims at the crosslink between FcγRIIB and the major cat allergen, Fel d1-specific IgE antibody bound to FcεRI. C) The bispecific antibody, Fab' anti-FcγRIIB-Fab' anti-IgE protein, that crosslinks IgE and FcγRIIB and inhibits IgE-induced histamine release of human mast cells and basophils.

of human mast cells and basophils and also Fel d1-induced systemic anaphylaxis. Although these two fusion proteins aim at the same strategy that is the induction of FcγRIIB-mediated inhibitory signaling, currently, there is no direct evidence whether Fel d1 can induce inhibitory signaling into the cells or not. In addition to these fusion proteins, a bispecific antibody against human IgE and FcγRII has been generated (Fig. 3C).[73] This bispecific antibody is composed of Fab' fragment of anti-human IgE and Fab' fragment of anti-human FcγRII. The Fab' anti-FcγRIIB-Fab' anti-IgE protein, which aims at the coengagement of IgE with FcγRIIB, inhibits IgE-induced histamine release of human mast cells and basophils.

Anti-Human FcγRIIB Antibody Therapy

As mentioned above, there has been no specific anti-human FcγRIIB antibody due to the high homology to FcγRIIA (96% homology). However, very recently, by immunizing human FcγRIIA-transgenic mice, Rankin CT et al have succeeded in generating high affinity mAb against human FcγRIIB that does not cross-react with FcγRIIA.[74]

Administration of the humanized anti-human FcγRIIB antibody, 2B6, can eliminate FcγRIIB-expressing B-cell lymphoma cell line in vitro or in vivo by ADCC. However, currently, it remains unclear whether 2B6 can induce inhibitory signaling. Moreover, unlike rituximab-targeting antigen CD20, FcγRIIB is expressed on various normal immune cells as well as on B-cell lymphomas, thus, further investigation will be required for the adaptation of clinical study.

IVIg—Possible FcγRIIB Targeting Therapy

Intravenous immunoglobulin (IVIg) has been used as the standard treatment for primary and secondary immunodeficient disorders such as hypo- and agammaglobulinemia for prevention of infectious disease. Currently, IVIg therapy has been established as an effective treatment for some autoimmune diseases including ITP,[75] Guillain-Barré syndrome,[76] multiple sclerosis,[77] myasthenia gravis,[78] and vasculitis.[79,80] Putative mechanisms of IVIg treatment are considered as follows:[81-83]

1. Interaction mediated by the antibody variable regions in the F(ab')₂ portion. An example is antagonistic effect against superantigens which could be related to the development of Kawasaki disease.
2. Effects mediated by the Fc portion.
3. Inhibitory effect against deposition of activated complement such as the interception of the conformation of complement membrane attack complex.
4. Effects mediated by immunoregulatory substances including soluble cytokine inhibitors other than immunoglobulin.

In particular, the Fc fragment plays an important role in these effects because intact IgG antibody has more suppressive effects than the F(ab')$_2$ antibody in the treatment of the disease model such as ITP.[33] It remains unclear whether the suppressive effect of the Fc fragment of IVIg interacts with FcγRIIB, however, in a murine model of ITP, FcγRIIB has been shown to play an essential role in the effects of IVIg.[84] In ITP, autoantibody-coated platelets bind to FcγRIII on macrophages, resulting in the FcγRIII cross-linking that triggers the phagocytosis of platelets. In a murine ITP, administration of the Fc fragment as well as IVIg to wild-type mice prevents platelet deletion induced by pathogenic anti-platelet antibody. However, IVIg treatment does not have a therapeutic effect in FcγRIIB-deficient mice or in mice treated with a blocking monoclonal antibody against FcγRIIB and FcγRIII, showing the requirement of FcγRIIB for the protection of ITP by IVIg.[84] In the nephrotoxic nephritis model induced by heterologus anti-glomerular basement membrane antiserum, IVIg down-regulates FcγRIV expression, while up-regulates the surface expression of FcγRIIB on macrophage infiltrating in the kidney and protects mice from fatal disease, suggesting that FcγRIV as well as FcγRIIB also contribute to the effect of IVIg.[85] Moreover, in the arthritis model induced by serum transfer from K/BxN mice, IVIg treatment is effective in protecting arthritis in wild-type mice but not in FcγRIIB-deficient mice.[86] The protective mechanism is partially associated with the increased population of FcγRIIB-expressing splenic macrophages, resulting in the induction of inhibitory signaling through FcγRIIB triggered by the cross-linking of FcγRIIB and FcγRIII or RIV by platelet-antibody ICs (Fig. 4A). These findings support the hypothesis that IVIg is a possible FcγRIIB-targeting therapy, however, it remains unclear whether the exact inhibitory signaling mediated by FcγRIIB involves the effect of IVIg treatment because mice deficient in SHIP, SHP-1 and Btk respond to the ameliorating effects of IVIg with the same kinetics as control mice.[87]

On the other hand, a recent report shows that the effect of IVIg in mouse ITP model seems to involve the acute interaction of activating FcγRs on dendritic cells, while FcγRIIB has a role in the late phase of IVIg action.[88] The adoptive transfer of IVIg-primed wild-type DCs ameliorated ITP, but donor cells from FcRγ chain-deficient mice did not inhibit ITP. Expression of FcγRIIB on donor DCs was not required for the amelioration of ITP, however, the effect of IVIg-treated DCs was observed in FcγRIIB-sufficient recipients but not FcγRIIB-deficient mice. These findings suggest that the interaction of IVIg with FcRγ-associating receptors, including activating FcγRI or paired immunoglobulin-like receptor (PIR)-A on DC, down-regulates the function of phagocytic macrophages (Fig. 4B).

In addition to the contribution of activating FcγR, it is reported that FcRn plays an essential role in the effects of IVIg in the arthritis model induced by serum transfer from K/BxN mice[89] and autoimmune skin blistering disease model induced by pathogenic IgG transfer.[90] FcRn-deficient mice show less susceptibility to both arthritis and skin blistering disease models and these mice are also resistant to the IVIg treatment. These findings show that the recycling of antibody by FcRn is critical for both the disease development and the therapeutic effect of IVIg in the pathogenic autoantibody-induced diseases. It is possible that IVIg blocks the recycling of autoantibodies by interacting with FcRn (Fig. 4C). These findings above suggest that both activating and inhibitory FcRs contribute to the effect of IVIg treatment at the various phases.

Conclusion

The deregulation of balancing by FcR-mediated activation or inhibition signaling affects the maintaining of peripheral tolerance, the prevention of allergic responses and the therapeutic effects by mAbs or IVIg, supporting the idea that FcR-mediated signaling pathways as well as FcR itself could be effective therapeutic targets. Recent accumulating structural analyses of FcR complexes provide detailed information of several peptides blocking the interaction between FcRs and immunoglobulin.[91-94] On the other hand, a molecule that targets FcR-mediated activation or inhibitory signaling has been hindered. If FcγRIIB-targeting mAbs or chimeric recombinant proteins can induce inhibitory signaling into immune or malignant cells, it will be launched as an effective therapeutic agent to treat allergy, autoimmune diseases and cancer.

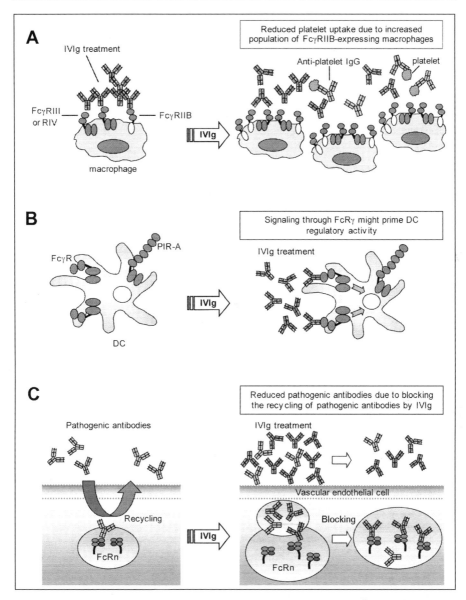

Figure 4. Possible mechanisms in IVIg treatment. A) In the murine model of ITP, anti-platelet antibody-coated platelets bind to FcγRIII or IV on macrophages, resulting in the FcγRIII or RIV cross-linking that triggers the phagocytosis of platelets. Administration of IVIg increases the population of FcγRIIB expressing splenic macrophages, resulting in the induction of inhibitory signaling through FcγRIIB triggered by the cross-linking of FcγRIIB and FcγRIII or RIV by platelet-antibody ICs. B) Interaction of IVIg with FcRγ-associating activating receptors including FcγR or PIR-A on DC, might down-regulate the function of phagocytic macrophage. C) The therapeutic saturation of FcRn by IVIg might contribute to blocking the recycling of pathogenic autoantibodies.

Acknowledgements

This work was supported by the CREST program of Japan Science and Technology Agency, a Grant-in-Aid from the Ministry of Education, Culture, Sports, Science and Technology of Japan, grants from the Mochida Memorial Foundation for Medical and Pharmaceutical Research, Kanzawa Medical Research Foundation, the Naito Foundation, Takeda Science Foundation and the 21st century COE program "Center for Innovative Therapeutic Development Towards the Conquest of Signal Transduction Diseases".

References

1. Heyman B. Regulation of antibody responses via antibodies, complement and Fc receptors. Annu Rev Immunol 2000; 18:709-737.
2. Ravetch JV, Bolland S. IgG Fc receptors. Annu Rev Immunol 2001; 19:275-290.
3. Takai T. Roles of Fc receptors in autoimmunity. Nat Rev Immunol 2002; 2:580-592.
4. Hogarth PM. Fc receptors are major mediators of antibody based inflammation in autoimmunity. Curr Opin Immunol 2002; 14:798-802.
5. Nimmerjahn F, Ravetch JV. Fcγ receptors: Old friends and new family members. Immunity 2006; 24:19-28.
6. Nimmerjahn F, Bruhns P, Horiuchi K et al. FcγRIV: A novel FcR with distinct IgG subclass specificity. Immunity 2005; 23:41-51.
7. Roopenian DC, Christianson GJ, Sproule TJ et al. The MHC class I-like IgG receptor controls perinatal IgG transport, IgG homeostasis, and fate of IgG-Fc-coupled drugs. J Immunol 2003; 170:3528-3233.
8. Sondermann P, Huber R, Oosthuizen V et al. The 3.2-Å crystal structure of the human IgG1 Fc fragment-FcγRIII complex. Nature 2000; 406:267-273.
9. Garman SC, Wurzburg BA, Tarchevskaya SS et al. Structure of the Fc fragment of human IgE bound to its high-affinity receptor FcεRIα. Nature 2000; 406:259-266.
10. Herr AB, Ballister ER, Bjorkman PJ. Insights into IgA-mediated immune responses from the crystal structures of human FcαRI and its complex with IgA1-Fc. Nature 2003; 423:614-620.
11. Radaev S, Motyka S, Fridman WH et al. The structure of a human type III Fcγ receptor in complex with Fc. J Biol Chem 2001; 276:16469-16477.
12. Woof JM, Burton DR. Human antibody-Fc receptor interactions illuminated by crystal structures. Nat Rev Immunol 2004; 4:89-99.
13. Maxwell KF, Powell MS, Hulett MD et al. Crystal structure of the human leukocyte Fc receptor, FcγRIIa. Nat Struct Biol 1999; 6:437-442.
14. Sondermann P, Huber R, Jacob U. Crystal structure of the soluble form of the human Fcγ-receptor IIb: A new member of the immunoglobulin superfamily at 1.7 A resolution. EMBO J 1999; 18:1095-1103.
15. Garman SC, Kinet JP, Jardetzky TS. Crystal structure of the human high-affinity IgE receptor. Cell 1998; 95:951-961.
16. Ding Y, Xu G, Yang M et al. Crystal structure of the ectodomain of human FcαRI. J Biol Chem 2003; 278:27966-27970.
17. Willcox BE, Thomas LM, Bjorkman PJ. Crystal structure of HLA-A2 bound to LIR-1, a host and viral major histocompatibility complex receptor Nat Immunol 2003; 4:913-919.
18. Chapman TL, Heikema AP, West AP Jr et al. Crystal structure and ligand binding properties of the D1D2 region of the inhibitory receptor LIR-1 (ILT2) Immunity 2000; 13:727-736.
19. Ghazizadeh S, Bolen JB, Fleit HB. Physical and functional association of Src-related protein tyrosine kinases with FcγRII in monocytic THP-1 cells. J Biol Chem 1994; 269:8878-8884.
20. Wang AV, Scholl PR, Geha RS. Physical and functional association of the high affinity immunoglobulin G receptor (FcγRI) with the kinases Hck and Lyn. J Exp Med 1994; 180:1165-1170.
21. Bolland S, Pearse RN, Kurosaki T et al. SHIP modulates immune receptor responses by regulating membrane association of Btk. Immunity 1998; 8:509-516.
22. Fluckiger AC, Li Z, Kato RM et al. Btk/Tec kinases regulate sustained increases in intracellular Ca^{2+} following B-cell receptor activation. EMBO J 1998; 17:1973-1985.
23. Tamir I, Stolpa JC, Helgason CD et al. The RasGAP-binding protein p62dok is a mediator of inhibitory FcγRIIB signals in B-cells. Immunity 2000; 12:347-358.
24. Hara I, Takechi Y, Houghton AN. Implicating a role for immune recognition of self in tumor rejection: passive immunization against the brown locus protein. J Exp Med 1995; 182:1609-1614.
25. Clynes R, Takechi Y, Moroi Y et al. Fc receptors are required in passive and active immunity to melanoma. Proc Natl Acad Sci USA 1998; 95:652-656.
26. Clynes RA, Towers TL, Presta LG et al. Inhibitory Fc receptors modulate in vivo cytoxicity against tumor targets. Nat Med 2000; 6:443-446.

27. Akiyama K, Ebihara S, Yada A et al. Targeting apoptotic tumor cells to FcγR provides efficient and versatile vaccination against tumors by dendritic cells. J Immunol 2003; 170:1641-1648.
28. Yada A, Ebihara S, Matsumura K et al. Accelerated antigen presentation and elicitation of humoral response in vivo by FcγRIIB- and FcγRI/III-mediated immune complex uptake. Cell Immunol 2003; 225:21-32.
29. Regnault A, Lankar D, Lacabanne V et al. Fcγ receptor-mediated induction of dendritic cell maturation and major histocompatibility complex class I-restricted antigen presentation after immune comple internalization. J Exp Med 1999; 189:371-380.
30. Rodriguez A, Regnault A, Kleijmeer M et al. Selective transport of internalized antigens to the cytosol for MHC class I presentation in dendritic cells. Nat Cell Biol 1999; 1:362-368.
31. Dhodapkar KM, Krasovsky J, Williamson B et al. Antitumor monoclonal antibodies enhance cross-presentation of cCellular antigens and the generation of myeloma-specific killer T-cells by dendritic cells. J Exp Med 2002; 195:125-133.
32. Kalergis AM, Ravetch JV. Inducing tumor immunity through the selective engagement of activating Fcγ receptors on dendritic cells. J Exp Med 2002; 195:1653-1659.
33. Bergtold A, Desai DD, Gavhane A et al. Cell surface recycling of internalized antigen permits dendritic cell priming of B-cells. Immunity 2005; 23:503-514.
34. Sinclair NR. Fc-signalling in the modulation of immune responses by passive antibody. Scand J Immunol 2001; 53:322-330.
35. Nakamura A, Akiyama K, Takai T. Fc receptor targeting in the treatment of allergy, autoimmune disease and cancer. Expert Opin Ther Targets 2005; 9:169-190.
36. Kinet JP. Atopic allergy and other hypersensitivities. Curr Opin Immunol 1999; 11: 603-605.
37. Weinberger M. Innovative therapies for asthma: anti-IgE—The future? Paediatr Respir Rev 2004; 5 Suppl A:S115-118.
38. Busse W, Neaville W. Anti-immunoglobulin E for the treatment of allergic disease. Curr Opin Allergy Clin Immunol 2001; 1:105-108.
39. Bruhns P, Fremont S, Daeron M. Regulation of allergy by Fc receptors. Curr Opin Immunol 2005; 17:662-669.
40. Strunk RC, Bloomberg GR. Omalizumab for asthma. N Engl J Med 2006; 354:2689-2695.
41. Goodnow CC. Pathways for self-tolerance and the treatment of autoimmune diseases. Lancet 2001; 357:2115-2121.
42. Scofield RH. Autoantibodies as predictors of disease. Lancet 2004; 363:1544-1546.
43. Olseni NJ, Stein CM. New drugs for rheumatoid arthritis. N Engl J Med 2004; 350:2167-2179.
44. Choy EH, Isenberg DA, Garrood T et al. Therapeutic benefit of blocking interleukin-6 activity with an anti-interleukin-6 receptor monoclonal antibody in rheumatoid arthritis: A randomized, double-blind, placebo-controlled, dose-escalation trial. Arthritis Rheum 2002; 46:3143-3150.
45. Nishimoto N, Yoshizaki K, Miyasaka N et al. Treatment of rheumatoid arthritis with humanized anti-interleukin-6 receptor antibody: A multicenter, double-blind, placebo-controlled trial. Arthritis Rheum 2004; 50:1761-1769.
46. Gorman C, Leandro M, Isenberg D. B-cell depletion in autoimmune disease. Arthritis Res Ther 2003; 5:S17-21.
47. Edwards JC, Szczepanski L, Szechinski J et al. Efficacy of B-cell-targeted therapy with rituximab in patients with rheumatoid arthritis. N Engl J Med 2004; 350:2572-2581.
48. Pescovitz MD. Rituximab, an anti-CD20 monoclonal antibody: History and mechanism of action. Am J Transplant 2006; 6:859-866.
49. Coiffier B. Monoclonal antibody as therapy for malignant lymphomas. CR Biol 2006; 329:241-254.
50. Gopal AK, Press OW. Clinical applications of anti-CD20 antibodies. J Lab Clin Med 1999; 134:445-450.
51. Uchida J, Hamaguchi Y, Oliver JA et al. The innate mononuclear phagocyte network depletes B-lymphocytes through Fc receptor-dependent mechanisms during anti-CD20 antibody immunotherapy. J Exp Med 2004; 199:1659-6169.
52. Anolik JH, Campbell D, Felgar RE et al. The relationship of FcγRIIIa genotype to degree of B-cell depletion by rituximab in the treatment of systemic lupus erythematosus. Arthritis Rheum 2003; 48:455-459.
53. Weng WK, Levy R. Two immunoglobulin G fragment C receptor polymorphisms independently predict response to rituximab in patients with follicular lymphoma. J Clin Oncol 2003; 21:3940-3947.
54. Glennie MJ, Johnson PW. Clinical trials of antibody therapy. Immunol. Today 2000; 21:403-410.
55. Gelderman KA, Tomlinson S, Ross GD et al. Complement function in mAb-mediated cancer immunotherapy. Trends Immunol 2004; 25:158-164.
56. Harris M. Monoclonal antibodies as therapeutic agents for cancer. Lancet Oncol 2004; 5:292-302.

57. Anderson KC, Bates MP, Slaughenhoupt BL et al. Expression of human B-cell-associated antigens on leukemias and lymphomas: A model of human B-cell differentiation. Blood 1984; 63:1424-1433.
58. Cartron G, Dacheux L, Salles G et al. Therapeutic activity of humanized anti-CD20 monoclonal antibody and polymorphism in IgG Fc receptor FcγRIIIa gene. Blood 2002; 99:754-758.
59. Zhang Z, Zhang M, Goldman CK et al. Effective therapy for a murine model of adult T-cell leukemia with the humanized anti-CD52 monoclonal antibody, Campath-1H. Cancer Res 2003; 63:6453-6457.
60. Slamon DJ, Godolphin W, Jones LA et al. Studies of the HER-2/neu proto-oncogene in human breast and ovarian cancer. Science 1989; 244:707-712.
61. Mosesson Y, Yarden Y. Oncogenic growth factor receptors: implications for signal transduction therapy. Semin Cancer Biol 2004; 14:262-270.
62. Sliwkowski MX, Lofgren JA, Lewis GD et al. Nonclinical studies addressing the mechanism of action of trastuzumab (Herceptin). Semin Oncol 1999; 4 Suppl 12:60-70.
63. zum Buschenfelde CM, Hermann C, Schmidt B et al. Antihuman epidermal growth factor receptor 2 (HER2) monoclonal antibody trastuzumab enhances cytolytic activity of class I-restricted HER2-specific T-lymphocytes against HER2-overexpressing tumor cells. Cancer Res 2002; 62: 2244-2247.
64. Kono K, Sato E, Naganuma H et al. Trastuzumab (Herceptin) enhances class I-restricted antigen presentation recognized by HER-2/neu-specific T cytotoxic lymphocytes. Clin Cancer Res 2004; 10:2538-2544.
65. Bolland S, Ravetch JV. Spontaneous autoimmune disease in FcγRIIB-deficient mice results from strain-specific epistasis. Immunity 2000; 13:277-285.
66. Yajima K, Nakamura A, Sugahara A et al. FcγRIIB deficiency with Fas mutation is sufficient for the development of systemic autoimmune disease. Eur J Immunol 2003; 33:1020-1029.
67. Kyogoku C, Dijstelbloem HM, Tsuchiya N et al. Fcγ receptor gene polymorphisms in Japanese patients with systemic lupus erythematosus. Arthritis Rheum 2002; 46:1242-1254.
68. Siriboonrit U, Tsuchiya N, Sirikong M et al. Association of Fcγ receptor IIb and IIIb polymorphisms with susceptibility to systemic lupus erythematosus in Thais. Tissue Antigens 2003; 61:374-383.
69. Floto RA, Clatworthy MR, Heilbronn KR et al. Loss of function of a lupus-associated FcγRIIb polymorphism through exclusion from lipid rafts. Nat Med 2005; 11:1056-1058.
70. Kono H, Kyogoku C, Suzuki T et al. FcγRIIB Ile232Thr transmembrane polymorphism associated with human systemic lupus erythematosus decreases affinity to lipid rafts and attenuates inhibitory effects on B-cell receptor signaling. Hum Mol Genet 2005; 14:2881-2892.
71. Zhu D, Kepley CL, Zhang M et al. A novel human immunoglobulin Fcγ-Fcε bifunctional fusion protein inhibits FcεRI-mediated degranulation. Nat Med 2002; 8:518-521.
72. Zhu D, Kepley CL, Zhang K et al. A chimeric human-cat fusion protein blocks cat-induced allergy. Nat Med 2005; 11:446-449.
73. Tam SW, Demissie S, Thomas D et al. A bispecific antibody against human IgE and human Fcgamma-RII that inhibits antigen-induced histamine release by human mast cells and basophils. Allergy 2004; 59:772-780.
74. Rankin CT, Veri MC, Gorlatov S et al. CD32B, the human inhibitory Fcγ receptor IIB, as a target for monoclonal antibody therapy of B-cell lymphoma. Blood 2006; 108:2384-2391.
75. Cines DB, Blanchettei VS. Immune thrombocytopenic purpura. N Engl J Med 2002; 346:995-1008.
76. Yuki N. Infectious origins of and molecular mimicry in, Guillain-Barre and Fisher syndromes. Lancet Infect Dis 2001; 1:29-37.
77. Durelli L, Isoardo G. High-dose intravenous immunoglobulin treatment of multiple sclerosis. Neurol Sci 2002; 23:S39-48.
78. Latov N, Chaudhry V, Koski CL et al. Use of intravenous γ globulins in neuroimmunologic diseases. J Allergy Clin Immunol 2001; 108:S126-132.
79. Wiles CM, Brown P, Chapel H et al. Intravenous immunoglobulin in neurological disease: A specialist review. J Neurol Neurosurg Psychiatry 2002; 72:440-448.
80. Burns JC. Kawasaki disease. Adv Pediatr 2001; 48:157-177.
81. Sewell WA, Jolles S. Immunomodulatory action of intravenous immunoglobulin. Immunology 2002; 107:387-393.
82. Simon HU, Spath PJ. IVIG—Mechanisms of action. Allergy 2003; 58:543-552.
83. Bayry J, Thirion M, Misra N et al. Mechanisms of action of intravenous immunoglobulin in autoimmune and inflammatory diseases. Transfus Clin Biol 2003; 10:165-169.
84. Samuelsson A, Towers TL, Ravetch JV. Anti-inflammatory activity of IVIG mediated through the inhibitory Fc receptor. Science 2001; 291:484-486.
85. Kaneko Y, Nimmerjahn F, Madaio MP et al. Pathology and protection in nephrotoxic nephritis is determined by selective engagement of specific Fc receptors. J Exp Med 2006; 203:789-797.
86. Bruhns P, Samuelsson A, Pollard JW et al. Colony-stimulating factor-1-dependent macrophages are responsible for IVIG protection in antibody-induced autoimmune disease. Immunity 2003; 18:573-581.

87. Crow AR, Song S, Freedman J et al. IVIg-mediated amelioration of murine ITP via FcγRIIB is independent of SHIP1, SHP-1 and Btk activity. Blood 2003; 102:558-560.
88. Siragam V, Crow AR, Brinc D et al. Intravenous immunoglobulin ameliorates ITP via activating Fcγ receptors on dendritic cells. Nat Med 2006; 12:688-692.
89. Akilesh S, Petkova S, Sproule TJ et al. The MHC class I-like Fc receptor promotes humorally mediated autoimmune disease. J Clin Invest 2004; 113:1328-1333.
90. Li N, Zhao M, Hilario-Vargas J et al. Complete FcRn dependence for intravenous Ig therapy in autoimmune skin blistering diseases. J Clin Invest 2005; 115:3440-3450.
91. McDonnell JM, Beavil AJ, Mackay GA et al. Structure based design and characterization of peptides that inhibit IgE binding to its high-affinity receptor. Nat Struct Biol 1996; 5:419-426.
92. Marino M, Ruvo M, De Falco S et al. Prevention of systemic lupus erythematosus in MRL/lpr mice by administration of an immunoglobulin-binding peptide. Nat Biotechnol 2000; 18:735-739.
93. Uray K, Medgyesi D, Hilbert A et al. Synthesis and receptor binding of IgG1 peptides derived from the IgG Fc region. J Mol Recognit 2004; 17:95-105.
94. Medgyesi D, Uray K, Sallai K et al. Functional mapping of the FcγRII binding site on human IgG1 by synthetic peptides. Eur J Immunol 2004; 34:1127-1135.

CHAPTER 18

Therapeutic Blockade of T-Cell Antigen Receptor Signal Transduction and Costimulation in Autoimmune Disease

Joseph R. Podojil, Danielle M. Turley and Stephen D. Miller*

Abstract

CD4+ T-cell-mediated autoimmune diseases are initiated and maintained by the presentation of self-antigen by antigen-presenting cells (APCs) to self-reactive CD4+ T-cells. According to the two-signal hypothesis, activation of a naïve antigen-specific CD4+ T-cell requires stimulation of both the T-cell antigen receptor (signal 1) and costimulatory molecules such as CD28 (signal 2). To date, the majority of therapies for autoimmune diseases approved by the Food and Drug Administration primarily focus on the global inhibition of immune inflammatory activity. The goal of ongoing research in this field is to develop antigen-specific treatments which block the deleterious effects of self-reactive immune cell function while maintaining the ability of the immune system to clear nonself antigens. To this end, the signaling pathways involved in the induction of CD4+ T-cell anergy, as apposed to activation, are a topic of intense interest. This chapter discusses components of the CD4+ T-cell activation pathway that may serve as therapeutic targets for the treatment of autoimmune disease.

Introduction

An important goal of current research is to develop new therapies for autoimmune diseases by specifically inhibiting and/or tolerizing self-reactive immune cells. While the present chapter focuses on the regulation of T-cell antigen receptor (TCR) and costimulatory molecule signaling pathways in one particular autoimmune disease, multiple sclerosis (MS) and its mouse model experimental autoimmune encephalomyelitis (EAE), similar approaches are ongoing in other autoimmune diseases as well as in tissue transplantation. Approximately 350,000 people in the United States of America have MS, a T-cell mediated demyelinating disease hypothesized to be triggered by an initiating event, possibly an infectious one. In these subjects, myelin-specific autoreactive CD4+ T-cells damage central nervous system (CNS) myelin. MS is characterized by perivascular CD4+ T-cell and mononuclear cell infiltration with subsequent primary demyelination of axonal tracks leading to progressive paralysis.[1] While CD4+ T-cells can discriminate between specific peptide antigens in the context of MHC II in an antigen-specific and MHC II haplotype-restricted manner through use of the TCR,[2] the TCR in and of itself is not intrinsically able to distinguish the difference between self- and nonself-peptides. Therefore, during thymic CD4+ T-cell selection, the majority of self-reactive T-cells are clonally deleted subsequent to presentation of self-antigens on

*Corresponding Author: Stephen D. Miller—Department of Microbiology-Immunology, Northwestern University Feinberg School of Medicine, Tarry 6-718, 303 E. Chicago Ave., Chicago, IL 60611, USA. Email: s-d-miller@northwestern.edu

Multichain Immune Recognition Receptor Signaling: From Spatiotemporal Organization to Human Disease, edited by Alexander B. Sigalov. ©2008 Landes Bioscience and Springer Science+Business Media.

thymic antigen-presenting cells (APCs).[3,4] Self-reactive CD4+ T-cells that escape thymic negative selection maintain the capability to respond to self-antigens presented by activated peripheral APCs. The ability to control peripheral activation of self-reactive T-cells is dependent on the level of costimulatory molecules expressed on the surface of APCs. In turn, the level of costimulatory molecule expression and cytokine production of APCs is regulated by the presence or absence of inflammation, infectious agents and other pathologic conditions. Thus, self-tolerance in the periphery is maintained, in part, by presentation of self-peptides on immature APCs that lack expression of costimulatory molecules resulting in anergy induction in self-reactive CD4+ T-cells. Costimulation blockade also represents a putative therapeutic strategy for treatment of established autoimmune diseases as a means to re-establish self-tolerance. A following section elaborates on this topic.

MS is an autoimmune disease characterized by T-cell responses to a variety of myelin proteins including myelin basic protein (MBP), myelin proteolipid protein (PLP) and/or myelin-oligo-dendrocyte glycoprotein (MOG).[5] There are four courses of clinical disease in MS: (1) relaps-ing-remitting, (2) secondary progressive, (3) primary progressive and (4) progressive-relapsing (Fig. 1). Correspondingly, there are relapsing-remitting and chronic mouse EAE models of MS. Relapsing-remitting EAE (R-EAE) is characterized by transient ascending hind limb paralysis, perivascular mononuclear-cell infiltration and fibrin deposition in the brain and spinal cord with adjacent areas of acute and chronic demyelination.[6] The facts that the inducing antigen (Ag) in MS has not been identified and that CD4+ T-cell responses to multiple epitopes on a number of myelin proteins activated via epitope spreading are probably responsible for chronic disease progres-sion make the use of antigen-specific tolerance-based immunotherapies problematic at this time. Furthermore, in human MS, a pathological role for epitope spreading is difficult to verify because the initiating antigen is not known. In contrast, animal models, such as EAE, have the advantage of a known initiating antigen. For example, in the SJL model of disease in which mice are primed with PLP$_{139-151}$ in complete Freund's adjuvant (CFA), peripheral PLP$_{139-151}$-specific CD4+ T-cell reactivity is maintained throughout the disease. However, prior to the first relapse PLP$_{178-191}$-specific CD4+ T-cell reactivity arises (intramolecular epitope spreading) and during the second relapse T-cells specific for a myelin basic protein epitope, MBP$_{84-104}$, arise due to intermolecular epitope spreading (Fig. 2). While Ag-specific tolerance can be induced in this experimental model and the self-peptides have been well characterized, this is not true for humans with MS. Therefore, the development of more efficacious and focused antigen nonspecific immunosuppressive therapies is currently favored.

T-Cell Activation: Target for Treatment of Disease

As first proposed by Lafferty and Cunningham,[7] activation of naïve T-cells requires two signals. The first signal received by a naïve CD4+ T-cell comes from the Ag-specific TCR interacting with an antigenic peptide presented in the context of MHC II on the APC surface. The second set of signals includes secretory products, such as cytokines, that are produced by either the APC or the activated CD4+ T-cell itself and costimulatory molecules that are expressed on the cell surface of activated APC. For example, CD80 (B7-1) and CD86 (B7-2) expressed on the APC surface interact with the coreceptor CD28 that is constitutively expressed on the surface of CD4+ T-cells.[8,9] The overall effect of CD28 ligation is to increase the level of proliferation and cytokine production, promote cell survival and enhance expression of CD40 ligand (CD40L) and adhe-sion molecules necessary for trafficking, such as VLA-4.[10-12] The costimulatory molecule pairs, CD28-CD80/CD86 and CD40-CD40L and cellular adhesion molecules represent putative therapeutic targets for blockade of autoreactive CD4+ T-cell activation and trafficking to inflam-matory sites. In addition to CD28 costimulation, the production of interferon-gamma (IFN-γ) or interleukin 4 (IL-4) by activated CD4+ T-cells or release of IL-12 by activated macrophages, dendritic or B-cells directs the local population of naïve CD4+ T-cells to differentiate toward the IFN-γ-producing Th1 cell or IL-4-producing Th2 cell phenotypes, respectively.[12] Recently, a third population of CD4+ effector T-cells that secrete IL-17 has been identified. The Th17 cell secretes

IL-17, IL-6, IL-21 and TNF-α and is hypothesized to differentiate from a common naïve CD4+ T-cell precursor cell that has been activated in the presence of TGF-β and IL-6. Furthermore, APC-secreted IL-23 is thought to maintain the survival of Th17 cells in vivo[13,14] and Th17 cells are critical for the development and maintenance of EAE.[15,16] Therefore, the development of an immune-mediated therapy may work through one of three possible mechanisms either alone or in combination: (1) induction of anergy in self-reactive CD4+ T-cells; (2) deletion of self-reactive CD4+ T-cells by apoptosis; or (3) immune deviation.

Previously tested immunotherapeutic strategies have been shown to work at least in part through the alteration of signal 1 and/or the inhibition of costimulatory molecule stimulation (signal 2). In this manner, CD4+ T-cell anergy is hypothesized to be induced in T-cells undergoing activation at the time of treatment via a short-term blockade of CD28-CD80/CD86 interactions. CD28-CD80/CD86 inhibitory reagents are currently being tested in phase I/II clinical trials in various autoimmune diseases. The goal for treatment of autoimmune diseases, such as MS, is to re-establish tolerance to self-antigens. The difficulty in the development of these therapies lies in maintaining the ability of the patient to normally recognize and react to nonself-antigens. While these therapies are still under development, current therapies for MS approved by the Food and Drug Administration (FDA) focus on immune deviation or nonspecific immunosuppression. For example, the administration of interferon-β is used to decrease the severity and frequency of disease relapses. Secondly, systemic or mucosal administration of antigens or altered peptide ligands has been tested with mixed success. Copaxone (Glatiramer Acetate) is a random mixture of peptides of various lengths composed of glutamine, lysine, alanine and tyrosine. This mixture is administered via daily subcutaneous injections to treat relapsing-remitting MS. The mechanism of action is believed to be the elicitation of suboptimal TCR signaling in the absence of costimulatory molecule signaling (signal 1 in the absence of signal 2). Treatment with Copaxone is hypothesized to induce a low level of TCR stimulation, thus inducing immune deviation toward a Th2 phenotype (disease-regulatory) as compared to Th1/Th17 phenotypes (disease-promoting). In an attempt to further test the two-signal hypothesis, several groups have investigated the therapeutic potential of anti-CD3 monoclonal antibody (mAb) treatment of various autoimmune diseases. However, treatment with an unaltered anti-CD3 mAb is potentially a double-edged sword: while treatment eliminates pathogenic autoreactive CD4+ T-cells and thereby ameliorates autoimmune disease progression, it may also induce serious nonspecific side effects through bystander activation of T-cells. For example, the induction of general immunosuppression increases the patient's susceptibility to opportunistic infection and the common occurrence of high-dose syndrome in which treatment recipients suffer severe side effects due to the nonspecific production of inflammatory cytokines such as TNF-α. Furthermore, crosslinking of CD3 may in some cases initiate a signal of sufficient strength that eliminates the need for a costimulatory molecule-induced reduction in the signal threshold required for T-cell activation.

Due to the aforementioned complications associated with the use of an unaltered anti-CD3 mAb, modifications to the anti-CD3 mAb have been made so that the deleterious side effects are avoided by reducing/eliminating the antibody ability to bind to Fc receptors and thus decreasing the ability to efficiently crosslink the TCR. The regulatory properties of nonmitogenic anti-CD3 mAb treatment are believed to be due to the lower levels of TCR-mediated signaling since the nonmitogenic anti-CD3 mAb is not stabilized by binding to Fc receptors on the surface of the APCs to allow for efficient TCR crosslinking. In this manner, nonmitogenic anti-CD3 mAb is hypothesized to favor T-cell differentiation into a Th2 cell phenotype and the development of regulatory T-cells (Treg).[17,18] Therefore, the possibility exists that treatments induce immune deviation. In this scenario, the T-cell-mediated immune response is changed from a Th1/Th17-like (disease-promoting) response to a Th2-like (disease-regulating) response. In support of this hypothesis, findings from numerous studies suggest that cytokines mediate a protective effect in nonmitogenic anti-CD3 mAb treatment. However, there is currently debate concerning the exact contribution of cytokines to the underlying mechanisms of treatment.[19,20] Furthermore, activated Th1 cells, but not naïve CD4+ T-cells, appear to become unresponsive to subsequent restimulation

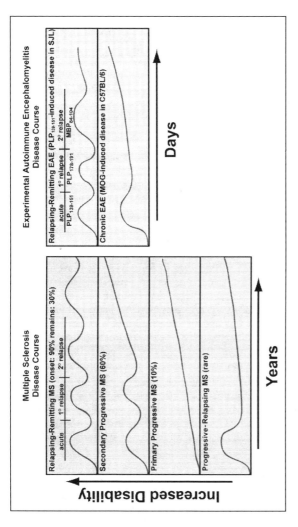

Figure 1. Clinical disease courses in multiple sclerosis (MS) and its mouse model experimental autoimmune encephalomyelitis (EAE). The clinical disease course of MS is classified according to the characteristics and severity of disease progression over time. The most common disease course of MS is Relapsing-Remitting MS (RRMS). This disease course is characterized by a defined acute attack (increase in disability) followed by a full recovery and subsequent attacks over time. Secondary Progressive MS (SPMS) is similar to RRMS, but instead of full recovery during remission, residual deficit is maintained. SPMS is characterized by less recovery during remission following attacks and fewer attacks as the disease course switches from a relapsing-remitting disease course to a more progressive disease course. Primary Progressive MS (PPMS) is a disease course characterized by a progressive increase in disability over time in the absence of well-defined relapses and/or remissions. Progressive-Relapsing MS (PRMS) is the least common of the disease courses characterized by a progressive disability from the onset of disease. PRMS contains clear relapses in disease severity in the absence or presence of full recovery. In the SJL mouse, relapsing-remitting model of MS, i.e., R-EAE, the dominate spread epitope for each consecutive relapse is well characterized. During the acute phase of the disease the majority of activated CD4+ T-cells are specific for the dominate proteolipid protein epitope, PLP$_{139-151}$ used to induce disease. The dominate epitope during the primary relapse is PLP$_{178-191}$ (intramolecular epitope spreading) and during the secondary relapse this epitope is MBP$_{84-104}$ (intermolecular epitope spreading). In contrast, epitope spreading in the chronic disease model has been suggested, but the consecutive dominant epitopes have not been identified.

following treatment with nonmitogenic anti-CD3 mAb.[21] To gain a better understanding of the potential Th cell subset specificity of nonmitogenic anti-CD3 mAb treatment, the efficacy of nonmitogenic anti-CD3 mAb treatment to induce tolerance has been compared in Th1 and Th2 cells revealing that tolerance is induced in Th1 cells as determined by proliferation and IL-2 production.[22] In contrast, no effect on the Th2 cell phenotype and activity was seen. These findings lend support to the theory that nonmitogenic anti-CD3 mAb may specifically downregulate Th1 cell function. Treatment of disease with a nonmitogenic anti-CD3 mAb provides a therapy that potentially blocks or induces a suboptimal signal 1 in the absence of costimulatory signals (signal 2). This therapy is hypothesized to represent a treatment by which only activated immune cells are affected at the time of treatment, allowing the maintenance of host defense against nonself-antigens at times post nonmitogenic anti-CD3 treatment.

Besides the administration of antigens in a tolerogenic form for the treatment of autoimmune diseases in humans,[23] adhesion molecule and costimulatory molecule blockade are currently being tested. For example, the use of Tysabri, a monoclonal antibody able to block the interaction of the adhesion molecule VLA-4 with its target ligand VCAM-1 expressed by endothelial cells, has been re-approved for the treatment of patients who have inadequate responses to other approved MS therapies. Clinical trials are also ongoing to study the therapeutic effect of a CD28-CD80/CD86 blockade by the use of the extracellular portion of cytotoxic T-lymphocyte-associated antigen 4 (CTLA-4) covalently linked to the Fc portion of an immunoglobulin molecule (CTLA-4-Ig). CTLA-4-Ig treatment is thought to block CD28-CD80/CD86 interactions as CTLA-4-Ig binds with high affinity to CD80/CD86 expressed on activated APCs. In this manner, the autoreactive CD4+ T-cell would receive signal 1 in the absence of signal 2. CD80 and CD86 have been shown to have differential roles in T-cell activation and differentiation.[24] Thus, conflicting results have been obtained using anti-CD80 and anti-CD86 mAbs to regulate autoimmune disease. Treatment with anti-CD80 mAb surrounding autoantigen priming has been shown to block EAE development induced with suboptimal concentrations of $PLP_{139-151}$ or MBP_{84-104} in SJL mice, whereas anti-CD86 mAb treatment has been reported to either exacerbate disease[25] or to have no effect.[26] In contrast, treatment with intact anti-CD86 mAb initiated during the remission following acute disease does not affect disease progression (relapses) in $PLP_{139-151}$-induced R-EAE.[27] Treatment with monovalent, noncrosslinking anti-CD80 Fab fragments during EAE remission blocks clinical relapses and epitope spreading to the $PLP_{178-191}$ epitope,[27] whereas treatment with intact anti-CD80 mAb leads to a profound exacerbation of disease relapses concomitant with accelerated epitope spreading.[28] Likewise, treatment of mice with a small molecule inhibitor of CD28 during disease decreases disease severity and proliferation of myelin-specific CD4+ T-cells upon ex vivo activation and increases CD4+ T-cell apoptosis.[29] Thus, a short-term blockade of CD28-CD80 interactions may represent a therapy which would predominantly target activated T-cells during the treatment period, thus allowing maintenance of host defense against infection.

Coupled-Cell Tolerance: Antigen-Specific Induction of Tolerance to a Self-Antigen

While the therapeutic treatments mentioned above are antigen nonspecific, a variety of current autoimmune therapies use antigen-specific approaches and are currently under development. For example, the intravenous injection of ethylene carbodiimide (ECDI)-antigen-coupled splenocytes (Ag-SP) is an efficient method of promoting clonal anergy of antigen-specific CD4+ T-cells both in vivo and in vitro.[30-33] Ag-SP promotes T-cell tolerance in many animal models of autoimmune and inflammatory disease including experimental autoimmune thyroditis,[34] uveitis[35] and neuritis,[36] as well as the non-obese diabetic (NOD) mouse model of diabetes (Kohm, AP, Miller SD, unpublished observations) and transplant survival.[37] In EAE, Ag-SP induces a long-lasting, antigen-specific tolerance in both the active-priming and adoptive transfer models of EAE regardless of whether the treatment is administered prior to or following disease initiation.[33,38-40] Ag-SP also appears to be nontoxic and well tolerated by treated animals at all stages of disease. In contrast, i.v. tolerance induced by intravenous administration of soluble peptides can induce severe anaphylactic responses and depending on the antigen result in the death of treated animals.[41,42] Since Ag-SP can induce long-lasting, Ag-specific tolerance in CD4[+] T-cells in the absence of any negative side effects, Ag-SP has significant therapeutic potential for future autoimmune therapy.

The mechanism of Ag-SP-induced self-tolerance is not completely understood. The two-signal hypothesis is presumed to play an active role in the induction of CD4[+] T-cell tolerance. Syngeneic donor spleen cells are fixed with ECDI in the presence of antigen. ECDI fixes antigen to the cells by crosslinking the free amino and carboxyl groups of the peptides to the donor cell surface proteins. This produces peptide-coated cells that function as potent tolerance-inducing carriers. The mechanisms of Ag-SP-induced tolerance was initially hypothesized to be mediated by direct interactions between MHC II-peptide complexes and the TCR expressed by target CD4[+] T-cells (signal 1).[43,44] Furthermore, the level of costimulatory molecule expression by the donor cells (signal 2) is also believed to be an important factor in the ability of Ag-SP to render cells anergic.[32] For example, lipopolysaccharide (LPS)-preactivated-coupled cells with high CD80/CD86 expression are not capable of inducing tolerance, suggesting that successful tolerance induction is dependent upon the lack of costimulatory signals coming from the APC.[43,44] CTLA-4 ligation during the secondary antigen encounter also appears to be important for the maintenance of the tolerized state.[43,44]

Alternative mechanisms may also contribute to the induction of functional tolerance by Ag-SP. In addition to peptide antigens, both whole protein and mouse spinal cord homogenate (MSCH) efficiently induce tolerance in CD4[+] T-cells when coupled to ECDI-fixed spleen cells.[45-47] Ag-SP is also effective when multiple encephalitogenic peptides are coupled to cells allowing the simultaneous targeting of multiple myelin-associated antigens[48] and effectively blocking possible spread epitopes (Figs. 1 and 2). The efficiency of Ag-SP is independent of de novo antigen processing by the donor-coupled cells since the inclusion of antigen-processing inhibitors during fixation do not inhibit tolerance induction.[49] On the other hand, the antigen must be physically attached to the donor cells for tolerance induction to occur. Donor spleen cells from MHC II-deficient mice that are ECDI-coupled to myelin peptides are able to ameliorate clinical disease,[50] but twice the number of MHC II-lacking Ag-SP donor cells is required to induce the level of protection equivalent to that of syngeneic-derived donor cells. Tolerance in this case is hypothesized to occur through the reprocessing of the donor-coupled cells by host APCs that are then able to represent antigen to host T-cells.[50] There is still much to be learned about the mechanism of coupled cell tolerance induction. However, taken together, the findings suggest that Ag-SP is an efficient method to restore Ag-specific self-tolerance during autoimmune disease. Ag-SP also lacks many of the safety concerns that accompany other methods of tolerance induction such as the anaphylactic responses associated with tolerance induced by intravenous injection of soluble peptides. In light of these findings, the use of peptide-coupled APCs holds therapeutic promise as a potential therapy for MS and other autoimmune diseases.

Immune Synapse: Activation of the TCR in Lipid Rafts

As mentioned above, CD4[+] T-cells respond to their environment by the use of a multichain immune recognition receptor (MIRR), i.e., the peptide-specific, MHC II-restricted TCR. While TCR-mediated recognition of a peptide bound to the MHC II molecules on the surface of the APC is necessary for T-cell activation (signal 1), the cytoplasmic tails of the TCR alpha/beta chains do not have inherent kinase activity. Signaling through the TCR is achieved through the associated accessory proteins that contain immunoreceptor tyrosine-based activation motifs (ITAMs). Following crosslinking of the TCR, the intracellular signaling cascade is initiated by phosphorylation of ITAM tyrosines by Src family kinases. While events that induce the association of Src family kinases with TCR remain undetermined, specialized cholesterol- and sphingolipid-rich membrane domains known as lipid rafts appear to function as platforms for the interaction (Fig. 3).[51] Due to the biochemical properties of cholesterol and sphingolipids, the lipids are tightly packed together, including specific membrane-associated proteins and excluding others. It is currently hypothesized that Src family kinases preferentially associate with lipid rafts and that upon the recognition of an antigenic peptide the TCR translocates into the lipid rafts where the Src family kinases

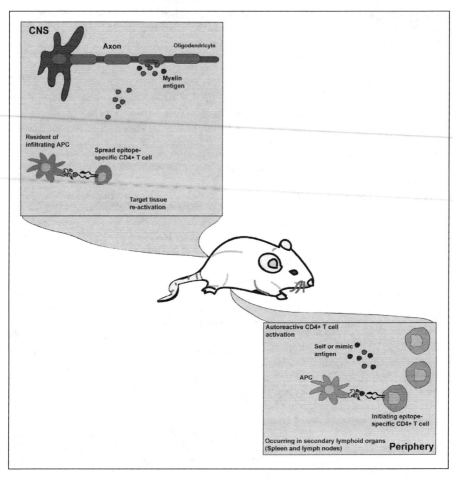

Figure 2. Epitope spreading. Animal models of multiple sclerosis (MS) have helped to identify putative mechanisms by which epitope spreading occurs. In R-EAE, the activation of the autoreactive CD4+ T-cells specific for the initiating antigen epitope (blue CD4+ T-cell) occurs in the draining lymph node. Upon activation, the activated CD4+ T-cells enter circulation and extravasate into the central nervous system (CNS). Once in the CNS, the autoreactive CD4+ T-cells initiate myelin destruction and activate resident and infiltrating APCs. The activated infiltrating immune cells secrete cytokines and chemokines that not only recruit immune cells into the CNS but also help to open the blood-brain barrier (BBB). Myelin antigens not only reactivate the CD4+ T-cells specific for the initiating antigen, but are also released, phagocytized, processed and presented by APCs to CD4+ T-cells. For example, in PLP$_{139-151}$-induced R-EAE in SJL mice, the initiating epitope is PLP$_{139-151}$ and this population of CD4+ T-cells is responsible for the initial acute phase of the disease. During the acute phase of the disease, the destruction of myelin allows for the release of both PLP and MBP. Due to antigen availability, the activation of the secondary population of CD4+ T-cells specific for PLP$_{178-191}$ (red CD4+ T-cell) occurs prior to the primary relapse, e.g., intramolecular spread epitope. In the case of R-EAE, the activation of the spread epitope-specific CD4+ T-cells has been shown to occur within the CNS. During the secondary relapse, CD4+ T-cells specific for MBP$_{84-104}$ are activated, e.g., intermolecular epitope spreading. A color version of this figure is available online at www.eurekah.com.

phosphorylate ITAMs on the cytoplasmic tail of the TCR.[52] Therefore, TCR signaling may be regulated by the ability of the TCR to associate with lipid rafts upon crosslinking.

Based on the ability of lipid rafts to exclude or include specific proteins, lipid raft-associated proteins are modified to allow for inclusion. For example, the Src family kinases Fyn, Lyn and Lck, which initiate TCR ITAM phosphorylation, are myristoylated and palmitoylated.[53,54] A list of the TCR signaling components that associate with lipid rafts is presented in Table 1. Another transmembrane protein involved in CD4+ T-cell activation is LAT which is palmitylated upon TCR crosslinking.[55] LAT mutants that cannot be palmitylated are not able to associate with lipid rafts, thereby altering TCR signaling.[55] The function of LAT as it pertains to T-cell activation versus anergy is discussed in a following section. The overall physical outcome of ligating the TCR is formation of the immune synapse. The immune synapse is a highly ordered membrane structure in which the TCR, associated signaling proteins, cytoskeleton and cellular adhesion molecules are concentrated to allow for sufficient intercellular protein-protein interactions.[56] The TCR-APC immune synapse is a dynamic structure containing a central cluster of TCRs ringed by adhesion molecules. On the cytoplasmic side, it contains signaling molecules such as Src family kinases and protein kinase C and the integrin-associated cytoskeleton proteins including talin.[57,58] Following crosslinking of the TCR, the immune synapse persists for more than an hour in a cytoskeleton-dependent manner, thereby allowing the TCR to be stimulated multiple times.[56] Stimulation of CD28 has also been shown to enhance the recruitment of lipid rafts to the immune synapse.[59] In this manner, the organization of the T-cell plasma membrane during T-cell-APC interaction not only contributes to the inclusion of the necessary signaling molecules but also allows for sufficient and sustained TCR signaling.

NFAT: Regulation of T-Cell Activation and Anergy

The previous sections were focused on defining the known immune mechanisms involved in an autoimmune disease and pointed out possible therapeutic targets. The following sections discuss the signaling pathways that may play a putative role in the induction of self-reactive CD4+ T-cell anergy. First and foremost, it is probable that several forms of anergy exist that

---

Table 1. Lipid raft-associated components of the TCR signaling complex

TCR-associated signaling molecules included in lipid rafts before TCR crosslinking
    Lck
    Fyn
    Itk
    Syk
    Ras
    Cbl-b
    CD4
    Actin
TCR-associated signaling molecules included in lipid rafts after TCR crosslinking
    Zap-70
    Slp-76
    Vav
    Grd-2
    PLC-γ1
    PKC
    LAT
    TCR
TCR-associated signaling molecule excluded from lipid rafts after TCR crosslinking
    CD45

have yet to be completely characterized biochemically. Part of the confusion may arise from the multiple costimulatory molecules that modulate T-cell responses following stimulation of the TCR. This discussion focuses on anergy induced by blocking cell cycle progression. The most consistent properties of anergic CD4+ T-cells is the decreased production of IL-2 and decreased proliferation.[60] Anergy has also been defined as an unresponsive state which is reversible by IL-2.[61-63] By this definition, therefore, one can conclude that an anergic CD4+ T-cell has been previously activated to the extent that it expresses the high-affinity IL-2 receptor and that the anergic T-cell is unresponsive but not nonviable. A critical point is that anergy is only a relative measure of the immune response. For example, although substantial decreases in responsiveness can be achieved in vitro, that level of responsiveness may cause significant effects in vivo.

Initial characterization of anergy in vitro in which TCR engagement (signal 1) occurred without costimulation (signal 2) demonstrated that T-cell clones were unable to proliferate or produce IL-2 under these conditions. These studies initiated a flurry of investigations into proposed intrinsic signaling defects that suggested that a myriad of deficiencies—such as a lack of mitogen-activated protein kinase (MAPK) signaling, Ras activation or the upregulation of dominant "anergic" factors—gave rise to the anergy phenotype.[60] As a result, a coherent model for the molecular mechanism of anergy induction was difficult to develop, in part due to the varied model systems used to induce T-cell anergy, including oral administration of soluble peptide or superantigen treatment in vivo and crosslinking of the CD3 complex in the absence of costimulation in vitro. A more recent model system for induced T-cell unresponsiveness utilizes prolonged nuclear factor of activated T-cells (NFAT) occupancy of anergy-associated gene promoters in the absence of MAPK signaling induced by treatment with the potent Ca^{2+} ionophore ionomycin. This causes the upregulation of a unique set of genes responsible for the induction of this form of T-cell tolerance by the NFAT/NFAT homodimer.[64] Anergy induction upregulates the expression of several ubiquitin E3 ligases, including Cbl-b (Casitas B-lineage lymphoma B), Itch and GRAIL (gene related to anergy in lymphocytes), leading to degradation of key signaling proteins in T-cell activation.[65] Cbl-b promotes the conjugation of ubiquitin to phosphatidylinositol 3-kinase (PI3 kinase) and modulates its recruitment to CD28 and TCR–CD3 complexes, thereby regulating the activation of Vav (Fig. 3). In support of this, the increased tyrosine phosphorylation of Vav in Cbl-b$^{-/-}$ T-cells and the enhanced T-cell proliferation and IL-2 production have been shown to be reversed by PI3 kinase inhibitors.[66] While controversy still exists as to whether PI3 kinase plays a crucial role in T-cell activation, particularly in CD28-mediated signaling,[67] recent data show that ligation of CD28 induces the formation of a grb-2-associated binder 2 (grb-2)/SRC homology phosphatase-2/PI3 kinase complex.[68] This suggests the induction of CD4+ T-cell tolerance is regulated by altered NFAT transcriptional activity leading to expression of anergy-associated genes rather than activation-associated genes.

Of the multiple signaling pathways that are upregulated during T-cell activation, Ca^{2+} signaling is critical for the first step of anergy induction. Lack of CD4+ T-cell costimulation through the interaction of CD28 with CD80/CD86 expressed on the surface of the APC correlates with an unbalanced partial form of signaling in which TCR-mediated Ca^{2+} influx predominates. While CD28 ligation is not directly coupled to Ca^{2+} mobilization, CD28 signaling potentiates TCR signals that do not involve Ca^{2+} influx. Experimentally, this is shown by the fact that anergy induced in CD4+ T-cells activated with Ca^{2+} ionophores is closely related to that induced in the absence of CD28 costimulation following TCR stimulation. As mentioned above, Ca^{2+}-induced anergy is mediated primarily by NFAT. NFAT is a transcription factor regulated by the protein phosphatase calcineurin and both NFAT activation and anergy induction are blocked by calcineurin inhibitors such as cyclosporin A (CsA).[69] NFAT was initially identified as an inducible nuclear factor that could bind the IL-2 promoter in activated T-cells.[70] However, when all proteins of the known NFAT family were isolated and characterized, it became clear that their expression is not limited to T-cells (Table 2). At least one NFAT family member is expressed by almost every cell type that has been examined. The NFAT family consists of five members: NFAT1 (also known as NFATp or NFATc2), NFAT2 (also known as NFATc or NFATc1), NFAT3 (also known as NFATc4),

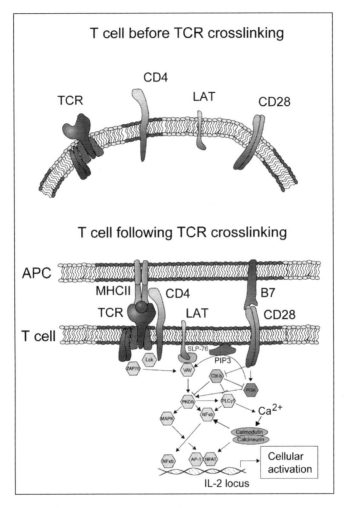

Figure 3. Signal transduction pathways involved in T-cell anergy. Signals delivered by the engagement of the TCR (signal 1) and costimulatory molecules, such as CD28, (signal 2) induce different signaling pathways that result in the activation of multiple transcription factors. Prior to crosslinking of the TCR, LAT and CD28 are located outside of lipid rafts (red portion of the lipid bilayer). Ligation of the TCR by peptide—MHC on an APC triggers the recruitment of the TCR and signaling elements, e.g., phospholipase C-1 (PLC-1) for the Ca^{2+} influx—nuclear factor of activated T-cells (NFAT) pathway and PKC-θ for the NF-κB and AP-1 pathway, which control nuclear transcriptional and gene activation to the lipid rafts. Prior to crosslinking the TCR, the necessary components for TCR signaling are distributed throughout the T-cell surface. Following crosslinking, the TCR and its accessory signaling proteins are recruited to lipid rafts. As a consequence, the TCR is a central component of the immune synapse in which sufficient TCR stimulation (signal 1) and costimulatory molecule stimulation (signal 2) occur. In the nucleus, NFAT cooperates with AP-1 and other transcription factors to induce a program of gene expression leading to IL-2 production. TCR engagement (signal 1) in the absence of costimulation (signal 2) results in induction NFAT proteins without concomitant AP-1 activation. In the absence of cooperative binding to AP-1 (FOS and JUN), NFAT transcriptionally regulates a distinct set of anergy-inducing genes, e.g., *Cbl-b*. Anergy-associated factors inhibit T-cell function at different levels leading to a state of T-cell unresponsiveness. A color version of this figure is available online at www.eurekah.com.

NFAT4 (also known as NFATx or NFATc3) and NFAT5 (also known as TonEBP or OREBP). All NFAT proteins have a highly conserved DNA-binding domain that is structurally related to the DNA-binding domain of the REL family of transcription factors. Four of the NFAT family members are regulated by Ca^{2+} signaling, while NFAT1, 2 and 4 are the predominate members expressed in $CD4^+$ T-cells. Since Ca^{2+} is activated following TCR stimulation, the NFAT family members are activated within a $CD4^+$ T-cell, but the outcome of NFAT activation depends on whether the $CD4^+$ T-cell receives costimulatory signals.

The molecular regulation of tolerance induction is an emerging area of study in which the complexity of intracellular signaling is beginning to be identified. The regulation of T-cell activation is regulated by NFAT through its interaction with AP-1.[71] The discovery that NFAT proteins can also form transcriptional complexes with other partners and even be transcriptionally active by themselves has introduced the possibility of defining new roles for NFAT proteins in T-cells.[65] In the two-signal hypothesis for T-cell activation, stimulation of the TCR (signal 1) and costimulatory molecule stimulation, i.e., CD28-CD80/CD86 (signal 2), are both required for full activation, whereas signaling through the TCR only induces T-cell anergy. In the absence of costimulatory molecule-enhanced AP-1 and the CD28-induced stabilization of the immune synapse, NFAT regulates transcription of a specific program of genes involved in the negative regulation of TCR signaling (Fig. 3). In this model, costimulatory signals push the TCR-induced signaling above the threshold level, allowing for cellular activation. For example, in the absence of the CD28-induced signaling pathway there is an unbalanced activation of the $CD4^+$ T-cell resulting in an altered set of transcription factors present in the nucleus due to the decreased activation of RAS—MAPK, protein kinase C (PKC) or inhibitor of NF-κB (IκB) kinases (IKK) signaling.[72] To illustrate the dependence of anergy-associated gene expression on NFAT, treatment of *Nfat1*$^{-/-}$ T-cells and wildtype $CD4^+$ T-cells with the calcineurin inhibitor CsA inhibits the activation of NFAT, blocks the expression of these anergy-associated genes and impairs induction of anergy in treated cells.[72] Since NFAT proteins control two opposing aspects of T-cell function, activation and anergy, it is likely that the availability of transcriptional partners in response to activating or anergizing stimuli determines which set of genes is activated. Among the proteins expressed by anergic T-cells there is a group of E3 ubiquitin ligases, i.e., Itch, Cbl-b and GRAIL.[65,73-75] Alterations in the molecules that negatively regulate TCR signaling such as Cbl-b have been shown to be involved in the initiation of autoimmune disease.[76,77] For example, the loss of Cbl-b expression allows for the hyper-reactive signaling through the TCR. Interestingly, T-cells from Cbl-b$^{-/-}$ mice are characterized by a lower threshold of activation following TCR-mediated signaling that results in hypersensitivity upon TCR engagement and activation of downstream signaling pathways without the normal requirement for coreceptor stimulation.[76] These two signaling pathway intermediates represent two putative candidates for the mechanism by which Ag-SP-induced tolerance of $CD4^+$ T-cells occurs.

NFAT Inhibitors: Putative Therapeutics

Given the important role of NFAT proteins in the control of T-cell activation, NFAT is considered to be an optimal target for therapeutic approaches aimed at regulating T-cell-mediated immune responses. For example, inhibitors of calcineurin, such as CsA and FK506, block the downstream activation of NFAT. These compounds are extensively used as immunosuppressive agents.[78] While the mechanism of action of these inhibitors is through the blockade of calcineurin and the inhibition of NFAT activation, they are not specific for NFAT. Thus, the caveat exists that nonNFAT-associated effects on T-cell function are involved in regulating T-cell function.[79] To test this possibility, several studies have begun to identify the protein-protein interaction site between NFAT and calcineurin. By developing an inhibitory peptide that blocks the interaction between NFAT and calcineurin, the specificity of the inhibitor can be significantly increased. For example, the interaction of NFAT proteins with calcineurin is mapped to their N-terminal regulatory domain, allowing for the development of a more specific NFAT inhibitor, i.e., a high affinity calcineurin-binding peptide, MAGHP**VIVIT**GPHEE.[80,81] A cell-permeable version of the VIVIT peptide, which is able to selectively inhibit calcineurin-mediated NFAT dephosphorylation,[81] has

Table 2. *NFAT family members*

NFAT Family Member	Alternative Names	Regulation	Expression	Immune Phenotype of Knockout Mice	Reference
NFAT1	NFATc2, NFATp	Ca²⁺/calcineurin	T-cells; vascular endothelial cells; skeletal muscle cells; chondrocytes; adipocytes; pancreatic islet-cells	• Enhanced B- and T-cell responses; • Th2 skewing with decreased IFN-γ production and increased IL-4 expression; • Allergic responses; • Suppression of chrondrogenesis	71,72,92-95
NFAT2	NFATc, NFATc1	Ca²⁺/calcineurin	T-cells; cardiac muscle cells, skeletal muscle cells	• Embryonic lethal due to defect in cardiac valve; • in RAG complementation system: impaired Th2 cell responses and IL-4 production	71,72,92-95
NFAT3	NFATc4	Ca²⁺/calcineurin	Perivascular tissue cells; adipocytes; cardiac muscle cells	Not reported yet	71,72,92-95
NFAT4	NFATc3, NFATx	Ca²⁺/calcineurin	T-cells; skeletal muscle cells; keratinocytes	• Decreased number of single-positive CD4 and CD8 cells due to increased apoptosis of double positive thymocytes	71,72,92-95
NFAT5	TonEBP; OREBP	Osmotic stress	T-cells; most cell types	• Decreased cellularity in thymus and spleen; • Impaired T-cell function in hyperosmotic conditions.	71,93

been successfully used to prolong graft survival in an experimental system of islet-cell transplantation in mice.[82] While these peptides are much more selective than CsA, they maintain the ability to inhibit the interaction of calcineurin with other substrates that use similar PXIXIT motifs to interact with calcineurin, e.g., calcineurin-binding protein 1 (CABIN1) or A-kinase anchor protein (AKAP79).[83] Recently, two additional regions of calcineurin have been found to be necessary for NFAT binding.[83,84] Therefore, it has been hypothesized that the amino acids flanking the PXIXIT motif may provide specific targets for future therapeutics.

The therapeutic use of peptide inhibitors is still limited by problems associated with the route of administration and half-life/stability of the inhibitor in vivo. As an alternative, the use of small organic molecules may overcome the peptide inhibitor limitations, since limitless structural changes can be designed to improve the specificity, stability, delivery and distribution of these molecules. Recently, several small organic molecules were identified that specifically inhibit calcineurin-induced NFAT activation, blocking NFAT-dependent cytokine production by T-cells.[85] While the current small molecule inhibitors are efficient at blocking NFAT-dependent transcription and are able to potentiate CsA effects, these molecules still act upstream of calcineurin.[86] Therefore, for these agents the same nonspecific side effects as known for CsA and FK506 may exist for in vivo use. The ability of small molecules to selectively inhibit calcineurin and NFAT protein—protein interactions points to the possibility of using them to modulate specific NFAT-regulated functions. Differential interactions between various NFAT family members and specific transcriptional partners might underlie the ability of NFAT to integrate multiple signaling pathways and control diverse cellular functions. If the protein—protein contact surfaces are specific for different interactions, molecules could be designed to block NFAT-regulated functions without affecting other calcineurin-regulated functions. Such molecules will most likely be therapeutically useful, with notably improved specificity and greatly reduced toxicity.

LAT: An Alternative Component of TCR Signaling

In addition to the regulation by transcription factors, components of the TCR signaling complex are also implicated in the regulation of CD4+ T-cell anergy. For example, the adaptor molecule LAT is a transmembrane protein that facilitates the formation of a multisubunit signaling complex with other signaling molecules such as phospholipase Cγ1, Gads-SLP-76, Grb2 and PI3 kinase[87] (Fig. 3). LAT is essential for TCR signaling. Upon TCR stimulation, phosphorylation of LAT is necessary for activation of MAPK cascades, Ca^{2+} flux and activation of the transcription factor AP-1. As mentioned in the previous section, the signaling cascade activated by the costimulatory molecule CD28 also activates AP-1. In this way, TCR and CD28 signaling pathways are coordinated to increase the level of AP-1 present within the T-cell, allowing for the regulation of activation-associated genes by the NFAT/AP-1 heterodimer. In the absence of AP-1, NFAT homodimerizes and regulates the expression of a cohort of ubiquitin E3 ligases, including Cbl-b, Itch and GRAIL.[65,73] Although Cbl-b is upregulated at both the mRNA and protein levels in ionomycin-anergized cells, ionomycin-treated Cbl-b-deficient CD4+ T-cells are also defective in LAT phosphorylation. Since there is an increased steady-state level of the LAT protein present in the cellular lysates of Cbl-b$^{-/-}$ T-cells, it is possible that Cbl-b regulates LAT steady-state protein amounts and thus more ionomycin is required to overcome this enlarged pool of LAT. To determine if the TCR signaling pathway is altered in anergic CD4+ T-cells, anergized antigen-specific transgenic T-cells were compared to control transgenic T-cells. While the immediate phosphorylation of TCR ζ-chain and ZAP-70 is normal in anergized T-cells, the adaptor protein LAT and its downstream target PLCγ1 are hypophosphorylated. The kinetics of both LAT and ZAP-70 activation is also decreased in anergic T-cells due to decreased recruitment of the p85 regulatory subunit of PI(3)K by LAT.[88] Interestingly, normal activation of the CD28 pathway was noted in ionomycin-anergized T-cells upon restimulation, demonstrating that the costimulatory cascade, which itself contributes to LAT phosphorylation, is unaltered. Inhibition of LAT activation may thus serve as a viable target for the induction of CD4+ T-cell anergy.

As discussed previously, a critical step in the activation of a naïve CD4⁺ T-cell is the costimulatory signal provided by the APC, where duration and strength of signal parameters help determine the outcome of the T-cell-APC interaction.[89] It has been demonstrated that although the number of conjugates between T-cells and APCs is not altered, the percentage of synapses that contains LAT is markedly decreased in ionomycin-anergized T-cells.[88] These results are consistent with previous work suggesting that ionomycin-anergized T-cells form unstable immune synapses. In order for LAT to be phosphorylated by active ZAP-70, it must first be palmitylated to facilitate trafficking to lipid rafts. In stimulated cells pretreated with ionomycin, there is diminished LAT localization to lipid rafts in addition to the lack of LAT phosphorylation compared to control cells stimulated without ionomycin treatment (Fig. 3). Because CD4 and Fyn localize to lipid rafts in both sets of stimulated cells, this argues that there is not a global defect in the constituents of lipid rafts. Based on these findings, many intriguing questions for future investigations into the control of T-cell activation can be formulated. LAT is located at the crucial juncture between recognition of an incoming signal from the TCR and its dissemination into multiple intracellular signaling cascades and second messengers. Attenuation of LAT localization and phosphorylation severely cripples T-cell activation.

Conclusions

It is clear that the T-cell repertoire in an autoimmune response, such as peptide-induced relapsing-remitting EAE, is dynamic and CD4⁺ T-cell responses to the initiating epitope play the dominant pathologic role during the acute disease episode but not in disease relapses. Understanding the mechanisms underlying spontaneous disease remission is critical to the ultimate design of therapeutic modalities. Current therapies for the re-establishment of self-tolerance in autoimmune disease focus on the inhibition of signal 1 and/or signal 2. For example, the blockade and/or provision of subthreshold levels of signal 1 in an antigen-specific therapy includes ECDI-antigen-coupled APC treatment and treatment by nonmitogenic anti-CD3 mAb to induce nonspecific signaling in activated T-cells. The blockade of signal 2 using CTLA-4-Ig to block CD28-CD80/CD86 interactions is currently used in the ongoing clinical trials. The molecular regulation of tolerance induction is an emerging area of study in which alterations in intracellular signaling pathways are beginning to be identified. As presented in this chapter, the regulation of T-cell activation appears to be controlled by NFAT through its interaction with AP-1.[71] Besides the positive regulation of transcription when dimerized with AP-1, NFAT also forms homodimers and complexes with other transcription factors, directly regulating transcription of anergy-associated genes.[65] For example, alterations in the molecules that negatively regulate TCR signaling, such as Cbl-b, have been shown to be involved in the initiation of autoimmune disease by allowing for hyperactive TCR signaling.[76,77] A characteristic feature of T-cells from Cbl-b⁻/⁻ mice is a lower threshold of the TCR-mediated activation, resulting in hypersensitivity following TCR engagement and activation of downstream signaling pathways without the normal requirement for coreceptor stimulation.[76] Furthermore, the dysregulation of TCR signaling cascade associated with T-cell survival, such as the PI3 kinase pathway, is associated with the loss of self-tolerance and the development of autoimmune disease.[90] This chapter has illustrated that three potential points of intervention exist for the induction of CD4⁺ T-cell anergy and may serve as potential therapeutic targets for regulation of autoimmune disease: (1) increasing the frequency of NFAT/NFAT homodimers as opposed to NFAT/AP-1 heterodimers; (2) increasing anergy-associated signaling pathway intermediates, such as Cbl-B and GRAIL; and (3) downregulation of the TCR signaling complex components, e.g., LAT.

Acknowledgements

The authors wish to acknowledge the support of grants from the National Institutes of Heath (NS034819, NS-026543 and NS-026543) and the Myelin Repair Foundation to SDM as well as a postdoctoral fellowship grant to JRP from the National Multiple Sclerosis Society (FG 1667A1/2-A-1).

References

1. Steinman L, Martin R, Bernard C et al. Multiple sclerosis: Deeper understanding of its pathogenesis reveals new targets for therapy. Annu Rev Neurosci 2002; 25:491-505.
2. Zinkernagel RM. H-2 restriction of virus-specific T-cell-mediated effector functions in vivo. II. Adoptive transfer of delayed-type hypersensitivity to murine lymphocytic choriomeningitis virus is restricted by the K and D region of H-2. J Exp Med 1976; 144:776-87.
3. Von Boehmer H. T-cell development and selection in the thymus. Bone Marrow Transplant 1992; 9 Suppl 1:46-48.
4. Rocha B, Vassalli P, Guy-Grand D. The extrathymic T-cell development pathway. Immunol Today 1992; 13:449-54.
5. Sospedra M, Martin R. Immunology of multiple sclerosis. Annu Rev Immunol 2005; 23:683-747.
6. Paterson PY, Swanborg RH. Demyelinating diseases of the central and peripheral nervous systems.Immunological Diseases, Vol. 4. In: Sampter M, Talmage DW, Frank MM et al, eds. Boston: Little, Brown and Co, 1988:1877-916.
7. Lafferty KJ, Cunningham AJ. A new analysis of allogeneic interactions. Aust J Exp Biol Med Sci 1975; 53:27-42.
8. Damle NK, Klussman K, Linsley PS et al. Differential costimulatory effects of adhesion molecules B7, ICAM-1, LFA-3 and VCAM-1 on resting and antigen-primed CD4 + T-lymphocytes. J Immunol 1992; 148:1985-92.
9. Gross JA, Callas E, Allison JP. Identification and distribution of the costimulatory receptor CD28 in the mouse. J Immunol 1992; 149:380-88.
10. Harding FA, McArthur J, Gross JA et al. CD28 mediated signalling costimulates murine T-cells and prevents induction of anergy in T-cell clones. Nature 1992; 356:607-09.
11. Norton SD, Zuckerman L, Urdahl KB et al. The CD28 ligand, B7, enhances IL-2 production by providing a costimulatory signal to T-cells. J Immunol 1992; 149:1556-61.
12. Seder RA, Germain RN, Linsley PS et al. CD28-mediated costimulation of interleukin 2 (IL-2) production plays a critical role in T-cell priming for IL-4 and interferon gamma production. J Exp Med 1994; 179:299-304.
13. Mangan PR, Harrington LE, O'Quinn DB et al. Transforming growth factor-beta induces development of the T(H)17 lineage. Nature 2006; 441:231-34.
14. Veldhoen M, Hocking RJ, Atkins CJ et al. TGFbeta in the context of an inflammatory cytokine milieu supports de novo differentiation of IL-17-producing T-cells. Immunity 2006; 24:179-89.
15. Chen Y, Langrish CL, McKenzie B et al. Anti-IL-23 therapy inhibits multiple inflammatory pathways and ameliorates autoimmune encephalomyelitis. J Clin Invest 2006; 116:1317-26.
16. Langrish CL, Chen Y, Blumenschein WM et al. IL-23 drives a pathogenic T-cell population that induces autoimmune inflammation. J Exp Med 2005; 201:233-40.
17. Herold KC, Burton JB, Francois F et al. Activation of human T-cells by FcR nonbinding anti-CD3 mAb, hOKT3gamma1(Ala-Ala). J Clin Invest 2003; 111:409-18.
18. Chatenoud L. CD3-specific antibody-induced active tolerance: From bench to bedside. Nat Rev Immunol 2003; 3:123-32.
19. Plain KM, Chen J, Merten S et al. Induction of specific tolerance to allografts in rats by therapy with nonmitogenic, nondepleting anti-CD3 monoclonal antibody: Association with TH2 cytokines not anergy. Transplantation 1999; 67:605-13.
20. Tran GT, Carter N, He XY et al. Reversal of experimental allergic encephalomyelitis with nonmitogenic, nondepleting anti-CD3 mAb therapy with a preferential effect on T(h)1 cells that is augmented by IL-4. Int Immunol 2001; 13:1109-20.
21. Smith JA, Tso JY, Clark MR et al. Nonmitogenic anti-CD3 monoclonal antibodies deliver a partial T-cell receptor signal and induce clonal anergy. J Exp Med 1997; 185:1413-22.
22. Smith JA, Tang Q, Bluestone JA. Partial TCR signals delivered by FcR-nonbinding anti-CD3 monoclonal antibodies differentially regulate individual Th subsets. J Immunol 1998; 160:4841-49.
23. Peng J, Liu C, Liu D et al. Effects of B7-blocking agent and/or CsA on induction of platelet-specific T-cell anergy in chronic autoimmune thrombocytopenic purpura. Blood 2003; 101:2721-6.
24. Schweitzer AN, Sharpe AH. Studies using antigen-presenting cells lacking expression of both B7-1 (CD80) and B7-2 (CD86) show distinct requirements for B7 molecules during priming versus restimulation of Th2 but not Th1 cytokine production. J Immunol 1998; 161:2762-71.
25. Kuchroo VK, Das MP, Brown JA et al. B7-1 and B7-2 costimulatory molecules differentially activate the Th1/Th2 developmental pathways: Application to autoimmune disease therapy. Cell 1995; 80:707-18.
26. Perrin PJ, Scott D, Davis TA et al. Opposing effects of CTLA4-Ig and anti-CD80 (B7-1) plus anti-CD86 (B7-2) on experimental allergic encephalomyelitis. J Neuroimmunol 1996; 65:31-39.
27. Miller SD, Vanderlugt CL, Lenschow DJ et al. Blockade of CD28/B7-1 interaction prevents epitope spreading and clinical relapses of murine EAE. Immunity 1995; 3:739-45.

28. Vanderlugt CL, Karandikar NJ, Lenschow DJ et al. Treatment with intact anti-B7-1 mAb during disease remission enhances epitope spreading and exacerbates relapses in R-EAE. J Neuroimmunol 1997; 79:113-18.
29. Srinivasan M, Gienapp IE, Stuckman SS et al. Suppression of experimental autoimmune encephalomyelitis using peptide mimics of CD28. J Immunol 2002; 169:2180-8.
30. Wetzig R, Hanson DG, Miller SD et al. Binding of Ovalbumin to mouse spleen cells with and without carbodiimide. J Immunol Methods 1979; 28:361-68.
31. Miller SD, Wetzig RP, Claman HN. The induction of cell-mediated immunity and tolerance with protein antigens coupled to syngeneic lymphoid cells. J Exp Med 1979; 149:758-73.
32. Jenkins MK, Schwartz RH. Antigen presentation by chemically modified splenocytes induces antigen-specific T-cell unresponsiveness in vitro and in vivo J Exp Med 1987; 165:302-19.
33. Miller SD, Tan LJ, Pope L et al. Antigen-specific tolerance as a therapy for experimental autoimmune encephalomyelitis. Int Rev Immunol 1992; 9:203-22.
34. Braley-Mullen H, Tompson JG, Sharp GC et al. Suppression of experimental autoimmune thyroiditis in guinea pigs by pretreatment with thyroglobulin-coupled spleen cells. Cell Immunol 1980; 51:408-13.
35. Dua HS, Gregerson DS, Donoso LA. Inhibition of experimental autoimmune uveitis by retinal photoreceptor antigens coupled to spleen cells. Cell Immunol 1992; 139:292-305.
36. Gregorian SK, Clark L, Heber-Katz E et al. Induction of peripheral tolerance with peptide-specific anergy in experimental autoimmune neuritis. Cell Immunol 1993; 150:298-310.
37. Elliott C, Wang K, Miller SD et al. Ethylcarbodiimide as an agent for induction of specific transplant tolerance. Transplantation 1994; 58:966-68.
38. Kennedy MK, Tan LJ, Dal Canto MC et al. Regulation of the effector stages of experimental autoimmune encephalomyelitis via neuroantigen-specific tolerance induction. J Immunol 1990; 145:117-26.
39. Kennedy KJ, Smith WS, Miller SD et al. Induction of antigen-specific tolerance for the treatment of ongoing, relapsing autoimmune encephalomyelitis—A comparison between oral and peripheral tolerance. J Immunol 1997; 159:1036-44.
40. Vandenbark AA, Vainiene M, Ariail K et al. Prevention and treatment of relapsing autoimmune encephalomyelitis with myelin peptide-coupled splenocytes. J Neurosci Res 1996; 45:430-38.
41. Smith CE, Eagar TN, Strominger JL et al. Differential induction of IgE-mediated anaphylaxis after soluble vs cell-bound tolerogenic peptide therapy of autoimmune encephalomyelitis. Proc Natl Acad Sci USA 2005; 102:9595-600.
42. Pedotti R, Mitchell D, Wedemeyer J et al. An unexpected version of horror autotoxicus: anaphylactic shock to a self-peptide. Nat Immunol 2001; 2:216-22.
43. Eagar TN, Karandikar NJ, Bluestone J et al. The role of CTLA-4 in induction and maintenance of peripheral T-cell tolerance. Eur J Immunol 2002; 32:972-81.
44. Eagar TN, Turley DM, Padilla J et al. CTLA-4 regulates expansion and differentiation of Th1 cells following induction of peripheral T-cell tolerance. J Immunol 2004; 172:7442-50.
45. Kennedy MK, Dal Canto MC, Trotter JL et al. Specific immune regulation of chronic-relapsing experimental allergic encephalomyelitis in mice. J Immunol 1988; 141:2986-93.
46. Kennedy MK, Tan LJ, Dal Canto MC et al. Inhibition of murine relapsing experimental autoimmune encephalomyelitis by immune tolerance to proteolipid protein and its encephalitogenic peptides. J Immunol 1990; 144:909-15.
47. Tan LJ, Kennedy MK, Miller SD. Regulation of the effector stages of experimental autoimmune encephalomyelitis via neuroantigen-specific tolerance induction. II. Fine specificity of effector T-cell inhibition. J Immunol 1992; 148:2748-55.
48. Smith CE, Miller SD. Multi-peptide coupled-cell tolerance ameliorates ongoing relapsing EAE associated with multiple pathogenic autoreactivities. J Autoimmunity 2006; 27:218-31.
49. Pope L, Paterson PY, Miller SD. Antigen-specific inhibition of the adoptive transfer of experimental autoimmune encephalomyelitis in Lewis rats. J Neuroimmunol 1992; 37:177-90.
50. Turley DM, Miller SD. Peripheral tolerance Induction using ethylenecarbodiimide-fixed APCs uses both direct and indirect mechanisms of antigen presentation for prevention of experimental autoimmune encephalomyelitis. J Immunol 2007; 178:2212-20.
51. Simons K, Ikonen E. Functional rafts in cell membranes. Nature 1997; 387:569-72.
52. Langlet C, Bernard AM, Drevot P et al. Membrane rafts and signaling by the multichain immune recognition receptors. Curr Opin Immunol 2000; 12:250-5.
53. Vidalain PO, Azocar O, Servet-Delprat C et al. CD40 signaling in human dendritic cells is initiated within membrane rafts. EMBO J 2000; 19:3304-13.
54. Shenoy-Scaria AM, Gauen LK, Kwong J et al. Palmitylation of an amino-terminal cysteine motif of protein tyrosine kinases p56lck and p59fyn mediates interaction with glycosyl-phosphatidylinositol-anchored proteins. Mol Cell Biol 1993; 13:6385-92.

55. Zhang W, Trible RP, Samelson LE. LAT palmitoylation: its essential role in membrane microdomain targeting and tyrosine phosphorylation during T-cell activation. Immunity 1998; 9:239-46.
56. Bromley SK, Burack WR, Johnson KG et al. The immunological synapse. Annu Rev Immunol 2001; 19:375-96.
57. Grakoui A, Bromley SK, Sumen C et al. The immunological synapse: A molecular machine controlling T-cell activation. Science 1999; 285:221-7.
58. Monks CR, Freiberg BA, Kupfer H et al. Three-dimensional segregation of supramolecular activation clusters in T-cells. Nature 1998; 395:82-6.
59. Viola A, Schroeder S, Sakakibara Y et al. T-lymphocyte costimulation mediated by reorganization of membrane microdomains. Science 1999; 283:680-2.
60. Schwartz RH. T-cell anergy. Annu Rev Immunol 2003; 21:305-34.
61. Jenkins MK, Mueller D, Schwartz RH et al. Induction and maintenance of anergy in mature T-cells. Adv Exp Med Biol 1991; 292:167-76.
62. Schwartz RH. A cell culture model for T-lymphocyte clonal anergy. Science 1990; 248:1349-56.
63. Schwartz RH, Mueller DL, Jenkins MK et al. T-cell clonal anergy. Cold Spring Harb Symp Quant Biol 1989; 54:605-10.
64. Borde M, Barrington RA, Heissmeyer V et al. Transcriptional basis of lymphocyte tolerance. Immunol Rev 2006; 210:105-19.
65. Heissmeyer V, Rao A. E3 ligases in T-cell anergy—Turning immune responses into tolerance. Sci STKE 2004; 2004:pe29.
66. Fang D, Liu YC. Proteolysis-independent regulation of PI3K by Cbl-b-mediated ubiquitination in T-cells. Nat Immunol 2001; 2:870-5.
67. Ward SG, Cantrell DA. Phosphoinositide 3-kinases in T-lymphocyte activation. Curr Opin Immunol 2001; 13:332-8.
68. Parry RV, Whittaker GC, Sims M et al. Ligation of CD28 stimulates the formation of a multimeric signaling complex involving grb-2-associated binder 2 (gab2), SRC homology phosphatase-2 and phosphatidylinositol 3-kinase: evidence that negative regulation of CD28 signaling requires the gab2 pleckstrin homology domain. J Immunol 2006; 176:594-602.
69. Diehn M, Alizadeh AA, Rando OJ et al. Genomic expression programs and the integration of the CD28 costimulatory signal in T-cell activation. Proc Natl Acad Sci USA 2002; 99:11796-801.
70. Goodnow CC. Pathways for self-tolerance and the treatment of autoimmune diseases. Lancet 2001; 357:2115-21.
71. Macian F, Lopez-Rodriguez C, Rao A. Partners in transcription: NFAT and AP-1. Oncogene 2001; 20:2476-89.
72. Crabtree GR, Olson EN. NFAT signaling: Choreographing the social lives of cells. Cell 2002; 109 Suppl:S67-79.
73. Jeon MS, Atfield A, Venuprasad K et al. Essential role of the E3 ubiquitin ligase Cbl-b in T-cell anergy induction. Immunity 2004; 21:167-77.
74. Seroogy CM, Soares L, Ranheim EA et al. The gene related to anergy in lymphocytes, an E3 ubiquitin ligase, is necessary for anergy induction in CD4 T-cells. J Immunol 2004; 173:79-85.
75. Anandasabapathy N, Ford GS, Bloom D et al. GRAIL: An E3 ubiquitin ligase that inhibits cytokine gene transcription is expressed in anergic CD4+ T-cells. Immunity 2003; 18:535-47.
76. Naramura M, Kole HK, Hu RJ et al. Altered thymic positive selection and intracellular signals in Cbl-deficient mice. Proc Natl Acad Sci USA 1998; 95:15547-52.
77. Bachmaier K, Krawczyk C, Kozieradzki I et al. Negative regulation of lymphocyte activation and auto-immunity by the molecular adaptor Cbl-b. Nature 2000; 403:211-6.
78. Gremese E, Ferraccioli GF. Benefit/risk of cyclosporine in rheumatoid arthritis. Clin Exp Rheumatol 2004; 22:S101-7.
79. Kiani A, Rao A, Aramburu J. Manipulating immune responses with immunosuppressive agents that target NFAT. Immunity 2000; 12:359-72.
80. Aramburu J, Garcia-Cozar F, Raghavan A et al. Selective inhibition of NFAT activation by a peptide spanning the calcineurin targeting site of NFAT. Mol Cell 1998; 1:627-37.
81. Aramburu J, Yaffe MB, Lopez-Rodriguez C et al. Affinity-driven peptide selection of an NFAT inhibitor more selective than cyclosporin A. Science 1999; 285:2129-33.
82. Noguchi H, Matsushita M, Okitsu T et al. A new cell-permeable peptide allows successful allogeneic islet transplantation in mice. Nat Med 2004; 10:305-9.
83. Li H Rao A, Hogan PG. Structural delineation of the calcineurin-NFAT interaction and its parallels to PP1 targeting interactions. J Mol Biol 2004; 342:1659-74.
84. Rodriguez A, Martinez-Martinez S, Lopez-Maderuelo MD et al. The linker region joining the catalytic and the regulatory domains of CnA is essential for binding to NFAT. J Biol Chem 2005; 280:9980-4.

85. Roehrl MH, Kang S, Aramburu J et al. Selective inhibition of calcineurin-NFAT signaling by blocking protein-protein interaction with small organic molecules. Proc Natl Acad Sci USA 2004; 101:7554-9.
86. Venkatesh N, Feng Y, DeDecker B et al. Chemical genetics to identify NFAT inhibitors: Potential of targeting calcium mobilization in immunosuppression. Proc Natl Acad Sci USA 2004; 101:8969-74.
87. Houtman JC, Houghtling RA, Barda-Saad M et al. Early phosphorylation kinetics of proteins involved in proximal TCR-mediated signaling pathways. J Immunol 2005; 175:2449-58.
88. Hundt M, Tabata H, Jeon MS et al. Impaired activation and localization of LAT in anergic T-cells as a consequence of a selective palmitoylation defect. Immunity 2006; 24:513-22.
89. Friedl P, den Boer AT, Gunzer M. Tuning immune responses: diversity and adaptation of the immunological synapse. Nat Rev Immunol 2005; 5:532-45.
90. Ohashi PS. T-cell signalling and autoimmunity: Molecular mechanisms of disease. Nat Rev Immunol 2002; 2:427-38.
91. McMahon EJ, Bailey SL, Castenada CV et al. Epitope spreading initiates in the CNS in two mouse models of multiple sclerosis. Nat Med 2005; 11:335-39.
92. Rao A, Luo C, Hogan PG. Transcription factors of the NFAT family: Regulation and function. Annu Rev Immunol 1997; 15:707-47.
93. Macian F. NFAT proteins: key regulators of T-cell development and function. Nat Rev Immunol 2005; 5:472-84.
94. Horsley V, Pavlath GK. NFAT: Ubiquitous regulator of cell differentiation and adaptation. J Cell Biol 2002; 156:771-4.
95. Masuda ES, Imamura R, Amasaki Y et al. Signalling into the T-cell nucleus: NFAT regulation. Cell Signal 1998; 10:599-611.

CHAPTER 19

MHC and MHC-Like Molecules:
Structural Perspectives on the Design of Molecular Vaccines

Vasso Apostolopoulos,* Eliada Lazoura and Minmin Yu

Abstract

Major histocompatibility complex (MHC) molecules bind and present short antigenic peptide fragments on the surface of antigen presenting cells (APCs) to T-cell receptors. Recognition of peptide-MHC complexes by T-cells initiates a cascade of signals in T-cells and activated cells either destroy or help to destroy the APC. The MHCs are divided into three subgroups: MHC class I, MHC class II and MHC class III. In addition, nonclassical MHC molecules and MHC-like molecules play a pivotal role in shaping our understanding of the immune response. In the design of molecular vaccines for the treatment of diseases, an understanding of the three-dimensional structure of MHC, its interaction with peptide ligands and its interaction with the T-cell receptor are important prerequisites, all of which are discussed herein.

Introduction

The Major Histocompatibility Complex (MHC) is a set of molecules displayed on cell surfaces that is responsible for antigen presentation to lymphocytes. The MHC molecules control the immune response through "self" and "non self" recognition and consequently serve as targets in transplantation rejection. More than two decades of intense research were required to determine the function of MHC molecules in antigen presentation. In the early 1970s, neither the function of the T-cell receptor (TCR) nor its ligand were known. The interaction between the MHC and antigen was also unclear. MHC restriction was only established in the early 1970s with the discovery that MHC molecules control T helper cells and B-cells,[1] T helper cells and macrophages,[2] as well as CD8 T-cell recognition of target cells.[3] It was demonstrated that T-cells recognize antigenic peptides in the context of MHC; and consequently, T-cells were said to be MHC restricted.

The MHC multigene clusters are located on chromosome 17 in mice and chromosome 6 in humans. The MHC molecules are divided into three subgroups: MHC class I, MHC class II and MHC class III. The MHC class I locus encodes heterodimeric peptide-binding proteins as well as antigen processing molecules, such as TAP and tapasin. The MHC class II locus encodes heterodimeric peptide-binding proteins and proteins that modulate peptide loading onto MHC class II proteins in the lysosomal compartment, such as MHC class II DM, -DQ and -DP. MHC class I proteins are expressed on all nucleated cells and MHC class II are found on only a few specialized cell types; B-cells, neutrophils, dendritic cells and thymic epithelial cells and can be induced on macrophages and human T-cells. The MHC class III locus encodes for other immune

*Corresponding Author: Vasso Apostolopoulos—Burnet Institute at Austin, Kronheimer Building, Studley Road, Heidelberg, VIC 3084, Australia. Email: vasso@burnet.edu.au

Multichain Immune Recognition Receptor Signaling: From Spatiotemporal Organization to Human Disease, edited by Alexander B. Sigalov. ©2008 Landes Bioscience and Springer Science+Business Media.

components, i.e., complement components (C2, C4, factor B), cytokines (TNF-α and TNF-β) and HSP70.

Stimulation of CD8 T Cells

MHC, first described more than 70 years ago to control transplant rejection in mice, was initially termed H-2. There are three gene families for murine MHC class I: H-2K, H-2D and H-2L and for human MHC: HLA-A, HLA-B and HLA-C. The first step in CD8[+] T-cell generation is the uptake and presentation of peptides by antigen presenting cells through their MHC molecules. Peptides bound to MHC class I are usually endogenous and cytosolic although exogenous peptides may also be presented by MHC class I molecules.[4,5] Exogenous antigens are taken up by antigen presenting cells (APCs), primarily dendritic cells, into phagosomes and early and late endosomes and presented to MHC class II molecules. Numerous reports have demonstrated that in early endosomes some antigens can either degrade or escape out of the endosome into the cytosol and enter the proteasome.[4,5] Likewise, endogenous peptides are primarily generated in the cytosol by the proteasome. From this point, exogenous- and endogenous-derived peptides follow the same pathway. The proteasome consists of 24 subunits, half of which contain proteolytic activity. The proteasome degrades antigens (proteins) into small peptides which are released into the cytosol. The peptides are transported from the cytosol into the endoplasmic reticulum (ER) via the transporter associated with antigen processing (TAP1 and TAP2) by ATP. In the ER, peptides bind to the newly synthesized MHC class I molecules following the formation of a large multimeric complex which involves TAP, tapasin, calreticulin, calnexin and ER60. From the ER, the peptide-MHC class I complex is transported to the surface of APCs through the secretory pathway (Golgi) where the complex undergoes several posttranslational modifications. Then the peptide-MHC class I complexes expressed at the APC surface interact with the T-cell antigen receptors (TCRs) of CD8 T-cells. In the late 1980s and early 1990s, the first reported crystal structures of human and murine MHC class I molecules with bound peptides provided structural insights on how the MHC specifically binds and presents antigenic peptides to T-cells.

MHC Class I Molecules

The first X-ray structure of an MHC class I molecule, the human HLA-A2, was determined in 1987.[6] To date more than 150 peptide-bound MHC structures are available, including human HLA-A1, HLA-A2, HLA-A3, HLA-A11, HLA-B27, HLA-A31, HLA-Aw68, HLA-B35, HLA-B53, HLA-B44 and HLA-57, as well as murine H-2K[b], H-2D[b], H-2L[d] and H-2K[k]. The H-2 and HLA molecules consist of a glycosylated 45 kDa (340 amino acids) heavy α chain which is noncovalently associated with the nonglycosylated 11.6 kDa (96 amino acids) MHC light chain, $β_2$-microglobulin (Fig. 1). The heavy and light chains exist in the MHC in a 1:1 ratio. The heavy chain anchors the MHC complex into the cell membrane and is divided into three extracellular domains ($α_1$, $α_2$, $α_3$; 90 amino acids each), a hydrophobic transmembrane region (40 amino acids) and a short cytoplasmic tail (30 amino acids). The $α_3$ domain is membrane bound and interacts with CD8. Both the α chain and the $β_2$-microglobulin are members of the Ig superfamily and share a disulfide-bonded domain structure with antibody. Peptides are closely associated with the MHC by specific interactions in the peptide binding groove (eight-stranded β-pleated sheet floor) which is located between the $α_1$ and $α_2$ helices, forming a cleft (Fig. 1).[7-12] A long groove between the helices constitutes the binding site for processed peptides.[6] The side chains of peptides, i.e., anchor residues, fit into specificity pockets that extend along the floor of the groove. The peptide-binding groove is subdivided into various pockets (A-F).[13] The amino acid sequence between the $α_1$ and $α_2$ helices varies from allele to allele, thus, changing the specificity of the peptide binding groove. The cleft is closed at the ends, limiting the size of suitable peptides to 8-10 amino acids.[14] Peptides are held by hydrogen bonds at the N- and C-termini and by binding to the specificity pockets which anchor the peptide in the groove (Fig. 1). The anchor residues slightly vary between MHC alleles, whereas non-anchor amino acids vary considerably, thus allowing numerous peptides to be presented by a few MHC class I alleles. The non-anchor amino acids are known to interact with the

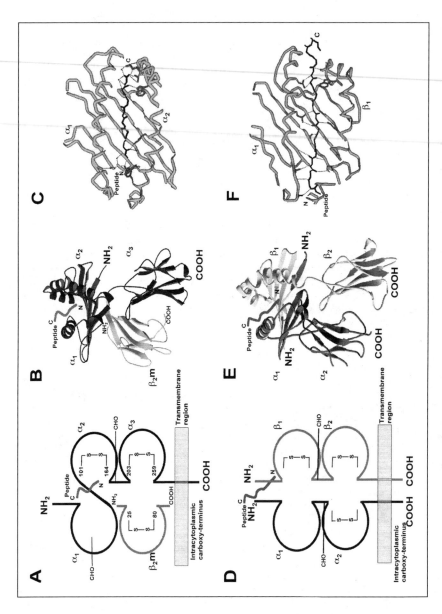

Figure 1. Structure of MHC class I and class II molecules. Schematic representation (A and D) and crystallographic structures of peptide-MHC class I (B and C) and class II (E and F) molecules. The peptide is bound between the α_1 and α_2 helices of MHC class I and the α_1 and β_1 helices of MHC class II. Top view of the peptide bound in the peptide binding groove is also shown (C and F). Complexes of H-2K^b with ovalbumin peptide SIINFEKL and I-A^b with ovalbumin CD4 peptide are depicted for MHC class I and class II, respectively.

TCR. From peptide elution and pool sequencing studies it was found that MHC molecules can have a specific preference for the type of amino acid at certain positions along the peptide.[14-16] The structural information is very important for interpretation of immunological data and it is of great benefit in drug design and development of novel peptide-based vaccines against many diseases.

Noncanonical Features of Peptide-Binding to MHC Class I Molecules

Peptides presented by MHC class I molecules for recognition by CD8 T-cells are usually cytosolic 8-10-mers. Cytoplasmic proteins undergo a number of different posttranslational modifications such as cytosolic O-β-linked glycosylation of serine and threonine residues with N-acetylglucosamine (GlcNAc). Synthetic glycopeptides with O-β-GlcNAc substitution at serine residues have been reported to be efficiently transported by TAP into the ER and presented by MHC class I molecules to induce CD8 T-cells.[17] These glycopeptides constitute up to 0.1% of the total bound peptides on MHC class I in vivo. In addition, glycopeptide-specific CD8 T-cells can also be generated after immunization of mice with a synthetic peptide conjugated to GlcNAc or galabiose disaccharide. Crystal structures of H-2K^b and H-2D^b in complex with glycopeptides have been identified, demonstrating that the carbohydrate residue makes contact with the TCR for activation.[18,19] Furthermore, we have demonstrated that the central GalNAc residue in a MUC1 peptide acts as an anchor, allowing high affinity binding.[20] Phosphorylated oncogene protein peptides and virus phosphoprotein peptides (phosphopeptides) have also been shown to be efficiently transported from the cytosol to the ER by TAP and presented by MHC class I to induce effective CD8 T-cell responses.[21] Such peptides are likely to constitute a subset of the MHC class I restricted CD8 T-cells in vivo. The effect of phosphopeptides binding to class I MHC has been shown to have neutral, negative, or even positive effects on the interaction with the TCR.

As noted above, MHC class I molecules have specificity pockets which bind to specific side chains of the bound peptide. However, numerous peptides without an appropriate anchor motif, have been shown to bind to MHC class I and still result in cell lysis.[22,23] Although 8- to 10-mer peptides bind preferentially to MHC class I, shorter[24] and longer[25] peptides have been shown to bind to MHC class I with following recognition by T-cells. Numerous studies have identified unusually long peptides in complex with MHC class I. The first crystal structure has been reported for the rat MHC class I molecule in complex with a 13-mer peptide anchored into position in a canonical manner, via its N- and C-termini.[26,27] However, the peptide has been shown bulge in the center and two different bulge conformations could be adopted by the same peptide.[26,27] A rigid, centrally bulged 13-mer viral peptide identified to bind to HLA-B3508 generates a biased T-cell response.[28] This peptide in complex with HLA-B3508 and TCR as well as a bulged 11-mer peptide from EBV in complex with HLA-B3501 and TCR have also been recently crystallized.[29,30] Other 11- to 14-mer peptides in complex with MHC class I have been identified for H2-D^b, H-2L^d, HLA-A01, HLA-A201, HLA-A03, HLA-A11, HLA-A3101, HLA-A6801, HLA-B0702, HLA-B2702, HLA-B44 and HLA-B5703 (reviewed by ref. 31). For these peptides, extensions have also been demonstrated at either the N- or C- termini. For example, a 15-mer peptide bound to H-2L^d has six amino acids which are extended at the N-terminus.[32] The crystal structure of a 10-mer peptide bound to HLA-A2 demonstrated that the glycine residue is extended from the C-terminus.[33] A 12-mer peptide presented by HLA-A2 extends at the carboxyl terminus out of the class I binding site.[34] Four residues could be added at the C- but not the N-terminus of VSV8 peptide (RGYVYQL) bound to H-2K^b (RGYVYQGL-KSGN)[25] (also reviewed by ref. 7). Furthermore, short, less than 8-mer peptides have also been demonstrated to bind to MHC class I and induce T-cell response. Mice immunized with MUC1 conjugated to mannan generate CTLs which recognize MUC1-9 (SAPDTRPAP) and MUC1-8 (SAPDTRPA) peptides in complex with the H-2K^b class I molecule.[35] Deletions of the peptide from the C-terminus (7-mer SAPDTRP; 6-mer SAPDTR and 5-mer SAPDT) can also be presented by H-2K^b and recognized by CTLs.[35] Other short immunogenic peptides which bind to MHC class I, H-2L^d and are recognized by CTL include the 3-mer (QNH), 4-mers (QNHR, ALDL, PFDL) and 5-mers (RALDL, HFMPT).[24,36,37]

Recently, the crystal structure of a 5-mer peptide from SEV9 representing N-terminal deletions has been determined in complex with H-2D^b.molecule.[38]

These observations are clearly outside the normal structural guidelines for tight binding of class I peptides, as deduced from the crystal structures of many peptide-MHC complexes. To complicate matters further, a distinction between the affinity a peptide has for a given MHC and the stability it can provide to the MHC-peptide complex has been suggested.[39] It has been found that the peptide YEA9 (SRDHSRTPM), lacking anchor motifs, still has high affinity at 25 °C, but provides low thermal stability at 37 °C.[39] We have noted that overlapping peptides from the MUC1 (TSAPDTRPA, SAPDTRPAP, APDTRPAPG presented by HLA-A2 or H-2K^b and RPAPGSTAP, PAPGSTAPG, APGSTAPGS presented by H-2D^b) are of low affinity and do not contain the anchor motifs, but can induce high avidity CTL.[22,23] One explanation for the low affinity of the peptides is that these peptides may bind transiently in the cleft, being anchored at only one main position while the remainder of the peptide may oscillate in and out of the cleft. Another explanation is that the central residues P3 to P6 of the peptide loop are out of the groove. Evidence supporting this hypothesis is that antibodies have been raised which can recognize the peptide component in the MHC-peptide complex.[40,41] Despite the lack of a hydrophobic anchor at P6, the high affinity peptide YEA9 binds to H-2K^b utilizing a new pocket, pocket E.[39] The Arg-P2 anchor of the peptide utilizes the B pocket and both Arg-P2 and Arg-P6 are required for high affinity binding. Supporting these findings, Arg-P5 from the SSYRRPVGI, the peptide from Influenza A PB1$_{703-711}$, has been also shown to bind in the E pocket.[42] Furthermore, for a low affinity peptide from MUC1 which binds to H-2K^b and induces CD8 T-cells, the crystal structure demonstrated that the mode of its binding to MHC class I is similar to that of high affinity peptides.[35] These findings broaden our understanding of the requirements and limitations of peptide binding and of epitope selection criteria for the MHC class I molecule. Immunological and structural information about canonical and noncanonical modes of peptide binding to MHC class I molecules will aid in the design of alternative and improved peptide-based vaccines for many diseases, such as cancer.

Stimulation of CD4 T Cells

Exogenous proteins are usually presented as peptides by MHC class II molecules. Exogenous proteins are endocytosed by antigen presenting cells into early endosomes where they remain in late endosomes/lysosomes and are digested into small peptide fragments. MHC class II molecules are assembled in the ER by association of the α and β chains with the invariant chain.[43,44] The invariant chain occupies the peptide-binding groove and, as such, avoids peptide binding. In addition, the invariant chain transports the MHC class II heterodimer to the late endosome/lysosome. In the late endosome/lysosome, the invariant chain is degraded by proteases/cathepsins until CLIP (class II associated invariant chain peptide) remains associated with MHC class II. CLIP is removed from MHC class II by HLA-DM peptide exchange factor[45] and peptide fragments from the exogenous protein associate with MHC class II molecules. These peptide-MHC class II complexes are transported to the cell surface where they interact with CD4 T-cells.

MHC Class II Molecules

MHC class II molecules share a peptide antigen binding fold that is very similar to MHC class I molecules despite the differences in domain organization (Fig. 1). MHC class II genes encode I-A and I-E products for mouse and HLA-DP, HLA-DQ and HLA-DR for human (HLA-DPA1, -DPB1, -DQA1, -DQB1, -DRA, -DRB1). The first crystal structures of a MHC class II molecule were determined for the human HLA-DR1 complexed with different peptides (1993) and with an influenza virus peptide (1994).[46,47] Quickly thereafter, the structures of HLA-DR3 and HLA-DR4[48,49] and murine I-E^k, I-A^d, I-A^k and I-A^{g7} complexes[50,51] have been reported. MHC class II molecule is a noncovalently bonded heterodimer composed of α and β chains (α$_1$, α$_2$, β$_1$ and β$_2$) (Fig. 1). The 230-residue α chain and the 240-residue β chain are glycosylated, resulting in molecular weights of 33 kDa and 28 kDa, respectively. A β-pleated sheet floor between the α$_1$

and β_1 helices forms the peptide binding cleft which is open, thus accommodating longer, usually 13 to18-residue long-peptides than that of MHC class I. The α_2 and β_2 domains are membrane bound and CD4 interacts with α_2. Peptides are not anchored or fixed into pockets at the N- and C- termini, but rather they bind by pockets throughout the binding groove and hydrogen bonding interactions which extend throughout the entire peptide backbone (Fig. 1). The anchoring amino acids within the peptide vary considerably between MHC alleles.

MHC Class III Molecules

MHC class III molecules are encoded by numerous genes, some of which are related to the immune system and some are not. Included are genes for complement proteins (C2, C4a, C4b and Bf), cytokines (TNF-α, TNF-β, and lymphotoxin), enzymes required for steroid synthesis, heat shock proteins and many unidentified proteins. MHC class III genes are important in immune regulation and inflammation. MHC class III molecules do not participate in binding antigenic peptides.

MHC Gene Mutations

Mutation rates of MHC genes have primarily been detected by skin graft rejections by inbred and hybrid mice. Mutations have been found in the H-2 K, D, L and I-A^b loci of the MHC in mice. Several features of these mouse H-2 mutants are also shared by human HLA class I alleles. Aside from the single allele at a single locus, the mutation rates are equivalent to non MHC genes and other mouse genes. The H-2K^b gene accounts for more than half of all reported H-2 mutations and mutates at a rate of 2×10^{-4} per gene per generation (non H-2 MHC genes mutate between 10^{-6}–10^{-5} per gene per generation). A number of H-2 mutations have been identified, including, but not limited, to: H-2K$^{bm1-bm22}$ where most of the mutations are within residues 116 and 121 (Fig. 2); H-2D$^{bm, 13\,bm24}$ with mutations at residue 114; H-2^{bm12} with a mutation in the MHC class II α chain at residues 67, 70 and 71; H-2^{dm2} has the L^d gene deleted; H-2K$^{dm4,\,dm5}$ have mutations in H-2K^d in amino acids 114 and 158, respectively; and H-2D^{dm6} with a mutation in H-2D^d at residue 133.[52]

The mutations lead to major differences in biological responses to the MHC mutants. For example, the H-2D^{bm14} mouse becomes a non CD8 T-cell responder to Moloney murine leukemia virus, whereas the H-2D^{bm13} mouse has an increased CD8 T-cell activity. Similar differences have been demonstrated in H-2K^b mutants. The H-2K^b mutants are clustered in the α_1 and α_2 domains and also affect peptide binding by the K^b molecule, TCR recognition or both (Fig. 2). For example, H-2K^{bm1} and H-2K^{bm8} are unable to present the MHC class I derived peptide from ovalbumin (OVA8 peptide, SIINFEKL) to CD8$^+$ T-cells. P2-Ile of OVA8 binds in the B-pocket of H-K^b and the H-2K^{bm8} variant has mutations at residues 22, 23, 24 and 30 on the floor of the groove, thus affecting the shape and chemical characteristics of the B-pocket. H-2K^{bm1} mutations are clustered in the N-terminal region of the second helical segment of the α_2 domain and exposed and, hence, are likely to alter contacts in the MHC-TCR interface (Fig. 2). Structural studies of H-2K^{bm3}-dEV8 peptide in complex with 2C TCR revealed that only two amino acid mutations in H-2K^{bm3} result in a dramatic effect on MHC function.[53] This structure offers a unique opportunity to understand allo-reactivity at the molecular level. Mutational studies of MHC class I reveal that, in the immune response, micro-recombination at the genetic level results in profound changes at the protein level. Such a mechanism can be a driving force in modulating CD8 T-cell immunity during evolution.

Nonclassical MHC Class I Molecules

For many years it was believed that T-cells recognize peptides only via classical MHC class I or class II molecules. However, studies over the last 10 years have demonstrated that a variety of different antigens such as the nonclassical MHC molecules can be also recognized by the TCR. The MHC encoded class I molecules are subdivided into two families: (i) the classical MHC class I molecules (class Ia) and (ii) the nonclassical MHC class 1 molecules (class 1b). While the MHC class Ia molecules are encoded by highly polymorphic mouse or human genes, the MHC

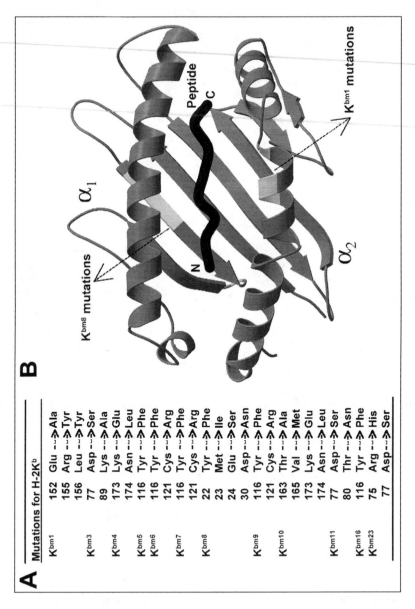

Figure 2. MHC class I, H-2K^b mutations. A) A list of H-2K^b mutations. B) Crystallographic structure of MHC class I molecule, H-2K^b, with highlighted positions of H-2K^{bm1} and H-2K^{bm8} mutations.

class Ib molecules show limited allelic variation. The detailed structure and function of MHC class Ia molecules have been extensively studied, whereas MHC class Ib molecules are less well understood. However, there is accumulating evidence that like the MHC class Ia molecules, certain MHC class Ib molecules play vital roles in immunological recognition events. MHC class Ia (classical) molecules H-2K, H-2D and H-2L in mice and HLA-A, HLA-B and HLA-C in humans present antigenic peptides to the αβ TCR on T-cells. The MHC class Ib (nonclassical) family members include H-2M, H-2Q and H-2T in mice and HLA-E, HLA-F, HLA-G and HFE (HLA-H) in humans. These molecules have been shown to play a major role in innate immunity and, more recently, in adaptive immunity regulating immunity to viruses, bacteria, tumors and self antigens.[54]

HLA-E in humans and Qa-I^b in mice have 73% amino acid identity and regulate natural killer (NK) cells, acting as a ligand for CD94/NKG2 receptors.[55] Since CD94/NKG2 receptors on NK cells inhibit NK cell cytotoxicity, the presence of an intact class I leader peptide/HLA-E complex serves to protect cells from NK cell lysis. CD94/NKG2 receptors are also expressed on CD8 T-cells,[56] thus, CD8 T-cells can interact with HLA-E/Qa-I^b molecules. Indeed, HLA-E/Qa-I^b has been demonstrated to induce CD8 T-cell responses to virus- and pathogen-derived peptides.[57] HLA-E is expressed by most cells. HLA-G also regulates activation of NK cells acting as a ligand for killer-cell immunoglobulin (Ig)-like receptor (KIR) 2DL4. In addition, HLA-G interacts with high affinity with the Ig-like transcripts (ILTs), ILT2 and ILT4 inhibitory receptors, which are expressed on numerous cell types, such as monocytes, macrophages, CF8 T-cells and NK cells.[58] HLA-G is expressed primarily on fetal trophoblasts in the maternal endometrium during placenta formation. Although the exact role of HLA-G in the endometrium during pregnancy is unclear, it is believed that it has a role in maternal tolerance of the fetus, protecting it against deleterious effects of maternal NK cells, CD8 T-cells, macrophages and mononuclear cells. Also, HLA-G is expressed in the thymus and on various dendritic cell subsets.[59] Like HLA-G, HLA-F is also expressed in the trophoblast[60] and also interacts with ILT2 and ILT4 receptors. In general, little is known about HLA-F. HLA-F is known to bind a restricted subset of peptides derived from the leader peptides of other MHC class I molecules. HLA-F exhibits few polymorphisms. There is convincing evidence that HLA-E presents peptides to CD8 T-cells; however, it is not clear whether HLA-F and HLA-G are also involved in activating T-cells. H-2Q9 plays a role in the adaptive immune response and in anti-tumor immune responses. Q9 induces restricted CD8 T-cells which are able to recognize tumors.[61] Like MHC class Ia, it also induces CD8 memory T-cells. Likewise, Qa-2 also induces tumor-specific CD8 T-cells; however, the peptides presented by Qa-2 molecules are still unknown.

The crystal structures for HLA-E, HLA-G, H-2Q9, H-2M3 and HFE have been determined and the basic architectural framework closely resembles that of the MHC class Ia molecules (Fig. 3) (reviewed by ref. 62). The MHC class Ib molecules consist of a heavy, 276-residue long, chain containing α$_1$, α$_2$ and α$_3$ helices and the 99-residue β$_2$ microglobulin. Peptides bind between the α$_1$ and α$_2$ helices in the heavy chain. The mode of peptide binding retains many of the standard features observed in MHC class Ia complexes but additional, novel features are also noted. For example, HLA-E and its murine homologue Qa allow high affinity binding to a specific, hydrophobic peptide. The specificity is conferred by hydrophobic pockets accommodating side chain anchors from peptide residues 2, 3, 6, 7 and 9, with additional binding affinity provided by hydrogen bonds to the peptide main chain throughout its length. The unique characteristics of the HLA-E peptide binding groove are consistent with a function requiring formation of highly stable, conserved complexes with MHC class Ia leader sequences. Thus, HLA-E is involved in the antigen processing pathway to CD94/NKG2 receptor-bearing NK cells. The HLA-G peptide repertoire is very restricted with a single peptide accounting for 15% of all eluted ligands. The crystal structure of HLA-G gave insights into its high affinity interaction with ILT2 and ILT4 receptors and the restricted peptide repertoire.[63] In complex with an endogenous peptide from histone H2A protein, the structure of the HLA-G revealed a remarkable constrained binding mode similar to that of the HLA-E molecule.[63] This peptide binds in the hydrophobic peptide binding groove formed by

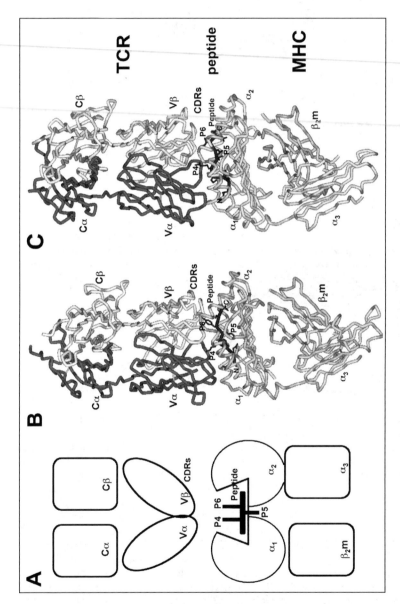

Figure 3. TCR-peptide-MHC class I complex. A) Schematic representation of MHC-peptide in complex with the TCR. Crystallographic structures of α 2C TCR in complex with H2K^b-dEV8 peptide (B) and αβ KK50.4 TCR in complex with HLA-E- CMV leader peptide (C).

the α_1 and α_2 helices with a β-sheet floor, involving 2 salt bridges, 17 hydrogen bonds, 16 water mediated hydrogen bonds and a large number of van der Waals interactions. Like in the classical peptide-MHC class I complexes, the peptide is anchored at the N- and C-termini with a central bulge. Peptide residues 2, 3, 6, 7 and 9 interact with well-defined pockets in the peptide binding groove of HLA-G and residues 4, 5 and 8 point up towards the TCR. Similar to MHC class Ia molecules, H-2Q9 binds a large array of peptides.[64] The crystal structure of H-2Q9 revealed an unusual peptide binding motif with 2 anchor residues located at the C-terminus of the peptide. There are limited contacts of the peptide bound in the hydrophobic and shallow peptide binding grooves and the central region of the peptide is bulged.[65] H-2M3 presents short formylated hydrophobic peptides from bacterial origin to CD8 T-cells. The crystal structure of H-2M3 is similar to most classical MHC class I molecules.[66] The main difference is that H-2M3 has a shorter antigen binding groove lined with mainly neutral amino acids. Only five hydrogen bonds anchor the peptide backbone in the binding groove.[66] Nevertheless, it is likely that the main features of the H-2M3 interaction with the TCR are similar to standard TCR-peptide-MHC class I complexes. In mice, (HFE) is encoded outside the MHC gene region and could be classified as a MHC class I-like molecule. HFE interacts with the transferrin receptor and plays a role in iron metabolism. The crystal structure revealed that HFE is similar to classical MHC class I molecules; however, the peptide binding groove is very narrow and does not allow binding of peptides.[67] Even though HFE does not present peptides, the antigen receptors of CD8 T-cells have been demonstrated to interact with HFE in MHC class Ia deficient mice. The precise role of HFE in the immune response is still unclear.

Further functional and structural studies are required to define the role of MHC class Ib molecules in the immune response. The ability of CD8 T-cells to recognize both MHC class Ia and class Ib molecules suggests a common ancestry between these two molecules. Since it appears that in evolution HLA-E and HLA-F preceded that of HLA-A, HL-B and HLA-C, structural analysis of MHC class Ib in complex with the TCR might give insights in the origins and basis of MHC restriction. Moreover, it is plausible to develop vaccines based on MHC class Ia- and class Ib-restricted responses. The advantage of such a vaccine remains to be determined.

Nonclassical MHC Class II Molecules

The nonclassical MHC class II molecules include HLA-DM and HLA-DO in humans and H-2DM and H-2-O in mice. HLA-DM functions as a chaperone for classical MHC class II molecules in endosomal and lysosomal loading compartments since it stabilizes the empty MHC class II peptide binding groove until appropriate peptide ligands are available to bind. Also, HLA-DM allows the release of low stability CLIP peptides in exchange for high stability peptides.[68] HLA-DM is expressed in all APCs expressing MHC class II. Conversely, HLA-DO is expressed mainly in B-cells where it binds tightly to HLA-DM, modulating its activity. The association HLA-DO with HLA-DM is essential for the intracellular transport of HLA-DO and these two molecules remain associated in the endosomal system.[68] The physiological relevance of this interaction is not clearly understood. Nonetheless, both HLA-DM and HLA-DO are critical factors in shaping the MHC class II-associated self or foreign peptide repertoire of APCs and, hence, govern initiation or prevention of the immune response.

MHC-Peptide Interaction with the TCR

The structural analyses of single peptide-MHC class I complexes identified which peptide side chains point up towards the TCR. Peptides which bind to MHC class I are fixed at the N- and C-termini and bulged in the central region which interacts with the TCR. However, in peptide-MHC class II complexes, peptides bind deeper in the binding groove and are less accessible for interaction with the TCR, although amino acids at the N-terminal extension (P-4 to P-1) can still play a major role in this interaction.

The TCR is structurally and functionally similar to the B-cell antigen receptor. TCR is composed of disulfide-linked α and β polypeptide chains, each having separate constant and variable

Table 1. Crystal structures of MHC-peptide-TCR complexes

MHC	Peptide	TCR αβ	Year Published
Mouse MHC class I			
H-2K^b	dEV8	2C	1996
H-2K^b	dEV8	2C	1998
H-2K^b	SIYR	2C	2000
H-2K^b	pBM1	scBM3.3	2000
H-2K^b	pKB1	scKB5-C20	2002
H-2K^{bm3}	dEV8	2C	2002
H-2K^b	VSV8	scBM3.3	2003
Human MHC class I			
HLA-A2	Tax	A6	1996
HLA-A2	Tax	B7	1998
HLA-A2	TaxP6A	A6	1999
HLA-A2	TaxV7R	A6	1999
HLA-A2	TaxY8A	A6	1999
HLA-A2	MP$_{58-66}$	JM22	2003
HLA-A2	p1049	AHIII 12.2	2003
HLA-B8	FLR	LC13	2003
HLA-A2	ESO9V	IG4	2005
HLA-A2	ESO9C	IG4	2005
HLA-B3508	LPEP	SB27	2005
Mouse MHC class II			
I-A^k	CA	scD10	1999
I-A^u	MBP	sc172.10	2005
Human MHC class II			
HLA-DR1	HA	HA1.7	2000
HLA-DR4	HA	HA1.7	2002
HLA-DR2b	MBP	Ob.1A12	2005
HLA-DR2a	MBP	3A6	2005
Nonclassical MHC class Ib molecule			
HLA-E	CMV leader	KK50.4	2006
MHC-like molecule			
T22	—	γδTCR-G8	2005

domains much like immunoglobulins. The variable domain contains three hypervariable regions that are responsible for antigen recognition (Fig. 3). The crystal structures of the Vα domain of the αβ TCR and the β chain of the αβ TCR were determined in 1995. The contact surface between Vβ and the constant region Cβ domains led to the conclusion that the β chain is involved in antigen recognition. In 1996, the first crystal structures—MHC class I-peptide-TCR complexes, 2C TCR bound to H-2K^b-dEV8 peptide and A6 TCR bound to HLA-A2-Tax peptide—were determined (Table 1).[69,70] These structures demonstrated that the TCR is oriented diagonally across the upper face of the MHC. It was also noted that the complementarity-determining region (CDR) α$_1$ and β$_1$ loops interact with the N- and C-termini of the peptide and with the MHC α-helixes. The CDR α$_2$ is oriented directly over the MHC α$_2$ helix and CDR β$_2$ is located above the MHC α$_1$ helix, whereas the CDR3 loops bind to the amino acids P4-P6 of the peptide.[69-71] Further structural studies revealed that the diagonal orientation of the TCR across the MHC has some variation, particularly with respect to placement of the TCR Vβ.

In 1999, the first crystal structure of the TCR bound to MHC class II I-A^k indicated more restricted relative orientation in the TCR-MHC class II complex.[72] Today, the structures of 24 MHC class I and class II-peptide-TCR complexes are determined and demonstrate that there is substantial degree of structural variability in the MHC-peptide-TCR recognition.

In 2006, the first crystal structure of the αβ TCR in complex with an MHC class Ib molecule was determined. In the complex of the KK50.4 TCR with HLA-E bound to a CMV-derived leader peptide (Table 1), the TCR adopts a diagonal orientation above the α_1 and α_2 helices of HLA-E and the CDRs interact to peptide-HLA-E complex similarly, like TCR recognizes MHC class Ia molecules (Fig. 3).[73] The CDR2 β uses 30% of the buried surface area at the interface and all 3 CDR β loops interact with residue 8 of the bound peptide, indicating that this residue determines self from nonself peptides.[73] The TCR interacts with four unique amino acids in the HLA-E molecule and the interaction between the TCR and HLA-E is of low affinity.[73]

MHC Class I-like Molecules

It is becoming evident that MHC class I-like molecules are a class of cell surface receptors which interact with CD8 T-cells, NK cells, or γδ T-cells. Structural and immunological studies of the MHC class I-like molecules (T10, T22, FcRn, CD1, as well as the stress-induced MICA, MICB, ULBP, Rae1 and H60 molecules) have given insights of their role and importance in the immune response.

The MHC class I-like molecules, mouse encoded T10 and closely related human T22 (94% identity), are recognized by γδ T-cells without the requirement of a peptide. The crystal structure of T10 revealed a 12-amino acid deletion in the α_1 domain and a 3 residue deletion in the α_2 domain, thus explaining the inability of T10 molecules to bind peptides. In addition, T22 has a 13-amino acid deletion in the α_2 domain resulting in a partial unfolding of the α_2-helix and exposure of the β-sheet floor of the $\alpha_1 \alpha_2$ domain. The primary structures of T10 and T22 are only 40% identical to that of HLA-A2. Since the differences between MHC class I molecules are large, it is assumed that the interaction of MHC class I-like molecules with the TCR would be strikingly different compared to the MHC class I complexes. Indeed, the structure of T22 in complex with γδ TCR determined in 2005 (Table 1), revealed that the TCR predominantly uses germline-encoded residues of its CDR3 δ chain loop to bind T22 in an orientation substantially different from that seen in αβ TCR-peptide-MHC complexes.[74]

The neonatal Fc receptor (FcRn) is also an MHC-like molecule which does not bind peptides or other ligands. FcRn has a very similar fold to that of MHC class I molecules but has a much narrower groove which cannot accommodate a bound ligand. The crystal structure of FcRn/Fc complex showed that the Fc binding site is distinct from the standard peptide and TCR binding site of MHC class I molecule,[75] thus indicating that the MHC fold can be utilized in a completely different context within the immune system.

Another example of MHC class I-like molecules that also lack ligand binding is a group of cell surface receptors: MICA, MICB and ULBP in human and Rae1 and H60 in mice. Their expression on cells is low and these receptors are expressed on fibroblasts, epithelial cells, dendritic cells and endothelial cells in response to stress, such as oxidative stress, thermal stress, bacterial infection and tumor growth. The crystal structures of MICA and Rae-1β showed an absence of any binding groove due to a reduced distance between the α_1 and α_2 helices.[76] The receptor for MICA is present on most γδ T-cells, CD8$^+$ αβ T-cells and NK cells—NKG2D, a fourth member of the CD94/NKG2 family of NK cell receptors.[76] MIC proteins do not associate with β_2 m and the murine Rae1 and H60 exist only as an isolated $\alpha_1 \alpha_2$ platform.

A significant and exciting new advance in our understanding of T-cell biology and cellular immunity has come from the demonstration that nonpeptide antigens can also be recognized by the TCR when bound in the context of nonMHC encoded proteins, called CD1 antigens. CD1 is a family of cell surface glycoproteins that are noncovalently associated with β_2 m and encoded outside the classical MHC class I and class II loci. CD1 closely resembles MHC class I with α_1, α_2, α_3 and β_2m domains. The expression of CD1 molecules is independent of TAP; however, it localizes in MHC class II compartments (late endosomes/lysosomes) where CD1 molecules bind to exogenous antigens. The CD1 family is divided into two groups in humans: group I comprises CD1a, CD1b and CD1c isotypes and group II comprises CD1d. CD1e has an intermediate isotype. Members of the CD1 family are all encoded by genes clustered on chromosome 1. The CD1 locus in mice is found on chromosome 3. It is less complex and contains only two closely related genes, CD1d1 and CD1d2 which are members of group II subfamily. The CD1d isotype is completely preserved in human, mouse, rabbit and rat. CD1 group I molecules are primarily expressed on dendritic cells and are upregulated by GM-CSF, IL-3, or IL-4, suggesting an active role in inflammation. CD1 group II molecules are more widely expressed on B-cells, immature cortical thymocytes and epithelial cells [reviewed by 9,76]. Twelve years ago, human CD1b was demonstrated to interact with CD4$^-$/CD8$^-$ $\alpha\beta$ T-cells. The nonpeptide antigens which bind to CD1b are bacterial components such as mycolic acid, lipoarabinomannan and phosphatidylinositol mannosides.[77] It was puzzling for some time how antigens of bacterial origin were presented by CD1 molecules until the first crystal structure of a CD1 molecule was determined nine years ago. The mouse CD1d1 crystal structure provided new insights into the mechanism by which CD1 could present lipid antigens to T-cells.[78] The overall structure of murine CD1d1 is very similar to that of MHC class I, although the binding groove of CD1d1 is significantly narrower, deeper, more hydrophobic and consists of two large pockets. The lipid tails of glycolipids and lipopeptides are bound in the groove and their polar moieties are presented to T-cells.[78] The crystal structures of human CD1a, CD1b and CD1d have also been determined, suggesting possible molecular mechanisms of presentation of such antigens in the immune system.

Future Prospects

Induction of cellular immunity seems to be even more complex than we thought 15 years ago. Increasing structural information and our recent progress in studies of cellular immune responses help us better understand how the immune system presents antigenic peptides and how T-cells recognize them. This is very important in peptide-based vaccine design. Recent advances in the field have significantly increased our knowledge of the structural immunology of the MHC superfamily. Knowing the structures of classical and nonclassical MHC class I and class II molecules and MHC-like molecules has helped to interpret immunological data and benefited drug design and development of novel ligand-based vaccines against many diseases such as cancer.

Acknowledgements

VA is supported by an RD Wright Research Fellowship (223316) and a project grant (223310) from the National Health and Medical Research Council of Australia.

References

1. Kindred B, Shreffler DC. H-2 dependence of co-operation between T and B cells in vivo. J Immunol 1972; 109:940-3.
2. Rosenthal AS, Shevach EM. Function of macrophages in antigen recognition by guinea pig T-lymphocytes. I. Requirement for histocompatible macrophages and lymphocytes. J Exp Med 1973; 138:1194-212.
3. Zinkernagel RM, Doherty PC. Restriction of in vitro T-cell-mediated cytotoxicity in lymphocytic choriomeningitis within a syngeneic or semiallogeneic system. Nature 1974; 248:701-2.
4. Harding CV, Song R. Phagocytic processing of exogenous particulate antigens by macrophages for presentation by class I MHC molecules. J Immunol 1994; 153:4925-33.
5. Zwickey HL, Potter TA. Antigen secreted from noncytosolic Listeria monocytogenes is processed by the classical MHC class I processing pathway. J Immunol 1999; 162:6341-50.
6. Bjorkman PJ, Saper MA, Samraoui B et al. Structure of the human class I histocompatibility antigen, HLA-A2. Nature 1987; 329:506-12.
7. Apostolopoulos V, Lazoura E. Noncanonical peptides in complex with MHC class I. Expert Rev Vaccines 2004; 3:151-62.
8. Apostolopoulos V, McKenzie IF, Wilson IA. Getting into the groove: Unusual features of peptide binding to MHC class I molecules and implications in vaccine design. Front Biosci 2001; 6:D1311-20.
9. Apostolopoulos V, Yu M, McKenzie IF et al. Structural implications for the design of molecular vaccines. Curr Opin Mol Ther 2000; 2:29-36.

10. Lazoura E, Apostolopoulos V. Insights into peptide-based vaccine design for cancer immunotherapy. Curr Med Chem 2005; 12:1481-94.
11. Lazoura E, Apostolopoulos V. Rational Peptide-based vaccine design for cancer immunotherapeutic applications. Curr Med Chem 2005; 12:629-39.
12. Pietersz GA, Pouniotis DS, Apostolopoulos V. Design of peptide-based vaccines for cancer. Curr Med Chem 2006; 13:1591-607.
13. Garrett TP, Saper MA, Bjorkman PJ et al. Specificity pockets for the side chains of peptide antigens in HLA-Aw68. Nature 1989; 342:692-6.
14. Rammensee HG, Friede T, Stevanoviic S. MHC ligands and peptide motifs: First listing. Immunogenetics 1995; 41:178-228.
15. Falk K, Rotzschke O, Stevanovic S et al. Allele-specific motifs revealed by sequencing of self-peptides eluted from MHC molecules. Nature 1991; 351:290-6.
16. Rammensee HG. Chemistry of peptides associated with MHC class I and class II molecules. Curr Opin Immunol 1995; 7:85-96.
17. Haurum JS, Hoier IB, Arsequell G et al. Presentation of cytosolic glycosylated peptides by human class I major histocompatibility complex molecules in vivo. J Exp Med 1999; 190:145-50.
18. Glithero A, Tormo J, Haurum JS et al. Crystal structures of two H-2Db/glycopeptide complexes suggest a molecular basis for CTL cross-reactivity. Immunity 1999; 10:63-74.
19. Speir JA, Abdel-Motal UM, Jondal M et al. Crystal structure of an MHC class I presented glycopeptide that generates carbohydrate-specific CTL. Immunity 1999; 10:51-61.
20. Apostolopoulos V, Yuriev E, Ramsland PA et al. A glycopeptide in complex with MHC class I uses the GalNAc residue as an anchor. Proc Natl Acad Sci USA 2003; 100:15029-34.
21. Andersen MH, Bonfill JE, Neisig A et al. Phosphorylated peptides can be transported by TAP molecules, presented by class I MHC molecules and recognized by phosphopeptide-specific CTL. J Immunol 1999; 163:3812-8.
22. Apostolopoulos V, Haurum JS, McKenzie IF. MUC1 peptide epitopes associated with five different H-2 class I molecules. Eur J Immunol 1997; 27:2579-87.
23. Apostolopoulos V, Karanikas V, Haurum JS et al. Induction of HLA-A2-restricted CTLs to the mucin 1 human breast cancer antigen. J Immunol 1997; 159:5211-8.
24. Reddehase MJ, Rothbard JB, Koszinowski UH. A pentapeptide as minimal antigenic determinant for MHC class I-restricted T-lymphocytes. Nature 1989; 337:651-3.
25. Horig H, Young AC, Papadopoulos NJ et al. Binding of longer peptides to the H-2Kb heterodimer is restricted to peptides extended at their C terminus: Refinement of the inherent MHC class I peptide binding criteria. J Immunol 1999; 163:4434-41.
26. Rudolph MG, Stevens J, Speir JA et al. Crystal structures of two rat MHC class Ia (RT1-A) molecules that are associated differentially with peptide transporter alleles TAP-A and TAP-B. J Mol Biol 2002; 324:975-90.
27. Speir JA, Stevens J, Joly E et al. Two different, highly exposed, bulged structures for an unusually long peptide bound to rat MHC class I RT1-Aa. Immunity 2001; 14:81-92.
28. Tynan FE, Burrows SR, Buckle AM et al. T-cell receptor recognition of a 'super-bulged' major histocompatibility complex class I-bound peptide. Nat Immunol 2005; 6:1114-22.
29. Miles JJ, Elhassen D, Borg NA et al. CTL recognition of a bulged viral peptide involves biased TCR selection. J Immunol 2005; 175:3826-34.
30. Tynan FE, Borg NA, Miles JJ et al. High resolution structures of highly bulged viral epitopes bound to major histocompatibility complex class I. Implications for T-cell receptor engagement and T-cell immunodominance. J Biol Chem 2005; 280:23900-9.
31. Burrows SR, Rossjohn J, McCluskey J. Have we cut ourselves too short in mapping CTL epitopes? Trends Immunol 2006; 27:11-6.
32. Samino Y, Lopez D, Guil S et al. A long N-terminal-extended nested set of abundant and antigenic major histocompatibility complex class I natural ligands from HIV envelope protein. J Biol Chem 2006; 281:6358-65.
33. Collins EJ, Garboczi DN, Wiley DC. Three-dimensional structure of a peptide extending from one end of a class I MHC binding site. Nature 1994; 371:626-9.
34. Chen Y, Sidney J, Southwood S et al. Naturally processed peptides longer than nine amino acid residues bind to the class I MHC molecule HLA-A2.1 with high affinity and in different conformations. J Immunol 1994; 152:2874-81.
35. Apostolopoulos V, Yu M, Corper AL et al. Crystal structure of a noncanonical low-affinity peptide complexed with MHC class I: A new approach for vaccine design. J Mol Biol 2002; 318:1293-305.
36. Eisen HN, Sykulev Y, Tsomides TJ. Antigen-specific T-cell receptors and their reactions with complexes formed by peptides with major histocompatibility complex proteins. Adv Protein Chem 1996; 49:1-56.

37. Gillanders WE, Hanson HL, Rubocki RJ et al. Class I-restricted cytotoxic T-cell recognition of split peptide ligands. Int Immunol 1997; 9:81-9.
38. Glithero A, Tormo J, Doering K et al. The crystal structure of H-2D(b) complexed with a partial peptide epitope suggests a major histocompatibility complex class I assembly intermediate. J Biol Chem 2006; 281:12699-704.
39. Apostolopoulos V, Yu M, Corper AL et al. Crystal structure of a noncanonical high affinity peptide complexed with MHC class I: a novel use of alternative anchors. J Mol Biol 2002; 318:1307-16.
40. Apostolopoulos V, Chelvanayagam G, Xing PX et al. Anti-MUC1 antibodies react directly with MUC1 peptides presented by class I H2 and HLA molecules. J Immunol 1998; 161:767-75.
41. Chelvanayagam G, Apostolopoulos V, McKenzie IF. Milestones in the molecular structure of the major histocompatibility complex. Protein Eng 1997; 10:471-4.
42. Meijers R, Lai CC, Yang Y et al. Crystal structures of murine MHC Class I H-2 D(b) and K(b) molecules in complex with CTL epitopes from influenza A virus: Implications for TCR repertoire selection and immunodominance. J Mol Biol 2005; 345:1099-110.
43. Pieters J. MHC class II restricted antigen presentation. Curr Opin Immunol 1997; 9:89-96.
44. Watts C. Capture and processing of exogenous antigens for presentation on MHC molecules. Annu Rev Immunol 1997; 15:821-50.
45. Kropshofer H, Hammerling GJ, Vogt AB. How HLA-DM edits the MHC class II peptide repertoire: Survival of the fittest? Immunol Today 1997; 18:77-82.
46. Brown JH, Jardetzky TS, Gorga JC et al. Three-dimensional structure of the human class II histocompatibility antigen HLA-DR1. Nature 1993; 364:33-9.
47. Stern LJ, Brown JH, Jardetzky TS et al. Crystal structure of the human class II MHC protein HLA-DR1 complexed with an influenza virus peptide. Nature 1994; 368:215-21.
48. Dessen A, Lawrence CM, Cupo S et al. X-ray crystal structure of HLA-DR4 (DRA*0101, DRB1*0401) complexed with a peptide from human collagen II. Immunity 1997; 7:473-81.
49. Ghosh P, Amaya M, Mellins E et al. The structure of an intermediate in class II MHC maturation: CLIP bound to HLA-DR3. Nature 1995; 378:457-62.
50. Fremont DH, Hendrickson WA, Marrack P et al. Structures of an MHC class II molecule with covalently bound single peptides. Science 1996; 272:1001-4.
51. Scott CA, Peterson PA, Teyton L et al. Crystal structures of two I-Ad-peptide complexes reveal that high affinity can be achieved without large anchor residues. Immunity 1998; 8:319-29.
52. Melvold RW, Wang K, Kohn HI. Histocompatibility gene mutation rates in the mouse: A 25-year review. Immunogenetics 1997; 47:44-54.
53. Luz JG, Huang M, Garcia KC et al. Structural comparison of allogeneic and syngeneic T-cell receptor-peptide-major histocompatibility complex complexes: A buried alloreactive mutation subtly alters peptide presentation substantially increasing V(beta) Interactions. J Exp Med 2002; 195:1175-86.
54. Rodgers JR, Cook RG. MHC class Ib molecules bridge innate and acquired immunity. Nat Rev Immunol 2005; 5:459-71.
55. Borrego F, Ulbrecht M, Weiss EH et al. Recognition of human histocompatibility leukocyte antigen (HLA)-E complexed with HLA class I signal sequence-derived peptides by CD94/NKG2 confers protection from natural killer cell-mediated lysis. J Exp Med 1998; 187:813-8.
56. Braud VM, Allan DS, O'Callaghan CA et al. HLA-E binds to natural killer cell receptors CD94/NKG2A, B and C. Nature 1998; 391:795-9.
57. Pietra G, Romagnani C, Mazzarino P et al. HLA-E-restricted recognition of cytomegalovirus-derived peptides by human CD8+ cytolytic T-lymphocytes. Proc Natl Acad Sci USA 2003; 100:10896-901.
58. Allan DS, Colonna M, Lanier LL et al. Tetrameric complexes of human histocompatibility leukocyte antigen (HLA)-G bind to peripheral blood myelomonocytic cells. J Exp Med 1999; 189:1149-56.
59. Mallet V, Blaschitz A, Crisa L et al. HLA-G in the human thymus: A subpopulation of medullary epithelial but not CD83(+) dendritic cells expresses HLA-G as a membrane-bound and soluble protein. Int Immunol 1999; 11:889-98.
60. Ishitani A, Sageshima N, Lee N et al. Protein expression and peptide binding suggest unique and interacting functional roles for HLA-E, F and G in maternal-placental immune recognition. J Immunol 2003; 171:1376-84.
61. Chiang EY, Stroynowski I. A nonclassical MHC class I molecule restricts CTL-mediated rejection of a syngeneic melanoma tumor. J Immunol 2004; 173:4394-401.
62. Sullivan LC, Hoare HL, McCluskey J et al. A structural perspective on MHC class Ib molecules in adaptive immunity. Trends Immunol 2006.
63. Clements CS, Kjer-Nielsen L, Kostenko L et al. Crystal structure of HLA-G: A nonclassical MHC class I molecule expressed at the fetal-maternal interface. Proc Natl Acad Sci USA 2005; 102:3360-5.

64. Joyce S, Tabaczewski P, Angeletti RH et al. A nonpolymorphic major histocompatibility complex class Ib molecule binds a large array of diverse self-peptides. J Exp Med 1994; 179:579-88.

65. He X, Tabaczewski P, Ho J et al. Promiscuous antigen presentation by the nonclassical MHC Ib Qa-2 is enabled by a shallow, hydrophobic groove and self-stabilized peptide conformation. Structure 2001; 9:1213-24.

66. Wang CR, Castano AR, Peterson PA et al. Nonclassical binding of formylated peptide in crystal structure of the MHC class Ib molecule H2-M3. Cell 1995; 82:655-64.

67. Lebron JA, Bennett MJ, Vaughn DE et al. Crystal structure of the hemochromatosis protein HFE and characterization of its interaction with transferrin receptor. Cell 1998; 93:111-23.

68. Alfonso C, Karlsson L. Nonclassical MHC class II molecules. Annu Rev Immunol 2000; 18:113-42.

69. Garboczi DN, Ghosh P, Utz U et al. Structure of the complex between human T-cell receptor, viral peptide and HLA-A2. Nature 1996; 384:134-41.

70. Garcia KC, Degano M, Stanfield RL et al. An alphabeta T-cell receptor structure at 2.5 A and its orientation in the TCR-MHC complex. Science 1996; 274:209-19.

71. Garcia KC, Degano M, Pease LR et al. Structural basis of plasticity in T-cell receptor recognition of a self peptide-MHC antigen. Science 1998; 279:1166-72.

72. Reinherz EL, Tan K, Tang L et al. The crystal structure of a T-cell receptor in complex with peptide and MHC class II. Science 1999; 286:1913-21.

73. Hoare HL, Sullivan LC, Pietra G et al. Structural basis for a major histocompatibility complex class Ib-restricted T-cell response. Nat Immunol 2006; 7:256-64.

74. Adams EJ, Chien YH, Garcia KC. Structure of a gammadelta T-cell receptor in complex with the nonclassical MHC T22. Science 2005; 308:227-31.

75. Burmeister WP, Huber AH, Bjorkman PJ. Crystal structure of the complex of rat neonatal Fc receptor with Fc. Nature 1994; 372:379-83.

76. Rudolph MG, Stanfield RL, Wilson IA. How TCRs bind MHCs, peptides and coreceptors. Annu Rev Immunol 2006; 24:419-66.

77. Chatterjee D, Khoo KH. Mycobacterial lipoarabinomannan: An extraordinary lipoheteroglycan with profound physiological effects. Glycobiology 1998; 8:113-20.

78. Zeng Z, Castano AR, Segelke BW et al. Crystal structure of mouse CD1: An MHC-like fold with a large hydrophobic binding groove. Science 1997; 277:339-45.

CHAPTER 20

SCHOOL Model and New Targeting Strategies

Alexander B. Sigalov*

Abstract

Protein-protein interactions play a central role in biological processes and thus are an appealing target for innovative drug design and development. They can be targeted by small molecule inhibitors, peptides and peptidomimetics, which represent an alternative to protein therapeutics that carry many disadvantages.

In this chapter, I describe specific protein-protein interactions suggested by a novel model of immune signaling, the Signaling Chain HOmoOLigomerization (SCHOOL) model, to be critical for cell activation mediated by multichain immune recognition receptors (MIRRs) expressed on different cells of the hematopoietic system. Unraveling a long-standing mystery of MIRR triggering and transmembrane signaling, the SCHOOL model reveals the intrareceptor transmembrane interactions and interreceptor cytoplasmic homointeractions as universal therapeutic targets for a diverse variety of disorders mediated by immune cells. Further, assuming that the general principles underlying MIRR-mediated transmembrane signaling mechanisms are similar, the SCHOOL model can be applied to any particular receptor of the MIRR family. Thus, an important application of the SCHOOL model is that global therapeutic strategies targeting key protein-protein interactions involved in MIRR triggering and transmembrane signal transduction may be used to treat a diverse set of immune-mediated diseases. This assumes that clinical knowledge and therapeutic strategies can be transferred between seemingly disparate disorders, such as T-cell-mediated skin diseases and platelet disorders, or combined to develop novel pharmacological approaches. Intriguingly, the SCHOOL model unravels the molecular mechanisms underlying ability of different human viruses such as human immunodeficiency virus, cytomegalovirus and severe acute respiratory syndrome coronavirus to modulate and/or escape the host immune response. It also demonstrates how the lessons learned from viral pathogenesis can be used practically for rational drug design.

Application of this model to platelet collagen receptor signaling has already led to the development of a novel concept of platelet inhibition and the invention of new platelet inhibitors, thus proving the suggested hypothesis and highlighting the importance and broad perspectives of the SCHOOL model in the development of new targeting strategies.

Introduction

Specific protein-protein interactions are responsible for the function of numerous processes in the cell and constitute the foundation for the majority of cell recognition, proliferation, growth,

*Alexander B. Sigalov—Department of Pathology, University of Massachusetts Medical School, 55 Lake Avenue North, Worcester 01655, Massachusetts, USA.
Email: alexander.sigalov@umassmed.edu

Multichain Immune Recognition Receptor Signaling: From Spatiotemporal Organization to Human Disease, edited by Alexander B. Sigalov. ©2008 Landes Bioscience and Springer Science+Business Media.

differentiation, programmed cell death and signal transduction in health and disease.[1-4] It seems that almost every important pathway includes and is critically influenced by protein-protein interactions.[1] Because of the ubiquitous nature of these interactions and the knowledge that inappropriate protein-protein binding can lead to disease, the specific and controlled inhibition and/or modulation of these interactions provides a promising novel approach for rational drug design, as revealed by recent progress in the design of inhibitory antibodies, peptides and small molecules. A number of recent reviews have addressed this topic.[5-12] Thus, revealing information about specific protein-protein interactions in any particular pathway (i.e., transmembrane signaling) can provide targets for a generation of new drugs.

Long-Standing Mystery of MIRR Triggering and Transmembrane Signaling

Multichain immune recognition receptors (MIRRs) recognize foreign antigens and initiate a variety of biological responses. Examples of MIRRs include the T-cell receptor (TCR) complex, the B-cell receptor (BCR) complex, Fc receptors (e.g., FcεRI, FcαRI, FcγRI and FcγRIII), NK receptors (e.g., NKG2D, CD94/NKG2C, KIR2DS, NKp30, NKp44 and NKp46), immunoglobulin (Ig)-like transcripts and leukocyte Ig-like receptors (ILTs and LIRs, respectively), signal regulatory proteins (SIRPs), dendritic cell immunoactivating receptor (DCAR), myeloid DNAX adapter protein of 12 kD (DAP12)-associating lectin 1 (MDL-1), novel immune-type receptor (NITR), triggering receptors expressed on myeloid cells (TREMs) and the platelet collagen receptor, glycoprotein VI (GPVI). MIRR-mediated transmembrane (TM) signal transduction plays an important role in health and disease[13-21] making these receptors attractive targets for rational intervention in a variety of immune disorders. Thus, future therapeutic strategies depend on our detailed understanding of the molecular mechanisms underlying the MIRR triggering and subsequent TM signal transduction.

All members of the MIRR family are multisubunit complexes formed by the association of recognition subunits with signal-transducing subunits that contain in their cytoplasmic (CYTO) domains the immunoreceptor tyrosine-based activation motif (ITAM) or the YxxM motif, found in the DAP-10 CYTO domain (see Chapter 12). This association in resting cells is mostly driven by the noncovalent TM interactions between recognition and signaling components and plays a key role in receptor assembly and integrity (see also Chapters 1-5).[18,21-26] Crosslinking of the receptors after ligand binding results in phosphorylation of the ITAM/YxxM tyrosines, which triggers the elaborate intracellular signaling cascade. The extracellular (EC) recognition of an antigen/ligand and the sequence of biochemical events that ensues after the phosphorylation of ITAMs/YxxM are understood in significant detail. However, the molecular mechanism linking EC antigen/ ligand-induced clustering of MIRR ligand-binding subunits to intracellular phosphorylation of signaling subunits has been a long-standing unsolved mystery. It was also unknown how this putative mechanism can explain the intriguing ability of immune cells to discern and differentially respond to slightly different ligands. This impeded our advance understanding of the immune response, the development of novel pharmacological approaches and even more important, the potential transfer of clinical knowledge, experience and therapeutic strategies between seemingly disparate immune disorders.

Despite numerous models of MIRR-mediated TM signal transduction suggested for particular MIRRs (e.g., TCR, BCR, FcRs, NK receptors, etc.), no current model fully explains how ligand-induced TM signal transduction commences at the molecular level. As a consequence, these models are mostly descriptive and do not reveal clinically important potential points of therapeutic intervention. In addition, no general model of MIRR-mediated immune cell activation has been suggested, thus preventing the potential transfer of therapeutic strategies between seemingly disparate immune disorders.

A recently developed novel mechanistic model, the SCHOOL model,[27-30] describes the crucial protein-protein interactions underlying the molecular mechanism of MIRR triggering and TM signaling (Fig. 1, see also Chapter 12). In this chapter, I describe these specific interactions as new

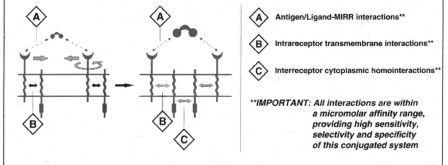

SCHOOL model and new therapeutic targets

MIRR assembly

Antigen recognition subunit
Signal-transducing subunit

A. SCHOOL model of MIRR signaling*

Antigen/Ligand

Signaling oligomers — Signaling oligomers — Signaling oligomers — Signaling oligomers

Activation signal — Internalization — Activation signal — Signal amplification — Internalization — Signal amplification

① → ② → ③ → ④ → ⑤

* Antibody-stimulated MIRR signaling is not shown.

B. SCHOOL model: Major driving forces in MIRR triggering and signaling

Ⓐ Antigen/Ligand-MIRR interactions**

Ⓑ Intrareceptor transmembrane interactions**

Ⓒ Interreceptor cytoplasmic homointeractions**

***IMPORTANT: All interactions are within a micromolar affinity range, providing high sensitivity, selectivity and specificity of this conjugated system*

C. Key protein-protein interactions of MIRR signaling as new therapeutic targets

Target I. Transmembrane interactions

Maintain receptor integrity in unstimulated cells and dictate the overall geometry and topology of MIRRs

Target II. Cytoplasmic interactions

Lead to formation of competent signaling subunit oligomers and drive MIRR triggering/signaling

Figure 1, legend viewed on following page.

Figure 1, viewed on previous page. Structural assembly of MIRRs (the inset), the signaling chain homooligomerization (SCHOOL) model of MIRR signaling (A,B) and new therapeutic targets revealed by the model (C). The model proposes that formation of competent MIRR signaling subunit oligomers driven by the homooligomerization of signaling subunits is necessary and sufficient to trigger the receptors and induce transmembrane (TM) signal transduction and downstream sequence (see Chapter 12 for detail). All interchain interactions in this intermediate are shown by light gray arrows reflecting their transition state. Immunoreceptor tyrosine-based activation motifs (ITAMs) are shown as gray rectangles. Circular arrow indicates ligand-induced receptor reorientation. Phosphate groups are shown as gray circles. Small solid black arrows indicate specific intersubunit hetero- and homointeractions between TM and cytoplasmic (CYTO) domains, respectively. Within the model, MIRR triggering and signaling is an outcome of the ligand-induced interplay between three key protein-protein interactions: antigen/ligand-MIRR interactions, intrareceptor TM interactions and interreceptor CYTO homointeractions (B). Two of these interactions can be considered as new therapeutic targets (C): 1) TM interactions between MIRR antigen-recognizing and signal-transducing subunits (target I) that play an important role in receptor assembly and integrity on resting cells; and 2) CYTO homointeractions between MIRR signaling subunits (target II) that represent a main driving force of MIRR triggering/signaling.

therapeutic targets revealed by the model for the treatment of diverse immune and other disorders mediated by MIRRs. Assuming that the similar structural architecture of the MIRRs dictates similar mechanisms of MIRR triggering and subsequent TM signal transduction, the model suggests that these targets are similar in seemingly unrelated diseases. This builds the structural basis for the development of novel pharmacological approaches as well as the transfer of clinical knowledge, experience and therapeutic strategies between various immune disorders. In addition, it significantly improves our understanding of the immunomodulatory activity of human viruses such as human immunodeficiency virus (HIV), human cytomegalovirus (CMV) and severe acute respiratory syndrome coronavirus (SARS-CoV), human T-cell leukemia type 1 virus (HTLV-1) and assumes that the lessons learned from viral pathogenesis can be used for the development of new therapeutic approaches. An important application of this hypothesis is that a general pharmaceutical approach may be used to treat diverse immune-mediated diseases.

SCHOOL Model of MIRR Triggering and Signaling: Basic Concept, Major Driving Forces, Restraints and Advantages

Basic Concept

Recently, a novel biophysical phenomenon, the homointeractions of intrinsically disordered CYTO domains of ITAM-containing MIRR signaling subunits, has been discovered.[31] It demonstrates that intrinsically disordered proteins do not necessarily undergo a transition between disordered and ordered states upon interaction,[32,33] a finding that opposes the generally accepted view on the behavior of natively unfolded proteins. Interestingly, this homooligomerization is best described by a two-step monomer-dimer-tetramer fast dynamic equilibrium with dissociation constants in the micromolar affinity range.[31,33] The overall binding affinity between proteins is known to depend on the function of the protein complex. For example, obligate homodimers have been reported to associate strongly with nano- or picomolar binding affinity[34] while, in contrast, proteins that associate and dissociate in response to changes in their environment, such as the majority of signal transduction mediators, tend to bind more weakly. In this context, micromolar binding affinities, in combination with a rapid association and dissociation kinetics,[31] make the homotypic CYTO interactions between MIRR signaling subunits a valid candidate for involvement in MIRR-mediated signal transduction.

Hypothesizing a crucial physiological role of these unique homointeractions, the SCHOOL model suggests that formation of competent MIRR signaling subunit oligomers is necessary and sufficient to trigger the receptors and induce TM signal transduction and the downstream signaling sequence (Fig. 1A, see also Chapter 12).[27-29] Within the model, MIRR engagement by multivalent antigen or anti-MIRR antibodies (e.g., anti-CD3ε and anti-TCRβ for TCR or anti-Igβ antibodies for BCR) leads to receptor clustering coupled with a multi-step structural reorganization driven by the homooligomerization of MIRR signaling subunits (Fig. 1A). Ligand-induced MIRR clustering leads to receptor reorientation and formation of a dimeric/oligomeric intermediate in which signaling chains from different receptor units start to trans-homointeract and form signaling oligomers (Fig. 1A, stages 1 and 2). Upon formation of signaling oligomers, protein tyrosine kinases phosphorylate the tyrosine residues in the ITAMs located on the CYTO tails of MIRR signaling subunits, leading to the generation of intracellular activation signal(s), dissociation of signaling oligomers and internalization of the engaged MIRR ligand-binding subunits (Fig. 1A, stages 2 and 3). Then, signaling oligomers interact with the signaling subunits of nonengaged receptors resulting in formation of higher-order signaling oligomers, thus propagating and amplifying the activation signal and resulting in internalization of the non-engaged MIRR recognition subunits (Fig. 1A, stages 4 and 5).

Major Driving Forces

Introducing the homotypic interactions between MIRR signaling subunits as one of the key interactions involved in MIRR triggering and TM signaling, the plausible and easily testable SCHOOL model defines this process as an outcome of the interplay between three major driving forces (Table 1, Fig. 1B):

1) Antigen/ligand-MIRR interactions. These interactions cluster two or more MIRRs in sufficient proximity and correct (permissive) relative orientation to initiate homointeractions between particular MIRR signaling subunits.

2) Intrareceptor TM interactions. These interactions stabilize and maintain receptor integrity in resting cells and balance opposing interactions, the interreceptor CYTO homointeractions, in stimulated cells, thus helping to discriminate ligands/antigens in their functional ability to trigger MIRRs and induce a cellular activation signal.

3) Interreceptor homointeractions. These interactions between the CYTO domains of MIRR signaling subunits lead to the formation of oligomeric signaling structures, thus triggering phosphorylation of ITAMs and initiating the signaling cascade.

Thus, the SCHOOL model reveals the last two key interactions of MIRR triggering/signaling as new therapeutic targets (Fig. 1C).

Antigen/ligand-MIRR interactions are generally of low affinity (micromolar range) and have rapid association and dissociation kinetics (reviewed, for example, for TCR in 35). This low affinity binding in combination with fast kinetics allows immune cells to recognize and discriminate a variety of antigens/ligands with high specificity, selectivity and sensitivity in order to respond with a variety of biological responses. Considering that EC and TM regions of MIRRs are well-ordered receptor segments while MIRR signaling CYTO domains have been recently shown to represent a novel class of intrinsically disordered proteins,[31-33] an important and intriguing question is raised: how do MIRRs transduce highly ordered information about antigen recognition/discrimination from outside the cell through the cell membrane into intracellular biochemical events, thus triggering specific pathways and resulting in a specific functional outcome?

Despite intensive studies of MIRR-mediated TM signal transduction, the only model that can answer this question and even more important, mechanistically explain how this signaling starts, is the SCHOOL model (see also Chapter 12).[27-29] Intriguingly, all three protein-protein interactions, namely antigen/ligand-MIRR EC interactions as well as intrareceptor TM heterointeractions and interreceptor CYTO homointeractions (Fig. 1B, Table 1), fall within the similar micromolar affinity range and are characterized by relatively rapid kinetics.[31,35-41] This conjugated and well-balanced system of interprotein interactions provides the ideal basis to explain the molecular mechanisms of the ability of MIRRs to transduce the extracellular information about recognition of different ligands/antigens through the cell membrane and translate it into different activation signals, thus triggering different intracellular pathways and resulting in different cell responses. Within the model, the MIRR-generated intracellular activation signals are combinatorial in nature and

Table 1. *Major driving forces in MIRR triggering and transmembrane signaling as revealed by the SCHOOL model*

Protein-Protein Interactions	Interaction Milieu	Role in MIRR Triggering/Signaling	Affinity Range
Between antigen/ligand and MIRR recognition subunit(s)	EC	Cluster MIRRs in sufficient interreceptor proximity and correct (permissive) orientation relative to each other to promote the interreceptor CYTO homointeractions between MIRR signaling subunits, resulting in formation of competent signaling oligomers and thus initiating the downstream signaling cascade	µM
Between MIRR recognition and MIRR signaling subunits*	TM	Define the overall rigid geometry and topology of the MIRR. Maintain the integrity of a functional receptor in resting cells. Balance opposing interactions, the CYTO homointeractions, thus helping to discriminate ligands/antigens in their functional ability to cluster MIRRs in sufficient interreceptor proximity and correct (permissive) orientation relative to each other to promote formation of competent signaling subunit oligomers	µM
Homointeractions between MIRR signaling subunit(s)*	CYTO	Lead to formation of competent signaling subunit oligomers, thus initiating the downstream signaling cascade	µM

*Within the SCHOOL model, these TM and CYTO interactions represent the opposing forces that balance resting and differently triggered patterns of MIRR receptor triggering and signaling. Abbreviations: CYTO, cytoplasmic; EC, extracellular; MIRR, multichain immune recognition receptor; SCHOOL model, signaling chain homooligomerization model; TM, transmembrane.

involve multiple components such as formation of different competent MIRR signaling subunit oligomers (see also Chapter 12)[27-30] and different ITAM Tyr phosphorylation patterns.[42-54] This system also explains mechanistically high specificity, selectivity and sensitivity of immune cells in recognition and discrimination of different antigens/ligands and how this recognition/discrimination results in different functional outcomes. This is particularly important for the TCR[55] that has four different signaling subunits, namely ζ and CD3ε, CD3δ and CD3γ, known to play different roles in T-cell biology (see Chapters 1 and 12). In addition, in contrast to other MIRR signaling subunits, ζ has three ITAMs that can provide differential tyrosine phosphorylation patterns in response to different ligands, initiating different intracellular signaling pathways. Thus, within the model, TCR-mediated signaling and cell activation has the highest combinatorial potential as compared to other MIRRs, explaining a high variability of distinct TCR-triggered intracellular signaling pathways and therefore distinct T-cell functional responses depending on the nature of the stimulus (see also Chapter 12).[27-30]

Restraints

Interactions between TM helices of recognition and signaling MIRR subunits maintain receptor integrity in unstimulated cells and determine the relative positions of these subunits in the receptor complex (angles, distances, etc.), thus dictating the overall geometry and topology of MIRRs. Within the SCHOOL model, the overall structural architecture (i.e., geometry and topology) of MIRRs that is dictated and maintained by TM interactions between MIRR recognition and signaling subunits (Fig. 1, see also Chapter 12),[27-30] in combination with the requirement to initiate interreceptor homointeractions between MIRR signaling subunits (Fig. 1), impose several restraints for multivalent antigen/ligand-induced MIRR triggering (Table 2, see also Chapter 12):[27-30]

- sufficient interreceptor proximity in MIRR dimers/oligomers
- correct (permissive) relative orientation of the receptors in MIRR dimers/oligomers
- long enough duration of the MIRR-ligand interaction that generally correlates with the strength (affinity/avidity) of the ligand
- sufficient lifetime of an individual receptor in MIRR dimers/oligomers

The importance of these factors for productive MIRR triggering and TM signaling is strongly supported by a growing body of evidence and described in detail in Chapter 12 of this book. Briefly, it should be noted that the restraints imposed by the model play an especially important role during the first stage of MIRR triggering (Fig. 1). At this point, these spatial, structural and temporal requirements (correct relative orientation, sufficient proximity, long enough duration of the MIRR-ligand interaction and lifetime of MIRR dimers/oligomers) should be fulfilled to favor initiation of trans-homointeractions between MIRR signaling subunits and formation of competent signaling subunit oligomers. If these requirements are not fulfilled at this "final decision-making" point, the formed MIRR dimers/oligomers may dissociate from the ligand and remain signaling-incompetent and/or break apart to its initial monomeric receptor complexes. Also, at this stage, slightly different ligands may bring two or more MIRRs in different relative orientations that favor homointeractions between different signaling subunits and result in formation of different signaling oligomers or their combinations, thus initiating distinct signaling pathways. This mechanism can explain the ability of MIRRs to differentially activate a variety of signaling pathways depending on the nature of the stimulus.

Advantages

The SCHOOL model is fundamentally different from those numerous models that have been previously suggested for particular MIRRs and has several important advantages (see also Chapter 12):[27-30]

- This is the first general mechanistic model for all MIRRs known to date, including TCR, BCR, Fc receptors, NK receptors, ILTs, LIRs, SIRPs, DCAR, MDL-1, NITR, TREMs, GPVI and others and for those that will be discovered in the future. Assuming the general principles underlying MIRR triggering and TM signaling mechanisms are similar for all MIRRs, the SCHOOL model can easily be applied to any particular receptor of the MIRR family,
- This is the first model that is based on specific protein-protein interactions—biochemical processes that can be influenced and controlled[2,10-12,56]—and specific inhibition and/or modulation of these interactions provides a promising novel approach for rational drug design, as revealed by recent progress in the design of inhibitory antibodies, peptides and small molecules.[1,3-8,12]
- Introducing the CYTO homointeractions between MIRR signaling subunits as one of the key elements of MIRR triggering and signaling, the SCHOOL model imposes functionally important restraints (Table 2, see also Chapter 12) and suggests molecular mechanisms for the vast majority of unexplained immunological observations accumulated to date (see also Chapter 12).[27-30]
- Unraveling the molecular mechanisms underlying MIRR triggering and subsequent TM signaling, the model suggests unique and powerful tools to study the immune response and a means to control and/or modulate it (see also Chapter 12).[27-30,57]

Table 2. *Selected main restraints for MIRR signaling imposed within the SCHOOL model by the overall structural architecture and topology of MIRRs in combination with the major driving forces in MIRR triggering and transmembrane signaling*

Restraints	Functional Significance
Sufficient interreceptor proximity in MIRR dimers/oligomers	Two or more antigen/ligand-clustered MIRRs should be in sufficient proximity to each other to initiate CYTO homointeractions between signaling subunits with subsequent formation of competent signaling subunit oligomers
Correct (permissive) relative orientation of the receptors in MIRR dimers/oligomers	Within two or more antigen/ligand-clustered MIRRs, particular MIRR signaling subunit(s) should be in correct orientation relative to each other to initiate CYTO homointeractions between these signaling subunits with subsequent formation of competent signaling subunit oligomers
Long enough duration of the MIRR-ligand interaction that generally correlates with the strength (affinity/avidity) of the ligand	Main protein-protein interactions involved in MIRR triggering and TM signaling (Table 1) fall into a similar low/moderate (micromolar) affinity range. For this reason, the multivalent antigen/ligand-receptor contact should last long enough to bring two or more MIRRs in sufficient proximity and correct relative orientation toward each other and hold them together to promote the interreceptor CYTO homointeractions between MIRR signaling subunits, resulting in formation of competent signaling subunit oligomers and thus initiating the downstream signaling cascade
Sufficient lifetime of an individual receptor in MIRR dimers/oligomers	Similarly to a restraint on duration of antigen/ligand-MIRR contact, in order to initiate the downstream signaling cascade, a lifetime of an individual receptor in antigen/ligand-clustered MIRRs should be sufficient to promote the interreceptor CYTO homointeractions between MIRR signaling subunits

Abbreviations: CYTO, cytoplasmic; MIRR, multichain immune recognition receptor; SCHOOL model, signaling chain homooligomerization model; TM, transmembrane.

- Based on specific protein-protein interactions, the SCHOOL model reveals new therapeutic targets (Fig. 1) for the treatment of a variety of disorders mediated by immune cells.[27-30,57,58]
- An important application of the SCHOOL model is that similar therapeutic strategies targeting key protein-protein interactions involved in MIRR triggering and TM signal transduction may be used to treat diverse immune-mediated diseases. This assumes that clinical knowledge, experience and therapeutic strategies can be transferred between seemingly disparate immune disorders or used to develop novel pharmacological approaches and that a general pharmaceutical approach may be used to treat diverse immune disorders.

SCHOOL Model: New Intervention Points for MIRR-Mediated Immune Disorders

As mentioned previously (Table 1, Fig. 1B), the SCHOOL model defines MIRR triggering and subsequent TM signaling as an outcome of the interplay between three crucially important interactions (Table 1, Fig. 1B): (1) antigen/ligand-MIRR EC interactions, (2) intrareceptor TM interactions and (3) interreceptor CYTO homointeractions. The SCHOOL model reveals these specific protein-protein interactions as points of intervention to inhibit and/or modulate MIRR-mediated TM signaling, thus inhibiting and/or modulating the immune response. While antigen/ligand-receptor interactions are a well-known target for drug design and development (see also Chapters 15, 17-19),[59-73] the last two protein-protein interactions that are critically involved in MIRR triggering/signaling, represent promising novel therapeutic targets as revealed by the model (Fig. 1C).[27-30,57,58] As suggested by the model, controlled inhibition/modulation of these particular interactions represents a means to inhibit/modulate MIRR-mediated TM signaling and specific downstream signaling pathways, thus inhibiting/modulating the immune response. This can be used in rational drug design and the development of novel strategies for the treatment of a variety of diseases and medical conditions that involve MIRR-mediated signaling. Importantly, unraveling the molecular basis of MIRR triggering and signaling and revealing specific protein-protein interactions that play a critical role in MIRR-mediated TM signal transduction and cell activation, the SCHOOL model suggests invaluable and unique powerful tools to dissect mechanisms of the related cell functional outcomes in response to antigen/ligand and to study many important aspects of viral pathogenesis (see also Chapter 22).[27-30,57,58]

In this Chapter, I demonstrate how the SCHOOL model, together with the lessons learned from viral pathogenesis, can be used practically for rational drug design and the development of new therapeutic approaches to treat a variety of seemingly unrelated disorders, such as T-cell-mediated skin diseases and platelet disorders.

Transmembrane Interactions as Immunotherapeutic Targets

Main Concept

Since it was first published in 2004,[28] the SCHOOL model has revealed intra-MIRR TM interactions as important therapeutic targets as well as points of great interest to study the molecular mechanisms underlying the MIRR-mediated cell response in health and disease (Figs. 1 and 2).[27-30,57,58] Notably, the model has provided a mechanistic explanation at the molecular level for specific processes behind "outside-in" MIRR signaling that were unclear (see also Chapter 12).[27-30,57,58] Examples include molecular mechanisms of action of the therapeutically important TCR TM peptides[74-77] first introduced by Manolios et al in 1997[78] and the mechanism underlying HIV-1 fusion peptide (FP)-induced inhibition of antigen-dependent T-cell activation.[79] The relevance of the latter mechanism has since been confirmed experimentally.[80]

Within the SCHOOL model, upon antigen/ligand stimulation, the intra-MIRR TM interactions balance opposing interactions, the inter-MIRR CYTO homointeractions and represent one of three major driving forces of MIRR triggering that helps to discriminate ligands/antigens in their functional ability to trigger MIRRs and induce a cellular activation signal (Table 1, Fig.

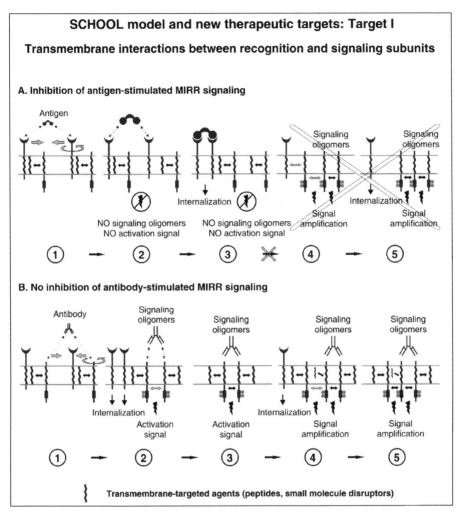

SCHOOL model and new therapeutic targets: Target I

Transmembrane interactions between recognition and signaling subunits

A. Inhibition of antigen-stimulated MIRR signaling

B. No inhibition of antibody-stimulated MIRR signaling

Transmembrane-targeted agents (peptides, small molecule disruptors)

Figure 2. Target I. Transmembrane interactions between MIRR recognition and signaling subunits. This is a simplified graphical illustration of the molecular mechanisms underlying proposed intervention by transmembrane-targeted agents (peptides and their derivatives, small molecule disruptors of protein-protein interactions, etc). Within the SCHOOL model, specific blockade of transmembrane interactions between recognition and signaling subunits is proposed to result in "predissociation" of the receptor complex, thus preventing formation of competent signaling oligomers and inhibiting antigen-dependent immune cell activation (A). In contrast, stimulation of these "predissociated" MIRRs with cross-linking antibodies to signaling subunit should not effect, according to the model, on receptor triggering and cell activation (B). It is noteworthy that the proposed strategies can be used not only to inhibit but also to modulate MIRR-mediated transmembrane signal transduction, thus modulating the immune response (see main text for details). Abbreviations and symbols as in Figure 1. Reprint from Trends Pharmacol Sci, 27, Sigalov AB, Immune cell signaling: a novel mechanistic model reveals new therapeutic targets, 518-524, copyright 2006 with permission from Elsevier.

1B). As suggested by the model (Figs. 1C and 2),[27-30,57,58] specific blockade or disruption of the TM interactions between MIRR recognition and signaling subunits causes a physical and functional disconnection of the subunits. Peptides and their derivatives, small molecule disruptors of protein-protein interactions, site-specific mutations and other similar agents/modifications can be used to affect the MIRR TM interactions. It should be noted that in this context, a physical disconnection means "predissociation" rather than full dissociation of the subunits because in the absence of stimulus, they can still remain together. Antigen/ligand stimulation of these "predissociated" receptors leads to reorientation and clustering of the recognition but not signaling subunits. As a result, signaling oligomers are not formed, ITAM Tyr residues do not become phosphorylated and the signaling cascade is not initiated (Fig. 2A). In contrast, this "predissociation" does not prevent the formation of signaling oligomers when signaling subunits are clustered by specific antibodies that trigger cell activation, e.g., anti-MIRR signaling antibodies (Fig. 2B) such as anti-CD3 for TCR and anti-Igβ antibodies for BCR, or anti-TCRβ antibodies for TCR (not illustrated).

Our current understanding of the MIRR structure and the nature and specificity of TM interactions between receptor recognition and signaling subunits not only allows us to block or disrupt these protein-protein interactions but also to modulate the interactions by sequence-based approach with using corresponding peptides and/or their derivatives. Strengthening/weakening and/or selective disruption of the association between particular recognition and signaling subunits might allow us not to inhibit, but rather to modulate the ligand-induced cell response. In addition, selective functional disconnection of particular signaling subunits from their recognition partner represents an invaluable tool in studies of MIRR-mediated TM signaling and cell activation. It should also be noted that methods of computational design, synthesis and optimization of TM peptides and peptidomimetics, as well as high-throughput screening techniques to search for the relevant TM mutations or small molecule disruptors, are currently developed and well-established,[1-11,34,76,81-90] making the proposed powerful approach both feasible and of great fundamental and clinical value.

I suggest that the TM interactions between recognition and signaling MIRR subunits represent extremely important points of control in MIRR triggering and cell activation. Since we can now use the SCHOOL model to design the TM-targeted agents effective in inhibition and/or modulation of MIRR-mediated TM signaling (Figs. 1 and 2, see also Chapter 12),[27-30,57,58] we have a powerful and well-controlled influence upon MIRR-mediated cell activation and control the immune response. The relevant TM-targeted agents for any particular member of MIRR family can be readily designed using the SCHOOL model and our knowledge about structural organization of this receptor. Examples include the TM peptides of TCR,[74-76,78,81] NK receptors[91] and GPVI[58] tested to inhibit/modulate the relevant receptor-mediated cell response. Importantly, the SCHOOL model unravels the TM-targeted molecular mechanisms underlying ability of different human viruses such as human immunodeficiency virus, cytomegalovirus and severe acute respiratory syndrome coronavirus to modulate and/or escape the host immune response.[29,30,57] It also demonstrates how the lessons learned from viral pathogenesis can be used practically for rational drug design.[30,58] These and other examples that successfully prove the main concept of the SCHOOL model-driven TM strategy are considered in detail below.

Obviously, allowing us to effectively control MIRR signaling and therefore the immune response, the MIRR intrareceptor TM interactions represent an important target of pharmacological intervention as first revealed and suggested by the SCHOOL model in 2004.[28] It further assumes that a general therapeutic strategy aiming to disrupt/modulate these interactions in the MIRRs may be used in the existing and future treatment of seemingly unrelated immune diseases. In other words, according to the main concept of the SCHOOL model, specific therapeutic agent(s) that target particular MIRR(s) involved in pathogenesis of the relevant immune disorder can be readily designed using basic principles of structural assembly of this receptor and the SCHOOL model as applied to this particular member of MIRR family.

Table 3. Selected agents reported to modulate the immune cell response and suggested or predicted by the SCHOOL model to affect MIRR transmembrane interactions

Agent	MIRR	Action	Mechanism as Suggested by the SCHOOL Model	Potential Clinical Use
TCR CP	TCR	Selectively inhibits antigen-stimulated TM signal transduction[75,76,78,81] Efficiently abrogates T-cell-mediated immune responses in mice and man in vitro and in vivo[74,77]	Disrupts TCRα-CD3δε and TCRα-ζ TM interactions resulting in disconnection/predissociation of these signaling subunits from the remaining complex and thus preventing the formation of signaling oligomers upon antigen stimulation and, consequently, inhibiting T-cell activation (Figs. 2 and 3, see also Chapter 12)[27-30]	T-cell-mediated immune disorders: dermatoses, arthritis, etc. Anti-tumor therapy
CD3δ-CP CD3ε-CP CD3γ-CP	TCR	CD3δ-CP and CD3γ-CP do not inhibit antigen-stimulated T-cell proliferation and IL-2 secretion.[77]CD3δ-CP, CD3ε-CP and CD3γ-CP prevent disease development and progression in rats with adjuvant-induced arthritis[77]	Disrupt TCRα-CD3δ (CD3δ-CP), TCRα-CD3ε (CD3ε-CP), TCRβ-CD3ε (CD3ε-CP) and TCRβ-CD3γ (CD3γ-CP) TM interactions, resulting in selective disconnection/predissociation of the particular signaling subunits from the remaining receptor complex, thus modulating the cell response[27-30]	T-cell-mediated immune disorders: dermatoses, arthritis, etc. Anti-tumor therapy
NK-CP ζ-CP	NKp44 NKp46 NKp30 NKG2D NKG2C KIR2DS	Inhibit NK cell cytolytic activity[91]	Disrupt the TM interactions between NK receptor ligand-binding subunits and associated homodimeric signaling subunits, such as ζ-ζ, γ-γ or DAP-12 (Fig. 2)[27-30]	NK cell-mediated diseases
NK-CP	NKG2D	Predicted to inhibit NKG2D signaling pathway critically involved in CD4+ T-cell-mediated colitis progression*	Predicted to disrupt the TM interactions between NK2D receptor ligand-binding subunits and associated homodimeric DAP-10 signaling subunit (Fig. 2)[27-30]	Inflammatory bowel diseases[210]
GPVI-CP	Platelet GPVI	Inhibits collagen-induced platelet activation and aggregation[58]	Disrupts the TM interactions between collagen-binding GPVI subunit and associated homodimeric γ-chain (Fig. 4)[58,108]	Platelet-mediated diseases and conditions

*To be proved in the future.

Abbreviations: CP, core peptide; DAP-12, DNAX activation protein 12; GPVI, glycoprotein VI; MIRR, multichain immune recognition receptor; NK cells, natural killer cells; TCR, T-cell antigen receptor; TM, transmembrane.

An exciting and promising example of using the SCHOOL model-driven TM approach for both fundamental and clinical applications has been recently demonstrated by Collier et al,[77] as covered in more detail below.

Direct and Indirect Evidence: Transmembrane Peptides and Immune Cell Activation

Direct Evidence

The SCHOOL model is the first model to clearly explain molecular mechanisms of action of TCR TM peptides (see also Chapter 16) and extend the concept of their action through these mechanisms to any other TM peptides of MIRRs and to the MIRR-mediated processes involved in viral pathogenesis.[27-30,57,58] Selected agents suggested or predicted by the SCHOOL model to affect MIRR TM interactions, thus inhibiting or modulating MIRR-mediated immune cell activation, are listed in Table 3.

TM peptides capable of inhibiting MIRR-mediated cell activation were first reported in 1997 for antigen-stimulated TCR-mediated T-cell activation by Manolios et al.[78] Since that time, despite extensive basic and clinical studies of these and several other TM peptides (see also Chapter 16),[74,75,77,92-101] the molecular mechanisms of action of these clinically relevant peptides have not been elucidated until 2004 when the SCHOOL model was first introduced.[28]

The vast majority of basic and clinical findings were reported for the TCR TM core peptide (TCR CP), or TCR mimic peptide, which represents a synthetic peptide corresponding to the sequence of the TM region of the ligand-binding TCRα chain critical for TCR assembly and function. This TM region has been shown to interact with the TM domains of the signaling CD3δε and ζ subunits,[22,23] thus maintaining the integrity of the TCR in resting T-cells.

Briefly, as suggested by the SCHOOL model (Figs. 2 and 3, Table 3, see also Chapter 12),[27-30,57] the TCR CP competes with the TCRα chain for binding to CD3δε and ζ hetero- and homodimers, respectively, thus resulting in disconnection/predissociation of the signaling subunits from the remaining receptor complex (Fig. 3). This leads to inhibition of antigen- but not antibody-mediated TCR triggering and cell activation (Figs. 2 and 3). It should be highlighted that the proposed mechanism is the only mechanism consistent with all experimental and clinical data reported up to date for TCR and other MIRR TM peptides and their lipid and/or sugar conjugates.[58,74,75,77,92-100]

Recently, new experimental evidence supporting the proposed mechanism of inhibitory action of TCRα CP has been reported.[100] This study has clearly shown that this peptide does not affect TCR assembly and cell surface expression.[100] Most strikingly, Kurosaka et al[100] have demonstrated that TCRα CP coprecipitates with CD3δε. This finding perfectly fits the molecular explanation of its inhibitory action suggested for the first time in 2004 by applying the SCHOOL model[28] and later developed further.[27-30,57] Again, within the model, competing with the TCRα chain, TCRα CP binds to CD3δε and ζ signaling subunits, preventing an antigen-induced formation of the relevant competent signaling oligomers and thus inhibiting an antigen-dependent T-cell response (Figs. 2 and 3, Table 3, see also Chapter 12).

The SCHOOL model predicts that the same mechanisms of inhibitory action can be applied to MIRR TM peptides corresponding to the TM regions of not only the MIRR recognition subunits but to the corresponding signaling subunits as well.[27-30] This was recently confirmed experimentally[77,91] by showing that the synthetic peptides corresponding to the sequences of the TM regions of the signaling CD3 (δ, ε, or γ) and ζ subunits are able to inhibit the immune response in vivo (CD3 TM peptides) and NK cell cytolytic activity in vivo (ζ TM peptide) (Table 3).

Interestingly, the model suggests a molecular explanation for the intriguing phenomenon recently reported by Collier et al[77] and interpreted by the authors as a discrepancy in CD3 TM peptide activity between in vitro and in vivo T-cell inhibition. It has been shown that the CD3δ and CD3γ TM peptides do not impact T-cell function in vitro (the CD3ε TM peptide has not been used in the reported in vitro experiments because of solubility issues) but that all three CD3 TM peptides decrease signs of inflammation in the adjuvant-induced arthritis rat model in vivo and inhibit an

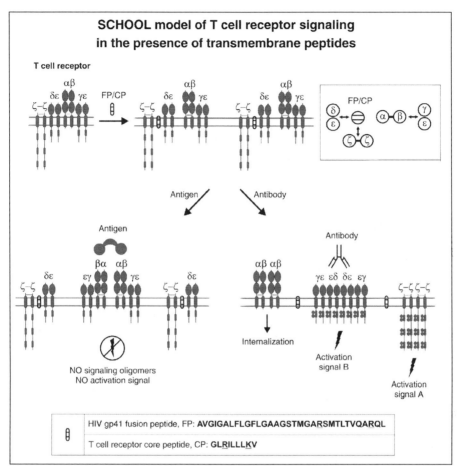

Figure 3. A proposed molecular mechanism of action of the T-cell receptor core peptide (CP) and HIV-1 gp41 fusion peptide (FP). Considering the close similarity in patterns of inhibition of T-cell activation and immunosuppressive activity observed for CP and FP, the SCHOOL model reasonably suggests a similar molecular mechanism of action for both peptides. Within the SCHOOL model, these peptides compete with the TCRα chain for binding to the CD3δε and ζ signaling subunits, thus disrupting the transmembrane (TM) interactions between these subunits and resulting in disconnection and predissociation of the relevant signaling subunits from the remaining receptor complex (also shown in the inset as a simplified axial view). This prevents formation of signaling oligomers upon multivalent antigen stimulation, thus inhibiting antigen-mediated T-cell activation. In contrast, stimulation of these "predissociated" MIRRs with cross-linking antibodies to signaling subunit should still lead to receptor triggering and cell activation. The model predicts that the same mechanisms of inhibitory action can be applied to TCR TM peptides corresponding to the TM regions of not only the TCRαβ recognition subunits but the corresponding CD3ε, CD3δ, CD3γ and ζ signaling subunits as well. In addition, similar mechanisms are proposed to be used by other viruses, such as cytomegalovirus and severe acute respiratory syndrome-associated coronavirus, in their pathogenesis to modulate the host immune response.

immune response.[77] Within the SCHOOL model, these data do not reveal any discrepancy between in vivo and in vitro experiments. Instead, they can be considered, in fact, as the first direct experimental evidence of our ability to selectively modulate the MIRR-mediated TM signaling and the immune response, as predicted by the model.[27-30] In this context, the CD3δ and CD3γ TM peptides disconnect the corresponding signaling subunits (CD3δ and CD3γ, respectively) from the remaining receptor complex (Table 3). Therefore, antigen stimulation does not result in formation of the relevant competent CD3δ or CD3γ signaling oligomers and phosphorylation of their ITAM tyrosine residues, preventing initiation of the corresponding signaling pathways and cell responses. Further, in their in vitro experiments, the authors[77] used an interleukin 2 (IL-2) production assay and T-cell proliferation as markers of T-cell activation. However, the previously reported in vitro activation studies with T-cells lacking CD3γ and/or CD3δ cytoplasmic domains clearly indicate that antigen-stimulated induction of cytokine secretion and T-cell proliferation in these cells are intact,[102-105] explaining the absence of inhibitory effect of the CD3δ and CD3γ TM peptides in the in vitro activation assays used.[77] However, in vivo deficiency either of CD3δ or CD3γ results in severe immunodeficiency disorders.[106,107] This can explain the inhibitory effect observed in the in vivo studies for all three CD3 TM peptides.[77]

These experimental data[77] successfully proved that our ability to selectively physically disconnect specific signaling subunits using the MIRR TM peptides in line with the SCHOOL model can result in their selective functional disconnection and thus provide a powerful tool to study MIRR functions and immune cell signaling.[27-30] Even more importantly, it also confirms that as predicted using the SCHOOL model,[27-30] agents targeted specific intra-MIRR TM interactions can be designed not only to inhibit but also specifically modulate the immune response and therefore result in the development of novel therapeutic strategies for a variety of immune disorders.

Similar molecular mechanisms of action are suggested by the SCHOOL model for other MIRR TM peptides and describe and/or predict their inhibitory/modulatory effect on MIRR-mediated cell activation (Table 3). Recently, the SCHOOL model-driven TM-targeted strategy has been successfully applied to develop a novel concept of platelet inhibition and resulted in the invention of a new class of platelet inhibitors (Fig. 4, Table 3, see also Chapter 12).[58,108] This issue will be covered in more detail below.

In summary, considering the high therapeutic potential of the MIRR TM peptides illustrated by the clinical results for the TCR CP (Table 4) and the promising results for other peptides (Table 3),[58,108] the SCHOOL model represents an invaluable tool in further development of this novel pharmacological approach targeting MIRR TM interactions.

Indirect Evidence

In contrast to MIRRs, single-chain receptors (SRs) can be characterized in the structural context as receptors with extracellular recognition domains and intracellular signaling domains located on the same protein chain. Examples include receptor tyrosine kinases (RTKs) that are TM glycoproteins consisting of a variable extracellular N-terminal domain, a single membrane spanning domain and a large cytoplasmic portion composed of a juxtamembrane domain, the highly conserved tyrosine kinase domain and a C-terminal regulatory region. Ligand binding is believed to stimulate monomeric receptor dimerization and trans-autophosphorylation at defined tyrosine residues through intrinsic kinase activity.[109-111] Further, the basic principles of SR signaling, namely the ligand-induced receptor dimerization/oligomerization and trans-autophosphorylation

Table 4. Effect of TCR core peptide on T-cell-mediated dermatoses in man*

Diagnosis	Number of Patients	Cure	Improvement	No Effect
Atopic dermatitis	5	3	2	-
Lichen planus	2	1	1	-
Psoriasis	2	-	1	1

*Adapted from [Gollner GP, Muller G, Alt R et al. Therapeutic application of T-cell receptor mimic peptides or cDNA in the treatment of T-cell-mediated skin diseases. Gene Ther 2000; 7:1000-1004].

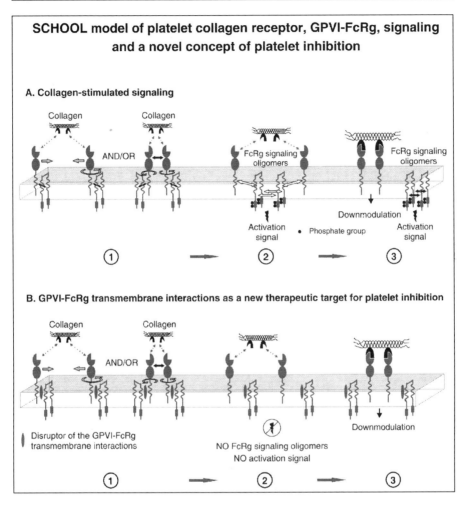

Figure 4. Novel concept of platelet inhibition. A) The signaling chain homooligomerization (SCHOOL) model of collagen-stimulated GPVI-FcRγ transmembrane (TM) signaling, proposing that the homooligomerization of the FcRγ signaling subunit plays a central role in triggering the GPVI-FcRγ receptor complex. The model also assumes that not only is sufficient proximity of the receptor units in formed and/or preformed receptor dimers/oligomers required to trigger MIRRs but also a correct interunit relative orientation and geometry. Small solid black arrows indicate specific intersubunit hetero- and homointeractions between TM and cytoplasmic domains, respectively. Circular arrows indicate collagen-induced receptor reorientation. All interchain interactions in a dimeric intermediate are shown by large white arrows reflecting their transition state. Immunoreceptor tyrosine-based activation motifs are shown as dark gray rectangles. Phosphate groups are shown as black circles. B) Specific disruption of the GPVI-FcRγ TM interactions results in "predissociation" of the GPVI-FcRγ receptor complex, thus preventing formation of FcRγ signaling oligomers and inhibiting collagen-dependent platelet activation and aggregation. Reproduced with permission from Wiley-Blackwell Publishing Ltd. Sigalov AB. More on: glycoprotein VI oligomerization: a novel concept of platelet inhibition. J Thromb Haemost 2007; 5:2310-2312.

of Tyr residues in CYTO signaling domains, are considered to represent common mechanisms of triggering and TM signal transduction for the vast majority of various receptors.[12,83,112-123]

Dimerization of SRs is known to be mostly driven by the homointeractions between receptor TM domains.[83,113,117,119-122,124] These findings reveal the interreceptor TM protein-protein interactions as an attractive point of pharmacological intervention. At present, there is a growing line of experimental evidence indicating that an application of TM-targeted strategy to inhibit/modulate SR signaling might represent a promising therapeutic strategy.[83,117,121,124-130]

It should be noted that despite apparent similarities in using TM peptides in both SR- and MIRR-targeted TM strategies, the basic principles of the molecular mechanism underlying inhibition and/or modulation of SR and MIRR signaling by using the TM agents are totally different. As established in the field of SR signaling, the SR-targeted TM peptides/agents block/disrupt/modulate interreceptor TM interactions crucial for antigen/ligand-induced receptor oligomerization. This prevents the formation of competent receptor oligomers, whereas MIRR-targeted TM peptides/agents, as suggested by the SCHOOL model (see also Chapter 12)[27-30,57,58] affect intra-MIRR TM interactions between recognition and signaling subunits, thus preventing upon antigen/ligand stimulation, formation of competent MIRR signaling oligomers but not MIRR oligomers/clusters in terms of MIRR recognition subunits and those signaling subunits that are not affected by the TM agents. Because of this fundamental difference in molecular mechanisms of action of SR- and MIRR-targeted TM peptides/agents, I consider the SR-related findings as indirect evidence for the fundamental and clinical relevance of the MIRR TM-targeted strategy suggested by the SCHOOL model. Several examples of using TM peptides to inhibit SR signaling are described in more detail below.

Ligand binding-induced association of the TM domains has been proposed to favor productive dimerization of intracellular kinase domains to promote trans-autophosphorylation.[124] Studies with the epidermal growth factor (EGF) and ErbB2 receptors have shown that synthetic peptides encompassing the TM domains of these receptors inhibit the autophosphorylation and signaling pathway of their cognate receptor.[124,129] These peptides are thought to block/disrupt specific TM interactions, thereby inhibiting receptor dimerization and activation.[124,129]

Using differential epitope tagging, it has been demonstrated that β2-adrenergic receptors form homodimers and TM domain VI of the receptor may represent part of an interface for receptor dimerization.[125] As shown, a peptide derived from this domain inhibits both dimerization and β-adrenergic agonist-promoted stimulation of adenylyl cyclase activity.[125] In contrast, a peptide based on the sequence of transmembrane domain 6 of the D1 dopamine receptor (D1DR) has been found to specifically inhibit D1DR binding and function without affecting receptor oligomerization.[126] One possible explanation for this finding is that in addition to ligand-stimulated dimerization of receptors, the correct (permissive) relative orientation in the receptor dimers formed can also play an important role in D1DR signaling. The importance of the relative orientation has been shown for other SRs such as, for example, EGF receptors,[131] Epo receptor,[114,132-134] toll-like receptors (TLRs)[135] and the integral membrane receptor LuxPQ.[136] The presence of the TM peptide bound to the D1DR TM domain is likely to prevent ligand-induced formation of receptor dimers with correct intermolecular orientation, thus preventing generation of the activation signal.

Another example of SR-targeted TM inhibitory peptides, the short peptide sequences corresponding to the Neu RTK TM domain, have been also reported to independently fold in membranes, interact with the full-length receptor and inhibit transformation of cells in vitro and in vivo.[137]

G-protein-coupled receptors (GPCR) are characterized by the presence of seven TM domains and represent a superfamily of proteins that mediate the function of neurotransmitters and peptide hormones and are involved in viral entry and perception of light, smell and taste. Structural analogs of individual TM domains of GPCRs have been reported to serve as potent and specific receptor inhibitors.[128] Peptide sequences corresponding to the TM domains of chemokine receptors, CXCR4, also called fusin, an alpha-chemokine receptor specific for stromal-derived-factor-1 and CCR5, the chemokine receptor which HIV uses as a coreceptor to gain entry into macrophages, have been demonstrated to specifically inhibit receptor signaling and the in vitro replication of HIV-1.[128] Similarly, peptides mimicking the TM domains of cholecystokinin receptor A, have been found to abolish ligand binding and signaling through the receptor.[128]

Thus, the sequence-based blockade of the interreceptor TM protein interactions as applied to SR signaling provides indirect evidence for the importance and clinical significance of the intra-MIRR TM-targeted strategy suggested by the SCHOOL model.

Transmembrane Interactions and Viral Pathogenesis

In general terms, viral pathogenesis is the process by which viral infection leads to disease. The consequences of a viral infection depend on a number of viral and host factors that affect pathogenesis. Infection of host cells by enveloped viruses requires fusion of the viral membrane with the host cell membrane. This fusion is mediated by viral glycoproteins (gp), the proteins that are anchored to the viral membrane. The fusion glycoproteins of enveloped viruses, typically type-I integral membrane proteins, are known to contain in their sequences a short region called the "fusion peptide" (FP), which is required for mediating membrane fusion.[138,139] This region interacts with the host cell membrane at an early stage of the membrane fusion process. Despite advances in our understanding of the major principles of viral fusion mediated by the fusion glycoproteins,[138-143] little is known about their role in functional modulation of MIRR-mediated TM signal transduction.

In this section, I focus on MIRR signaling-related immunomodulatory activity recently reported for HIV and CMV. As suggested by the SCHOOL model, the molecular mechanisms underlying this activity affect MIRR TM interactions and can be also used by other viruses. To illustrate this point, I describe an application of the model in the pathogeneses of two other viruses, SARS-CoV and HTLV-1. I also demonstrate how the SCHOOL model-driven TM strategy, together with the lessons learned from viral pathogenesis, can be used practically for rational drug design and the development of new therapeutic approaches.

HIV Pathogenesis

CD4+ T-cells are the main targets of HIV-1 in the host. The magnitude of viral replication in these cells is closely linked to their activation state. In activated memory CD4+ T-cells, HIV-1 readily undergoes multiple rounds of replication, whereas resting helper T-cells are largely refractory to productive infection.[144,145] Indeed, several steps in the life cycle of HIV-1 have been identified where potent blocks in virus propagation occur when ample T-cell activation is lacking.

Fusion Peptide

The FP found in the N terminus of the HIV envelope glycoprotein gp41 functions together with other gp41 domains to fuse the virion with the host cell membrane.[146,147] Surprisingly, this peptide has been recently shown to have not only a fusogenic activity but also a T-cell-targeted immunomodulatory activity: it colocalizes with CD4 and TCR molecules, coprecipitates with the TCR and inhibits antigen-stimulated T-cell proliferation and proinflammatory cytokine secretion in vitro.[79] These effects are specific, T-cell activation via PMA/ionomycin or mitogenic antibodies to CD3 is not affected by FP and FP does not interfere with antigen-presenting cell function.[79] In mice, HIV FP shows immunosuppressive activity, inhibiting the activation of arthritogenic T-cells in the autoimmune disease model of adjuvant arthritis and reducing the disease-associated interferon-γ (IFN-γ) response.[79] The close match between these findings[79] and the experimental data generated for TCR CP[75,78,98] suggests a mechanistic similarity underlying the TCR-targeted HIV FP and TCR CP activities.

However, as with TCR CP, despite ongoing studies of HIV gp41 FP,[80] the molecular mechanisms of immunomodulatory action of this peptide have not been elucidated until 2006 when the SCHOOL model was first applied to this area.[29] Considering the close similarity in patterns of inhibition of T-cell activation and immunosuppressive activity observed for FP[79] and CP,[75,78,98] the SCHOOL model reasonably suggests a similar molecular mechanism of action for TCR TM peptides and HIV gp41 FP (Fig. 3, Tables 3 and 5).[27-30,57] Primary sequence analysis of these two peptides (Table 6) shows different primary sequences but a similarity in charged or polar residue distribution patterns with two positively charged residues spaced apart by 4 (CP) or 8 (FP) amino acids. For CP, Arg and Lys residues are known to mediate the interaction between recognition TCRα subunit and signaling CD3δε and ζ subunits.[23] Importantly, for FP, both arginines are located in the C-terminal half, suggesting that this sequence could be important for the interaction with the TCR. Figure 3 shows a potential mode of action of CP and FP as proposed by the SCHOOL model (see also Tables 3 and 5). Briefly, CP and FP compete with the TCRα chain for binding to CD3δε and ζ hetero- and homodimers, respectively, thus resulting in TM disconnection/predissociation of the signaling subunits from the remaining receptor complex (Fig. 3). This mechanism of FP action suggests the existence of an interaction interface in the C-terminal half of the peptide. Within the model,[29,30,57] the peptide prevents formation of CD3δε and ζ signaling oligomers and thus inhibits antigen-dependent T-cell activation (Fig. 3, Table 5), acting similarly in this respect to TCR CP (Fig. 3, Tables 3 and 4).[27-30,75] However, stimulation with anti-CD3 antibodies of these "predissociated" TCRs still should result[27-30,57] and results[79,80] in receptor triggering and cell activation. The model suggests that clinically relevant antibodies (OKT3) could be used to modulate the affected T-cell response during HIV infection. Recently, OKT3 antibodies have been used successfully in HIV therapy to augment immune activation.[148] More recent studies[80] have confirmed the predicted molecular mechanism of immunomodulatory activity of the HIV FP. Finally, it should be noted that the proposed mechanism is the only mechanism consistent with all experimental data on immunomodulatory action of HIV gp41 FP reported up to date.[79,80]

A highly specific natural inhibitor of HIV-1 gp41 FP has been recently reported to block HIV-1 entry.[149,150] This agent that has been isolated from human hemofiltrate and designated VIRus Inhibitory Peptide (VIRIP),[150] represents a 20-residue peptide, corresponding to the C-proximal region of α1-antitrypsin. Importantly, it has been shown that VIRIP directly interacts with the gp41 FP and a few amino acid changes increase its antiretroviral activity potency by two orders of magnitude, thus demonstrating the usability and efficiency of rational peptide design approaches.[150]

According to the SCHOOL model, the TCR TM interactions represent not only important therapeutic targets for immune-mediated diseases but also a point of HIV intervention. The molecular mechanisms revealed by the model can be used in rational antiviral drug design and the development of novel antiviral therapies.

HIV Nef Protein

Another application of the SCHOOL model to HIV pathogenesis is related to the molecular mechanisms of action of the HIV pathogenicity factor Nef, a key protein in viral replication and progression of disease. Several studies have shown that this protein interacts with the TCR ζ chain and mediates downmodulation of TCR—CD3 complex.[151-153] Notably, Nef lowers the threshold of CD4+ T-cell activation.[154,155] Other study showed that Nef induces transcription of an array of genes almost identical to that triggered upon exogenous stimulation of TCR.[156] Nef has been also reported to affect T-cell activation events through its interactions within the lipid raft microenvironment,[157] induce signal transduction via the recruitment of a signaling machinery, thereby mimicking a physiological cellular mechanism to initiate the TCR cascade[158] and, finally, to form a signaling complex with the TCR, which bypasses the requirement of antigen to initiate T-cell activation.[159] Thus, the extent of T-cell activation imprinted by expression of Nef is a matter of controversy. In addition, although we know that Nef binds the TCR ζ chain,[152,160] the role of this interaction and the mechanism used by Nef to modulate T-cell activation remain unknown.

Importantly, similar to ζ,[31,33] Nef exists in several discrete oligomeric species, namely monomers, dimers and trimers.[161] Within the model,[27-29] natively oligomeric Nef may crosslink homodimeric ζ chains, leading to the formation of multivalent TCR complexes that have been shown to be responsible for sensing low concentrations of antigen.[162] This mechanism could explain the observed activation sensitization in T-cells by Nef.[154,155] On the other hand, Nef dimers may crosslink ζ homodimers in the "permissive" relative orientation and promote formation of competent signaling ζ oligomers, generating an activation signal A (Fig. 3, see also Chapter 12)[27-29] and resulting in

Table 5. *Selected viral agents reported to modulate the immune cell response and suggested or predicted by the SCHOOL model to affect MIRR transmembrane interactions*

Agent	MIRR	Action	Mechanism as Suggested by the SCHOOL Model
HIV gp41 FP	TCR	Colocalizes with CD4 and TCR molecules, coprecipitates with the TCR and inhibits antigen-specific T-cell proliferation and proinflammatory cytokine secretion in vitro[79] Blocks the TCR/CD3 TM interactions needed for antigen-triggered T-cell activation[80]	Similarly to the TCR CP, disrupts TCRα-CD3δε and TCRα-ζ TM interactions resulting in dissociation of these signaling subunits from the remaining complex and thus preventing the formation of signaling oligomers upon antigen stimulation and, consequently, inhibiting T-cell activation[29,30,57]
CMV pp65	NKp30	Interacts directly with NKp30, leading to dissociation of the linked ζ subunit and, consequently, to reduced killing[193]	Affects the NKp30-ζ TM interactions resulting in dissociation of the ζ signaling subunit from the remaining complex and thus preventing the formation of ζ signaling oligomers upon antigen stimulation and, consequently, inhibiting NK cell cytolytic activity[29,30]
SARS-CoV FP	TCR*	Not reported yet	Similarly to the TCR CP and HIV gp41 FP, disrupts TCRα-CD3δε and TCRα-ζ TM interactions resulting in dissociation of these signaling subunits from the remaining complex and thus preventing the formation of signaling oligomers upon antigen stimulation and, consequently, inhibiting T-cell activation*

*As predicted by the SCHOOL model.
Abbreviations: CMV, human cytomegalovirus; CP, core peptide; FP, fusion peptide; gp, glycoprotein; MIRR, multichain immune recognition receptor; NK cells, natural killer cells; pp65, 65 kDa phosphoprotein; SARS-CoV, severe acute respiratory syndrome coronavirus; TCR, T-cell antigen receptor; TM, transmembrane.

Table 6. *Primary sequences of MIRR transmembrane domains involved in viral pathogenesis and viral fusion proteins and peptides suggested or predicted by the SCHOOL model to affect MIRR transmembrane interactions*

MIRR	Description	Sequence*
TCR	TCRα TMD	VIGF**R**ILLL**K**VAGFNLLMTL
	TCRα CP	GL**R**ILLL**K**V**
	SARS-CoV FP	MY**K**TPTL**K**YFGGFNFSQIL
	HIV gp41 FP	AVGIGALFLGFLGAAGSTMGA**RS**MTLTVQARQL
	HTLV-1 gp21***	AVPVAVWLVSALAMGAGVAGGITGSMSLASG**KS**LLHEVD**K**D
MIRR?	LASV FP	GTFTWTLSDSEG**K**DTPGGYCLT**R**WMLIEAEL**K**CFGNTAV
	LCMV FP	GTFTWTLSDSSGVENPGGYCLT**K**WMILAAEL**K**CFGNTAV
	MOPV FP	GLFTWTLSDSEGNDMPGGYCLT**RS**MLIGLDL**K**CFGNTAI
	TACV FP	AFFSWSLTDPLGNEAPGGYCLE**K**WMLVASEL**K**CFGNTAI
NKp30	NKp30 TMD	GTVLLL**R**AGFYAVSFLSVAVG
	ζ subunit TMD	LCYLL**D**GILFIYGVILTALFL*****
	CMV pp65***	ME**SR**G**RR**CP**E**MISVLGPISGHVL**K**AVFS**R**G**D**TPVLPHET**R**LLQTGIHV**R**VSQPS_LIVSQYTP**D**STPCH**R*****

*Basic amino acid residues are indicated in bold.
**Corresponds to the predetermined assembly TM sequence of murine TCRα.[78]
***N-terminal end.
****Acidic amino acid residues are underlined.

Abbreviations: CMV, human cytomegalovirus; CP, core peptide; FP, fusion peptide; gp, glycoprotein; HTLV-1, human T-lymphotropic virus type 1; LASV, Lassa virus; LCMV, lymphocytic choriomeningitis virus; MOPV, Mopeia virus; MIRR, multichain immune recognition receptor; NK cells, natural killer cells; pp65, 65 kDa phosphoprotein; SARS-CoV, severe acute respiratory syndrome coronavirus; TACV, Tacaribe virus; TCR, T-cell antigen receptor; TCRα, TCR alpha chain; TM, transmembrane; TMD, transmembrane domain.

dissociation of the ζ signaling oligomers from the remaining receptor complex with its subsequent internalization. The SCHOOL model suggests that the oligomer interfaces of ζ and/or Nef are involved in the molecular mechanisms underlying the immunomodulatory effects of Nef.[27-29] As recently shown,[163] a Nef mutant carrying a mutation targeted to the conserved residue D123, in addition to losing the ability to oligomerize, is defective for major histocompatibility complex class I (MHC-I) downmodulation and enhancement of viral infectivity, suggesting that the oligomerization of Nef may be critical for its multiple functions.

In this regard, I suggest that both proposed mechanisms may take place in vivo and selection between these two alternative pathways may possibly depend on the type of cells infected and/or on the cell membrane lipid content. Thus, CYTO heterointeractions at the Nef-ζ interface and CYTO homointeractions in Nef and ζ oligomers may represent attractive targets for the design of antiviral agents.

The usability and efficiency of this SCHOOL model-driven CYTO approach have been later demonstrated for Nef-mediated internalization of surface CD80 or CD86 that is dependent on the binding of Nef to the CYTO domain of the target CD80 or CD86 molecule, respectively.[164] This issue will be covered in more detail below.

SARS-CoV Pathogenesis

The coronavirus SARS CoV is the etiological agent of SARS that represents the life-threatening disease associated with a mortality of about 10%.[165] In recent studies, in which a total of 38 patients with SARS were enrolled, have shown that CD4+ and CD8+ T-lymphocyte levels were reduced in 100% and 87% of patients, respectively.[166] Thus, one can suggest that the virus can have an immunomodulatory activity and this activity is TCR-targeted.

In the traditional view of HIV disease course, acute HIV infection is characterized by massive and rapid CD4+ T-cell loss, whereas chronic infection is characterized by persistent immune activation that drives viral replication and further CD4+ T-cell depletion.[167,168] Thus, HIV infection has been thought of as a relatively indolent disruption of CD4+ T-cells eventually leading to collapse of immune function. As with SARS CoV,[166] this notion has been largely based on measurements of CD4+ T-cell counts in peripheral blood.[167,168]

Despite the lack of direct evidence, it is reasonable to suggest that SARS-CoV has a TCR-targeted immunomodulatory activity. More specifically, as with HIV, this activity might be especially important during virus entry to suppress the host response to virus infection. Like other enveloped viruses encoding class I viral fusion proteins such as HIV[169] and Ebola and avian sarcoma viruses,[170] SARS-CoV is presumed to use membrane fusion mechanisms for viral entry.[171-173] It has been shown that the SARS-CoV viral spike (S) protein 2 (S2) is a class I viral fusion protein and is responsible for driving viral and targeT-cell membrane fusion.[174] Recently, inhibitory peptides derived from the membrane-proximal heptad repeat region (HR2) of the S2 protein have been suggested as an attractive basis for the development of therapeutics for SARS.[175] The putative SARS-CoV FP has also been identified at the N terminus of the SARS-CoV S2 subunit.[176] As shown by using synthetic peptides,[176] the fusogenic activity of the SARS-CoV FP appears to be dependent on its amino acid sequence, as scrambling the peptide renders it unable to partition into large unilamellar vesicles (LUVs), assume a defined secondary structure, or induce both fusion and leakage of LUV.

Primary sequence analysis of the SARC-CoV FP and TCRα TM domain (or TCR CP) shows different primary sequences but reveals a similarity in charged or polar residue distribution patterns with two positively charged residues spaced apart by 4 amino acids (Table 6). These two positively charged residues are critical for TCR assembly and function. Within the SCHOOL model,[27-30,57] the RILLLK and RSMTLTVQAR motifs in TCR CP and HIV FP (Table 6), respectively, play an important role in mimicking the TCRα TM region and therefore in an inhibitory activity of these peptides (Fig. 3, Tables 3 and 5). Intriguingly, SARS CoV FP has a structural motif KTPTLK that is strikingly similar to that of TCR CP (Table 6). Considering the common structural features of three peptides (TCR CP, SARS-CoV FP and HIV FP) as well as functional similarities between HIV FP and SARS FP in the context of their T-cell-targeted activities, I suggest that like TCR

CP and HIV FP, SARS-CoV FP should mimic the TCRα TM domain and therefore exhibit an inhibitory effect on the antigen-mediated TCR TM signaling (Table 5). In the context of the SCHOOL model, molecular mechanisms of this inhibitory action of SARS-CoV FP are similar to those suggested for TCR CP and HIV FP (Fig. 3, Tables 3 and 5).[27-30,57]

As hypothesized, the TCR TM interactions might represent a point of SARS-CoV intervention. If true, the molecular mechanisms revealed can be used in rational antiviral drug design and the development of novel antiviral therapies. I believe that future studies will experimentally prove this hypothesis.

HTLV-1 Pathogenesis

HTLV-1 is a type C complex retrovirus. It infects and immortalizes human CD4+ T-cells in vitro and is associated with the development of adult T-cell leukemia/lymphoma (ATL).[177-180] Recent observations demonstrate an immunomodulatory ability of the HTLV-1 regulatory protein p12[181,182] and suggest roles that T-cell activation may play in the pathogenesis of HTLV-1-induced disease.[177,181-183]

Below, I consider similarities between the HIV gp41 and HTLV-1 gp21 FPs and the Nef and HTLV-1 p12 proteins, respectively and describe my structural and functional predictions related to potential TCR-targeted activities of HTLV-1 FP and p12. Currently, there is no experimental evidence for these predicted activities. Despite this, I believe that future studies will prove the hypotheses made by using the SCHOOL model.

Fusion Peptide

Similarly to HIV gp41 protein,[146,147] the ectodomain of HTLV-1 TM protein (gp21) contains an N-terminally located fusion peptide, a sequence that inserts into target cellular membranes and is well-known to be critical for membrane fusion activity.[184,185] However, in contrast to the HIV FP, there has been no report to date of an immunomodulatory activity of the HTLV-1 FP.

Primary sequence analysis of these two FPs (Table 6) indicates different sequences but reveals an interesting similarity in charged or polar residue distribution patterns with two positively charged residues spaced apart by 8 (HIV FP) or 7 (HTLV-1 FP) amino acids. Considering the structural similarities of both FPs and the fact that T-cells are main target for both viruses, it is reasonable to suggest a TCR-targeted immunomodulatory activity for the HTLV-1 FP. As proposed by the SCHOOL for HIV FP and TCR CP (Fig. 3, Tables 3 and 5),[27-30,57] a potential mode of action of HTLV-1 can involve the TM competition with the TCRα subunit for binding to CD3ε and ζ subunits, thus resulting in TM disconnection/predissociation of the signaling subunits from the remaining receptor complex (Fig. 3, Table 3 and 5). As with the HIV FP, this mechanism of HTLV-1 FP action suggests the existence of an interaction interface in the C-terminal half of the peptide. Within the model, the peptide should prevent formation of signaling-competent CD3ε and ζ oligomers and thus inhibit antigen-dependent T-cell activation, acting similarly in this respect to both TCR CP and HIV FP (Fig. 3, Tables 3, 4 and 5).[27-30,57,75] However, stimulation with anti-CD3 antibodies of these "predissociated" TCRs should still result (Fig. 3)[27-30,57] in receptor triggering and cell activation. As with HIV infection, the model suggests that clinically relevant antibodies (i.e., OKT3) could be used to modulate the affected T-cell response during HTLV-1 infection.

In summary, I propose a new hypothesis that considers the largely unexplored immunomodulatory role of the FP in the HTLV-1 infection and pathogenesis of ATL. If true, this hypothesis will generate new therapeutic targets and opportunities. I also suggest that our current and future clinical knowledge, experience and therapeutic strategies can be potentially transferred in this respect between the HIV- and HTLV-1-related medical conditions.

HTLV-1 p12 Protein

The p12 protein of HTLV-1 is a small oncoprotein that has been shown to have multiple functions. Expression of p12 has been demonstrated to induce nuclear factor of activation of T-cells (NF-AT), increase calcium release and transcriptional factor Stat 5 activation in T-cells suggesting that p12 may alter T-cell signaling.[186-188] Interestingly, p12 is important for viral infectivity in quiescent human peripheral blood lymphocytes (PBLs) and the establishment of persistent infection in rabbits.[189,190] Despite the distinct structures, both retroviral accessory proteins HTLV-1 p12 and HIV Nef are able to modulate TCR-mediated signaling and play a critical role in enhancing viral infectivity in primary lymphocytes and infected animals. It has been recently reported that p12 could complement for effects of Nef on HIV-1 infection of Magi-CCR5 cells, which express CD4, CXCR4 and CCR5 on the surface, or macrophages.[182] Also, the clones of Jurkat cells expressing the highest levels of p12 have been found to exhibit a more rapid rate of cell proliferation than the parental cells.[182] Similarly to HIV Nef, the p12 protein, upon engagement of the TCR, relocalizes to the interface between T-cells and antigen-presenting cells, defined as the immunological synapse (IS).[181] Both Nef and p12 are recruited to the IS, but Nef potentiates TCR signaling[191] while p12 dampens it.[181]

In summary, targeting TCR-mediated signaling seems to be a shared feature of both HIV and HTLV-1 viruses, reflecting probably their similar evolutionary pathway towards their adaptation to the host immune response. Thus, it is possible that similar molecular mechanisms may be involved in TCR-targeting strategies used by Nef and p12 to modulate TCR-mediated signaling pathways. If true, this hypothesis will generate new therapeutic targets (i.e., protein-protein interactions at the interface of p12 and its potential TCR-related partners) and opportunities, similar to those suggested for HIV Nef.

CMV Pathogenesis

To escape from NK cell-mediated surveillance, human CMV interferes with the expression of NKG2D ligands in infected cells. In addition, the virus may keep NK inhibitory receptors engaged by preserving human leukocyte antigen (HLA) class I molecules that have a limited role in antigen presentation.[192] Despite considerable progress in the field, a number of issues regarding the involvement of NK receptors in the innate immune response to human CMV remain unresolved.

Recently, a direct interaction between the human CMV tegument protein pp65 and the NK cell activating receptor NKp30 has been reported.[193] It has been shown that the binding of pp65 to NKp30 is specific and functional. Surprisingly, the recognition of pp65 by NKp30 does not lead to NK cell activation but instead results in a general inhibition mediated by the dissociation of the signaling ζ subunit from the NKp30-ζ receptor complex.[193] This results in the diminishing of activating signals and loss in the ability of NK cells to kill normal, tumor and virus-infected cells.[193]

Within the context of SCHOOL model,[27-30] the reported action of the human CMV pp65 protein may be due to its potential impact on the TM interactions between NKp30 and ζ, leading to disconnection and dissociation of the ζ subunit.[30] This would prevent the formation of signaling-competent ζ oligomers upon ligand stimulation and consequently, inhibit NK cell cytolytic activity (Fig. 2, Table 5) in a manner similar in this respect to the inhibitory action of TCR CP (Fig. 3, Table 3). Primary sequence analysis of the N-terminal end of pp65 shows the existence of multiple positively and negatively charged amino acid residues (Table 6). This pp65 region possibly contains the sequence that mimics the NKp30 or ζ TM domain with the Arg or Asp residues, respectively, that are known to mediate the interaction between recognition NKp30 chain and signaling ζ subunit (see also Chapter 4).[18] However, further experimental studies are needed to confirm the proposed mechanism.

Lessons from Viral Pathogenesis

General issues related to viral pathogenesis in the context of MIRR TM signaling are covered in more detail in Chapter 22. In this section, I briefly describe several important lessons that we can learn from the SCHOOL model-revealed similarity of the molecular mechanisms underlying viral pathogenesis and MIRR signaling-targeted immunomodulatory viral activity important for viral immune escape. I also consider the striking similarities of the molecular mechanisms and basic structural principles that are suggested by the model to explain immunomodulatory effects of viral fusion and accessory proteins and synthetic agents affecting intra- or inter-MIRR protein-protein interactions in the TM or CYTO milieus, respectively.

It seems that in general, viruses use TM-targeted immunomodulatory activity of their fusion proteins mostly during virus entry to suppress the host immune response, whereas modulation of CYTO interactions by using accessory proteins such as HIV Nef and HTLV-1 p12 plays a role in viral replication and enhancing viral infectivity in the host. Thus, our improved understanding of MIRR signaling-targeted immunomodulatory viral activity might allow us to reveal novel targets at these stages of viral pathogenesis.

I believe that lessons that we can learn from viral pathogenesis in the context of the SCHOOL model of immune signaling are very important for our further understanding of the molecular mechanisms used by viruses to infect the host and escape its immune response. I also believe that these lessons are of both fundamental and clinical value. Why?

1. Now we know the molecular mechanisms of inhibitory action of MIRR TM peptides such as TCR TM peptides,[74,77,78,81,93,95-97] NK TM peptides[91] and GPVI TM peptide,[58] as suggested by the SCHOOL model.[27-30,58] We also know that the same mechanisms are very likely to be used in vivo by HIV gp41 FP and also, as predicted by the SCHOOL model, by fusion proteins of other viruses, such as SARS-CoV and HTLV-1 FPs, to suppress the host immune response.[29,30,57] Considering the high specificity and efficiency of viral agents in inhibition of immune receptors in combination with our current knowledge of the protein-protein interactions underlying this process, we can now use modern well-established computational, bioinformatic and synthetic methodologies[82-90] to design and produce highly specific and effective TM-targeted agents that are able to affect specific TM interactions of a targeted MIRR and suppress and/or modulate the MIRR-mediated immune response. These agents would be of great fundamental and clinical value.

   Similar conclusions can be drawn for our ability to use the SCHOOL model of immune signaling and the lessons learned from our current knowledge of the CYTO-targeted viral strategies to design and produce efficient and specific CYTO-targeted agents.

2. According to the SCHOOL model, TCR CP, HIV gp41 FP and, as predicted, SARS-CoV FP and HTLV-1 FP, affect similar TCR TM interactions (Fig. 3, Tables 3 and 5).[27-30,57] Primary sequence analysis of these peptides (Table 6) shows different primary sequences but a striking similarity in charged or polar residue distribution patterns, suggesting that a computational approach combined with the molecular mechanisms of action of these peptides revealed by the SCHOOL model, can and should be used in the rational design of effective immunomodulatory TM-targeted peptides. General well-known principles of designing TM peptides with an ability to insert into the membrane might be readily used at this stage.[86]

3. As suggested by the SCHOOL model (Fig. 2), TM-targeted agents should inhibit MIRR-mediated cell activation induced only by antigen/ligand but not antibodies to MIRR signaling subunits. Indeed, it has been shown for TCR CP[75] and HIV FP[79] that these TM peptides inhibit only antigen-mediated T-cell activation, whereas stimulation with anti-CD3 antibodies in the presence of the peptides still results in functional cell response. Entirely similar considerations can be applied for other viral FPs such as SARS-CoV FP and HTLV-1 FP. Thus, the SCHOOL model suggests that antibodies to MIRR signaling subunits can be used as immunotherapeutics to modulate the affected immune cell response during viral infection.

4. For TCR, considering our selective ability to physically and more importantly, functionally disconnect any particular CD3 and/or ζ signaling subunits from the remaining receptor by using the relevant TM peptides and basic principles of the SCHOOL model of TCR signaling (Fig. 3), we can design, synthesize and use these peptides as a powerful tool to dissect fine molecular mechanisms of viral pathogenesis in the context of TCR signaling.

5. Two unrelated enveloped viruses, HIV and human CMV, use a similar mechanism to modulate the host immune response mediated by two functionally different MIRRs—TCR and NKp30. As predicted by the SCHOOL model, SARS-CoV and HTLV-1 can also use similar mechanisms during virus infection. Intriguingly, as shown in Table 6, similar positively charged residue distribution pattern with two Arg and/or Lys residues spaced apart by 8 amino acids is observed for the FPs of seemingly unrelated viruses such as HIV, Lassa virus (LASV),[194] lymphocytic choriomeningitis virus (LCMV),[194] Mopeia virus (MOPV)[194] and Tacaribe virus (TACV).[194] Thus, it is very likely that similar general immunomodulatory mechanisms can be or are used by other viral and possibly nonviral pathogens (see also Chapter 22). In addition, as with HIV gp41 FP,[149,150] it is promising to apply a similar strategy to block viral entry by using the agents able to interact directly with FPs of other viruses.

Novel Concept of Platelet Inhibition

Damage to the integrity of the vessel wall results in exposure of the subendothelial extracellular matrix, which triggers platelet adhesion and aggregation.[195,196] The consequence of this process is the formation of a thrombus, which prevents blood loss at sites of injury or leads to occlusion and irreversible tissue damage or infarction in diseased vessels.[196] Despite intensive research efforts in antithrombotic drug discovery and development, uncontrolled hemorrhage still remains the most common side effect associated with antithrombotic drugs that are currently in use.

The major physiological function of platelets is hemostasis, prevention of bleeding and the effect of aspirin has established that they are also involved in its pathological variant, thrombosis.[20] Platelets also play a critical role in coronary artery disease and stroke, as evidenced by the well-documented benefits of antiplatelet therapy.[197]

Platelet adhesion, aggregation and activation induced by collagen is critically dependent upon the engagement and clustering of GPVI, a type I transmembrane platelet glycoprotein of about 62 kDa and the major signalling receptor for collagen on platelets (see also Chapter 5).[21,196,198-200] GPVI has no intrinsic signaling capacity and signaling is achieved through the association with its signaling partner, the FcRγ chain.[21] The selective inhibition of GPVI and/or its signaling is thought by most experts in the field to inhibit thrombosis without affecting hemostatic plug formation, thus providing new therapeutic strategies to fight platelet-mediated diseases (see also Chapter 5).[21,201-204] In contrast to antithrombotic drugs that are currently in use, GPVI receptor-specific inhibitors represent an ideal class of clinically suitable antithrombotics. However, despite intensive studies of the GPVI-FcRγ receptor complex,[21,199,205,206] the mechanism of GPVI signaling was not known until very recently when the SCHOOL model was introduced and applied to GPVI triggering and TM signal transduction.[27-30,58] This resulted in the development of a novel concept of platelet inhibition and the invention of new platelet inhibitors within this promising antithrombotic strategy.[58,108] The invented inhibitors are proposed to be useful in the prevention/treatment of thrombosis and other medical conditions involving collagen-induced platelet activation and aggregation as well as in the production of drug-coated medical devices.[58,108]

Within the SCHOOL model, GPVI-mediated platelet activation is a result of the interplay between GPVI-FcRγ TM interactions, the association of two TM Asp residues in the FcRγ homodimer with the TM Arg residue of GPVI,[207] that maintain receptor integrity in platelets under basal conditions and homointeractions between FcRγ subunits, leading to initiation of a signaling response (Table 3, Fig. 4A). Binding of the multivalent collagen ligand to two or more GPVI-FcRγ receptor complexes pushes the receptors to cluster, rotate and adopt an appropriate orientation relative to each other (Fig. 4A, step 1), at which point the trans-homointeractions between FcRγ molecules are initiated. Upon formation of FcRγ signaling oligomers, the Src-family kinases Fyn or Lyn phosphorylate the tyrosine residues in the FcRγ ITAM that leads to TM transduction of the activation signal (Fig. 4A, step 2) and dissociation of FcRγ oligomers and downmodulation of the engaged GPVI subunits (Fig. 4A, step 3). Later, the dissociated oligomeric FcRγ chains can interact with FcRγ subunits of the non-engaged GPVI-FcRγ complexes, resulting in formation of higher-order signaling oligomers and their subsequent phosphorylation, thus providing lateral signal propagation and amplification (not shown).

For the preformed oligomeric receptor complexes described by Berlanga et al,[208] this model suggests that under basal conditions, the overall geometry of the receptor dimer keeps FcRγ chains apart, whereas stimulation by collagen results in breakage of GPVI-GPVI extracellular interactions and reorientation of signaling FcRγ homodimers, thus bringing them into a close proximity and an appropriate relative orientation permissive of initiating the FcRγ homointeractions (Fig. 4A). Thus, the SCHOOL model highlights a striking similarity between the data on the coexistence of mono- and multivalent TCRs[162] or GPVIs[208] in resting T-cells or nonstimulated platelets, respectively and suggests a similar molecular explanation to answer an important and intriguing question raised in these studies: why does the observed basal TCR or GPVI oligomerization not lead to receptor triggering and subsequent T-cell or platelet activation, respectively, whereas agonist-induced receptor crosslinking/clustering does? See also Chapter 12.[27-30,58]

Suggesting how binding to collagen triggers the GPVI-mediated signal cascade at the molecular level, the SCHOOL model of collagen-induced GPVI signaling reveals GPVI-FcRγ TM interactions as a novel therapeutic target for the prevention and treatment of platelet-mediated thrombotic events.[29,30,58] Specific blockade or disruption of these interactions causes a physical and functional disconnection of the subunits (Fig. 4B, Table 3). Antigen stimulation of these "predissociated" receptor complexes leads to clustering of GPVI but not FcRγ subunits. As a result, FcRγ signaling oligomers are not formed, ITAM Tyr residues do not become phosphorylated and the signaling cascade is not initiated. Agents that target GPVI-FcRγ TM interactions may thus represent a novel class of platelet inhibitors. These include, but are not limited to, peptides, peptide derivatives and compositions and nonpeptide small molecule inhibitors. Preliminary experimental results[58,108] provided support for this novel concept of platelet inhibition and demonstrated that incubation of whole blood samples with a peptide corresponding to the TM domain of GPVI (Gly-Asn-Leu-Val-Arg-Ile-Cys-Leu-Gly-Ala-Val) at a final concentration of 100 μM prior to addition of collagen (10 and 20 μg/ml) or convulxin (10 ng/ml) leads to a 30-60% reduction in both the percentage of P-selectin-positive platelets and the expression of the platelet activation markers, P-selectin and PAC-1 (Sigalov AB, Barnard, MR, Frelinger AL, Michelson AD, unpublished results). This effect is specific: platelet activation via ADP (20 μM) is not affected by the peptide. As assumed by the SCHOOL model, this peptide penetrates the platelet membrane and competitively binds to the FcRγ TM domain, thus replacing GPVI receptor from its interaction with the signaling FcRγ subunit and resulting in "predissociation" of the GPVI—FcRγ receptor complex (Fig. 4B). Notably, a control peptide containing a single amino acid substitution (Arg to Ala) does not display inhibitory activity, a phenomenon predicted by the SCHOOL model since this peptide cannot compete with GPVI for binding with FcRγ in the TM milieu.

In conclusion, a combination of basic principles of the SCHOOL model with a recently reported computational design of peptides that target TM helixes in a sequence-specific manner[85] and other well-established techniques to search for the relevant TM mutations or small molecule disruptors as well as to synthesize and optimize TM peptides and peptidomimetics[1-12,34,76,81-90] opens up a new avenue for designing novel platelet inhibitors, making the proposed strategy both feasible and of great fundamental and clinical value. Combining breakthrough scientific ideas and advances in different fields[28-31,33,57,58] and the high market potential,[209] the suggested technology opens new perspectives in innovative antithrombotic drug discovery and development.

Inflammatory Bowel Diseases: a Novel Treatment Strategy

As another interesting application of the SCHOOL model that challenges its predictive power, I describe my prognosis related to the use of TM- and possibly CYTO-targeted agents as therapeutics to treat inflammatory bowel diseases (IBDs). Briefly, intestinal inflammation in colitic severe combined immunodeficiency (SCID) mice has been recently shown to be characterized by significant increase of CD4+NKG2D+ T-cells.[210] As also demonstrated,[210] neutralizing anti-NKG2D mAb treatment prevents or ameliorates the development of colitis primarily by inhibiting the expansion and/or infiltration of pathogenic T-cells in the colon and secondarily by inhibiting the development of pathogenic Th1 cells. The authors concluded that targeting of NKG2D signaling in NKG2D-expressing pathogenic CD4+ T-cells may be a useful strategy for the treatment of Th1-mediated chronic intestinal inflammation such as Crohn's disease.[210]

I suggest that TM-targeted agents such as the TM peptides designed by using the SCHOOL model-driven TM-targeted strategy should have specific NKG2D-inhibitory activity and therefore can be used as promising therapeutics to prevent and/or treat IBDs (Table 4). Future studies will prove or disprove this hypothesis.

Cytoplasmic Homointeractions as Immunotherapeutic Targets

Main Concept

As mentioned above, the CYTO domains of the vast majority of MIRR signaling subunits, namely, CD3e, CD3d, CD3g, ζ, Igα, Igβ and FcRγ, have been recently shown to represent a new class of intrinsically disordered proteins (IDPs, see also Chapter 12).[31-33] By definition, IDPs (or natively unfolded, or intrinsically unstructured) are proteins that lack a well-defined ordered structure under physiological conditions in vitro, i.e., neutral pH and room temperature.[211] A highly flexible, random coil-like conformation is the native and functional state for many proteins known to be involved in cell signaling.[212-214]

In addition, intrinsically disordered regions of human plasma membrane proteins have been very recently demonstrated to preferentially occur in the cytoplasmic segment.[215] Finally, it has been suggested that protein phosphorylation, one of the critical and obligatory events in cell signaling, occurs predominantly within intrinsically disordered protein regions.[216] My major assumption is that a flexible, random-coil conformation of the MIRR signaling subunit CYTO domains plays an important role in MIRR triggering and TM signaling.[27-29,31-33] I also suggest that the CYTO domains of those MIRR signaling subunits that have not been studied so far (e.g., DAP12, DAP10 and FcεRIβ), are IDPs as well. Future studies will prove or disprove this hypothesis.

Surprisingly, all intrinsically disordered CYTO domains studied exist under physiological conditions as specific oligomers (mostly, dimers), as I discovered in 2001 and published in 2004[31] and even more interestingly, these IDPs do not undergo a transition between disordered and ordered states upon dimerization.[31-33] This specific dimerization is distinct from nonspecific aggregation behavior seen in many systems. These findings oppose the generally accepted view on the behavior of IDPs, providing first evidence for the existence of specific dimerization interactions for IDP species and thus opening a new line of research in this new and quickly developing field of IDPs. The unusualness and uniqueness of the discovered biophysical phenomenon that was found to be a general phenomenon with all CYTO domains studied in this work,[31] led me to hypothesize that the homointeractions between MIRR signaling subunits represent the missing piece in the puzzle of MIRR triggering and TM signal transduction and to develop the SCHOOL model (see also Chapter 12).[28-30,57,58]

Since it was first published in 2004,[28] the SCHOOL model has revealed inter-MIRR CYTO homointeractions as important therapeutic targets as well as points of great interest to study molecular mechanisms underlying the MIRR-mediated cell response in health and disease (Figs. 1 and 5).[27-30,57,58] Within the model, upon antigen/ligand stimulation, these interactions represent one of three major driving forces of MIRR triggering signal (Table 1, Fig. 1B, see also Chapter 12). As suggested by the SCHOOL model, specific blockade of the interreceptor CYTO homointeractions between MIRR signaling subunits by CYTO-targeted agents or site-specific point mutations within the dimerization/oligomerization interfaces prevents formation of competent signaling oligomers (Figs. 1 and 5) and initiation of a MIRR-mediated cell response. Similar to Target I, the intra-MIRR TM interactions, modulation of the inter-MIRR homointeractions between particular signaling cytoplasmic domains might allow us to modulate the ligand-induced cell response. In addition, our ability to selectively prevent the formation of signaling oligomers of particular subunit(s) might also prove to be an important tool in functional studies of MIRRs. Peptides and their derivatives, small molecule disruptors of protein-protein interactions, site-specific mutations and other similar agents/modifications can be used to affect the MIRR CYTO interactions. As mentioned above,

SCHOOL model and new therapeutic targets: Target II

Homointeractions between cytoplasmic domains of signaling subunits

A. Antigen-stimulated MIRR signaling   B. Antibody-stimulated MIRR signaling

NO homointeractions
NO signaling oligomers
NO activation signal

NO homointeractions
NO signaling oligomers
NO activation signal

■ **Cytoplasmic-targeted agents (peptides, small molecule disruptors) and site-specific mutations**

Figure 5. Target II. Cytoplasmic homointeractions between MIRR signaling subunits. This is a simplified graphical illustration of the molecular mechanisms underlying proposed intervention by cytoplasmic-targeted agents (peptides and their derivatives, small molecule disruptors of protein-protein interactions, etc) and site-specific mutations. Specific blockade of homointeractions between signaling subunits is proposed to prevent formation of signaling oligomers, thus inhibiting antigen-dependent immune cell activation (A). Within the model, in contrast to transmembrane-targeted agents (Target I, Fig. 2), stimulation of MIRRs with cross-linking antibodies to signaling subunit in the presence of cytoplasmic-targeted agents should not result in receptor triggering and cell activation (B). The proposed cytoplasmic-targeted strategies can be used not only to inhibit but also to modulate MIRR-mediated transmembrane signal transduction, thus modulating the immune response (see main text for details). Abbreviations and symbols as in Figure 1.

methods of computational design, synthesis and optimization of peptides and peptidomimetics as well as high-throughput screening techniques to search for the relevant mutations or small molecule disruptors are currently developed and well-established,[1-11,56,217-221] thus making the proposed CYTO-targeted approach both feasible and of great fundamental and clinical value.

Importantly, in contrast to TM-targeted agent-affected MIRRs that can be still activated by specific antibodies (Fig. 2B), stimulation of CYTO-targeted agent-affected MIRRs with specific antibodies that trigger cell activation should not result in MIRR triggering and generation of the activation signal (Fig. 5).

Thus, I suggest that like intra-MIRR TM interactions, the interreceptor CYTO homointeractions between MIRR signaling subunits represent extremely important points of control in MIRR triggering and cell activation. Since now we can use the SCHOOL model to design the CYTO-targeted agents effective in inhibition and/or modulation of MIRR-mediated TM signaling (Figs. 1 and 5, see also Chapter 12)[27-29] and have a powerful and well-controlled influence upon MIRR-mediated cell activation, thus controlling the immune response. The relevant

CYTO-targeted agents for any particular member of MIRR family can be readily designed using the SCHOOL model and our knowledge about structural organization of this receptor.

Evidence: Cytoplasmic Agents and Immune Cell Activation

Since homooligomerization of the MIRR signaling subunit CYTO domains was discovered[31] and these CYTO homointeractions were suggested to represent an important therapeutic target,[27-29] no direct experimental evidence has been reported to support this hypothesis. However, there is a growing line of indirect evidences indicating the importance of CYTO domains in functionally relevant homooligomerization of other receptors in vivo and demonstrating that the SCHOOL model-driven MIRR CYTO-targeted strategy using a variety of CYTO-targeted agents and/or mutations (Fig. 5) is technologically feasible and can be readily applied in both fundamental and clinical applications. These findings are mostly related to the field of SR triggering and TM signaling and will be described below.

Mutations

Fas (CD95, APO-1, TNFRSF6) is a tumor necrosis factor (TNF) receptor superfamily member that directly triggers apoptosis and contributes to the maintenance of lymphocyte homeostasis and prevention of autoimmunity.[222] Although Fas-associated death domain (FADD) and caspase-8 have been identified as key intracellular mediators of Fas signaling, it is not clear how recruitment of these proteins to the Fas death domain (DD) leads to activation of caspase-8 in the receptor signaling complex.[222,223] Recently, ligand-induced formation of surface receptor oligomers has been reported for Fas receptor.[224] A cytoplasmic DD of this SR, upon ligand stimulation, binds to the homologous DD of the adaptor protein FADD and homooligomerizes, thus initiating the caspase signaling cascade (Fig. 6A). Interestingly, an autoimmune lymphoproliferative syndrome-linked mutation in Fas cytoplasmic domain (T225K) impairs receptor oligomerization and inhibits Fas-mediated signaling but retains the ability to interact with FADD (Fig. 6A).[224] This suggests that homointeractions between signaling cytoplasmic tails themselves play an important role in ligand-induced surface receptor oligomerization and subsequent signaling.

This interesting finding supports the proposed MIRR CYTO-targeted strategy and provides a promising direction for future research. One can also hypothesize that similar mutations located in the CYTO domains of MIRR signaling subunits might occur naturally in MIRR-mediated disorders and disturb the homooligomerization interface(s), thus preventing formation of competent signaling subunit oligomers and MIRR triggering.

Cytoplasmic Peptides and Peptidomimetics

There is growing line of evidence indicating that CYTO peptides and peptidomimetics can be successfully used to target CYTO hetero- or homointeractions between entire protein molecules or the CYTO domains of TM proteins.[164,225-230] This means that once we can identify a new promising therapeutic CYTO target, it is technologically feasible to design, synthesize and use the relevant peptide-based agents, peptidomimetics and small molecules (or screen for the appropriate agents by using high throughput screening assays). Selected examples of CYTO-targeted agents used to inhibit CYTO protein-protein interactions, thus modifying the functional response, are considered in more detail below.

Myeloid differentiation factor 88 (MyD88) is a critical adaptor protein that recruits signaling proteins to TLR/IL-1 receptor (IL-1R) superfamily and thus plays a crucial role in the signaling pathways triggered by these receptors in innate host defense.[231,232] A critical event in MyD88-trgiggered signaling pathway is homodimerization of MyD88 mediated by its TLR/IL-1R translation initiation domain (TIR) that is able to heterodimerize with the receptor and homodimerize with another MyD88 molecule (Fig. 6B).[228,229,232] Dimerization of MyD88 favors the recruitment of downstream signaling molecules such as two IL-1R-associated kinases (IRAKs): IRAK1 and IRAK4 (Fig. 6B). Recently, eptapeptides that mimic the BB-loop region of the conserved TIR domain of MyD88, have been shown to effectively inhibit homodimerization with either the isolated TIR or full-length MyD88 (Fig. 6B).[229] The authors also demonstrated

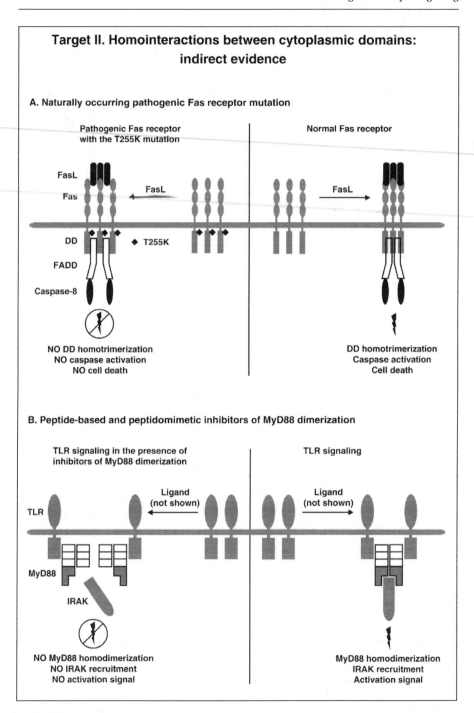

Target II. Homointeractions between cytoplasmic domains: indirect evidence

A. Naturally occurring pathogenic Fas receptor mutation

Pathogenic Fas receptor with the T255K mutation

Normal Fas receptor

FasL

Fas

FasL

FasL

DD   ◆ T255K

FADD

Caspase-8

NO DD homotrimerization
NO caspase activation
NO cell death

DD homotrimerization
Caspase activation
Cell death

B. Peptide-based and peptidomimetic inhibitors of MyD88 dimerization

TLR signaling in the presence of inhibitors of MyD88 dimerization

TLR signaling

TLR

Ligand (not shown)

Ligand (not shown)

MyD88

IRAK

NO MyD88 homodimerization
NO IRAK recruitment
NO activation signal

MyD88 homodimerization
IRAK recruitment
Activation signal

Figure 6, legend viewed on following page.

Figure 6, viewed on previous page. Indirect evidence for importance of the proposed cytoplasmic-targeted strategy. A) Fas apoptosis signaling by a normal Fas receptor and the receptor with the Fas T225K mutation that naturally occurs in patients with the autoimmune lymphoproliferative syndrome (ALPS). In contrast to all other ALPS-associated Fas DD mutations, this pathogenic mutation specifically disrupts homooligomerization of the cytoplasmic tails of the receptor but retains the ability to interact with FADD [Siegel RM, Muppidi JR, Sarker M et al. SPOTS: signaling protein oligomeric transduction structures are early mediators of death receptor-induced apoptosis at the plasma membrane. J Cell Biol 2004; 167:735-744]. As shown, the blockade of the cytoplasmic homointeractions does not allow full caspase-8 activation and apoptosis induction, thus revealing these protein-protein interactions as a therapeutic target. Abbreviations: FasL, Fas Ligand; DD, Death Domain; FADD, Fas-associated Death Domain protein. B) TLR signaling in the absence or presence of peptide-based and peptidomimetics inhibitors of MyD88 dimerization. As reported [Loiarro M, Capolunghi F, Fanto N et al. Pivotal Advance: Inhibition of MyD88 dimerization and recruitment of IRAK1 and IRAK4 by a novel peptidomimetic compound. J Leukoc Biol 2007; 82:801-810; Loiarro M, Sette C, Gallo G et al. Peptide-mediated interference of TIR domain dimerization in MyD88 inhibits interleukin-1-dependent activation of NF-κB. J Biol Chem 2005; 280:15809-15814], cell-permeable analogs of MyD88 peptides derived from the TIR domain of MyD88 as well as a synthetic peptidomimetic compound effectively inhibit homodimerization of MyD88 TIR domains, significantly reducing IL-1 signaling in vitro and dose-dependently inhibiting IL-1β-induced production of IL-6 in treated mice. This suggests that inhibition of MyD88 homodimerization in the cytoplasmic milieu may have therapeutic potential. Abbreviations: TLR, toll-like receptor; MyD88, myeloid differentiation factor 88; IRAK, interleukin-1 (IL-1) receptor-associated kinase; TIR, toll/IL-1 receptor domain.

that a cell permeable analog of MyD88 eptapeptide inhibits homodimerization of MyD88 TIR domains in an in vitro cell system and significantly reduces IL-1 signaling, indicating that the MyD88 homodimerization interface is a good target for specific inhibition of MyD88-mediated signaling in vivo.[229]

Importantly, a synthetic peptidomimetic compound modeled after the structure of a heptapeptide in the BB-loop of the MyD88-TIR domain has been shown very recently to inhibit MyD88 dimerization in coimmunoprecipitation experiments.[228] This effect is specific for homodimerization of the TIR domains and does not affect homodimerization of the DDs. The agent causes inhibition of IL-1β-mediated activation of NF-κB transcriptional activity.[228] After oral administration, the compound results in dose-dependent inhibition of IL-1β-induced production of IL-6 in treated mice.[228] In addition, it suppresses B-cell proliferation and differentiation into plasma cells in response to CpG-induced activation of TLR9, a receptor that requires MyD88 for intracellular signaling.[228] These data indicate that the peptidomimetic compound studied blocks IL-1R/TLR signaling by interfering with MyD88 homodimerization. This suggests that inhibition of MyD88 homodimerization in the CYTO milieu by peptide-based agents or peptidomimetics may have therapeutic potential in treatment of chronic inflammatory diseases.[228]

These findings strongly support basic principles of the SCHOOL model-suggested strategy with its new point of intervention to inhibit/modulate MIRR triggering and the immune response, the MIRR signaling subunit CYTO homointeractions. As with TM-targeted strategy, our current understanding of MIRR structure and the nature and specificity of antigen/ligand-induced homointeractions between receptor signaling subunits not only allows us to inhibit these protein-protein interactions but also to modulate the interactions by a sequence-based approach using corresponding peptides and/or their derivatives. Peptidomimetics and small molecules can be used for these purposes, as well. Strengthening/weakening and/or selective inhibition of the association between particular signaling subunits might allow us not to inhibit, but rather to modulate the ligand-induced cell response. In addition, selective functional inhibition of particular signaling subunits represents an invaluable tool in studies of MIRR-mediated TM signaling and cell activation.

As another example, the processes by which Nef mediates the redistribution of CD80 and CD86 in human monocytic cells can be considered.[164] The endocytic mechanism used to trigger internalization of CD80 and CD86 is known to involve Nef binding to the CYTO tails of these target proteins.[164] In an inhibition assay, synthetic peptides corresponding to the CYTO domains of CD80 or CD86 have been demonstrated to inhibit Nef binding to the same peptides immobilized on polystyrene plates.[164] Introduction of these CYTO peptides into Nef-expressing U937 cells using the Chariot reagent at 4°C causes substantial reduction in the loss of CD80 or CD86, respectively, from the cell surface of Nef-expressing cells,[164] thus proving the principal feasibility and the utility of the CYTO-targeted strategy suggested by the SCHOOL model.

Interestingly, unlike wild-type Nef, the Nef D123G mutant has been shown to lose its ability to mediate efficient internalization of cell-surface CD80 or CD86, or bind to the CYTO peptides of CD80 or CD86.[164] As mentioned before, mutation of a conserved D123 residue affects the ability of Nef to form dimers and results in impairment of Nef biological functions such as MHC class I downmodulation and enhancement of viral infectivity, indicating that the oligomerization of Nef may be critical for its multiple functions.[163] In this regard, I suggest that the impaired function of the Nef D123G mutant with regard to downmodulation of CD80/CD86 can be explained by its inability to form oligomers. If true, this means that the rational design of antiviral agents that are able to target CYTO homointeractions in Nef oligomers may represent an attractive target in the CYTO milieu, not only with regard to Nef-mediated modulation of TCR triggering and TM signaling, but also with respect to other Nef biological functions.

Peptide-based CYTO-targeted strategy has been also successfully applied to modulate outside-in TM signaling mediated by the platelet receptors such as GPIb/IX/V,[225] GPIIb[226] and the megakaryocyte- and platelet-specific integrin αIIbβ3.[227]

The platelet GPIb/IX/V receptor plays a key role in platelet adhesion at sites of vascular damage through its interaction with subendothelial-bound von Willebrand factor (VWF).[233,234] However, despite the crucial role that the GPIb/IX/V receptor complex plays in hemostasis, the molecular mechanisms of its signaling are not completely understood. The GPIb/IX/V complex consists of four subunits, namely, GPIbα, GPIbβ, GPIX and GPV. An amino acid sequence in the CYTO domain of the GPIbβ subunit between residues R151 and A161 has been shown to be highly conserved across species and plays an important physiological role.[225] It has been also reported[225] that a synthetic CYTO-targeted agent, the cell-permeable palmitylated peptide corresponding to this sequence, completely inhibits low-dose thrombin- and ristocetin-induced aggregation in washed platelets, significantly reduces thromboxane (TXA) production in platelets stimulated by thrombin compared with collagen, substantially decreases activation of the integrin αIIbβ3 in response to thrombin and significantly reduces the adhesion of washed platelets to VWF under static conditions and the velocity of platelets rolling on VWF. This demonstrates an effective impact of this peptide-based CYTO-targeted agent on platelet function in terms of rolling velocity, adhesion, spreading, signaling to αIIbβ3 and aggregation.

The integrin αIIbβ3 plays an important role in hemostasis mediating platelet adhesion, aggregation and bidirectional signaling.[235,236] Little is known about the molecular mechanisms underlying the regulation of αIIb-mediated outside-in signaling. Recently, it has been shown that this signaling is enhanced in platelets of a patient lacking the terminal 39 residues of the β3 CYTO domain, as detected by thromboxane production and granule secretion and requires ligand cross-linking of αIIbβ3 and platelet aggregation.[227] A synthetic CYTO-targeted agent, the cell-permeable palmitylated β3 peptide corresponding to the CYTO sequence R724-R734, has been demonstrated to effectively and specifically inhibit this outside-in signaling,[227] thus supporting basic principles and feasibility of the SCHOOL model-suggested CYTO-targeted strategy.

All integrin α subunits are known to contain a highly conserved KXGFFKR motif in their CYTO domains that plays a crucial role in the regulation of integrin affinity for their ligands.[226,237-239] A synthetic CYTO-targeted agent, the palmitylated peptide corresponding to the K989-R995 sequence of the CYTO domain of the platelet integrin GPIIb (αIIb) subunit has been shown to specifically induce platelet activation and aggregation equivalent to that of strong agonists such as thrombin.[226] The authors conclude that this lipid-modified peptide imitates the CYTO domain of GPIIb and, in a highly specific and effective manner, initiates parallel but independent signaling pathways, one leading to ligand binding and platelet aggregation and the other to intracellular signaling events such as TXA2 synthesis and secretion.[226]

An example of using a synthetic peptide to inhibit protein-protein homointeractions in the intracellular milieu has been recently reported in studies of Ebola virus (EBOV), a filovirus that causes sporadic outbreaks of a fatal hemorrhagic fever in Africa.[230,240] Viral protein 30 (VP30), one of seven structural proteins of this enveloped virus,[240] is the constituent of the nucleocapsid and represents an EBOV-specific transcription activation factor.[241] The essential role of homooligomerization for the function of VP30 and the significance of the self-assembly of VP30 for viral transcription and propagation have been recently reported.[230] Interestingly, it has been also shown that the homooligomerization of VP30 can be dose dependently inhibited by a 25-mer peptide derived from the presumed oligomerization interface region.[230] Importantly, when this peptide is transfected into EBOV-infected cells, the peptide inhibits viral replication, suggesting that inhibition of VP30 oligomerization represents a target for EBOV antiviral drugs.[230] This confirms that, as proposed by the SCHOOL model for MIRR-mediated TM signaling and cell activation,[27-29] protein-protein homodimerization/homooligomerization interface(s) can represent an important point of intervention in the CYTO milieu and be targeted by synthetic peptides, their derivatives and peptidomimetics.

Another potential application of the CYTO-targeted strategy involves the use of CYTO-targeted agents to modulate TLR4 signaling. This receptor is activated by monophosphoryl lipid A, derived from the active moiety (lipid A) of bacterial endotoxin (lipopolysaccharide, LPS). As recently demonstrated,[242] LPS binds to a secreted glycoprotein MD-2, which in turn binds to TLR4 and induces aggregation and signal transduction. It has been also shown that TLR4 can form homodimers.[243] Despite both TLR4 monomers and dimers are able to activate NF-κB, this activation is significantly enhanced upon homodimerization.[243] However, NF-κB activation by TLR4 monomer, but not homodimer, is completely inhibited by dominant negative MyD88, suggesting that TLR4 homodimers and monomers can activate NF-κB through different mechanisms.[243] Using the protein complementation assay, a novel method to detect protein-protein interactions in vivo,[244] the TLR4 homodimerization has been shown to be mediated by the TLR4 CYTO domain.[245] I suggest that, similar to other applications mentioned above, CYTO-targeted agents can be used to modulate TLR4-mediated signaling and cell activation, thus modulating the host immune response to LPS.

Conclusions

Despite numerous models of MIRR signaling suggested for particular MIRRs and a growing interest in targeting MIRR signaling as a potential treatment strategy for many immune disorders, the molecular mechanisms that underlie MIRR triggering and subsequent TM signal transduction were unknown for a long time, preventing our improved understanding of these fundamentally important processes and therefore the development of novel pharmacological approaches.

Discovery of an unusual and unique biophysical phenomenon, the existence of specific homointeractions between the intrinsically disordered CYTO domains of MIRR signaling subunits,[31,33] defined the last piece in the puzzle of MIRR triggering and TM signaling and led me to the development of a general model of MIRR-mediated immune cell activation, the SCHOOL model.[27-30,57,58] Suggesting MIRR triggering as an outcome of ligand-induced interplay between three major driving forces represented by well-defined protein-protein interactions that strikingly fall within the similar micromolar affinity range and are characterized by relatively rapid kinetics, the model finally unravels a long-standing mystery of MIRR-mediated TM signal transduction. Importantly, assuming that the molecular mechanisms underlying TM signaling and cell activation mediated by all receptors that belong to the MIRR family are similar, the SCHOOL model can be readily applied to any particular member of this receptor family. In doing so, the model suggests molecular mechanisms for the vast majority of unexplained immunological observations accumulated to date (see also Chapter 12) and reveals novel universal therapeutic targets for a diverse variety of disorders mediated by immune cells, thus opening new horizons in both fundamental and clinical research in different fields such as immunology, structural biology, virology, hematology and others.

My central hypothesis is that the similar structural architecture of the MIRRs dictates similar mechanisms of MIRR triggering and subsequent signaling and cell activation and therefore suggests the existence of similar therapeutic targets in seemingly unrelated diseases. This makes possible the development of novel pharmacological approaches as well as the transfer of clinical knowledge, experience and therapeutic strategies between various immune disorders. In addition, this hypothesis significantly improves our understanding of the immunomodulatory activity of many human viruses. Thus, the lessons learned from the SCHOOL model and viral pathogenesis indicate that a general drug design approach may be used to treat a variety of different and seemingly unrelated immune diseases. The model unraveled the striking similarity of the molecular mechanisms underlying immunomodulatory activities of the TCR TM peptides first introduced by Manolios et al in 1997[78] and viral fusion peptides that appeared to be used by different viruses not only to entry target cells but also to modulate and escape the host immune response. This suggests the possibility to design, synthesize and apply highly specific and effective therapeutic agents and strongly supports the feasibility, utility and both fundamental and clinical importance of the TM-targeted strategy suggested by the SCHOOL model.

Application of this model to the platelet collagen receptor GPVI has already resulted in the development of a novel concept of platelet inhibition and the invention of novel platelet inhibitors. Importantly, the similar basic principles based on our current knowledge of the structural assembly of MIRRs and the molecular mechanisms of MIRR signaling suggested by the SCHOOL model were used to explain immunomodulatory activity of TCR TM peptides and to design, synthesize and apply new GPVI-targeted platelet inhibitors. Again, within the model, a similar approach can be applied to any particular receptor of the MIRR family and therefore to any disease or medical conditions mediated by this receptor. This is not only a comprehensive example of the usability and predictive power of the SCHOOL model but also supports my central hypothesis in the context of our ability to develop general pharmacological approaches and transfer clinical knowledge, experience and therapeutic strategies between seemingly disparate immune-mediated diseases.

In summary, I would like to highlight that the SCHOOL model (described in Chapter 12 in more detail) provides a set of basic principles underlying MIRR-mediated signaling and indicates that a general drug design approach could be used to treat many different, seemingly unrelated, immune diseases. Considering the multiplicity and diversity of the MIRRs involved in the pathogenesis of numerous human diseases, the proposed model can contribute significantly to the improvement of existing therapies and the design of new therapeutic strategies for malignancies, diverse immune system disorders, including those with infections caused by various viruses and other MIRR-mediated medical conditions.

Acknowledgements

I would like to thank Walter M. Kim for his critical reading of this manuscript.

References

1. Fry DC. Protein-protein interactions as targets for small molecule drug discovery. Biopolymers 2006; 84:535-552.
2. Ryan DP, Matthews JM. Protein-protein interactions in human disease. Curr Opin Struct Biol 2005; 15:441-446.
3. Toogood PL. Inhibition of protein-protein association by small molecules: Approaches and progress. J Med Chem 2002; 45:1543-1558.
4. Fletcher S, Hamilton AD. Targeting protein-protein interactions by rational design: Mimicry of protein surfaces. J R Soc Interface 2006; 3:215-233.
5. Hershberger SJ, Lee SG, Chmielewski J. Scaffolds for blocking protein-protein interactions. Curr Top Med Chem 2007; 7:928-942.
6. Loregian A, Palu G. Disruption of protein-protein interactions: Towards new targets for chemotherapy. J Cell Physiol 2005; 204:750-762.
7. Sillerud LO, Larson RS. Design and structure of peptide and peptidomimetic antagonists of protein-protein interaction. Curr Protein Pept Sci 2005; 6:151-169.

8. Che Y, Brooks BR, Marshall GR. Development of small molecules designed to modulate protein-protein interactions. J Comput Aided Mol Des 2006; 20:109-130.
9. Berg T. Modulation of protein-protein interactions with small organic molecules. Angew Chem Int Ed Engl 2003; 42:2462-2481.
10. Archakov AI, Govorun VM, Dubanov AV et al. Protein-protein interactions as a target for drugs in proteomics. Proteomics 2003; 3:380-391.
11. Veselovsky AV, Ivanov YD, Ivanov AS et al. Protein-protein interactions: Mechanisms and modification by drugs. J Mol Recognit 2002; 15:405-422.
12. Pagliaro L, Felding J, Audouze K et al. Emerging classes of protein-protein interaction inhibitors and new tools for their development. Curr Opin Chem Biol 2004; 8:442-449.
13. Rudd CE. Disabled receptor signaling and new primary immunodeficiency disorders. N Engl J Med 2006; 354:1874-1877.
14. Takai T. Fc receptors and their role in immune regulation and autoimmunity. J Clin Immunol 2005; 25:1-18.
15. Takai T. Fc receptors: their diverse functions in immunity and immune disorders. Springer Semin Immunopathol 2006; 28:303-304.
16. Gomes MM, Herr AB. IgA and IgA-specific receptors in human disease: structural and functional insights into pathogenesis and therapeutic potential. Springer Semin Immunopathol 2006; 28:383-395.
17. Honda Z. Fcepsilon- and Fcgamma-receptor signaling in diseases. Springer Semin Immunopathol 2006; 28:365-375.
18. Biassoni R, Cantoni C, Falco M et al. Human natural killer cell activating receptors. Mol Immunol 2000; 37:1015-1024.
19. Moretta A, Bottino C, Vitale M et al. Activating receptors and coreceptors involved in human natural killer cell-mediated cytolysis. Annu Rev Immunol 2001; 19:197-223.
20. Clemetson KJ. Platelet receptors and their role in diseases. Clin Chem Lab Med 2003; 41:253-260.
21. Moroi M, Jung SM. Platelet glycoprotein VI: Its structure and function. Thromb Res 2004; 114:221-233.
22. Manolios N, Bonifacino JS, Klausner RD. Transmembrane helical interactions and the assembly of the T-cell receptor complex. Science 1990; 249:274-277.
23. Call ME, Pyrdol J, Wiedmann M et al. The organizing principle in the formation of the T-cell receptor-CD3 complex. Cell 2002; 111:967-979.
24. Michnoff CH, Parikh VS, Lelsz DL et al. Mutations within the NH2-terminal transmembrane domain of membrane immunoglobulin (Ig) M alters Ig alpha and Ig beta association and signal transduction. J Biol Chem 1994; 269:24237-24244.
25. Daeron M. Fc receptor biology. Annu Rev Immunol 1997; 15:203-234.
26. Borrego F, Kabat J, Kim DK et al. Structure and function of major histocompatibility complex (MHC) class I specific receptors expressed on human natural killer (NK) cells. Mol Immunol 2002; 38:637-660.
27. Sigalov A. Multi-chain immune recognition receptors: Spatial organization and signal transduction. Semin. Immunol 2005; 17:51-64.
28. Sigalov AB. Multichain immune recognition receptor signaling: Different players, same game? Trends Immunol 2004; 25:583-589.
29. Sigalov AB. Immune cell signaling: A novel mechanistic model reveals new therapeutic targets. Trends Pharmacol Sci 2006; 27:518-524.
30. Sigalov AB. Transmembrane interactions as immunotherapeutic targets: Lessons from viral pathogenesis. Adv Exp Med Biol 2007; 601:335-344.
31. Sigalov A, Aivazian D, Stern L. Homooligomerization of the cytoplasmic domain of the T-cell receptor zeta chain and of other proteins containing the immunoreceptor tyrosine-based activation motif. Biochemistry 2004; 43:2049-2061.
32. Sigalov AB, Aivazian DA, Uversky VN et al. Lipid-binding activity of intrinsically unstructured cytoplasmic domains of multichain immune recognition receptor signaling subunits. Biochemistry 2006; 45:15731-15739.
33. Sigalov AB, Zhuravleva AV, Orekhov VY. Binding of intrinsically disordered proteins is not necessarily accompanied by a structural transition to a folded form. Biochimie 2007; 89:419-421.
34. Jones S, Thornton JM. Principles of protein-protein interactions. Proc Natl Acad Sci USA 1996; 93:13-20.
35. Davis MM, Boniface JJ, Reich Z et al. Ligand recognition by alpha beta T-cell receptors. Annu Rev Immunol 1998; 16:523-544.
36. Bormann BJ, Engelman DM. Intramembrane helix-helix association in oligomerization and transmembrane signaling. Annu Rev Biophys Biomol Struct 1992; 21:223-242.

37. Finger C, Volkmer T, Prodohl A et al. The stability of transmembrane helix interactions measured in a biological membrane. J Mol Biol 2006; 358:1221-1228.
38. Andersen PS, Geisler C, Buus S et al. Role of the T-cell receptor ligand affinity in T-cell activation by bacterial superantigens. J Biol Chem 2001; 276:33452-33457.
39. Garcia KC, Tallquist MD, Pease LR et al. Alphabeta T-cell receptor interactions with syngeneic and allogeneic ligands: Affinity measurements and crystallization. Proc Natl Acad Sci USA 1997; 94:13838-13843.
40. Torigoe C, Inman JK, Metzger H. An unusual mechanism for ligand antagonism. Science 1998; 281:568-572.
41. Miura Y, Takahashi T, Jung SM et al. Analysis of the interaction of platelet collagen receptor glycoprotein VI (GPVI) with collagen. A dimeric form of GPVI, but not the monomeric form, shows affinity to fibrous collagen. J Biol Chem 2002; 277:46197-46204.
42. Pitcher LA, Mathis MA, Young JA et al. The CD3 gammaepsilon/deltaepsilon signaling module provides normal T-cell functions in the absence of the TCR zeta immunoreceptor tyrosine-based activation motifs. Eur J Immunol 2005; 35:3643-3654.
43. Pitcher LA, van Oers NS. T-cell receptor signal transmission: Who gives an ITAM? Trends Immunol 2003; 24:554-560.
44. Pike KA, Baig E, Ratcliffe MJ. The avian B-cell receptor complex: Distinct roles of Igalpha and Igbeta in B-cell development. Immunol Rev 2004; 197:10-25.
45. Storch B, Meixlsperger S, Jumaa H. The Ig-alpha ITAM is required for efficient differentiation but not proliferation of preB-cells. Eur J Immunol 2007; 37:252-260.
46. Gazumyan A, Reichlin A, Nussenzweig MC. Ig beta tyrosine residues contribute to the control of B-cell receptor signaling by regulating receptor internalization. J Exp Med 2006; 203:1785-1794.
47. Lin S, Cicala C, Scharenberg AM et al. The Fc(epsilon)RIbeta subunit functions as an amplifier of Fc(epsilon)RIgamma-mediated cell activation signals. Cell 1996; 85:985-995.
48. Sanchez-Mejorada G, Rosales C. Signal transduction by immunoglobulin Fc receptors. J Leukoc Biol 1998; 63:521-533.
49. Lysechko TL, Ostergaard HL. Differential Src family kinase activity requirements for CD3 zeta phosphorylation/ZAP70 recruitment and CD3 epsilon phosphorylation. J Immunol 2005; 174:7807-7814.
50. Kuhns MS, Davis MM. Disruption of extracellular interactions impairs T-cell receptor-CD3 complex stability and signaling. Immunity 2007; 26:357-369.
51. Chau LA, Bluestone JA, Madrenas J. Dissociation of intracellular signaling pathways in response to partial agonist ligands of the T-cell receptor. J Exp Med 1998; 187:1699-1709.
52. Jensen WA, Pleiman CM, Beaufils P et al. Qualitatively distinct signaling through T-cell antigen receptor subunits. Eur J Immunol 1997; 27:707-716.
53. Chae WJ, Lee HK, Han JH et al. Qualitatively differential regulation of T-cell activation and apoptosis by T-cell receptor zeta chain ITAMs and their tyrosine residues. Int Immunol 2004; 16:1225-1236.
54. Kesti T, Ruppelt A, Wang JH et al. Reciprocal Regulation of SH3 and SH2 Domain Binding via Tyrosine Phosphorylation of a Common Site in CD3{epsilon}. J Immunol 2007; 179:878-885.
55. Rudolph MG, Stanfield RL, Wilson IA. How TCRs bind MHCs, peptides and coreceptors. Annu Rev Immunol 2006; 24:419-466.
56. Arkin M. Protein-protein interactions and cancer: Small molecules going in for the kill. Curr Opin Chem Biol 2005; 9:317-324.
57. Sigalov AB. Interaction between HIV gp41 fusion peptide and T-cell receptor: Putting the puzzle pieces back together. FASEB J 2007; 21:1633-1634; author reply 1635.
58. Sigalov AB. More on: glycoprotein VI oligomerization: A novel concept of platelet inhibition. J Thromb Haemost 2007; 5:2310-2312.
59. Norman PS. Immunotherapy: 1999-2004. J Allergy Clin Immunol 2004; 113:1013-1023; quiz 1024.
60. Jackson SP, Schoenwaelder SM. Antiplatelet therapy: In search of the 'magic bullet'. Nat Rev Drug Discov 2003; 2:775-789.
61. Kepley CL. New approaches to allergen immunotherapy. Curr Allergy Asthma Rep 2006; 6:427-433.
62. Kraft S, Kinet JP. New developments in FcepsilonRI regulation, function and inhibition. Nat Rev Immunol 2007; 7:365-378.
63. McNicol A, Israels SJ. Platelets and anti-platelet therapy. J Pharmacol Sci 2003; 93:381-396.
64. Molloy PE, Sewell AK, Jakobsen BK. Soluble T-cell receptors: Novel immunotherapies. Curr Opin Pharmacol 2005; 5:438-443.
65. Pons L, Burks W. Novel treatments for food allergy. Expert Opin Investig Drugs 2005; 14:829-834.
66. Chatenoud L, Bluestone JA. CD3-specific antibodies: A portal to the treatment of autoimmunity. Nat Rev Immunol 2007; 7:622-632.
67. St Clair EW, Turka LA, Saxon A et al. New reagents on the horizon for immune tolerance. Annu Rev Med 2007; 58:329-346.

68. Hombach A, Heuser C, Abken H. The recombinant T-cell receptor strategy: Insights into structure and function of recombinant immunoreceptors on the way towards an optimal receptor design for cellular immunotherapy. Curr Gene Ther 2002; 2:211-226.

69. Luzak B, Golanski J, Rozalski M et al. Inhibition of collagen-induced platelet reactivity by DGEA peptide. Acta Biochim Pol 2003; 50:1119-1128.

70. O'Herrin SM, Slansky JE, Tang Q et al. Antigen-specific blockade of T-cells in vivo using dimeric MHC peptide. J Immunol 2001; 167:2555-2560.

71. Andrasfalvy M, Peterfy H, Toth G et al. The beta subunit of the type I Fcepsilon receptor is a target for peptides inhibiting IgE-mediated secretory response of masT-cells. J Immunol 2005; 175:2801-2806.

72. Cronin SJ, Penninger JM. From T-cell activation signals to signaling control of anti-cancer immunity. Immunol Rev 2007; 220:151-168.

73. Waldmann TA. Immune receptors: targets for therapy of leukemia/lymphoma, autoimmune diseases and for the prevention of allograft rejection. Annu Rev Immunol 1992; 10:675-704.

74. Enk AH, Knop J. T-cell receptor mimic peptides and their potential application in T-cell-mediated disease. Int Arch Allergy Immunol 2000; 123:275-281.

75. Wang XM, Djordjevic JT, Kurosaka N et al. T-cell antigen receptor peptides inhibit signal transduction within the membrane bilayer. Clin Immunol 2002; 105:199-207.

76. Amon MA, Ali M, Bender V et al. Lipidation and glycosylation of a T-cell antigen receptor (TCR) transmembrane hydrophobic peptide dramatically enhances in vitro and in vivo function. Biochim Biophys Acta 2006; 1763:879-888.

77. Collier S, Bolte A, Manolios N. Discrepancy in CD3-transmembrane peptide activity between in vitro and in vivo T-cell inhibition. Scand J Immunol 2006; 64:388-391.

78. Manolios N, Collier S, Taylor J et al. T-cell antigen receptor transmembrane peptides modulate T-cell function and T-cell-mediated disease. Nat Med 1997; 3:84-88.

79. Quintana FJ, Gerber D, Kent SC et al. HIV-1 fusion peptide targets the TCR and inhibits antigen-specific T-cell activation. J Clin Invest 2005; 115:2149-2158.

80. Bloch I, Quintana FJ, Gerber D et al. T-Cell inactivation and immunosuppressive activity induced by HIV gp41 via novel interacting motif. FASEB J 2007; 21:393-401.

81. Ali M, Salam NK, Amon M et al. T-Cell antigen receptor-alpha chain transmembrane peptides: Correlation between structure and function. Int J Pept Res Ther 2006; 12:261-267.

82. Melnyk RA, Partridge AW, Yip J et al. Polar residue tagging of transmembrane peptides. Biopolymers 2003; 71:675-685.

83. Smith SO, Smith C, Shekar S et al. Transmembrane interactions in the activation of the Neu receptor tyrosine kinase. Biochemistry 2002; 41:9321-9332.

84. Cunningham F, Deber CM. Optimizing synthesis and expression of transmembrane peptides and proteins. Methods 2007; 41:370-380.

85. Yin H, Slusky JS, Berger BW et al. Computational design of peptides that target transmembrane helices. Science 2007; 315:1817-1822.

86. Wimley WC, White SH. Designing transmembrane alpha-helices that insert spontaneously. Biochemistry 2000; 39:4432-4442.

87. Edwards RJ, Moran N, Devocelle M et al. Bioinformatic discovery of novel bioactive peptides. Nat Chem Biol 2007; 3:108-112.

88. Apic G, Russell RB. A shortcut to peptides to modulate platelets. Nat Chem Biol 2007; 3:83-84.

89. Ashish, Wimley WC. Visual detection of specific, native interactions between soluble and microbead-tethered alpha-helices from membrane proteins. Biochemistry 2001; 40:13753-13759.

90. Killian JA. Synthetic peptides as models for intrinsic membrane proteins. FEBS Lett 2003; 555:134-138.

91. Vandebona H, Ali M, Amon M et al. Immunoreceptor transmembrane peptides and their effect on natural killer (NK) cell cytotoxicity. Protein Pept Lett 2006; 13:1017-1024.

92. Ali M, De Planque MRR, Huynh NT et al. Biophysical studies of a transmembrane peptide derived from the T-cell antigen receptor. Letters in Peptide Science 2002; 8:227-233.

93. Huynh NT, Ffrench RA, Boadle RA et al. Transmembrane T-cell receptor peptides inhibit B- and natural killer-cell function. Immunology 2003; 108:458-464.

94. Bender V, Ali M, Amon M et al. T-cell antigen receptor peptide-lipid membrane interactions using surface plasmon resonance. J Biol Chem 2004; 279:54002-54007.

95. Gerber D, Quintana FJ, Bloch I et al. D-enantiomer peptide of the TCRalpha transmembrane domain inhibits T-cell activation in vitro and in vivo. FASEB J 2005; 19:1190-1192.

96. Gollner GP, Muller G, Alt R et al. Therapeutic application of T-cell receptor mimic peptides or cDNA in the treatment of T-cell-mediated skin diseases. Gene Ther 2000; 7:1000-1004.

97. Manolios N, Huynh NT, Collier S. Peptides in the treatment of inflammatory skin disease. Australas J Dermatol 2002; 43:226-227.

98. Ali M, Amon M, Bender V et al. Hydrophobic transmembrane-peptide lipid conjugations enhance membrane binding and functional activity in T-cells. Bioconjug Chem 2005; 16:1556-1563.

99. Wang XM, Djordjevic JT, Bender V et al. T-cell antigen receptor (TCR) transmembrane peptides colocalize with TCR, not lipid rafts, in surface membranes. Cell Immunol 2002; 215:12-19.

100. Kurosaka N, Bolte A, Ali M et al. T-cell antigen receptor assembly and cell surface expression is not affected by treatment with T-cell antigen receptor-alpha chain transmembrane Peptide. Protein Pept Lett 2007; 14:299-303.

101. Quintana FJ, Gerber D, Bloch I et al. A structurally altered D,L-amino acid TCRalpha transmembrane peptide interacts with the TCRalpha and inhibits T-cell activation in vitro and in an animal model. Biochemistry 2007; 46:2317-2325.

102. Buferne M, Luton F, Letourneur F et al. Role of CD3 delta in surface expression of the TCR/CD3 complex and in activation for killing analyzed with a CD3 delta-negative cytotoxic T-lymphocyte variant. J Immunol 1992; 148:657-664.

103. Luton F, Buferne M, Legendre V et al. Role of CD3gamma and CD3delta cytoplasmic domains in cytolytic T-lymphocyte functions and TCR/CD3 down-modulation. J Immunol 1997; 158:4162-4170.

104. Haks MC, Cordaro TA, van den Brakel JH et al. A redundant role of the CD3 gamma-immunoreceptor tyrosine-based activation motif in mature T-cell function. J Immunol 2001; 166:2576-2588.

105. Haks MC, Pepin E, van den Brakel JH et al. Contributions of the T-cell receptor-associated CD3gamma-ITAM to thymocyte selection. J Exp Med 2002; 196:1-13.

106. de Saint Basile G, Geissmann F, Flori E et al. Severe combined immunodeficiency caused by deficiency in either the delta or the epsilon subunit of CD3. J Clin Invest 2004; 114:1512-1517.

107. Roifman CM. CD3 delta immunodeficiency. Curr Opin Allergy Clin Immunol 2004; 4:479-484.

108. Sigalov AB. Inhibiting collagen-induced platelet aggregation and activation with reptide variants. US 12/001,258 and PCT PCT/US2007/025389 patent applications were filed on 12/11/2007 and 12/12/2007, respectively, claiming a priority to US provisional patent application 60/874,694 filed on 12/13/2006.

109. Heldin CH. Dimerization of cell surface receptors in signal transduction. Cell 1995; 80:213-223.

110. Hubbard SR. Structural analysis of receptor tyrosine kinases. Prog Biophys Mol Biol 1999; 71:343-358.

111. Weiss A, Schlessinger J. Switching signals on or off by receptor dimerization. Cell 1998; 94:277-280.

112. Klemm JD, Schreiber SL, Crabtree GR. Dimerization as a regulatory mechanism in signal transduction. Annu Rev Immunol 1998; 16:569-592.

113. Metzger H. Transmembrane signaling: the joy of aggregation. J Immunol 1992; 149:1477-1487.

114. Jiang G, Hunter T. Receptor signaling: When dimerization is not enough. Curr Biol 1999; 9:R568-571.

115. Mass RD. The HER receptor family: A rich target for therapeutic development. Int J Radiat Oncol Biol Phys 2004; 58:932-940.

116. Daniel PT, Wieder T, Sturm I et al. The kiss of death: Promises and failures of death receptors and ligands in cancer therapy. Leukemia 2001; 15:1022-1032.

117. Hernanz-Falcon P, Rodriguez-Frade JM, Serrano A et al. Identification of amino acid residues crucial for chemokine receptor dimerization. Nat Immunol 2004; 5:216-223.

118. Holler N, Tardivel A, Kovacsovics-Bankowski M et al. Two adjacent trimeric Fas ligands are required for Fas signaling and formation of a death-inducing signaling complex. Mol Cell Biol 2003; 23:1428-1440.

119. Lemmon MA, Schlessinger J. Regulation of signal transduction and signal diversity by receptor oligomerization. Trends Biochem Sci 1994; 19:459-463.

120. Marianayagam NJ, Sunde M, Matthews JM. The power of two: Protein dimerization in biology. Trends Biochem Sci 2004; 29:618-625.

121. Marmor MD, Skaria KB, Yarden Y. Signal transduction and oncogenesis by ErbB/HER receptors. Int J Radiat Oncol Biol Phys 2004; 58:903-913.

122. Mendrola JM, Berger MB, King MC et al. The single transmembrane domains of ErbB receptors self-associate in cell membranes. J Biol Chem 2002; 277:4704-4712.

123. Vandenabeele P, Declercq W, Beyaert R et al. Two tumour necrosis factor receptors: Structure and function. Trends Cell Biol 1995; 5:392-399.

124. Bennasroune A, Fickova M, Gardin A et al. Transmembrane peptides as inhibitors of ErbB receptor signaling. Mol Biol Cell 2004; 15:3464-3474.

125. Hebert TE, Moffett S, Morello JP et al. A peptide derived from a beta2-adrenergic receptor transmembrane domain inhibits both receptor dimerization and activation. J Biol Chem 1996; 271:16384-16392.

126. George SR, Lee SP, Varghese G et al. A transmembrane domain-derived peptide inhibits D1 dopamine receptor function without affecting receptor oligomerization. J Biol Chem 1998; 273:30244-30248.

127. Yin H, Litvinov RI, Vilaire G et al. Activation of platelet alphaIIbbeta3 by an exogenous peptide corresponding to the transmembrane domain of alphaIIb. J Biol Chem 2006; 281:36732-36741.
128. Tarasova NI, Rice WG, Michejda CJ. Inhibition of G-protein-coupled receptor function by disruption of transmembrane domain interactions. J Biol Chem 1999; 274:34911-34915.
129. Bennasroune A, Gardin A, Auzan C et al. Inhibition by transmembrane peptides of chimeric insulin receptors. Cell Mol Life Sci 2005; 62:2124-2131.
130. Bennett JS. Structure and function of the platelet integrin alphaIIbbeta3. J Clin Invest 2005; 115:3363-3369.
131. Zhang X, Gureasko J, Shen K et al. An allosteric mechanism for activation of the kinase domain of epidermal growth factor receptor. Cell 2006; 125:1137-1149.
132. Syed RS, Reid SW, Li C et al. Efficiency of signalling through cytokine receptors depends critically on receptor orientation. Nature 1998; 395:511-516.
133. Livnah O, Johnson DL, Stura EA et al. An antagonist peptide-EPO receptor complex suggests that receptor dimerization is not sufficient for activation. Nat Struct Biol 1998; 5:993-1004.
134. Ballinger MD, Wells JA. Will any dimer do? Nat Struct Biol 1998; 5:938-940.
135. Gay NJ, Gangloff M, Weber AN. Toll-like receptors as molecular switches. Nat Rev Immunol 2006; 6:693-698.
136. Neiditch MB, Federle MJ, Pompeani AJ et al. Ligand-induced asymmetry in histidine sensor kinase complex regulates quorum sensing. Cell 2006; 126:1095-1108.
137. Lofts FJ, Hurst HC, Sternberg MJ et al. Specific short transmembrane sequences can inhibit transformation by the mutant neu growth factor receptor in vitro and in vivo. Oncogene 1993; 8:2813-2820.
138. Durell SR, Martin I, Ruysschaert JM et al. What studies of fusion peptides tell us about viral envelope glycoprotein-mediated membrane fusion (review). Mol Membr Biol 1997; 14:97-112.
139. Pecheur EI, Sainte-Marie J, Bienven e A et al. Peptides and membrane fusion: towards an understanding of the molecular mechanism of protein-induced fusion. J Membr Biol 1999; 167:1-17.
140. Weissenhorn W, Hinz A, Gaudin Y. Virus membrane fusion. FEBS Lett 2007; 581:2150-2155.
141. Epand RM. Fusion peptides and the mechanism of viral fusion. Biochim Biophys Acta 2003; 1614:116-121.
142. Eckert DM, Kim PS. Mechanisms of viral membrane fusion and its inhibition. Annu Rev Biochem 2001; 70:777-810.
143. Teissier E, Pecheur EI. Lipids as modulators of membrane fusion mediated by viral fusion proteins. Eur Biophys J 2007; 36:887-899.
144. Fackler OT, Alcover A, Schwartz O. Modulation of the immunological synapse: A key to HIV-1 pathogenesis? Nat Rev Immunol 2007; 7:310-317.
145. Stevenson M. HIV-1 pathogenesis. Nat Med 2003; 9:853-860.
146. Bosch ML, Earl PL, Fargnoli K et al. Identification of the fusion peptide of primate immunodeficiency viruses. Science 1989; 244:694-697.
147. Gallaher WR. Detection of a fusion peptide sequence in the transmembrane protein of human immunodeficiency virus. Cell 1987; 50:327-328.
148. van Praag RM, Prins JM, Roos MT et al. OKT3 and IL-2 treatment for purging of the latent HIV-1 reservoir in vivo results in selective long-lasting CD4+ T-cell depletion. J Clin Immunol 2001; 21:218-226.
149. Blumenthal R, Dimitrov DS. Targeting the sticky fingers of HIV-1. Cell 2007; 129:243-245.
150. Munch J, Standker L, Adermann K et al. Discovery and optimization of a natural HIV-1 entry inhibitor targeting the gp41 fusion peptide. Cell 2007; 129:263-275.
151. Bell I, Ashman C, Maughan J et al. Association of simian immunodeficiency virus Nef with the T-cell receptor (TCR) zeta chain leads to TCR down-modulation. J Gen Virol 1998; 79(Pt 11):2717-2727.
152. Schaefer TM, Bell I, Fallert BA et al. The T-cell receptor zeta chain contains two homologous domains with which simian immunodeficiency virus Nef interacts and mediates down-modulation. J Virol 2000; 74:3273-3283.
153. Schaefer TM, Bell I, Pfeifer ME et al. The conserved process of TCR/CD3 complex down-modulation by SIV Nef is mediated by the central core, not endocytic motifs. Virology 2002; 302:106-122.
154. Keppler OT, Tibroni N, Venzke S et al. Modulation of specific surface receptors and activation sensitization in primary resting CD4+ T-lymphocytes by the Nef protein of HIV-1. J Leukoc Biol 2006; 79:616-627.
155. Schrager JA, Marsh JW. HIV-1 Nef increases T-cell activation in a stimulus-dependent manner. Proc Natl Acad Sci USA 1999; 96:8167-8172.
156. Simmons A, Aluvihare V, McMichael A. Nef triggers a transcriptional program in T-cells imitating single-signal T-cell activation and inducing HIV virulence mediators. Immunity 2001; 14:763-777.

157. Djordjevic JT, Schibeci SD, Stewart GJ et al. HIV type 1 Nef increases the association of T-cell receptor (TCR)-signaling molecules with T-cell rafts and promotes activation-induced raft fusion. AIDS Res Hum Retroviruses 2004; 20:547-555.
158. Krautkramer E, Giese SI, Gasteier JE et al. Human immunodeficiency virus type 1 Nef activates p21-activated kinase via recruitment into lipid rafts. J Virol 2004; 78:4085-4097.
159. Xu XN, Laffert B, Screaton GR et al. Induction of Fas ligand expression by HIV involves the interaction of Nef with the T-cell receptor zeta chain. J Exp Med 1999; 189:1489-1496.
160. Swigut T, Greenberg M, Skowronski J. Cooperative interactions of simian immunodeficiency virus Nef, AP-2 and CD3-zeta mediate the selective induction of T-cell receptor-CD3 endocytosis. J Virol 2003; 77:8116-8126.
161. Arold S, Hoh F, Domergue S et al. Characterization and molecular basis of the oligomeric structure of HIV-1 nef protein. Protein Sci 2000; 9:1137-1148.
162. Schamel WW, Arechaga I, Risueno RM et al. Coexistence of multivalent and monovalent TCRs explains high sensitivity and wide range of response. J Exp Med 2005; 202:493-503.
163. Liu LX, Heveker N, Fackler OT et al. Mutation of a conserved residue (D123) required for oligomerization of human immunodeficiency virus type 1 Nef protein abolishes interaction with human thioesterase and results in impairment of Nef biological functions. J Virol 2000; 74:5310-5319.
164. Chaudhry A, Das SR, Jameel S et al. A two-pronged mechanism for HIV-1 Nef-mediated endocytosis of immune costimulatory molecules CD80 and CD86. Cell Host Microbe 2007; 1:37-49.
165. Chen J, Subbarao K. The Immunobiology of SARS*. Annu Rev Immunol 2007; 25:443-472.
166. Cui W, Fan Y, Wu W et al. Expression of lymphocytes and lymphocyte subsets in patients with severe acute respiratory syndrome. Clin Infect Dis 2003; 37:857-859.
167. Douek D. HIV disease progression: immune activation, microbes and a leaky gut. Top HIV Med 2007; 15:114-117.
168. Hazenberg MD, Hamann D, Schuitemaker H et al. T-cell depletion in HIV-1 infection: How CD4+ T-cells go out of stock. Nat Immunol 2000; 1:285-289.
169. Gallaher WR, Ball JM, Garry RF et al. A general model for the transmembrane proteins of HIV and other retroviruses. AIDS Res Hum Retroviruses 1989; 5:431-440.
170. Gallaher WR. Similar structural models of the transmembrane proteins of Ebola and avian sarcoma viruses. Cell 1996; 85:477-478.
171. Xu Y, Lou Z, Liu Y et al. Crystal structure of severe acute respiratory syndrome coronavirus spike protein fusion core. J Biol Chem 2004; 279:49414-49419.
172. Zhu J, Xiao G, Xu Y et al. Following the rule: Formation of the 6-helix bundle of the fusion core from severe acute respiratory syndrome coronavirus spike protein and identification of potent peptide inhibitors. Biochem Biophys Res Commun 2004; 319:283-288.
173. Ingallinella P, Bianchi E, Finotto M et al. Structural characterization of the fusion-active complex of severe acute respiratory syndrome (SARS) coronavirus. Proc Natl Acad Sci USA 2004; 101:8709-8714.
174. Taguchi F, Shimazaki YK. Functional analysis of an epitope in the S2 subunit of the murine coronavirus spike protein: involvement in fusion activity. J Gen Virol 2000; 81:2867-2871.
175. Bosch BJ, Martina BE, Van Der Zee R et al. Severe acute respiratory syndrome coronavirus (SARS-CoV) infection inhibition using spike protein heptad repeat-derived peptides. Proc Natl Acad Sci USA 2004; 101:8455-8460.
176. Sainz B Jr, Rausch JM, Gallaher WR et al. Identification and characterization of the putative fusion peptide of the severe acute respiratory syndrome-associated coronavirus spike protein. J Virol 2005; 79:7195-7206.
177. Verdonck K, Gonzalez E, Van Dooren S et al. Human T-lymphotropic virus 1: Recent knowledge about an ancient infection. Lancet Infect Dis 2007; 7:266-281.
178. Hinuma Y, Nagata K, Hanaoka M et al. Adult T-cell leukemia: Antigen in an ATL cell line and detection of antibodies to the antigen in human sera. Proc Natl Acad Sci USA 1981; 78:6476-6480.
179. Poiesz BJ, Ruscetti FW, Gazdar AF et al. Detection and isolation of type C retrovirus particles from fresh and cultured lymphocytes of a patient with cutaneous T-cell lymphoma. Proc Natl Acad Sci USA 1980; 77:7415-7419.
180. Jones KS, Fugo K, Petrow-Sadowski C et al. Human T-cell leukemia virus type 1 (HTLV-1) and HTLV-2 use different receptor complexes to enter T-cells. J Virol 2006; 80:8291-8302.
181. Fukumoto R, Dundr M, Nicot C et al. Inhibition of T-cell receptor signal transduction and viral expression by the linker for activation of T-cells-interacting p12(I) protein of human T-cell leukemia/lymphoma virus type 1. J Virol 2007; 81:9088-9099.
182. Tsukahara T, Ratner L. Substitution of HIV Type 1 Nef with HTLV-1 p12. AIDS Res Hum Retroviruses 2004; 20:938-943.

183. Lin HC, Hickey M, Hsu L et al. Activation of human T-cell leukemia virus type 1 LTR promoter and cellular promoter elements by T-cell receptor signaling and HTLV-1 Tax expression. Virology 2005; 339:1-11.
184. Wilson KA, Bar S, Maerz AL et al. The conserved glycine-rich segment linking the N-terminal fusion peptide to the coiled coil of human T-cell leukemia virus type 1 transmembrane glycoprotein gp21 is a determinant of membrane fusion function. J Virol 2005; 79:4533-4539.
185. Wilson KA, Maerz AL, Poumbourios P. Evidence that the transmembrane domain proximal region of the human T-cell leukemia virus type 1 fusion glycoprotein gp21 has distinct roles in the prefusion and fusion-activated states. J Biol Chem 2001; 276:49466-49475.
186. Albrecht B, D'Souza CD, Ding W et al. Activation of nuclear factor of activated T-cells by human T-lymphotropic virus type 1 accessory protein p12(I). J Virol 2002; 76:3493-3501.
187. Ding W, Albrecht B, Kelley RE et al. Human T-cell lymphotropic virus type 1 p12(I) expression increases cytoplasmic calcium to enhance the activation of nuclear factor of activated T-cells. J Virol 2002; 76:10374-10382.
188. Nicot C, Mulloy JC, Ferrari MG et al. HTLV-1 p12(I) protein enhances STAT5 activation and decreases the interleukin-2 requirement for proliferation of primary human peripheral blood mononuclear cells. Blood 2001; 98:823-829.
189. Albrecht B, Collins ND, Burniston MT et al. Human T-lymphotropic virus type 1 open reading frame I p12(I) is required for efficient viral infectivity in primary lymphocytes. J Virol 2000; 74:9828-9835.
190. Collins ND, Newbound GC, Albrecht B et al. Selective ablation of human T-cell lymphotropic virus type 1 p12I reduces viral infectivity in vivo. Blood 1998; 91:4701-4707.
191. Fenard D, Yonemoto W, de Noronha C et al. Nef is physically recruited into the immunological synapse and potentiates T-cell activation early after TCR engagement. J Immunol 2005; 175:6050-6057.
192. Guma M, Angulo A, Lopez-Botet M. NK cell receptors involved in the response to human cytomegalovirus infection. Curr Top Microbiol Immunol 2006; 298:207-223.
193. Arnon TI, Achdout H, Levi O et al. Inhibition of the NKp30 activating receptor by pp65 of human cytomegalovirus. Nat Immunol 2005; 6:515-523.
194. Klewitz C, Klenk HD, ter Meulen J. Amino acids from both N-terminal hydrophobic regions of the Lassa virus envelope glycoprotein GP-2 are critical for pH-dependent membrane fusion and infectivity. J Gen Virol 2007; 88:2320-2328.
195. Weiss HJ. Platelet physiology and abnormalities of platelet function (first of two parts). N Engl J Med 1975; 293:531-541.
196. Nieswandt B, Brakebusch C, Bergmeier W et al. Glycoprotein VI but not alpha2beta1 integrin is essential for platelet interaction with collagen. EMBO J 2001; 20:2120-2130.
197. Michelson AD. Platelet inhibitor therapy: mechanisms of action and clinical use. J Thromb Thrombolysis 2003; 16:13-15.
198. Smethurst PA, Onley DJ, Jarvis GE et al. Structural basis for the platelet-collagen interaction: The smallest motif within collagen that recognizes and activates platelet Glycoprotein VI contains two glycine-proline-hydroxyproline triplets. J Biol Chem 2007; 282:1296-1304.
199. Gibbins JM, Okuma M, Farndale R et al. Glycoprotein VI is the collagen receptor in platelets which underlies tyrosine phosphorylation of the Fc receptor gamma-chain. FEBS Lett 1997; 413:255-259.
200. Nieswandt B, Watson SP. Platelet-collagen interaction: Is GPVI the central receptor? Blood 2003; 102:449-461.
201. Li H, Lockyer S, Concepcion A et al. The Fab fragment of a novel anti-GPVI monoclonal antibody, OM4, reduces in vivo thrombosis without bleeding risk in rats. Arterioscler Thromb Vasc Biol 2007; 27:1199-1205.
202. Lockyer S, Okuyama K, Begum S et al. GPVI-deficient mice lack collagen responses and are protected against experimentally induced pulmonary thromboembolism. Thromb Res 2006; 118:371-380.
203. Clemetson KJ. Platelet collagen receptors: A new target for inhibition? Haemostasis 1999; 29:16-26.
204. Massberg S, Konrad I, Bultmann A et al. Soluble glycoprotein VI dimer inhibits platelet adhesion and aggregation to the injured vessel wall in vivo. FASEB J 2004; 18:397-399.
205. Farndale RW. Collagen-induced platelet activation. Blood Cells Mol Dis 2006; 36:162-165.
206. Gawaz M. Role of platelets in coronary thrombosis and reperfusion of ischemic myocardium. Cardiovasc Res 2004; 61:498-511.
207. Feng J, Garrity D, Call ME et al. Convergence on a distinctive assembly mechanism by unrelated families of activating immune receptors. Immunity 2005; 22:427-438.
208. Berlanga O, Bori-Sanz T, James JR et al. Glycoprotein VI oligomerization in cell lines and platelets. J Thromb Haemost 2007; 5:1026-1033.
209. Collins B, Hollidge C. Antithrombotic drug market. Nat Rev Drug Discov 2003; 2:11-12.
210. Ito Y, Kanai T, Totsuka T et al. Blockade of NKG2D signaling prevents the development of murine CD4+ T-cell-mediated colitis. Am J Physiol Gastrointest Liver Physiol 2008; 294:G199-207.

211. Uversky VN, Gillespie JR, Fink AL. Why are "natively unfolded" proteins unstructured under physiologic conditions? Proteins 2000; 41:415-427.
212. Dyson HJ, Wright PE. Intrinsically unstructured proteins and their functions. Nat Rev Mol Cell Biol 2005; 6:197-208.
213. Dunker AK, Brown CJ, Lawson JD et al. Intrinsic disorder and protein function. Biochemistry 2002; 41:6573-6582.
214. Iakoucheva LM, Brown CJ, Lawson JD et al. Intrinsic disorder in cell-signaling and cancer-associated proteins. J Mol Biol 2002; 323:573-584.
215. Minezaki Y, Homma K, Nishikawa K. Intrinsically disordered regions of human plasma membrane proteins preferentially occur in the cytoplasmic segment. J Mol Biol 2007; 368:902-913.
216. Iakoucheva LM, Radivojac P, Brown CJ et al. The importance of intrinsic disorder for protein phosphorylation. Nucleic Acids Res 2004; 32:1037-1049.
217. Stockwell BR. Exploring biology with small organic molecules. Nature 2004; 432:846-854.
218. Bose M, Gestwicki JE, Devasthali V et al. 'Nature-inspired' drug-protein complexes as inhibitors of Abeta aggregation. Biochem Soc Trans 2005; 33:543-547.
219. Fry DC, Vassilev LT. Targeting protein-protein interactions for cancer therapy. J Mol Med 2005; 83:955-963.
220. Watt PM. Screening for peptide drugs from the natural repertoire of biodiverse protein folds. Nat Biotechnol 2006; 24:177-183.
221. Stoevesandt O, Elbs M, Kohler K et al. Peptide microarrays for the detection of molecular interactions in cellular signal transduction. Proteomics 2005; 5:2010-2017.
222. Thorburn A. Death receptor-induced cell killing. Cell Signal 2004; 16:139-144.
223. Wajant H. The Fas signaling pathway: More than a paradigm. Science 2002; 296:1635-1636.
224. Siegel RM, Muppidi JR, Sarker M et al. SPOTS: Signaling protein oligomeric transduction structures are early mediators of death receptor-induced apoptosis at the plasma membrane. J Cell Biol 2004; 167:735-744.
225. Martin K, Meade G, Moran N et al. A palmitylated peptide derived from the glycoprotein Ib beta cytoplasmic tail inhibits platelet activation. J Thromb Haemost 2003; 1:2643-2652.
226. Stephens G, O'Luanaigh N, Reilly D et al. A sequence within the cytoplasmic tail of GpIIb independently activates platelet aggregation and thromboxane synthesis. J Biol Chem 1998; 273:20317-20322.
227. Liu J, Jackson CW, Gruppo RA et al. The beta3 subunit of the integrin alphaIIbbeta3 regulates alphaIIb-mediated outside-in signaling. Blood 2005; 105:4345-4352.
228. Loiarro M, Capolunghi F, Fanto N et al. Pivotal Advance: Inhibition of MyD88 dimerization and recruitment of IRAK1 and IRAK4 by a novel peptidomimetic compound. J Leukoc Biol 2007; 82:801-810.
229. Loiarro M, Sette C, Gallo G et al. Peptide-mediated interference of TIR domain dimerization in MyD88 inhibits interleukin-1-dependent activation of NF-{kappa}B. J Biol Chem 2005; 280:15809-15814.
230. Hartlieb B, Modrof J, Muhlberger E et al. Oligomerization of Ebola virus VP30 is essential for viral transcription and can be inhibited by a synthetic peptide. J Biol Chem 2003; 278:41830-41836.
231. Akira S, Takeda K. Toll-like receptor signalling. Nat Rev Immunol 2004; 4:499-511.
232. O'Neill LA. The role of MyD88-like adapters in Toll-like receptor signal transduction. Biochem Soc Trans 2003; 31:643-647.
233. Berndt MC, Shen Y, Dopheide SM et al. The vascular biology of the glycoprotein Ib-IX-V complex. Thromb Haemost 2001; 86:178-188.
234. Savage B, Saldivar E, Ruggeri ZM. Initiation of platelet adhesion by arrest onto fibrinogen or translocation on von Willebrand factor. Cell 1996; 84:289-297.
235. Lefkovits J, Plow EF, Topol EJ. Platelet glycoprotein IIb/IIIa receptors in cardiovascular medicine. N Engl J Med 1995; 332:1553-1559.
236. Shattil SJ, Kashiwagi H, Pampori N. Integrin signaling: The platelet paradigm. Blood 1998; 91:2645-2657.
237. Coppolino M, Leung-Hagesteijn C, Dedhar S et al. Inducible interaction of integrin alpha 2 beta 1 with calreticulin. Dependence on the activation state of the integrin. J Biol Chem 1995; 270:23132-23138.
238. Hughes PE, Diaz-Gonzalez F, Leong L et al. Breaking the integrin hinge. A defined structural constraint regulates integrin signaling. J Biol Chem 1996; 271:6571-6574.
239. Weitzman JB, Pujades C, Hemler ME. Integrin alpha chain cytoplasmic tails regulate "antibody-redirected" cell adhesion, independently of ligand binding. Eur J Immunol 1997; 27:78-84.
240. Gonzalez JP, Pourrut X, Leroy E. Ebolavirus and other filoviruses. Curr Top Microbiol Immunol 2007; 315:363-387.
241. Muhlberger E, Weik M, Volchkov VE et al. Comparison of the transcription and replication strategies of marburg virus and Ebola virus by using artificial replication systems. J Virol 1999; 73:2333-2342.

242. Visintin A, Latz E, Monks BG et al. Lysines 128 and 132 enable lipopolysaccharide binding to MD-2, leading to Toll-like receptor-4 aggregation and signal transduction. J Biol Chem 2003; 278:48313-48320.
243. Zhang H, Tay PN, Cao W et al. Integrin-nucleated Toll-like receptor (TLR) dimerization reveals subcellular targeting of TLRs and distinct mechanisms of TLR4 activation and signaling. FEBS Lett 2002; 532:171-176.
244. Michnick SW, Remy I, Campbell-Valois FX et al. Detection of protein-protein interactions by protein fragment complementation strategies. Methods Enzymol 2000; 328:208-230.
245. Lee HK, Dunzendorfer S, Tobias PS. Cytoplasmic domain-mediated dimerizations of toll-like receptor 4 observed by beta-lactamase enzyme fragment complementation. J Biol Chem 2004; 279:10564-10574.

Chapter 21

Immune Receptor Signaling, Aging and Autoimmunity

Anis Larbi,* Tamas Fülöp and Graham Pawelec

Abstract

Aging is associated with a myriad of changes including alterations in glucose metabolism, brain function, hormonal regulation, muscle homeostasis and the immune system. Aged individuals, generally still defined as over 65 years old, differ from middle-aged or young donors in many features of the immune system. The major observation is that the elderly population is not able to cope with infections as well as younger adults and recovery generally takes longer. Moreover, some diseases first appear with advancing age and are likely associated with dysfunction of the immune system. Thus, Alzheimer's disease, atherosclerosis, type II diabetes and some autoimmune disorders are linked to changes in immune function. One major immune cell population implicated as being responsible for the initiation and chronicity of immune dysfunction leading to diseases or immunosuppression is the T-cell. Although many changes in B-cell and innate immune function in aging are associated with the appearance of disease, they are not as well studied and clearly demarcated as changes in the T-cell compartment. The adaptive immune system is coordinated by T-cells, the activation of which is required for the initiation, maintenance and termination of responses against pathogens. Changes in the expression and functions of the T-cell receptor (TCR) for antigen and its co-receptors are closely associated with immunosenescence. Certain similar changes have also been found in some other disease states, e.g., rheumatoid arthritis, systemic lupus erythematosus and cancer. In this chapter, we will summarize our knowledge about multichain immune recognition receptor signaling, mainly the TCR, in aging and autoimmune diseases.

Introduction

The percentage of individuals over 65 years old in the world is increasing, not only in developed countries but also to some extent in developing countries.[1] This phenomenon is largely due to improved medication and health care along with decreased malnutrition and death caused by common pathogens such as influenza.[2] However, the elderly population is particularly targeted for vaccination against pneumonia and influenza, because of the lower efficiency of their immune system and their difficulty to cope with infections.[3] A better understanding of aging and age-related diseases affecting the immune system is very important to keep these elderly individuals in the best of health. Public health services also benefit from a reduction in the high costs of maintaining institutionalized ill elderly people. Understanding physiological aging as well as age-related diseases

*Corresponding Author: Anis Larbi—Center for Medical Research (ZMF), Tübingen Aging and Tumour Immunology Group, Section for Transplantation Immunology and Immunohematology, University of Tübingen, Waldhörnlestrasse 22, D-72072 Tübingen, Germany. Email: anis.larbi@medizin.uni-tuebingen.de

Multichain Immune Recognition Receptor Signaling: From Spatiotemporal Organization to Human Disease, edited by Alexander B. Sigalov. ©2008 Landes Bioscience and Springer Science+Business Media.

would help to respond better to their specific requirements and to improve the quality of life for longer periods.[4] This is still challenging but several studies have demonstrated significant changes in immune system functions of elderly individuals when compared to their younger counterparts.[5] In the first part of this chapter, we will review which cells exhibit functional changes in aging and then consider receptor signaling in aging and autoimmune diseases such as rheumatoid arthritis (RA) and systemic lupus erythematosus (SLE). Finally, we will discuss possible interventions to modulate multichain immune recognition receptor (MIRR) signaling in order to restore normal immune function.

Immunosenescence

Overall, immune response dysregulation can be termed immunosenescence.[6] This phenomenon is very difficult to explain because it is multifactorial and may have different clinical consequences depending on the individual's health status and immunological history. Several causes for this age-related phenomenon have been put forward without explaining it entirely. There is the old paradigm, certainly still important, that age-related immune deficiency occurs with thymic involution.[7] This is based on the idea that T-cells are lost with time from the periphery but the thymus becomes less able to replace them, resulting in decreased numbers of naïve cells exported. There is a great deal of evidence for decreased, sometimes catastrophically decreased, naïve cells in the elderly.[8] Thymic atrophy is also thought to cause T-cell repertoire shrinkage that renders immunological protection incomplete, explaining the difficulty that elderly individuals encounter in overcoming infection, especially with pathogens that they have not previously encountered. More recently, longitudinal studies have associated immunosenescence with the accumulation of anergic T-cells, mainly CD8+ T-cells specific for antigens from cytomegalovirus (CMV).[9] These can represent >20% of the whole peripheral blood CD8 repertoire. Thus, aging is associated not only with changes in lymphocyte subsets but also with functional changes within these subsets. We will review this briefly now.

There is a consensus that T-cell functions are altered following TCR ligation. The main critical failure is the decrease in the production of interleukin-2 (IL-2) and consequently a reduced proliferative capacity even of the non-anergic cells.[10] This could help to explain decreased immune responses after antigen recognition. The most common changes demonstrated in T-cell functions and properties in aging are shown in Table 1. Several changes in surface marker expression occur in aging. Most consistently, the nonpolymorphic coreceptor CD28 is decreased leading to an increased number of CD28-negative cells, mostly in the CD8 compartment.[11] Mitogen-induced IL-2 production is severely impaired concomitantly with decreased IL-2 receptor (IL-2R) expression and proliferation.[12] Not only is the intensity of the response changed but also the type of response. Thus, there may be a shift towards Th2 responses with aging.[13] However, changes in T-cells with aging differ within the different T-cell subsets. Some reports have suggested differential susceptibility to apoptosis in CD8+ and CD4+ T-cells in aging.[14] Changes have also been demonstrated in membrane fluidity, DNA damage and telomere length.[15] All these changes lead to the impaired T-cell response with aging.

Elderly individuals may have a normal B-cell count and mount a good humoral response although low B-cell numbers have been described as part of the original immune risk phenotype (a cluster of immune parameters predicting mortality in longitudinal studies of a very elderly population.[9] However, the antibodies produced are commonly of low affinity, providing a less powerful response compared to young individuals.[16] B-cell lymphopoiesis is also reduced, which leads to an increase in the percentage of antigen-experienced cells when compared to newly-produced naïve B-cells.[17] This is analogous to the situation with T-cells described above.

Natural killer T(NKT)-cell cytotoxicity as well as interferon-gamma (IFN-γ) production decreases in aging. However the functionality of NKT-cells in aging is still controversial.[18] The percentage of CD3+Valpha24+ NKT-cells in peripheral blood from elderly donors was found to decrease and the majority of these cells are CD28-positive.[19] However the percentage of Valpha24+ NKT-cells is increased in the CD8+ compartment.

Table 1. Most significant changes in T-cell properties in aging

Decreased	Increased
Proliferation with mitogens	CD8+CD28- cells
IL-2 production	CD95 expression
Telomere length	CD45RO+ cells
Telomerase activity	DNA damage
Th1 response: IL-2. IFN-γ	IL-6, TNF-α secretion
Delayed-type hypersensitivity	Th2 response: IL-4, IL-5, IL-10, IL-12
TCR signal transduction	CD4+ T-cell apoptosis
Nuclear factor transcription activity	Anergic CMV-specific CD8+ T-cells
IL-2 receptor expression	
Membrane fluidity	
DNA repair	
CD8+ T-cell apoptosis	
Naïve CD4+ cells	
T-cell repertoire	
CD45RA+ cells	

While adaptive immunity has been clearly shown to be defective in aging, the role of cells from the innate immune system in age-related dysfunction is still a matter for debate. Nevertheless, we can state that the function of macrophages and neutrophils is impaired regarding Toll-like receptor function and expression.[20,21] It remains likely that delayed recovery in elderly individuals can also be caused by defects in innate immunity.

Receptor Signaling in Immunosenescence

Why T-cell functions are decreased in the elderly is still under debate. Defects in T-cell activation or subsequent thereto could be explained by extrinsic or intrinsic factors. It is known that the elderly often manifest a state of low-grade inflammation although this may be the case only for (the majority of) not perfectly healthy individuals, which is reflected by an increase in circulating pro-inflammatory cytokines (e.g., IL-6).[22]

It seems that the number of TCRs on the cell surface is not changed during aging. We will not discuss TCR assembly and signaling in detail here because the previous chapters of the present book cover specifically this area. In Table 2 we depict the TCR signaling alterations in aging in summary. The first step in TCR-mediated signaling is the activation of different tyrosine kinases, leading to the tyrosine phosphorylation of several downstream molecules. The level of tyrosine phosphorylation of p59fyn and ZAP-70 kinases is impaired in T-cells from old mice activated through the TCR/CD3 complex.[23] In human T-cells, an age-related defect is observed in tyrosine-specific protein phosphorylation after activation via TCR–CD3 complexes. In addition, a p59fyn and p56lck activity was recently shown to substantially decrease in T-cells of healthy elderly subjects.[24]

It is now well-documented that with aging other early events related to protein tyrosine phosphorylation following TCR activation are altered, such as the generation of myo-inositol 1,4,5-trisphosphate, intracellular free calcium mobilization and protein kinase C (PKC) translocation to the membrane.[25] It was shown that defects in translocation of PKC following TCR stimulation are present in T-cells of old humans and mice. Data are accumulating showing that more distal events, such as in the ras-mitogen activated protein kinase (MAPK) pathways, are

Table 2. TCR signaling alterations in aging

Intracellular free Ca²⁺	Lck activation
Myo-inositol 1,4,5-trisphosphate production	ITAM phosphorylation
Protein kinase C translocation	Linker of activated T-cells activation
CD69 expression	Fyn activation
CD25 expression	ZAP-70 activation
Membrane fluidity	Extracellular signal-regulated kinase activation
Cholesterol content	p38 activation
Raft-associated proteins	Proteasome activity
NF-AT distribution	NF-AT translocation
Regulation of cellular cholesterol	NF-κB relocalization
Raft coalescence	

also changed with aging. Whisler et al have shown that elderly subjects had a reduction in MAPK activation.[26] Because MAPK activation is correlated with IL-2 production, it is possible that the impaired signaling may represent the rate-limiting step for IL-2 production.

There is increasing experimental evidence that an appropriate balance between tyrosine kinase and phosphatase activities is essential for the regulation of cellular activation.[27] CD45 is a receptor-like protein tyrosine phosphatase expressed on all haematopoietic cells. It is a positive regulator of Src tyrosine kinases such as Lck by dephosphorylating their negative regulatory C-terminal residue.[28] Other phosphatases are expressed by T-cells but only a few studies tested the hypothesis that phosphatase dysregulation could be responsible for immunosenescence. Changes in the activity/localisation of transcription factors are the direct consequences of any impairment in the signaling cascade. As described above, calcium mobilization is deficient in T-cells from aged donors and the well-known decrease of NF-AT translocation to the nucleus is the direct consequence of this deficiency.[29] The other important transcription factor for IL-2 production is NF-kB, which is constitutively expressed in the cytoplasm and bound to an inhibitory protein, IkB, prior to activation. The decrease in NF-kB activation in mice and in humans is mainly due to a decreased inactivation of IkB by the proteasome.[30] Although there were differences in experimental groups in terms of age, experimental conditions or concentrations of stimuli in the different studies published, there is a general consensus that aging is associated with impairments in the activation of the TCR signaling cascade. One important question still to be solved is the following. Why are so many steps of the TCR signaling cascade shown to be altered, despite apparently unchanged TCR expression in aging? One answer may be that in the earlier studies, the analyses of TCR signaling were not sufficiently sophisticated to reveal subtle changes. This is not to say that the studies previously published are incorrect but that our knowledge in the field of signaling has recently progressed tremendously with the discovery of membrane microdomains which are the initiators of receptor signaling.[31] These will be considered next.

The Role of Membrane Rafts in TCR Signaling: The Aging Rafts

The concept of a spatial organization of signaling molecules in specialized cholesterol- and glycosphingolipid-enriched microdomains called membrane rafts has been introduced recently, suggesting that they provide a platform for lymphocyte signaling.[32] TCR ligation induces a redistribution of phosphorylated proteins into membrane rafts, which are highly compact and relatively small domains (20 to 200 nM) composed of saturated lipids and signaling molecules.[33] The saturation of the lipids as well as the enrichment in cholesterol allows the rafts to move through the membrane as discrete units. The role of membrane rafts is not limited to signal transduction but also to lipid transport, virus entry, cell movement, as well as cell-cell communication.[34] The accumulation or clustering of signaling molecules via membrane rafts initiates the formation of a signaling platform which increases the efficiency of signaling. The sustained T-cell activation via organised membrane raft signaling ultimately leads to the formation of a mature immune synapse needed to achieve full T-cell activation.[35] The organization and composition of the membrane will directly modulate the formation of such a signaling platform which ultimately influences cellular activation and functions.[36]

It is clear that there is an age-related alteration in the physico-chemical status of the plasma membrane of T-cells leading to decreased fluidity with in vivo aging as well as in in vitro models of senescence.[37] One should be attentive to the detrimental effects of such changes because membrane rafts and immune synapse formation are needed for the sustained activation of the cell. Because the activation of several steps of the signaling cascade is known to be impaired in aging, it is reasonable to test the hypothesis that downstream impairments are caused by upstream alterations. As the first events of the signaling cascade consist of movement of molecules recruited to or excluded from rafts, we will now review raft-associated changes noted in aging.

Although this paradigm has not been very extensively studied thus far, data are accumulating to support an important role for lipid rafts in age-related and diseases-related changes in signal transduction (reviewed by ref. 38). Simons et al were able to link membrane rafts with at least forty pathological cases and infections.[38] Miller et al were the first to demonstrate an alteration in several components of this signaling complex in naïve and memory T-cells in aging.[39] Because the latter accumulate with age, analyses of whole peripheral lymphocyte populations mostly reveal these changes. Several raft-associated or recruited proteins, such as the linker of activated T-cells (LAT), PKC and Vav fail to become activated in T-cells of aged mice. Moreover, LAT phosphorylation and redistribution to the T-cell-antigen-presenting cell (APC) immune synapse was impaired following TCR ligation.[39]

The cause of the increased rigidity of the membrane with aging is not known, but there are several possible explanations, of which the cholesterol hypothesis has been most thoroughly tested. Membrane rafts were originally found to float in fractions enriched in cholesterol and glyco-sphingolipids following centrifugation of Triton X-100-containing lysates. We have analysed the enrichment of cholesterol in membrane rafts of T-cells from young and elderly donors and found a 2-fold increase in the amount of cholesterol in membrane rafts in T-cells from the latter.[40] Such a change could have a detrimental effect on protein movement through the membrane and there-fore protein-protein interactions. This idea was also tested and it was found that T-cells exposed to anti-CD3 monoclonal antibodies (mAb) or a combination of anti-CD3 and anti-CD28 mAb induced significantly decreased membrane raft coalescence in T-cells of elderly subjects.[41]

The decrease in CD28 expression in T-cells with aging is a well-known phenomenon which can explain the decrease in raft coalescence, because membrane rafts are reported to be dependent on CD28 ligation for proper functioning.[42] How, then, can we explain the observed impaired coalescence with anti-CD3 stimulation alone? The signaling deficiencies shown in aging are linked to the TCR signaling cascade which is independent of CD28 signal transduction. When molecules associated with pathways of TCR-mediated signal transduction were assessed for acti-vation and localization into membrane rafts, we were able to show severe defects in Lck and LAT activation/localization in T-cells from the elderly following TCR ligation.[41] The age-associated alterations in the properties of membrane rafts include an increase in cholesterol content, impaired coalescence and selective differences in the recruitment of key proteins involved in TCR signaling. The localization of molecules through the membrane is dependent on posttranslational modifi-cations including acylation, farnesylation and palmitoylation.[43] Recently, it was demonstrated that LAT phosphorylation was not optimal in antigen-primed anergic CD4+ T-cells after TCR ligation.[44] More interestingly, LAT association with membrane rafts was defective in these CD4+ T-cells and this was partly explained by the impaired palmitoylation of LAT in these cells. Thus, the anergic state could be a consequence of changes in posttranslational modification of proteins which interact with membrane rafts. It is likely that anergic CD8+ cells are accumulated in aged humans. A large fraction of these cells are specific for viruses and become anergic for unknown reasons as a consequence of chronic antigenic stimulation, possibly resulting in exhaustion of their replicative potential (replicative senescence). Whether a change in the palmitoylation status of signaling molecules occurs during this process and provides useful information on possible interventions to modulate and restore cellular functions is not known but is under investigation in our group. It should be noted that TCR signaling alterations are different in different T-cell subsets, notably in CD4+ and CD8+ T-cells. Accumulating data show that CD4+ T-cells rely more on membrane rafts whereas CD8+ T-cells do not need such a process to signal properly.[45] This brings us to a discussion of raft heterogeneity in the immune system and its importance in the understanding of immunosenescence.

Heterogeneity in Membrane Rafts and T-Cell Subsets

Membrane rafts are not identical domains where the associated proteins are always the same. It is increasingly evident that a great deal of raft heterogeneity exists.[46] This became even clearer with the recent data from Douglass et al who demonstrated that signaling molecules are not restricted to one particular type of membrane raft.[47] Commonly, membrane rafts have been labelled and identified using the cholera toxin b subunit (CTxB) which targets the ganglioside M1, as a marker for membrane rafts. This study showed that signaling molecules such as LAT and Lck cocluster in domains of the membrane and that this was dependent on protein-protein interactions. This conclusion was based on the fact that Lck and LAT did not colocalize with CTxB fluorescence.[47] Now we know that membrane rafts are not restricted to those positive for this marker but other types such as GM3-rafts and flotillin-rafts also exist.[48-49] When CD4+ and CD8+ T-cells were analyzed for signaling molecule content in membrane rafts, we were able to demonstrate an age-associated decrease in LAT and Lck association and activation in the CD4+ subset while the CD8+ subset suffered less from aging in this respect.[45] There is a paradox here because it is the CD8+ not CD4+ compartment where CD28 expression is most markedly decreased with age and where anergic cells are more prominent. However, our study was limited to exploring molecules associated only with the GM1-rafts. It is possible that other raft domains are altered in the CD8+ compartment but these have simply not yet been investigated. Ongoing studies in this direction will help to better understand the role of membrane rafts in order to increase our knowledge of the spatiotemporal variables in TCR signaling in different T-cell subsets.

Protein-protein interactions direct signaling events but the lipid composition of the membrane facilitates protein movement from one raft to another to control protein-protein interactions. Recently, it was shown that inhibitory receptors such as the inhibitory killer immunoglobulin-like receptor 2DL2 (KIR2DL2) are excluded from membrane rafts and the immune synapse during the first steps of cellular activation but these receptors are recruited when cell functions needed to be down-regulated.[50] Our group has focussed on CMV-specific CD8+ cells which express the killer cell lectin-like receptor G1 (KLRG-1) and exhibit some dysfunctionalities. The expression of KLRG-1 and other important receptors of this type (CD161, KLRF-1 and NKG2A) and their localization in membrane rafts are under investigation. Results of these studies will help us to determine the role of such receptors in the inhibition of TCR signaling in aging.[51] Moreover, we have also shown that phosphatase activity in the membrane rafts of neutrophils could be responsible for age-related changes in susceptibility to apoptosis.[52] Although there is still little data, inhibitory signaling following TCR stimulation will need to be taken into account to explain changes in TCR signaling in aging.

MIRR Signaling and Autoimmunity

Rheumatoid Arthritis

Antibody production is a major step in the initiation and maintenance of autoimmune diseases. It has been shown that uncoupling of the B-cell antigen receptor (BCR) from calcineurin-dependent signaling pathways prevents self-antigen stimulation, such as by CpG DNA and thus prevents pro-liferation.[53] Concomitantly, a continuous activation of the extracellular signal-regulated kinase by self-antigen hinders cellular differentiation. In RA, B-cell tolerance is broken and auto-antibodies are secreted. RA patients suffer from a defect in central and peripheral B-cell tolerance.[54] Samuels et al showed that half of RA patients display unexpected immunoglobulin light chain repertoires.[54] Receptor editing and the regulation of recombination may be impaired and defective in RA. Because of this defect, BCR signaling will escape from control and assist in autoimmunity.

It was recently shown that MAPK activation in T-cells results in matrix metalloproteinases production by osteoclasts.[55] T-cells play a central role in the development and maintenance of RA mainly due to cytokine production.[56] Hence, the changes in TCR signaling have critical effects on activation of transcription factors and cytokine production. Synovial T-cells have a particular functional phenotype (hypo-responsiveness to TCR stimulation) which can be explained by the microenvironment of the synovial joint.[57] Chronic exposure of T-cells to tumour necrosis factor alpha (TNF-α) leads to disruption of TCR/CD3 assembly in the membrane which results in se-verely reduced calcium influx response.[58] In parallel, it was shown that TNF-α also down-regulated CD28 expression.[59] The link between increased TNF-α levels and the maintenance of RA is well-known. Thus, a change in TCR signaling caused by a differential MIRR spatial organization and coreceptor expression could be responsible for the hypo-responsiveness to TCR ligation in RA. This is not restricted to RA because elderly individuals commonly possess increased circulating TNF-α levels which can be seen as part of the "Inflam-Aging" process, a low-grade inflammation primarily caused by cytokines and oxidative stress.[60] Although there are some significant changes in TCR signal transduction, RA is mostly associated with changed B-cell functions as well as extrinsic changes including the cytokine environment. This might explain why changes in TCR signaling are mainly found in synovial T-cells.

Systemic Lupus Erythematosus

Recent studies have revealed an immunological disorder mainly in lymphocytes of SLE patients. SLE T-cells show an altered CD4:CD8 ratio, which is due to a decreased proportion of CD4+ T-cells and a concomitant increase in the proportion of CD8+ T-cells.[61] This parallels the inverted CD4:CD8 ratio which is a hallmark of the "immune risk phenotype" (IRP) predicting mortality in the very elderly.[62] T lymphocytes from SLE patients present abnormalities in TCR signaling which may also be to some extent similar to those found in the elderly. Abnormal expression of key signal-ing molecules and defective functions of T lymphocytes play a significant role in the pathogenesis of SLE. It is well-accepted that the expression of TCR zeta chain is defective in the majority of SLE patients.[63] However, TCR-mediated stimulation of SLE T-cells shows over-phosphorylation and different calcium response patterns when compared to healthy individuals but the outcome is the decreased IL-2 production. This phenomenon has been recently explained by Juang et al who demonstrated that cAMP response element modulator binds to the IL-2 promoter and sup-presses IL-2 production.[64] While it is not known why the zeta chain is down-regulated in SLE, there are several possibilities to explain the increased activation state of SLE T-cells that leads to their dysfunction upon stimulation. Of these, the altered structure of the receptors, the modula-tion of membrane clustering, the altered association of signaling molecules to membrane rafts or impaired inhibitory signaling may be important. For example, SLE T-cells have been shown to form greater amounts of GM1-rafts which are remarkably similar to those observed in T-cells of elderly donors. The alterations in the membrane raft signaling machinery represent an important mechanism that is responsible for the basal hyper-activated state of T-cells in SLE.[65] Jury et al were the first to describe a role for membrane rafts to explain the changes in SLE T-cell responses.[66] This study clearly showed that although there is a decrease in overall Lck expression at the cellular level, there is an increased activation of Lck in membrane rafts which explains the basal activation of SLE T-cells. This in turn leads to the inability to produce IL-2 upon stimulation to the same level as T-cells from control groups.[66] There is a correlation between aging and SLE since we previously described an increased basal phosphorylation of Lck in membrane rafts in T-cells from elderly donors which interferes with the proper signaling and IL-2 production.[41] Above, we discussed the putative role of phosphatases and inhibitory molecules in the changes of TCR signaling in aging. It was demonstrated that in SLE, CD45 is over-associated to membrane rafts which can explain changes of Lck activity by dephosphorylating Tyr505.[66] The localization of CD45 is important for its activity. Immunosenescence and SLE are very different phenomena. The first is a normal process with slow changes that results from immune remodelling over the lifespan, while the latter displays an earlier and more intense dysfunction of the immune system. However, despite this difference, it may be possible to increase our understanding of each by studying the other. The changes in TCR signaling in aging, RA and SLE, are described in Figure 1. Some investigators hypothesize that autoimmune diseases are premature amplifications of certain changes which occur more slowly during "normal" aging,[67] such as defects of T-cell selection, receptor functioning and apoptosis resistance. This notion can easily be extended to include changes in membrane properties and MIRR signaling. The modulation of such properties is of great interest for putative interventions to achieve improved immune functions or to control autoimmune diseases.

Therapeutic Strategies Targeting the MIRR Signaling

In order to influence cellular activation, the modulation of membrane rafts would be of importance not only for the TCR but all MIRRs. Using methyl β-cyclodextrin, a cholesterol-extracting molecule, it is possible to inhibit TCR signaling by modulating membrane raft properties. The activation of Lck and LAT and their association with membrane rafts are inhibited by methyl β-cyclodextrin.[40] It is of note that cyclodextrins are commonly used as vehicles for many drugs because they are functional excipients that increases drug solubility.[68] Some anti-fungal creams (Itraconazole),[69] hepatitis C drugs (PG301029),[70] anti-inflammatory drugs (Meloxicam®),[71] and many others[72] contain cyclodextrins. The use of cyclodextrins as vehicles can be valuable when T-cell function is specifically targeted. However, cyclodextrins are nonspecific polymers and one should take this into account for drug delivery as well as for side effects.

Protein-protein interactions as well as signaling molecule localization depend on posttranslational modifications of the proteins. The posttranslational modification of amino acids extends the range of functions of the protein by attaching other functional groups such as acetate, phosphate, lipids and carbohydrates. Palmitoylation, myristoylation, farnesylation and prenylation are very important for protein localization and interaction with adjacent molecules.[73] Hundt et al recently showed that LAT palmitoylation was defective in anergized CD4+ T-cells.[44] This can explain its altered association with membrane rafts and the central supra-molecular activation cluster (c-SMAC) of the immune synapse. The control of such a process could benefit cells which are hyper- or hypo-responsive. There is already some encouraging progress in this direction. For example, Garcia et al were able to reverse the age-related decrease in murine CD4+ T-cells using O-sialoglycoprotein endopeptidase.[74] The expression of activation markers such as CD25 and CD69, was restored using this approach.

MIRR assembly and signaling depend on the physico-chemical properties of the membrane. It has been shown that modulating the lipid composition of the extracellular milieu may lead to significant changes in membrane composition that ultimately result in perturbation of cellular functions.[75] We have seen previously that LAT can be displaced from membrane domains by changing its palmitoylation status. Stulnig et al have shown that the addition of polyunsaturated fatty acids to T-cell cultures in vitro leads to modifications of membrane rafts,[76,77] in particular, displacement of LAT.[76] This has direct effects on TCR signaling.[77] Therefore, we have here a real possibility to modulate T-cell and B-cell immune functions via their receptors by nutritional supplementation or by changes in food intake. Using this model, in our own study, healthy young donors were supplemented intravenously for 2 hours with a mixture of lipids (Intralipid 20%) which contains mainly palmitic, oleic and linoleic acids.[78] Blood samples were collected before and after injection and T-cells were isolated for further analysis. This study demonstrated that increases in lipid plasma levels have a direct effect on T-cell functions including signaling following TCR stimulation, IL-2 production and cell proliferation. This is of particular interest when we consider that this lipid

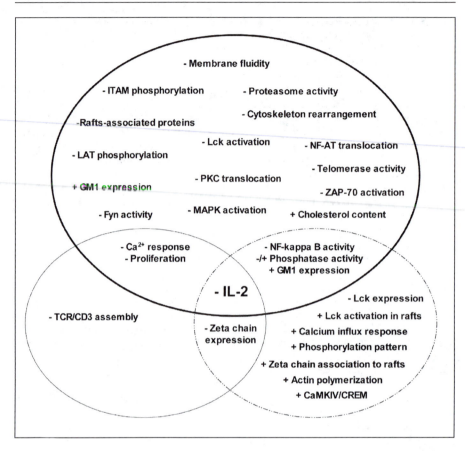

Figure 1. Comparative changes in TCR signaling in aging vs RA and SLE. The main changes in TCR signaling are depicted for T-cells from elderly donors (bold line), RA patients (continuous line) and SLE patients (discontinuous line). RA T-cells from the synovial joint mostly display resistance to activation. Peripheral T-cells from RA patients do not exhibit such changes. Decrease in the activation of molecules involved in TCR signal transduction is the hallmark of T-cells from elderly donors. SLE T-cells exhibit a very different pattern which consists of increased early signal transduction such as calcium influx and protein tyrosine phosphorylation but the end result is still a reduced capacity to produce IL-2. This is explained by CREM binding to the IL-2 promoter. Quantitatively, RA T-cells exhibit minor changes when compared to aging T-cells and SLE T-cells. The role of cytokines in autoimmune diseases is predominant, which explains why less attention has been paid to TCR signaling. Nevertheless, this pro-inflammatory environment will certainly influence TCR signaling. Increase in signaling events is represented by (+) while a decrease is represented by (-). Common changes are also mentioned. PKC, protein kinase C; ITAM, immunoreceptor tyrosine-based activation motif; LAT, linker of activated T-cells; MAPK, mitogen-activated protein kinase; GM1, gangliosides M1; CREM, cyclic adenosine monophosphate response element modulator; CaMKIV, Ca^{2+}/Calmodulin-dependent Protein Kinases IV.

supplementation is often given to hospitalized patients. One should reconsider the balance between beneficial and side effects of this supplementation in the case of immuno-depressed patients.[79] It is known that elderly individuals have a very different nutritional intake than young people.[80] Increased lifespan is due to better health services, vaccination and a better quality of life which includes food intake. Nevertheless, this can be improved even more because elderly individuals often have disturbed eating patterns which may not provide optimal nutrition.

Aging is associated with an increase in the number of anergic virus-specific CD8+ T-cells. Patients with certain autoimmune diseases or cancer may also display an increase in anergic and hypo-responsive T-cells. The possible causes of such phenomena have been described in this chapter focussing on the role of defects in TCR signaling. The accumulation of these anergic cells may influence the functioning of the other cells in several ways (e.g., competing for antigen, competing for cytokines, secreting suppressive cytokines, etc). Using tetramer technology, it is possible to detect antigen-specific cells and in combination with functional tests such as IFN-γ production[81] to assess which cells are responding and which belong to the anergic population. It may be also possible to selectively deplete the anergic population and thereby reconstitute immune function, however, it is not an easy task to perform on humans as yet.

Concerning other viral infections, it is worth mentioning that HIV, which enters T-cells via the TCR and TCR coreceptors such as CD4 causes a shift in phospholipid synthesis to neutral lipids and also causes polyunsaturated fatty acid peroxidation and deregulation of cytokine production.[82] We have the example here of a virus which is able to modulate the TCR environment in order to subvert cell proliferation according to its needs (viral replication), or for cell death initiation. How viruses can modulate TCR signaling is of great interest to open new windows for future strategies to modulate T-cell signaling and improve function in immuno-deficient individuals (elderly) as well as in pathological situations (auto-immune diseases, cancer).

Conclusion

TCR signaling changes with age as well as in autoimmunity and there are some resemblances between these two phenomena. The most important point is that protein-protein interactions and T-cell activation in an elderly population are very different from those in young healthy individuals and it helps to explain the changes in cellular function. Membrane rafting is critical for the assembly of the signaling platform for the TCR and BCR (and also for other receptors that we did not discuss here). The final outcome of protein rafting is the formation of the immunological synapse which is needed for sustained activation resulting in a complete immune response. We can document changes in molecular events with age and autoimmunity, but we are not yet able to explain these changes. Understanding the events that lead to changes in the TCR signaling cascade would be of great benefit considering the large number of diseases in which membrane raft dysfunction is thought to play a role.

Acknowledgements

The authors' own work was supported by the Deutsche Forschungsgemeinschaft (SFB 685) and the European Commission (QLK6-CT-2002-02283, "T-CIA"). The authors were all supported by the European Commission (EU contract 6FP-CT-2003-506850, "ZINCAGE"). Tamas Fulop was supported by the Canadian Institute of Health Research (No 63149), the Research Center on Aging and the University of Sherbrooke.

References

1. Crews DE, Zavotka S. Aging, disability and frailty: Implications for universal design. J Physiol Anthropol 2006; 25:113-8.
2. Wiet SG. Future of caring for an aging population: Trends, technology and caregiving. Stud Health Technol Inform 2005; 118:220-30.
3. Webster RG. Immunity to influenza in the elderly. Vaccine 2000; 18:1686-9.
4. Levy R. Costs and benefits of pharmaceuticals: The value equation for older Americans. Care Manag J 2002; 3:135-42.
5. Pawelec G, Adibzadeh M, Solana R et al. The T-cell in the ageing individual. Mech Ageing Dev 1997; 93:35-45.
6. Pawelec G. Immunosenescence and human longevity. Biogerontology 2003; 4:167-70.
7. Makinodan T. Nature of the decline in antigen-induced humoral immunity with age. Mech Ageing Dev 1980; 14:165-72.
8. Aspinall R Andrew D. Thymic involution in aging. J Clin Immunol 2000; 20:250-6.

9. Pawelec G, Akbar A, Caruso C et al. Human immunosenescence: Is it infectious? Immunol Rev 2005; 205:257-68.
10. Linton PJ, Haynes L, Tsui L et al. From naive to effector-alterations with aging. Immunol Rev 1997; 160:9-18.
11. Vallejo AN, Brandes JC, Weyand CM et al. Modulation of CD28 expression: Distinct regulatory pathways during activation and replicative senescence. J Immunol 1999; 162:6572-9.
12. Dennett NS, Barcia RN, McLeod JD. Age associated decline in CD25 and CD28 expression correlate with an increased susceptibility to CD95 mediated apoptosis in T-cells. Exp Gerontol 2002; 37:271-83.
13. Sandmand M, Bruunsgaard H, Kemp K et al. Is ageing associated with a shift in the balance between Type 1 and Type 2 cytokines in humans? Clin Exp Immunol 2002; 127:107-14.
14. Effros RB, Dagarag M, Spaulding C et al. The role of CD8+ T-cell replicative senescence in human aging. Immunol Rev 2005; 205:147-57.
15. Fulop T, Larbi A, Wikby A et al. Dysregulation of T-cell function in the elderly: Scientific basis and clinical implications. Drugs Aging 2005; 22:589-603.
16. van Dijk-Hard I, Soderstrom I, Feld S et al. Age-related impaired affinity maturation and differential D-JH gene usage in human VH6-expressing B lymphocytes from healthy individuals. Eur J Immunol 1997; 27:1381-6.
17. Miller JP, Allman D. Linking age-related defects in B lymphopoiesis to the aging of hematopoietic stem cells. Semin Immunol 2005; 17:321-9.
18. Mocchegiani E, Malavolta M. NK and NKT-cell functions in immunosenescence. Aging Cell 2004; 3:177-84.
19. DelaRosa O, Tarazona R, Casado JG et al. Valpha24+ NKT-cells are decreased in elderly humans. Exp Gerontol 2002; 37:213-7.
20. Sebastian C, Espia M, Serra M et al. MacrophAging: A cellular and molecular review. Immunobiology 2005; 210:121-6.
21. Fulop T, Larbi A, Douziech N et al. Signal transduction and functional changes in neutrophils with aging. Aging Cell 2004; 3:217-26.
22. Franceschi C, Bonafe M, Valensin S et al. Inflamm-aging. An evolutionary perspective on immunosenescence. Ann N Y Acad Sci 2000; 908:244-54.
23. Garcia GG, Miller RA. Single-cell analyses reveal two defects in peptide-specific activation of naive T-cells from aged mice. J Immunol 2001; 166:3151-7.
24. Fulop Jr T, Gagne D, Goulet AC et al. Age-related impairment of p56lck and ZAP-70 activities in human T lymphocytes activated through the TcR/CD3 complex. Exp Gerontol 1999; 34:197-216.
25. Kawanishi H. Activation of calcium (Ca)-dependent protein kinase C in aged mesenteric lymph node T and B-cells. Immunol Lett 1993; 35:25-32.
26. Whisler RL, Newhouse YG, Bagenstose SE. Age-related reductions in the activation of mitogen-activated protein kinases p44mapk/ERK1 and p42mapk/ERK2 in human T-cells stimulated via ligation of the T-cell receptor complex. Cell Immunol 1996; 168:201-10.
27. Mustelin T, Rahmouni S, Bottini N et al. Role of protein tyrosine phosphatases in T-cell activation. Immunol Rev 2003; 191:139-47.
28. Altin JG, Sloan EK. The role of CD45 and CD45-associated molecules in T-cell activation. Immunol Cell Biol 1997; 75:430-45.
29. Whisler RL, Beiqing L, Chen M. Age-related decreases in IL-2 production by human T-cells are associated with impaired activation of nuclear transcriptional factors AP-1 and NF-AT. Cell Immunol 1996; 169:185-95.
30. Ponnappan S, Uken-Trebilcock G, Lindquist M et al. Tyrosine phosphorylation-dependent activation of NFkappaB is compromised in T-cells from the elderly. Exp Gerontol 2004; 39:559-66.
31. Simons K, Ikonen E. Functional rafts in cell membranes. Nature 1997; 387:569-72.
32. Janes PW, Ley SC, Magee AI. Aggregation of lipid rafts accompanies signaling via the T-cell antigen receptor. J Cell Biol 1999; 147:447-61.
33. Hancock JF. Lipid rafts: Contentious only from simplistic standpoints. Nat Rev Mol Cell Biol 2006; 7:456-62.
34. Kusumi A, Suzuki K. Toward understanding the dynamics of membrane-raft-based molecular interactions. Biochim Biophys Acta 2005; 1746:234-51.
35. Balamuth F, Brogdon JL, Bottomly K. CD4 raft association and signaling regulate molecular clustering at the immunological synapse site. J Immunol 2004; 172:5887-92.
36. Manes S, Viola A. Lipid rafts in lymphocyte activation and migration. Mol Membr Biol 2006; 23:59-69.
37. Huber LA, Xu QB, Jurgens G et al. Correlation of lymphocyte lipid composition membrane microviscosity and mitogen response in the aged. Eur J Immunol 1991; 21:2761-5.

38. Simons K, Ehehalt R. Cholesterol, lipid rafts and disease. J Clin Invest 2002; 110:597-603.
39. Garcia GG, Miller RA. Single-cell analyses reveal two defects in peptide-specific activation of naive T-cells from aged mice. J Immunol 2001; 166:3151-7.
40. Larbi A, Douziech N, Khalil A et al. Effects of methyl-beta-cyclodextrin on T lymphocytes lipid rafts with aging. Exp Gerontol 2004; 39:551-8.
41. Larbi A, Douziech N, Dupuis G et al. Age-associated alterations in the recruitment of signal-transduction proteins to lipid rafts in human T lymphocytes. J Leukoc Biol 2004; 75:373-81.
42. Kovacs B, Parry RV, Ma Z et al. Ligation of CD28 by its natural ligand CD86 in the absence of TCR stimulation induces lipid raft polarization in human CD4 T-cells. J Immunol 2005; 175:7848-54.
43. Resh MD. Membrane targeting of lipid modified signal transduction proteins. Subcell Biochem 2004; 37:217-32.
44. Hundt M, Tabata H, Jeon MS et al. Impaired activation and localization of LAT in anergic T-cells as a consequence of a selective palmitoylation defect. Immunity 2006; 24:513-22.
45. Larbi A, Dupuis G, Khalil A et al. Differential role of lipid rafts in the functions of CD4+ and CD8+ human T lymphocytes with aging. Cell Signal 2006; 18:1017-30.
46. Pike LJ. Lipid rafts: heterogeneity on the high seas. Biochem J 2004; 378:281-92.
47. Douglass AD, Vale RD. Single-molecule microscopy reveals plasma membrane microdomains created by protein-protein networks that exclude or trap signaling molecules in T-cells. Cell 2005; 121:937-50.
48. Grauby-Heywang C, Turlet JM. Behavior of GM3 ganglioside in lipid monolayers mimicking rafts or fluid phase in membranes. Chem Phys Lipids 2006; 139:68-76.
49. Langhorst MF, Reuter A, Stuermer CA. Scaffolding microdomains and beyond: The function of reggie/flotillin proteins. Cell Mol Life Sci 2005; 62:2228-40.
50. Henel G, Singh K, Cui D et al. Uncoupling of T-cell effector functions by inhibitory killer immunoglobulin-like receptors. Blood 2006; 107:4449-57.
51. Ouyang Q, Wagner WM, Voehringer D et al. Age-associated accumulation of CMV-specific CD8+ T-cells expressing the inhibitory killer cell lectin-like receptor G1 (KLRG1). Exp Gerontol 2003; 38:911-20.
52. Fortin CF, Larbi A, Lesur O et al. Impairment of SHP-1 down-regulation in the lipid rafts of human neutrophils under GM-CSF stimulation contributes to their age-related, altered functions. J Leukoc Biol 2006; 79:1061-72.
53. Rui L, Vinuesa CG, Blasioli J et al. Resistance to CpG DNA-induced autoimmunity through tolerogenic B-cell antigen receptor ERK signaling. Nat Immunol 2003; 4:594-600.
54. Samuels J, Ng YS, Coupillaud C et al. Impaired early B-cell tolerance in patients with rheumatoid arthritis. J Exp Med 2005; 201:1659-67.
55. Rifas L, Arackal S. T-cells regulate the expression of matrix metalloproteinase in human osteoblasts via a dual mitogen-activated protein kinase mechanism. Arthritis Rheum 2003; 48:993-1001.
56. Brennan F, Foey A. Cytokine regulation in RA synovial tissue: Role of T-cell/macrophage contact-dependent interactions. Arthritis Res 2002; 4:S177-82.
57. Romagnoli P, Strahan D, Pelosi M et al. A potential role for protein tyrosine kinase p56(lck) in rheumatoid arthritis synovial fluid T lymphocyte hyporesponsiveness. Int Immunol 2001; 13:305-12.
58. Cope AP. Studies of T-cell activation in chronic inflammation. Arthritis Res 2002; 4:S197-211.
59. Lewis DE, Merched-Sauvage M, Goronzy JJ et al. Tumor necrosis factor-alpha and CD80 modulate CD28 expression through a similar mechanism of T-cell receptor-independent inhibition of transcription. J Biol Chem 2004; 279:29130-8.
60. Bruunsgaard H. Effects of tumor necrosis factor-alpha and interleukin-6 in elderly populations. Eur Cytokine Netw 2002; 13:389-91.
61. Pavon EJ, Munoz P, Navarro MD et al. Increased association of CD38 with lipid rafts in T-cells from patients with systemic lupus erythematosus and in activated normal T-cells. Mol Immunol 2006; 43:1029-39.
62. Hadrup SR, Strindhall J, Kollgaard T et al. Longitudinal studies of clonally expanded CD8 T-cells reveal a repertoire shrinkage predicting mortality and an increased number of dysfunctional cytomegalovirus-specific T-cells in the very elderly. J Immunol 2006; 176:2645-53.
63. Nambiar MP, Mitchell JP, Ceruti RP et al. Prevalence of T-cell receptor zeta chain deficiency in systemic lupus erythematosus. Lupus 2003; 12:46-51.
64. Juang YT, Wang Y, Solomou EE et al. Systemic lupus erythematosus serum IgG increases CREM binding to the IL-2 promoter and suppresses IL-2 production through CaMKIV. J Clin Invest 2005; 115:996-1005.
65. Krishnan S, Nambiar MP, Warke VG et al. Alterations in lipid raft composition and dynamics contribute to abnormal T-cell responses in systemic lupus erythematosus. J Immunol 2004; 172:7821-31.
66. Jury EC, Kabouridis PS, Flores-Borja F et al. Altered lipid raft-associated signaling and ganglioside expression in T lymphocytes from patients with systemic lupus erythematosus. J Clin Invest 2004; 113:1176-87.

67. Sibilia J. Novel concepts and treatments for autoimmune disease: Ten focal points. Joint Bone Spine 2004; 71:511-7.
68. Shimpi S, Chauhan B, Shimpi P. Cyclodextrins: Application in different routes of drug administration. Acta Pharm 2005; 55:139-56.
69. Groll AH, Wood L, Roden M et al. Safety, pharmacokinetics and pharmacodynamics of cyclodextrin itraconazole in pediatric patients with oropharyngeal candidiasis. Antimicrob Agents Chemother 2002; 46:2554-63.
70. Johnson JL, He Y, Jain A et al. Improving cyclodextrin complexation of a new antihepatitis drug with glacial acetic acid. AAPS PharmSciTech 2006; 7:E18.
71. Ghorab MM, Abdel-Salam HM, El-Sayad MA et al. Tablet formulation containing meloxicam and beta-cyclodextrin: mechanical characterization and bioavailability evaluation. AAPS Pharm Sci Tech 2004; 5:e.59.
72. Loftsson T, Masson M. Cyclodextrins in topical drug formulations: Theory and practice. Int J Pharm 2001; 225:15-30.
73. Magee T, Seabra MC. Fatty acylation and prenylation of proteins: What's hot in fat. Curr Opin Cell Biol 2005; 17:190-6.
74. Garcia GG, Miller RA. Age-related defects in CD4+ T-cell activation reversed by glycoprotein endopeptidase. Eur J Immunol 2003; 33:3464-72.
75. Cavaglieri CR, Nishiyama A, Fernandes LC et al. Differential effects of short-chain fatty acids on proliferation and production of pro- and anti-inflammatory cytokines by cultured lymphocytes. Life Sci 2003; 73:1683-90.
76. Zeyda M, Staffler G, Horejsi V et al. LAT displacement from lipid rafts as a molecular mechanism for the inhibition of T-cell signaling by polyunsaturated fatty acids. J Biol Chem 2002; 277:28418-23.
77. Stulnig TM, Berger M, Sigmund T et al. Polyunsaturated fatty acids inhibit T-cell signal transduction by modification of detergent-insoluble membrane domains. J Cell Biol 1998; 143:637-44.
78. Larbi A, Grenier A, Frisch F et al. Acute in vivo elevation of intravascular triacylglycerol lipolysis impairs peripheral T-cell activation in humans. Am J Clin Nutr 2005; 82:949-56.
79. Calder PC. n-3 fatty acids, inflammation and immunity—Relevance to postsurgical and critically ill patients. Lipids 2004; 39:1147-61.
80. Roberts SB, Rosenberg I. Nutrition and aging: Changes in the regulation of energy metabolism with aging. Physiol Rev 2006; 86:651-67.
81. Ouyang Q, Wagner WM, Wikby A et al. Compromised interferon gamma (IFN-gamma) production in the elderly to both acute and latent viral antigen stimulation: Contribution to the immune risk phenotype? Eur Cytokine Netw 2002; 13:392-4.
82. Raulin J. Human immunodeficiency virus and host cell lipids. Interesting pathways in research for a new HIV therapy. Prog Lipid Res 2002; 41:27-65.

CHAPTER 22

Viral Pathogenesis, Modulation of Immune Receptor Signaling and Treatment

Walter M. Kim and Alexander B. Sigalov*

Abstract

During the co-evolution of viruses and their hosts, the latter have equipped themselves with an elaborate immune system to defend themselves from the invading viruses. In order to establish a successful infection, replicate and persist in the host, viruses have evolved numerous strategies to counter and evade host antiviral immune responses as well as exploit them for productive viral replication. These strategies include those that target immune receptor transmembrane signaling. Uncovering the exact molecular mechanisms underlying these critical points in viral pathogenesis will not only help us understand strategies used by viruses to escape from the host immune surveillance but also reveal new therapeutic targets for antiviral as well as immunomodulatory therapy. In this chapter, based on our current understanding of transmembrane signal transduction mediated by multichain immune recognition receptors (MIRRs) and the results of sequence analysis, we discuss the MIRR-targeting viral strategies of immune evasion and suggest their possible mechanisms that, in turn, reveal new points of antiviral intervention. We also show how two unrelated enveloped viruses, human immunodeficiency virus and human cytomegalovirus, use a similar mechanism to modulate the host immune response mediated by two functionally different MIRRs—T-cell antigen receptor and natural killer cell receptor, NKp30. This suggests that it is very likely that similar general mechanisms can be or are used by other viral and possibly nonviral pathogens.

Introduction

Facing the destructive consequences of microbial infections, the human immune system has evolved two arms of host defense designed to discriminate foreign agents and mount appropriate effector responses: the innate and adaptive immune systems. Differing primarily in their receptors and receptor specificities, the innate immune system functions as the early and immediate defense mechanism and recognizes a broad set of conserved and invariant properties of nonself agents, such as viruses, through a diverse set of germ-line encoded pattern recognition receptors (PRRs), including members of the toll-like receptor (TLR) family and the retinoic acid inducible gene 1 (RIG-1)-like helicases.[1-3] In contrast, the adaptive arm of the immune system is the more slow-responding defense mechanism but the more pathogen-specific; infectious antigens are

*Corresponding Author: Alexander B. Sigalov—Department of Pathology, University of Massachusetts Medical School, 55 Lake Avenue North, Worcester, MA 01655, USA. Email: alexander.sigalov@umassmed.edu

Multichain Immune Recognition Receptor Signaling: From Spatiotemporal Organization to Human Disease, edited by Alexander B. Sigalov. ©2008 Landes Bioscience and Springer Science+Business Media.

processed in antigen-presenting cells (APCs), presented in the context of major histocompatibility complex (MHC) class I or II molecules and are recognized by somatically generated receptors on antigen-specific T-cells that are ultimately activated and perform effector functions. Collectively, the innate and adaptive immune systems work cooperatively to defend against infection, pathogenic proliferation and disease.

In order to persist in an immunocompetent host, viruses in particular have been described to have developed intricate strategies to evade the innate immune system.[4-11] Following viral infection and recognition of viral components by PRRs,[1,12-16] innate immune cells, such as dendritic cells and macrophages, normally respond robustly with secretion of type I interferons (IFNs), a group of pro-inflammatory cytokines that upregulate numerous interferon-stimulated genes (ISGs);[8,17-20] overexpression of ISGs initiates a series of antiviral, antiproliferative and immunoregulatory responses against the infected cell.[2,8,20-23] A number of viruses, including influenza and herpesvirus, employ diverse counteracting mechanisms to disrupt the IFN regulatory pathway at nearly every step, including blocking IFN induction/expression, intercepting binding of IFNs to their natural target receptors, modulating intracellular IFN-mediated signaling pathways and finally downregulation of ISG expression.[24-27] By disrupting the IFN regulatory pathway, viruses are able to attenuate the antiviral properties of type I IFNs and survive recognition by the innate immune system.

Because type I IFNs also upregulate expression of MHC class I and II proteins,[28-35] virus-mediated disruption of normal IFN activity has been suggested to not only interrupt innate immunity but adaptive immunity as well.[2] Other unrelated viruses, namely human immunodeficiency virus (HIV), human T-cell lymphotrophic virus (HTLV) and human cytomegalovirus (HCMV), have also developed strategies that modulate innate and adaptive immune processes, but do not involve type I IFNs nor IFN regulatory pathways. In contrast, HIV, HTLV and HCMV target members of the family of multichain immune recognition receptors (MIRRs) found on immune cells and either disrupt or surprisingly augment MIRR-mediated activation signaling as required for self-preservation. Predicted and explained by the signaling chain homooligomerization (SCHOOL) model (see also Chapters 12 and 20),[36] numerous unrelated viruses employ viral proteins either to (1) disrupt intermolecular transmembrane (TM) interactions between recognition and signaling subunits of MIRRs in an effort to disarm the receptor or (2) cluster the signaling subunits to activate or augment MIRR-triggered signaling. More interesting, these viruses have exquisitely incorporated targeting and manipulation of MIRR signaling in viral processes essential to the viral life cycle: viral entry, membrane targeting and viral escape and replication. By overlapping multiple functions in a single viral protein product, the virus is able to maintain a simple genome conducive to rapid replication but have the added benefit of diverse functionality.

In this chapter, we discuss an intriguing principle of convergence for a number of divergent viruses in their strategic choice to uniformly target MIRRs. Our investigation of how seemingly disparate viruses target a single family of membrane receptors exposes a redundancy in viral strategies exploiting the host innate and adaptive immune systems. MIRR-targeted strategies disrupting the MIRR TM architecture from the extracellular space as well as virus-induced clustering of MIRR signaling subunits from the cytoplasmic space (Fig. 1) will be described for a select group of viruses that are functionally disparate, target different host cells and differ in their replication strategies. We will also display the power of the SCHOOL model-guided primary sequence evaluation for a number of additional viruses and its ability to predict additional MIRR-targeting viral agents not previously conceived. Furthermore, by understanding the mechanisms viruses have developed over centuries of evolution to modulate MIRR-mediated triggering in the immune response, we gain insight into the fundamental details of the mechanisms underlying normal MIRR-mediated immune activation processes and can begin to learn how to take advantage of these optimized processes. Finally, the learned viral strategies and newly developed concepts of MIRR signaling can be translated towards new lines of rational drug design efforts targeting MIRRs and modulation of immune activation (see also Chapter 20). MIRR-targeted strategies stretch beyond the specific viruses discussed in this chapter and represent a surprising junction in viral strategies. Whether

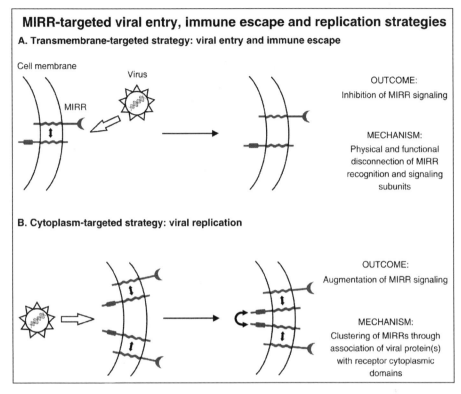

Figure 1. Targeting MIRRs: suggested immunomodulatory strategies used by viruses to entry target cells, survive and replicate. Transmembrane interactions between MIRR recognition and signaling subunits are shown by black arrows. Immunoreceptor tyrosine-based activation motifs (ITAMs) are shown as gray rectangles. Circular arrow indicates viral agent-induced receptor clustering.
Abbreviation: MIRR: multichain immune recognition receptor.

this strategy represents a convergence in evolution of disparate viruses or hints towards a similar evolutionary origin from which viruses have diverged remains to be determined.

Viruses: Classification and Pathogenesis

One of the quandaries encompassing virology and virologic discovery has been the difficulty in the classification or grouping of viruses. Although a single taxonomy governing the naming of viruses has been well-established, numerous classification methods have been suggested, highlighting similarities in virion structure, target organ systems or genomic composition. Here we describe the principles underlying the development of the Baltimore classification method and its application towards segregating viruses based on replication methods and pathogenesis. However, as a consequence of viral classification and the strict segregation of viruses from one other, universal viral strategies linking differentially classified viruses have been tragically overlooked. We postulate that a number of viruses that lie in different classifications are only seemingly different and that generic immunomodulatory strategies targeting MIRRs serve as a surprisingly common tactic shared by them.

Viral Classification

Viruses represent a collection of infectious, obligate intracellular parasites that require a living host cell to replicate. They are comprised of either DNA or RNA, a virion capsid comprised of proteins encoded by the viral nucleic acid and depending on the specific virus, a surrounding envelope. Due to the high genetic, morphologic and pathogenic variability found among different viruses, classification has proven difficult. Early attempts to organize viruses were based on their structural organization, highlighting differences in nucleic acid (DNA vs RNA), virion symmetry, presence of an envelope and number of capsomers.[37,38] For example, one system of viral classification[37,38] developed by Lwoff, Horne and Tournier, the LHT system, merged all viruses under one phylum, Vira, then divided into two subphyla, subphylum Deoxyvira (DNA viruses) and Ribovira (RNA viruses) which then divided into classes based on virion symmetry and finally segregated by number of capsomers present in the infecting virus.

The most recent and widely accepted virus classification system is based on functional characterization that differentiates viruses based on their replication strategies and chemical nature of its nucleic acid. Coined the "Baltimore classification",[39] viruses are grouped into seven groups or classes, termed the "Baltimore Classes I-VII" (Table 1). Each group of viruses uses a different replication strategy, such as exploitation of the host polymerases (Group I) or direct translation of injected positive-sense RNA (Group IV). Although each viral group contains viruses with the same type of nucleic acid (i.e., positive-sense single stranded (ss)RNA, double stranded (ds)DNA, etc.), there is remarkable variation in virion symmetry and presence or absence of an envelope surrounding the virus (Table 1). Therefore, viral architecture and morphology don't necessarily correlate with function and structurally different viruses unexpectedly share common functions and strategies. In this section, we further describe how three seemingly unrelated viruses (Table 1; HIV, HCMV, HTLV) share a common targeted approach in their mutual ability to modulate the immune system to enhance viral entry, replication and pathogenesis: uniform exploitation of the architecture and function of different MIRRs to directly suppress or augment immune activation (Fig. 1).

Viral Pathogenesis

Despite the vast diversity in viruses and target cells in the human host, there is a common sequence of processes that serve as the foundation for all viral infection. First, the infecting virus must migrate to the primary site of infection, usually through direct inoculation, or through the respiratory, gastrointestinal or genitourinary route. The virus then undergoes a process of viral entry, including attachment, a physical connection of the virus to the target cell through a viral cell recognition protein-host receptor interaction and penetration, exit from the extracellular space and entry into the cellular environment. Once inside the target cell, the virus particle uncoats and releases its viral contents, including its nucleic acid genome, in preparation of viral replication. Depending on the nature of the nucleic acid and the Baltimore group classification, viral genes may be translated directly by the host cell translation machinery (i.e., Group IV positive-sense ssRNA viruses) or incorporated into the host genome (i.e., Group I dsDNA viruses). Regardless of whether the expressing transcript originates from the viral particle itself or integrated viral genes, mRNA transcripts are translated, localize to the site of maturation and assemble into virion particles, encapsulating the viral genome in the process. Depending on the enveloped property of the infecting virus, the viral particle either surrounds itself in host membrane during budding and release (enveloped viruses) or releases without an envelope (non-enveloped viruses). Released viral progeny are then free to infect other host cells and proliferate in the host organism.

Collectively, these processes represent the fundamental stages in viral pathogenesis shared amongst members of virtually every group and class of viruses. However, inside the fine details of each stage lay intricate subprocesses that aid in enhancing viral persistence and virulence. Unexposed until recently (see also Chapter 20)[36,40,41] is the universal targeting of MIRRs that multiple viruses have surreptitiously concealed in several viral processes, including viral entry, membrane targeting and viral replication. In particular, HIV and HCMV specifically target different receptors within the MIRR family during viral entry through extracellular targeting mechanisms (Fig. 1A)

Table 1. Baltimore classification of viruses

Family	Capsid	Envelope	Genome Size (kb)	Representative Virus*	Primary Target Cell/Organ System
Group I: dsDNA					
Adenoviridae	Icosahedral	No	26-45	Adenovirus 1	Epithelial tight junctions: heart, pancreas, nervous system, prostate, testis, lung, liver, intestine
Herpesviridae	Icosahedral	Yes	125-240	Human cytomegalovirus (HCMV)	Fibroblasts
Papovaviridae	Icosahedral	No	7-8	BK Virus (polyomavirus)	Kidney epithelium, lymphocytes
Poxviridae	Ovoid	Yes	130-375	Vaccinia virus	Broad tropism
Group II: positive-sense ssDNA					
Circoviridae	Icosahedral	No	2	Transmitted transfusion virus (TTV)	Oral and intestinal mucosa
Parvoviridae	Icosahedral	No	4-6	Adeno-associated virus (AAV)	Broad tropism
Group III: dsRNA					
Bornaviridae	Icosahedral	No	5-6	Borna disease virus	Broad tropism, neuronal cells
Reoviridae	Icosahedral	No	19-32	Human rotavirus	Small intestine enterocytes
Group IV: positive-sense ssRNA					
Astroviridae	Isometric	No	6-7	Astrovirus 1	Jejunum, ileum
Caliciviridae	Icosahedral	No	7-8	Norwalk virus	Upper GI tract, jejunum
Coronaviridae	Helical	Yes	28-31	SARS coronavirus (SARS-CoV)	Upper airway, alveolar epithelial cells
Flaviviridae	Spherical	Yes	10-12	Hepatitis C virus	Hepatocytes

Continued on next page

Table 1. *Continued*

Family	Capsid	Envelope	Genome Size (kb)	Representative Virus*	Primary Target Cell/Organ System
Picornaviridae	Icosahedral	No	7-9	Hepatitis A virus	Hepatocytes, intestinal mucosa
Togaviridae	Icosahedral	Yes	10-12	Rubella virus	Nasopharynx, lymph nodes
Group V: negative-sense ssRNA					
Arenaviridae	Helical filaments	Yes	11	Lymphocytic choriomeningitis virus (LCMV)	Broad tropism, hilar lymph nodes, lung parenchyma
				Lassa virus (LASV)	Dendritic cells, macrophages and other immune cells, hepatocytes, endothelial cells
				Mopeia virus (MOPV)	Dendritic cells, macrophages, endothelial cells
				Tacaribe virus (TACV)	Dendritic cells, macrophages
Bunyaviridae	Helical filaments	Yes	11-19	Hantaan virus	Lung parenchyma, lymph nodes, hematopoietic cells
Filoviridae	Helical filaments	Yes	19	Ebola virus	Broad tropism, mononuclear phagocytic system, mucosa
				Zaire Ebola virus (ZEBOV)	Mononuclear phagocytic system
				Sudan Ebola virus (SEBOV)	Mononuclear phagocytic system
Orthomyxoviridae	Helical filaments	Yes	10-15	Influenza virus A	Upper and lower respiratory tract
Paramyxoviridae	Helical filaments	Yes	13-18	Parainfluenza virus 1	Lower respiratory tract epithelium

Continued on next page

Table 1. Continued

Family	Capsid	Envelope	Genome Size (kb)	Representative Virus*	Primary Target Cell/Organ System
Rhabdoviridae	Helical filaments	Yes	11-15	Vesicular stomatitis virus	Oral mucosa
Group VI: positive-sense ssRNA-RT					
Retroviridae	Spherical	Yes	7-13	Human immunodeficiency virus (HIV)	Intestinal mucosa, T-cells, dendritic cells, macrophages, microglia
				Human T-cell lymphotropic virus type 1 (HTLV-1)	T-cells
Group VII: dsDNA+RT					
Hepadnaviridae	Icosahedral	Yes	3-4 kb	Hepatitis B virus	Hepatocytes

*Underlined are the viruses discussed in this chapter.
Abbreviations: ds: double-stranded; kb: kilobase(s); RT: reverse transcriptase; SARS: severe acute respiratory syndrome; ss: single-stranded.

whereas HIV and HTLV target MIRRs from the intracellular environment (Fig. 1B) during viral replication. Although these viruses selectively inhibit or augment MIRR-mediated activation of the target cell during different viral stages, viral persistence is universally enhanced either through disarmament of the immune response or enhancement of the replicative environment. By overlapping multiple processes in each viral stage, viruses have demonstrated a remarkable efficiency in their life cycle that emphasizes their advanced evolution. Intriguingly, MIRR-targeted functions enacted by viral proteins seem to be present at multiple checkpoints in viral pathogenesis, coinciding with several viral stages. Therefore, it is not unreasonable to propose that MIRRs represent a key component in the host cell that multiple viruses have ubiquitously evolved to target, disrupt or activate as desired (Fig. 1).

Viral Entry and Membrane Targeting

In order for a virus to proliferate, it must first undergo a process of attachment to the target host cell and then penetration either through fusion or direct access; collectively, these two processes comprise viral entry and are often actuated by a single protein molecule. Viral attachment has been a subject of intense investigation and several details regarding the necessary specificity of viruses for their host cells have emerged. Interestingly, disparate viruses overlap in their specificities for their primary natural receptors. For example, members of the coronaviruses (OC43),[42] orthomyxoviruses (Influenza A, B)[43,44] and reoviruses (T3)[45-47] contain surface receptors that are specific for sialic acid residues found on the host cell receptor whereas members of the picornaviridae (rhinoviruses, polioviruses)[48-51] and retroviruses (HIV-1)[52-56] bind surface receptors that adopt the canonical immunoglobulin fold such as intercellular adhesion molecule-1 (ICAM-1), the immunoglobulin G (IgG) superfamily and CD4, respectively. Although there is little sequence or structural similarity in their envelope or capsid proteins, these viruses exhibit redundancy in receptor specificity.

Following attachment, the virus penetrates the host cell either through fusion in the case of enveloped viruses or direct entry for non-enveloped viruses. Although the steps and strategies non-enveloped viruses use to enter cells are largely unknown, the events leading to viral fusion have been studied in great detail. Membrane fusion of enveloped viruses is mediated by fusion proteins that exist primarily as homo- or heterodimeric type I integral membrane proteins found embedded in the surrounding envelope.[57,58] Concealed in the fusion protein is the fusion sequence or fusion peptide (FP), a short hydrophobic sequence ranging from 3-6 to 24-36 amino acids, that serves as the primary mediator of virus-host cell membrane anchoring. Depending on the location of the FP and the structural nature of the fusion protein, fusion proteins are segregated in three types. Type I fusion proteins found in such viruses as influenza are comprised of alpha-helix coiled-coil domains that contain FPs at the N-terminus. Type II fusion proteins contain primarily beta-sheet structures and contain internal FP sequences. The third group of fusion proteins do not fall in the type I and I classifications and are found in such viruses as coronaviruses and herpesviruses.

After translation in the host cell, type I and II fusion proteins are fusion-incompetent and require processing by viral proteases in order to be fusion-competent, or primed for fusogenic activity. Once the mature, processed and primed virus encounters a target cell, fusion events are mediated either by direct recognition and binding of the virus to its receptor on a target cell or a pH trigger commonly found in viruses that fuse within the endosome and not the outer membrane.[57-64] Once fusion is initiated, the fusion protein undergoes irreversible conformational changes that result in exposure of the FP. The hydrophobic peptide then embeds into the target host membrane, directly linking the virus and target cell. Previous investigation has attributed the embedding properties of the FP as a conclusion of predicted secondary sequences that FPs adopt amphipathic helices with hydrophobic residues on one face and polar residues on the opposing face.[65] However, recent work (see also Chapter 20)[41] has suggested that FPs from HIV and HCMV not only have generalized hydrophobic sequences, but sequences that specifically target host receptors, namely members of the MIRR family. If in fact MIRR-targeted strategies are conserved in a number of viruses and overlap with viral entry, sequence analysis of FPs from viruses other than HIV and HCMV should identify those viruses that share in their immunomodulatory specificities for MIRRs.

Human Immunodeficiency Virus

Viral entry of HIV is mediated by the product of HIV *env* expression, the type I fusion protein gp160, that is processed by HIV protease to yield the viral receptor gp120 (aa1-511) and fusion protein gp41 (aa512-684), found associated as heterohexameric complexes [(gp120)₃-(gp41)₃][66-68] on the surface of HIV particles. Following encounter of a target T-cell, gp120 first binds the CDR2 loop of the CD4 coreceptor. CD4 induces a conformational change in gp120 that enhances binding to a coreceptor, namely CXCR4 or CCR5, to form the ternary CD4-CXCR4/CCR5-gp120 complex.[52,54,69-75] Consequently, membrane fusion is initiated by ternary complex-induced conformational changes in the gp120-gp41 complex that release gp41 from its metastable state and allow for the FP (aa512-535) to integrate into the target host membrane. Once the adjoining membranes are anchored by gp41, fusion events mediated by both gp41 and gp120 occur, allowing for viral entry.

Until recently,[40,76] the function attributed to gp41 and namely the FP has been limited to anchoring of the infecting HIV particle to the target T-cell. However, it is becoming increasingly evident that the FP contributes much more to the viral pathogenesis than simply viral entry. Investigation of the primary sequence of HIV FP yields the presence of two positively charged arginines (Fig. 2C) that lie on the same face of a predicted alpha helix. Interestingly, the TM domain (TMD) of the T-cell receptor alpha chain (TCRα) also contains two positively charged residues (R, K) that lie on the same face as well, being separated by 4 residues (Fig. 2). Because the TMDs of other components of the TCR, namely the CD3δε and ζζ hetero- and homodimers, contain a negatively charged aspartate (D) residue, it is believed that electrostatic interactions drive TCR complex formation in the largely hydrophobic environment of the TM (Fig. 2A).[77] Therefore, by having similar electrostatic properties and distribution pattern of charged residues as the TCRα TMD, HIV FP may (1) specifically bind the electronegative components of the TCR complex in a transmembrane milieu and (2) physically and functionally disconnect the CD3δε and ζ signaling subunits from the remaining TCR complex by direct competition with the TCRα subunit (Fig. 2B).[40,41] This TCR-targeted functionality of the HIV FP adds a new dimension to the binding properties of the peptide and because of the adaptive immune function associated with the TCR, compounds an immunomodulatory role. These collective functions have been described in detail by the SCHOOL model[36,40,41,78,79] (see also Chapters 12 and 20) and are becoming increasingly substantiated by emerging experimental observation.

In in vitro coimmunoprecipitation and fluorescence resonance energy transfer (FRET) studies, HIV FP was demonstrated to specifically associate with TCR and the gp120 ligand, CD4 and to colocalize with TCR within 50Å.[80] Since neither gp120 nor the bulk of gp41 (aa535-684), which contains domains thought to also interfere with T-cell activation, were included in these experiments,[80] HIV FP must contain homing sequences that drive preferential localization and binding to the TCR without any extracellular contribution; the binding specificity is limited to the TM environment and is best explained by electrostatic interactions between the HIV FP and the TCR TMDs (Fig. 2B).[40]

Since gp120 is the primary HIV surface receptor that specifically binds CD4, HIV doesn't seemingly require gp41 or the FP particularly to serve as a binding partner for TCR or CD4. However, because the FP is heavily conserved amongst the divergent HIV subtypes, it must have other TCR-specific functions outside binding. In fact, FP was demonstrated to inhibit activation of primed lymph node cells and human T-cell lines in the presence of an activating antigen. However, in the presence of phorbol 12-myristate 13-acetate (PMA)/ionomycin or mitogenic antibodies to CD3, the inhibitory activity of HIV FP was abrogated.[80] These observations of FP closely mirror those of the recently studied TCR core peptide (CP) and are discussed in detail in Chapter 20. Briefly, TCR CP is a 9 amino acid peptide homologous to part of the TCRα TMD and contains the two electropositive residues (R, K) thought to be important for TCR complex formation (Fig. 2). TCR CP was also demonstrated to have immunosuppressive effects on T-cells in the presence of specific stimulating antigens, suggesting similar functionalities of TCR CP and

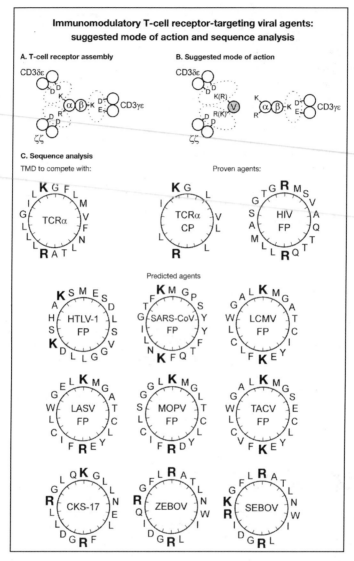

Figure 2. Suggested mode of action and sequence analysis of viral fusion protein regions and other domains proven and predicted to affect transmembrane interactions between T-cell receptor recognition and signaling subunits. A) Structural architecture of T-cell receptor is organized by three major assembly transmembrane forces, each involving one basic and two acidic amino acid residues. B) Within the SCHOOL model, viral agents (V) disrupt the transmembrane interactions between the ligand-binding TCRα chain and the CD3δε and ζζ signaling subunits which normally maintain the integrity of a functional T-cell receptor. This prevents formation of signaling oligomers upon multivalent antigen stimulation, thus inhibiting antigen-specific T-cell activation (see also Chapters 12 and 20). C) Helical wheel representations of proven and predicted immunomodulatory sequences of viral fusion protein regions and other domains. For illustrative purposes, the regions shown are restricted to 18 residues. As an ideal alpha helix consists of 3.6 residues per complete turn, the angle between two residues is chosen to be 100 degrees and thus there exists a periodicity after five turns and 18 residues. Positively charged residues are shown in bold. Legend continued on following page.

Figure 2, continued from previous page. Abbreviations: FP: fusion peptide; HIV: human immunodeficiency virus; HTLV-1: human T-cell lymphotropic virus type 1; LASV: Lassa virus; LCMV: lymphocytotic choriomeningitis virus; MOPV: Mopeia virus; SARS-CoV: severe acute respiratory syndrome coronavirus; SEBOV: Sudan Ebola virus; TACV: Tacaribe virus; TCR: T-cell receptor; V: viral agent; ZEBOV: Zaire Ebola virus.

HIV FP. However, those similarities were only described in retrospective analysis of the data, leading to assignments of novel functionality to the HIV FP.

As described by the SCHOOL model[36,41,78,79] (see also Chapters 12 and 20), both naturally-derived HIV FP and synthetically-designed TCR CP exploit their TM specificities for the CD3δε and ζζ components of the TCR to disrupt the TM interactions that hold the TCR complex together.[40,41] By disconnecting the recognition chains, TCRαβ, from the signaling chains, CD3δε and ζζ, HIV FP functionally disrupts the TCR complex and effectively disarms the MIRR. As a consequence, when TCRαβ recognizes and binds to its MHC-peptide partner on an APC, T-cell signaling is absent; the FP-associated signaling chains are unable to oligomerize and transduce the extracellular binding event (see Chapters 12 and 20).[40,41]

One of the defining features of the ability of HIV to replicate and proliferate is the low fidelity of HIV reverse transcriptase (RT) that leads to high mutability and sequence variability in HIV progeny during productive infection.[81] However, the HIV gp41 FP sequence is remarkably conserved among different HIV strains, suggesting a key role of not only the need for hydrophobic residues to embed in the target membrane and permit fusion but also the two electropositive residues that mediate binding to components of the TCR. As a result, HIV FP may not only serve as a fusogenic agent, but an immunosuppressive factor targeting the TCR as well, contributing to evasion of the adaptive immune response.

Human Cytomegalovirus

HCMV, a member of the betaherpesvirus subfamily of herpesviruses, is an enveloped virus characterized by a large genome (196 to 241 kbp) with the capacity to encode over 160 gene products. Existing as an opportunistic pathogen, HCMV proliferates during primary infection or reactivation of latent infection where an absence of effective immunity arises. Such conditions include modes where the immune system is compromised by other pathogenic agents (i.e., acquired immune deficiency syndrome, AIDS) or by prescribed immunosuppression (i.e., transplant recipients). However, the virus has also been demonstrated to replicate, reactivate and proliferate in environments where inflammation is markedly elevated.[82,83] Although the viral factors that mediate HCMV pathogenesis remain largely undetermined, three stages of HCMV pathogenesis have been described: (I) stimulation of a latently infected cell to differentiate and reactivate the latent virus to replicate by proinflammatory, cytokine-driven processes, (II) immunosuppression that allows amplification of productive viral replication, either systemically or locally and (III) direct or indirect viral or host immune-mediated damage that manifests as acute or chronic disease.[84] The immunosuppression or the ability of HCMV to evade and survive effector responses by innate and adaptive immune cells has been studied in great detail,[84] with novel mechanisms targeting disruption of MIRR signaling just now emerging.

Primarily infecting fibroblasts, HCMV has also been found to occupy professional APCs, namely macrophages and dendritic cells, following infection. Once inside the target host cell, HCMV prepares the cell for productive replication through two mechanisms: modulation of proinflammatory IFN cytokine production and reprogramming of cellular machinery. Immediately following entry, the tegument protein pp65, stored between the virion and surrounding envelope in the mature viral particle, is released and translocates to the nucleus, reducing the level of nuclear factor kappa B (NF-kB) production and blocking interferon regulatory factor-3 (IRF-3) activation.[85] Modulation of the IFN response is compounded by the activity of IE1-p72, a gene product expressed early after infection. By binding STAT1 and STAT2, IE1-p72 sequesters the signaling kinases and prevents their association with IRF-9, leading to the block of transcription of IFN-responsive genes.[86] HCMV also dramatically alters cellular gene expression and cell cycle progression immediately following infection, allowing for productive replication; the cell cycle is dysregulated and kept in a mitosis-like state, permitting early viral gene expression and productive replication of viral progeny before apoptosis occurs.

In addition to modulation of IFN signaling pathways in the infected cell, HCMV has been described extensively to have developed mechanisms of evading the natural killer cell (NK cell) arm of the innate immune system.[84] NK cells surveil the host environment and are able to discriminate normal cells from those under duress or infection by monitoring the differential surface expression of MHC molecules on cells through the killer cell immunoglobulin-like receptors (KIRs). Once downregulation of MHC expression is detected, ligation of the natural cytotoxic receptors (NCR) NKp46, NKp44 by viral hemagglutinin or NKp30 by unidentified ligands results in NK cell-mediated cytotoxicity and lysis of the affected cell. While production of MHC analogs by HCMV in an infected cell to conceal the infectious process has been described in great detail,[87] mechanisms of viral evasion targeting the NCR have not garnered much attention until recently.

NKp30 exists on the surface of NK cells as an NKp30-ζ receptor complex, comprised of the recognition subunit NKp30 associated with the immunoreceptor tyrosine activation motif (ITAM)-containing ζ signaling subunit homodimer to form a canonical MIRR. Supported by experimental evidence[88] and described by the SCHOOL model,[36,79] ligation of the recognition subunit NKp30 and subsequent oligomerization of the ζ signaling subunit results in full activation of the MIRR. Although natural ligands for NKp30 have yet to be extensively identified, recent studies have demonstrated that the tegument protein pp65 interacts specifically and directly with the NKp30 complex, thus representing one of the first molecules to be classified as a NKp30 ligand.[88] However, rather than induce activation of the targeted NK cell, pp65 exhibits deleterious effects and inhibits NK cell activation, resulting in the inability of the NK cell to kill normal, tumor and virus-infected cells. This inhibitory effect of pp65 is explained and described to be the consequence of dissociation of the signaling ζ chains from the recognition NKp30 receptor, which renders the MIRR nonfunctional.[88] However, until recent application of the SCHOOL model (see also Chapter 20),[36,41] the mechanism for how binding of the NKp30-ζ complex and dissociation of NKp30 from ζ results in the inhibition of NKp30 signaling was unknown.

Investigation of the primary sequence of the N-terminal domain of pp65 reveals the presence of several electronegative and more importantly, electropositive amino acid residues (see Chapter 20, Table 6) that may disrupt the TMD interactions between NKp30 and ζ and result in the inhibition of NK cell activation observed. By taking advantage of the presence of a negatively charged aspartate (D) in the TMD of ζ, the highly positively charged pp65 N-terminus may preferentially bind ζ through a TM interaction, effectively releasing NKp30 from its binding partner, similar to the described actions of HIV FP and TCR CP (Fig. 2). Experimental evidence substantiating this mechanism of defusion will need to be demonstrated, however it is evident that HCMV has developed specific mechanisms to target MIRRs redundant with other viral strategies, such as those previously described for HIV FP in this chapter.

While the primary function of pp65 has been attributed to immediate inhibition of NF-kB production and IRF-3 promoter-driven gene expression inside the infected cell,[85] pp65's effects on NK cell activity have been described as a result of extracellular exposure of pp65 to the NKp30-ζ complex[88]—a quandary that needs further investigation. Whether exogenous pp65's origins come from secretion of the protein or more likely release from apoptotic cells, the membrane targeting activity of pp65 may not be as disparate from HIV FP as one would imagine, despite the nonfusogenic activity of pp65 or the major classification differences between HIV and HCMV (Table 1). Demonstrated to specifically target the NKp30-ζ complex, pp65 may act identically to HIV FP in targeting an MIRR and disengaging the receptor to suppress the immune cell and permit viral persistence.

Prediction of MIRR-Targeting Viral Agents: HTLV-1 and Other Viruses

Like other retroviruses, HTLV-1 enters permissive cells by binding to cellular surface molecules such as heparin sulfate proteoglycans[89] and the ubiquitous glucose transporter GLUT1 that serves as a receptor for both HTLV-1 and HTLV-2 viruses,[90-92] followed by subsequent fusion of the viral and target cell membranes, thus releasing the viral core into the host cell cytoplasm.[90,93-96] This fusion is mediated by several viral envelope (Env) glycoproteins that are presented on the surface of virus or infected cell as a trimer of surface (SU) glycoprotein subunits anchored to a trimer of TM glycoproteins. Remarkably, infection with cell-free HTLV-1 virions remains inefficient because naturally infected lymphocytes produce very few cell-free virions and because, of the HTLV-1 virions that are released, only 1 in 10^5 to 10^6 is infectious.[91,94,95] The most efficient mode of HTLV-1 infection is cell-to-cell transmission that likely represents the sole mode of in vivo transmission for all retroviruses. Using confocal microscopy, the transfer of different HTLV-1 virion components from lymphocytes of infected patients to non-infected recipient lymphocytes has been directly visualized.[95]

Viral fusion results from a conformational change in the TM subunit of the Env protein, triggered by the SU/receptor interaction. This engagement exposes a FP located at the N terminus of the HTLV-1 TM protein gp21.[96,97] Similar to HIV gp41 FP, this sequence inserts into target cellular membranes and is well-known to be critical for membrane fusion activity.[98,99] However, in contrast to the HIV FP, there has been no report to date of an immunomodulatory activity of the HTLV-1 FP.

Because T-lymphocytes represent the major target cells for HTLV-1, it can be easily suggested that the TCR is a favorable target for inhibition at the viral entry stage. For these purposes, a TM-targeted strategy intended to physically and functionally disconnect TCR recognition and signaling subunits (Fig. 1A) might be effectively used by HTLV-1 as was described for HIV. The SCHOOL model[36,41,79] (see also Chapters 12 and 20) suggests that this "secret weapon" of HTLV-1 can be represented by the viral sequence that mimics the TMD of the TCR recognition subunit (for example, the TMD of TCRα chain) and is able to insert into the cell membrane where it competes with TCRα for binding to the CD3δε and ζ signaling chains in the TM milieu (Fig. 2B), thereby resulting in inhibition of antigen-induced T-cell activation as with HIV FP. Through helical wheel prediction (Fig. 2C) of the HTLV-1 FP, similarities in the location of electropositive residues previously described to be essential for the action of HIV FP, TCR CP and HCMV pp65 are revealed. Positioning of the charged lysine (K) residues in HTLV-1 FP is almost identical to those for the TCR CP and closely resemble those of the MIRR-disrupting viral agents HIV FP and HCMV pp65. Therefore, it is highly likely that HTLV-1 FP targets the TCR complex in a manner identical to HIV FP, TCR CP and HCMV pp65 and disrupts the TM interactions that hold the complex together, resulting in a defused TCR (Fig. 2B).

Intriguingly, analysis of other seemingly unrelated viruses has yielded similar correlations in primary structure and function. Earlier studies have reported an inhibitory effect of the CKS-17 peptide on lymphocyte proliferation, a synthetic 17-mer peptide with sequence corresponding to a highly conserved region of retroviral TM proteins of human and animal retroviruses including HTLV-1.[100] Later, the reported immunosuppression was further confirmed and further localized to a sequence essentially identical to the sequence present in the TM protein gp21 of HTLV-1,[101] supporting the hypothesis that this protein participates in the mechanism of immunosuppression previously reported for the TM proteins of feline leukemia virus and other animal retroviruses.

Interestingly, peptides corresponding to regions of HIV TM protein gp41 homologous to the highly conserved and immunosuppressive sequence contained within the TM proteins p15E and gp21 of animal and human retroviruses, respectively, have been also reported to inhibit lymphoproliferation.[101] Recently, filoviral 17-mer peptides corresponding to a 17 amino acid domain in filoviral glycoproteins that resembles an immunosuppressive motif in retroviral envelope proteins have been demonstrated to inhibit TCR-mediated T-cell activation and cell proliferation, providing new insights in the immunopathogenesis of Ebola and Marburg viruses.[102] In all these peptides (CKS-17; Zaire Ebola virus, ZEBOV and Sudan Ebola virus, SEBOV; Table 1), a striking similarity is observed between these peptides in charged or polar residue distribution patterns with positioning of the charged lysine (K) and/or arginine (R) residues almost identical to those for the HIV FP (Fig. 2C), suggesting again a similarity in the molecular mechanisms of their immunosuppressive action.

Based on the surprising conservation in positioning of the essential electropositive residues in the helical wheel predictions of HIV-1 FP and HTLV-1 FP and its similarity to those for the TCR CP, it is highly probable that proteins from other unrelated viruses that also participate in viral fusion would also target MIRRs on the surface of their target cell. Exploratory sequence investigation of FPs from severe acute respiratory syndrome coronavirus (SARS-CoV), Lassa virus (LASV), lymphocytic choriomeningitis virus (LCMV), Mopeia virus (MOPV) and Tacaribe virus (TACV) reveal evidence of such a hypothesis. As shown in Figure 2, there is striking similarity in the positioning of the electropositive residues on one face of the helix, despite the fact that the amino acid residues aren't necessarily conserved; for example, MOPV FP contains an arginine and lysine whereas TACV contains only lysine residues. This clearly demonstrates that viruses, despite their differences in virion structure, genomic composition or classification, have adopted similar mechanisms of specifically targeting MIRRs, disrupting their architecture and suppressing the immune system. Importantly, by virtue of the acquired insight into this conserved structural motif, expanded predictions, hypotheses and conclusions can be derived to begin answering the question of if shared MIRR-targeted strategies represent a conserved function or if they represent a convergent tactic of divergent viruses.

Viral Replication

Similar to viral entry, viruses have developed subprocesses targeting MIRRs that underlie other viral stages, namely viral replication, for enhancement of viral production and persistence. Following entry and uncoating in the target cell, viruses undergo an efficient and economical process of replication where copies of the viral genome are abundantly produced, viral genes are expressed and viral protein translations begin to assemble into competent viral particles. Due to the diversity in genomic structure found among the different viruses, there is also great diversity in the replication strategies they employ. Contrary to cellular genomes that are comprised uniformly of dsDNA, viral genomes span all possible structural organizations: dsDNA, dsRNA, positive-sense ssDNA, negative-sense dsDNA, positive-sense ssRNA, negative-sense ssRNA and mixed (ambisense) ssDNA or ssRNA. Consequently, viruses have developed unique replication strategies, used by the Baltimore classification method to group viruses, that require different host proteins as well as inclusion of different virally encoded proteins in their genomes. For example, group I dsDNA viruses, such as members of the adenoviral family, require host cell DNA polymerases to replicate their viral genomes and are therefore highly dependent on the replicative state of the cell; the target cell must be undergoing active replication and cell division where the cell's polymerases are most active. In contrast, group VI positive-sense ssRNA viruses, such as members of the retrovirus family, replicate their genomes by RNA-dependent DNA synthesis not by any host polymerases but by virus-encoded RT; the transcribed DNA is then used as the viral template for integration into the host genome and transcription. Because RT is not supplied by the target cell, it must be packaged with the viral progeny for further replication. Regardless of the structure and replication strategy of their genomes, all viruses express their genes as functional mRNAs early in infection and direct the cell's translational machinery to make viral proteins for eventual viral packaging.

Efficiency is essential to every viral stage but particularly to replication as it represents a pivotal point in virus production. Viruses have therefore optimized their replication strategies to exploit naturally occurring biological and cellular processes of their hosts, effectively hijacking the replication, transcription and translational machinery. However, replicative efficiency has its drawbacks; viruses are consequently dependent largely on the replicative capacity of their target cells and what functional state they are in during the infection. To overcome these limitations, several viruses have developed mechanisms of activating the infected target cell from within the cytoplasmic environment to enhance viral replication (Fig. 1B). In this section, we describe subprocesses within the realm of viral replication that two members of the retrovirus family enact by targeting a specific MIRR, namely the TCR, from the cytoplasmic environment. Coupled with the TM-targeted strategy employed by HIV-1 FP and HTLV-1 FP (Fig. 1A), the cytoplasmic-targeted strategy represents an interesting dichotomy of site of action and function that converge on the identity of the specific target.

Human Immunodeficiency Virus

Characterized by its positive polarity ssRNA genome and group VI classification, HIV shares a unique replicative process with other members of the retrovirus family that differs significantly from other viruses. Prior to replication, HIV virions attach to and enter T-lymphocytes following formation of the ternary HIV gp120-CD4-chemokine receptor CCR4/CXCR5 complex and direct membrane fusion mediated by HIV gp41, respectively.[52,54,69-75] Once inside the cell, the virion partially uncoats in the cytoplasm, releasing viral accessory proteins and the two copies of the positive-sense ssRNA genome housed inside the viral particle. HIV RT then initiates transcription of the viral genome, producing double-stranded cDNA transcription products that immediately associate with a number of viral (integrase, RT, matrix, Vpr)[103-105] and cellular (IN1, HMGA1, BAF, EED, LEDGF/p75)[106] proteins to form the preintegration complex (PIC). Due to the low fidelity of HIV RT[81] that results in 3×10^{-5} mutations per replication cycle in vivo,[107] HIV enjoys incredible genetic diversity during virus production that closely resembles evolution but in a rapid timescale. Viral particles that introduce mutations in their genomes that exhibit increased replicative capacity will propagate and dominate the infection whereas replication-deficient variants will cease to exist.

Once formed, the PIC migrates to the nucleus by the host nuclear import machinery that only actively translocates the PIC when the cell is arrested in the G1 phase of the cell cycle and nondividing. Following import into the nucleus, RT-transcribed viral cDNA is integrated into the host chromosome via HIV integrase, a hallmark event that is unique to HIV. Once integrated, HIV DNA is left untranscribed in a latent stage of infection until the infected T-lymphocyte is activated and coordinated interactions between HIV-encoded Tat protein, host NF-kB, Sp1 transcriptional transactivating proteins and the RNA polymerase II transcriptional complex facilitate production of high levels of viral RNA.[108] Newly transcribed mRNAs are exported from the nucleus to the cytoplasm by HIV Rev and then translated by host ER-associated and cytoplasmic ribosomes to yield gp120 Env and Gag/Gag-Pol polyproteins, respectively. Each viral protein species translocates to the cytoplasmic face of the plasma membrane where they associate with dimeric viral positive-sense ssRNA to form the premature viral bud that subsequently undergoes further processing, entering the final stages of viral assembly and release.

While much of the work investigating HIV replication has focused on the role of the viral regulatory protein Tat on HIV RNA transcription,[109-112] reports have suggested a key role of cellular activating factors in enhancing replication.[108] In order for HIV to emerge from latent infection where the HIV genome is transcriptionally silent, the infected T-lymphocyte must become activated and initiate a signaling cascade that ultimately results in the release of NF-kB from sequestration by IkB. Therefore, any mechanism that induces a state of activation within the infected cell would effectively enhance NF-kB activity and downstream replication of HIV. Recently, the viral accessory protein Nef has been described to affect the activation profile of CD4+ T-lymphocytes by reducing the threshold of T-cell activation[113,114] and also initiating a transcriptional program in Jurkat T-cells similar to that of a T-lymphocyte exogenously activated through the TCR.[115] Localization to the cytoplasmic face of the plasma membrane seems to be required for Nef-induced activation or augmentation of activation[116] and association with lipid rafts and cytoplasmic signaling proteins has been proposed to play a key role.[117] However, details of the specific mechanisms underlying Nef-mediated augmentation of activation or reduction in threshold for activation remain largely unknown.

Originally coined "negative factor" under reports that HIV Nef reduced replication by suppressing transcription of integrated HIV genes,[118] it is now evident that Nef mediates several processes that collectively enhance viral replication: (1) downmodulation of surface receptors, namely CD4,[119,120] MHC Class I proteins (HLA-A, B but not C or E),[121-123] CD28[124] and TCR in the context of simian immunodeficiency virus (SIV),[125] (2) enhancement of viral infectivity[126] and (3) modulation of signaling pathways. Among all of Nef's functions, downmodulation of the TCR remains the most controversial and intriguing. Because of its role in initiation of the signaling cascade in T-lymphocytes, TCR fills a strong potential role in Nef's reported effects on increasing the activation state of the cell. Interestingly, HIV-2 and SIV Nef have been reported to specifically interact with the ζ signaling chain of the TCR complex but additionally induce downregulation of surface TCR from the cell surface.[127-129] Functional mapping of SIV Nef has revealed that the C-terminal core domain, conserved among the different HIV-1 clades and strains, is responsible for specific ζ binding whereas the nonconserved N-terminal domain cooperatively binds AP-2 from the host thereby inducing downregulation of the bound TCR. Extrapolation of these results explains the lack of TCR downmodulation observed for several HIV-1 Nef variants,[130] considering the genetic variability in the N-terminal domain and strengthens the observed binding data surrounding the HIV-1 Nef-ζ interaction that has been previously disputed.[127,128]

Armed with the ability to form homooligomers on the one hand and specifically bind the signaling ζ chain of the TCR, on the other, HIV Nef can exert activating or augmenting effects on TCR-mediated stimulation, as described recently by the SCHOOL model (see also Chapter 20).[36,78,79] In contrast to extracellular targeting of the TCR by HIV FP as described earlier in this chapter, HIV Nef targets the TCR from the cytoplasmic environment and rather than inhibit TCR activation, enhances it. In the case of HIV FP, the signaling subunits of the TCR are physically and functionally disconnected from the recognition subunits through TMD interactions formed with HIV FP that effectively results in inhibition of antigen-mediated TCR signaling. HIV Nef may crosslink with the ζ signaling subunits through cytoplasmic interactions[36] (Fig. 1B; see also Chapters 12 and 20), cluster TCRs and instead of disengaging the receptor, activate it or prime it for activation. While a large component of the SCHOOL model requires the ability of the signaling chains of the TCR to homooligomerize in receptor clusters, HIV Nef has been reported to self-oligomerize, a property already described to be vital for function.[131-135] Therefore, through the combination of an interaction with ζ and self-oligomerization, HIV Nef may induce the formation of higher order receptor oligomers that directly activate the cell[115] or effectively reduce the threshold of stimulus required for full activation.[113,114] Recent studies have indeed demonstrated clustering of HIV Nef at the immunological synapse,[136] the interface between the infect T-lymphocyte and an APC, furthering supporting the notion that Nef interacts with cytoplasmic components of the TCR and likely participates in higher order oligomerization conducive to T-cell activation.

Interestingly, SIV seems to have developed additional methods of further exploiting the TCR-targeted augmentation of cellular activation Nef enacts. Characterized by rapid viral kinetics and the novel ability to replicate and proliferate in non-exogenously-stimulated macaque peripheral blood mononuclear cells (PBMC),[137] SIVsmPBj, a highly pathogenic strain of SIV, induces acute, destructive disease while exhibiting an augmented replicative state.[137,138] Underlying this disease is the presence of an ITAM sequence in SIVsmPBj Nef similar to that found in the signaling domains of the CD3δ, CD3ε, CD3γ and ζ components of the TCR. Therefore, upon localization to the inner leaflet of the plasma membrane and association with ζ during acute infection, SIVsmPBj Nef forms high order heterooligomeric Nef-ζ complexes with significantly increased numbers of ITAM domains as compared to nonSIVsmPBj variants. Consequently, the infected cell will be prone to not only clustering of the signaling chains of the TCR by binding Nef but additional induced activation by virtue of the supplied ITAM sequences present in the viral protein. By including ITAM sequences, SIVsmPBj effectively clusters viral ITAMs with host ITAMs to induce acute activation and replication.

Despite targeting the same receptor as HIV FP, HIV Nef has the complete opposite effect on its function; rather than inactivate the receptor as observed with HIV FP, HIV Nef activates it. Explained to be the result of a cytoplasmic-targeted strategy (Fig. 1B), it is intriguing that HIV developed two mechanisms of acting on the same receptor, but eliciting different outcomes depending on the viral stage and site of action. However, through those developed viral strategies, details on how MIRRs function and initiate the intracellular cascade are revealed and provide methods of studying immune regulation but also new avenues for development of novel immunomodulatory therapeutics.

Human T-Cell Lymphotropic Virus

There is a growing line of evidence that the accessory proteins of HTLV-1 are critically involved in viral transmission and propagation and may in fact be multifunctional proteins. Key among them is the p12 protein of HTLV-1, a small oncoprotein that is produced during the course of the natural infection in vivo and has been shown to have multiple functions. Analogous to the accessory HIV-1 Nef protein,[139,140] p12 is required for optimal viral infectivity in nondividing primary lymphocytes.[141-143] HTLV-1 viral infection of T-lymphocytes is known to induce T-cell activation.[138] As suggested, one mechanism involves activation of T-cells harboring the virus and is exemplified in vivo by infected, non-immortalized T-cell clones that display prolonged states of activation, whereas with a separate mechanism, virus-infected cells can induce activation of uninfected T-cells via T-cell-T-cell interactions.[138] In non-immortalized, HTLV-1-infected T-cells, spontaneous clonal proliferation is resistant to immunosuppression by transforming growth factor-β (TGF-β), a cytokine implicated in terminating T-cell activation, suggesting a potential role of HTLV-1 in a defense against TGF-β-induced immune suppression of the host cell.[144]

Spontaneous proliferation and virus production have been reported to increase in the presence of anti-CD3 and anti-TCR antibodies while addition of HLA class I antibodies, but not HLA class II or viral proteins, shut down virus production and cell proliferation.[145] These findings suggest that both virus and cell activation may occur through the TCR on the infected cell. Expression of p12 has been shown to induce nuclear factor of activation of T-cells (NFAT), enhance the production of interleukin-2 (IL-2), decrease MHC-I expression, increase cytoplasmic calcium and signal transducer and activator of transcription 5 (Stat 5) activation in T-cells further supporting the hypothesis that p12 may alter T-cell signaling.[143,146-150] Interestingly, p12 is important for viral infectivity in quiescent human peripheral blood lymphocytes (PBLs) and PBMCs and the establishment of persistent infection in vivo, suggesting a role for p12 in the activation of quiescent lymphocytes, a prerequisite for effective viral replication in vivo.[141,151] In this context, function of p12 in conditions where the majority of viral target cells are in quiescent states has been predicted to be similar to that of Nef.[141] HTLV-1 p12-expressing cells were reported to display a decreased requirement for IL-2 to induce proliferation during suboptimal stimulation with anti-CD3 and anti-CD28 antibodies.[149] HTLV-1 replication in infected lymphocytes has been also been reported to increase upon CD2 cross-linking.[152] This receptor is known to signal primarily through the associated CD3ε and ζ chains.[153,154] Studies have shown that the mitogenic activity of HTLV-1 viral particles is restricted to virus-producing T-cells, requires cell-to-cell contact and may be mediated through the lymphocyte-associated antigen 3 (LFA-3)/CD2 activation pathway and that HTLV-1 virions interfere mainly with activation of peripheral T-cells via CD2/ζ but not via the CD3/TCR complex.[155]

Overall, p12 seems to augment T-cell activation and facilitate viral replication. Thus, despite the distinct structures, both retroviral accessory proteins HTLV-1 p12 and HIV Nef are able to modulate TCR-mediated signaling and play a critical role in enhancing viral infectivity in primary lymphocytes and infected animals. Interestingly, it has been recently reported that p12 could complement for effects of Nef on HIV-1 infection of Magi-CCR5 cells, which express CD4, CXCR4 and CCR5 on the surface, or macrophages.[156] Also, Jurkat cell clones that express high levels of p12 have been found to exhibit a more rapid rate of cell proliferation than the parental cells.[156] Similarly to HIV Nef, the p12 protein, upon engagement of the TCR, localizes to the interface between T-cells and antigen-presenting cells, namely the immunological synapse.[157]

Intriguingly, similarly to HIV-1 Nef protein,[133] HTLV-1 p12 has also been shown to form dimers.[149] It can be suggested that homooligomerization of p12 contributes to p12-mediated augmentation of T-cell activation and that molecular mechanisms of this phenomenon are similar to those that have been suggested previously for Nef through application of the SCHOOL model of TCR signaling (see also Chapters 12 and 20).[36] If true, the homooligomerization interface(-s) of p12 represent potential therapeutic targets for antiviral treatment.

Translation of Redundant Viral Strategies into Disease Care

As depicted by members of the retroviridae and herpesviridae, namely HIV, HTLV and HCMV, a wide range of viruses has developed methods of targeting members of the MIRR family of surface receptors. However, depending on the needs of the virus and at which stage of viral replication the virus is in, MIRR-induced signaling is either disrupted or enhanced. More specifically, when HIV undergoes viral entry, MIRR-triggered activation is abrogated through disruption of TM interactions in TCR by HIV FP in order to evade immune activation. Similar function is required during persistence of HCMV infection where signaling through NKp30 is abrogated so as to inactivate the NK cell response and accompanying immune activation. However, where MIRR-triggered activation is needed for enhanced replication, exemplified by HIV and HTLV, viral proteins once again specifically target MIRRs, but in a concerted effort to induce triggering and subsequent cellular activation mechanisms conducive to viral production. Therefore, although viruses may be structurally different, contain different types of genomes and exhibit different replication strategies, many converge in their immune modulation strategies.

The combination of retrospective analysis of previous experiments investigating details of HIV, HTLV and HCMV pathogenesis and application of a novel model of MIRR triggering,[36,79] has revealed a couple of key features of MIRR triggering that viruses redundantly interfere to modulate the immune response: TM interactions between the recognition and signaling subunits of MIRRs and oligomeric clustering of signaling domains. Described as TM and cytoplasmic targets (Fig. 1), respectively, these two classes of interactions represent the foundations of MIRR triggering and provide avenues for novel but universal antiviral therapies and importantly, immunomodulatory treatment as well.

Current small molecule, antiviral research has focused on exploiting the differences between virus and host and selectively targeting a viral enzyme or process. However, due to the high mutation rate many viruses enjoy, therapies against protease or reverse transcriptase in HIV are being selected, resulting in drug resistant viral strains that exhibit even increased pathogenicity and necessitating the discovery of novel therapeutic targets. Our discussion of the specific targeting of MIRR signaling subunits, namely ζ, by HIV and HTLV provides that opportunity. Targeting of TCR-mediated signaling seems to be a shared feature of both HIV and HTLV-1 viruses and reflects a similar evolutionary pathway towards their adaptation to the host immune response that may also be shared with other unrelated viruses. Instead of inhibiting a specific enzymatic function, Nef and p12 functional targeting strategy would involve disrupting the protein-protein interface between the viral protein and the partner signaling chain to abrogate its activating function. In addition, the homointeractions between viral proteins may also emerge as a functional target since homooligomerization of viral proteins has also been shown to be essential for function. Careful investigation of the interacting surfaces on both the viral and MIRR may reveal unique features essential for binding, highlighting more rationalized drug targeting. Finally, extension of this protein-protein interaction disruption strategy should also be applied to other viruses to determine if there is increased redundancy in the processes outlined by Nef and p12. If so, MIRR-targeted antiviral research may provide a new line of generic but universal antiviral therapies.

An intriguing extension of the revealed strategies viruses redundantly use to target MIRRs is the application of them towards development of immunomodulatory agents. Viruses have evolved over thousands to millions of years and have optimized methods of disarming and evading the immune response for self-preservation. Therefore, investigation of how viruses have adapted to disarm the innate and adaptive immune system will prove invaluable in rational drug design efforts aiming to reduce immune activation or inflammation. One viral strategy, namely the disruption of TMD interactions between the signaling and recognition subunits in MIRRs suggested for HIV, HTLV, HCMV and other viruses here (Fig. 1A; see also Chapter 20), provides such an avenue for exquisite drug development that has the potential for rapid development.

Retrospective analysis of the primary sequence of HIV FP and HCMV pp65 revealed the presence of specific electropositive residues that mirror those found in the TMD of the TCRα signaling subunit (Fig. 2C; see also Chapter 20, Table 6). Combining that observation with functional data describing the inhibitory effect they have on TCR and NKp30 signaling, it is highly probable that they compete with TCRα for binding with its signaling binding partners, effectively disrupting the TCR complex and rendering it useless (Fig. 2B; see also Chapter 20). Therefore, membrane-targeted strategies mimicking those of HIV FP and pp65 and exploiting the binding contribution of electropositive amino acid residues will likely have similar effects and provide useful as therapies for immune disorders characterized by chronic inflammation. Coincidentally, one such avenue of TCR-targeting research has already undergone development with promising results. Derived from the primary sequence of the TCRα TMD region, synthetic hydrophobic peptides, coined the TCR core peptides or TCR CPs, were produced and exhibit inhibitory function in not only T-cells, but B cells and NK cells as well.[158] Further studies with a D-amino acid variant also show strong efficacy, suggesting that chirality plays little role in the function of the peptide, leaving sequence pattern and electrostatics as the only mediators of function.[159]

Although the TM-targeting strategy employed by TCR CP was not a prospective application based on learned viral strategies, it displays the intellectual and rational research power that can be attained by investigating what viruses and nature have already employed and optimized. Hence, we have begun to investigate the primary sequences (Fig. 2C; see also Chapter 20)[40,41] of several unrelated viruses and see a remarkable homology in primary sequence and sequence pattern of a number of viral proteins, highlighting the presence of electropositive residues that may also target MIRRs. Future collaborations in bioinformatics, biochemistry and virology will undoubtedly reveal new details of the viral immune evasion strategies that are shared amongst a number of viruses that may prove useful in developing rational approaches to immune therapy.

Conclusions and Perspectives

Viral infection and the resultant immune response form a violent interplay where host homeostasis is interrupted by a propagating virus seeking to proliferate and the immune system working to quell the infection. In many cases, the virus and human host have coevolved to exist symbiotically where the virus resides in a latent phase nonpathogenic to the host. However, as new viruses emerge or crossover from other species, they will need to replicate rapidly and efficiently so as to proliferate as quickly as possible. This poses the largest pathogenic threat to humans and incurs disease that defeats the immune system and results in death of the human host. Therefore, we are forced to develop novel strategies to target the infecting virus. However, rather than targeting virus-specific proteins or processes, it would be advantageous to transfer therapeutic strategies that target redundant processes found among a number of viruses. In this chapter, we have described the universal targeting of members of the MIRR family by a number of seemingly unrelated viruses that function through similar mechanisms. Therefore, it is possible to take advantage of these general processes in drug development; the tedious work of developing virus-specific therapies would be eliminated and powerful far-reaching agents could be conceived.

In addition to the antiviral lessons learned from investigating the role of MIRRs in viral pathogenesis, several details regarding normal MIRR structure-function relationships and therapeutic intervention can be extrapolated. As demonstrated by the similar function of natural HIV FP and synthetically derived TCR CP, viral immune evasion strategies can be transferred to therapeutic strategies that require similar functionalities. Viruses represent years of evolution and the efficiency and optimization that come along with it. Therefore, viral functions should not only be studied as foreign processes but as efficient strategies we can use in our own attempts at immune evasion or immunomodulation.

References

1. Medzhitov R. Recognition of microorganisms and activation of the immune response. Nature 2007; 449:819-826.
2. Pichlmair A, Reis e Sousa C. Innate recognition of viruses. Immunity 2007; 27:370-383.

3. Takeuchi O, Akira S. Recognition of viruses by innate immunity. Immunol Rev 2007; 220:214-224.
4. Schroder M, Bowie AG. An arms race: Innate antiviral responses and counteracting viral strategies. Biochem Soc Trans 2007; 35:1512-1514.
5. Loo YM, Gale M Jr. Viral regulation and evasion of the host response. Curr Top Microbiol Immunol 2007; 316:295-313.
6. Keller BC, Johnson CL, Erickson AK et al. Innate immune evasion by hepatitis C virus and West Nile virus. Cytokine Growth Factor Rev 2007; 18:535-544.
7. Coscoy L. Immune evasion by Kaposi's sarcoma-associated herpesvirus. Nat Rev Immunol 2007; 7:391-401.
8. Takaoka A, Yanai H. Interferon signalling network in innate defence. Cell Microbiol 2006; 8:907-922.
9. van Wamel WJ, Rooijakkers SH, Ruyken M et al. The innate immune modulators staphylococcal complement inhibitor and chemotaxis inhibitory protein of staphylococcus aureus are located on beta-hemolysin-converting bacteriophages. J Bacteriol 2006; 188:1310-1315.
10. Rajagopalan S, Long EO. Viral evasion of NK-cell activation. Trends Immunol 2005; 26:403-405.
11. Kosugi I, Kawasaki H, Arai Y et al. Innate immune responses to cytomegalovirus infection in the developing mouse brain and their evasion by virus-infected neurons. Am J Pathol 2002; 161:919-928.
12. Haller O, Weber F. Pathogenic viruses: Smart manipulators of the interferon system. Curr Top Microbiol Immunol 2007; 316:315-334.
13. Saito T, Gale M Jr. Principles of intracellular viral recognition. Curr Opin Immunol 2007; 19:17-23.
14. Kurt-Jones EA, Popova L, Kwinn L et al. Pattern recognition receptors TLR4 and CD14 mediate response to respiratory syncytial virus. Nat Immunol 2000; 1:398-401.
15. Pasare C, Medzhitov R. Toll-like receptors: Linking innate and adaptive immunity. Adv Exp Med Biol 2005; 560:11-18.
16. Janeway CA Jr. Approaching the asymptote? Evolution and revolution in immunology. Cold Spring Harb Symp Quant Biol 1989; 54(Pt 1):1-13.
17. Doly J, Civas A, Navarro S et al. Type I interferons: Expression and signalization. Cell Mol Life Sci 1998; 54:1109-1121.
18. Kunzi MS, Pitha PM. Interferon targeted genes in host defense. Autoimmunity 2003; 36:457-461.
19. Le Page C, Genin P, Baines MG et al. Interferon activation and innate immunity. Rev Immunogenet 2000; 2:374-386.
20. Ozato K, Tailor P, Kubota T. The interferon regulatory factor family in host defense: Mechanism of action. J Biol Chem 2007; 282:20065-20069.
21. Galligan CL, Murooka TT, Rahbar R et al. Interferons and viruses: signaling for supremacy. Immunol Res 2006; 35:27-40.
22. Perry AK, Chen G, Zheng D et al. The host type I interferon response to viral and bacterial infections. Cell Res 2005; 15:407-422.
23. Bonjardim CA. Interferons (IFNs) are key cytokines in both innate and adaptive antiviral immune responses—and viruses counteract IFN action. Microbes Infect 2005; 7:569-578.
24. Cebulla CM, Miller DM, Sedmak DD. Viral inhibition of interferon signal transduction. Intervirology 1999; 42:325-330.
25. Garcia-Sastre A. Mechanisms of inhibition of the host interferon alpha/beta-mediated antiviral responses by viruses. Microbes Infect 2002; 4:647-655.
26. Goodbourn S, Didcock L, Randall RE. Interferons: Cell signalling, immune modulation, antiviral response and virus countermeasures. J Gen Virol 2000; 81:2341-2364.
27. Levy DE, Garcia-Sastre A. The virus battles: IFN induction of the antiviral state and mechanisms of viral evasion. Cytokine Growth Factor Rev 2001; 12:143-156.
28. Yang I, Kremen TJ, Giovannone AJ et al. Modulation of major histocompatibility complex Class I molecules and major histocompatibility complex-bound immunogenic peptides induced by interferon-alpha and interferon-gamma treatment of human glioblastoma multiforme. J Neurosurg 2004; 100:310-319.
29. Agrawal S, Kishore MC. MHC class I gene expression and regulation. J Hematother Stem Cell Res 2000; 9:795-812.
30. Thomas HE, Parker JL, Schreiber RD et al. IFN-gamma action on pancreatic beta cells causes class I MHC upregulation but not diabetes. J Clin Invest 1998; 102:1249-1257.
31. Gruschwitz MS, Vieth G. Up-regulation of class II major histocompatibility complex and intercellular adhesion molecule 1 expression on scleroderma fibroblasts and endothelial cells by interferon-gamma and tumor necrosis factor alpha in the early disease stage. Arthritis Rheum 1997; 40:540-550.
32. Dhib-Jalbut SS, Xia Q, Drew PD et al. Differential up-regulation of HLA class I molecules on neuronal and glial cell lines by virus infection correlates with differential induction of IFN-beta. J Immunol 1995; 155:2096-2108.

33. Chang CH, Hammer J, Loh JE et al. The activation of major histocompatibility complex class I genes by interferon regulatory factor-1 (IRF-1). Immunogenetics 1992; 35:378-384.
34. Beniers AJ, Peelen WP, Debruyne FM et al. HLA-class-I and -class-II expression on renal tumor xenografts and the relation to sensitivity for alpha-IFN, gamma-IFN and TNF. Int J Cancer 1991; 48:709-716.
35. Giacomini P, Fisher PB, Duigou GJ et al. Regulation of class II MHC gene expression by interferons: insights into the mechanism of action of interferon (review). Anticancer Res 1988; 8:1153-1161.
36. Sigalov AB. Immune cell signaling: A novel mechanistic model reveals new therapeutic targets. Trends Pharmacol Sci 2006; 27:518-524.
37. Lwoff A, Tournier P. The classification of viruses. Annu Rev Microbiol 1966; 20:45-74.
38. Lwoff A, Horne R, Tournier P. A system of viruses. Cold Spring Harb Symp Quant Biol 1962; 27:51-55.
39. Baltimore D. Expression of animal virus genomes. Bacteriol Rev 1971; 35:235-241.
40. Sigalov AB. Interaction between HIV gp41 fusion peptide and T-cell receptor: Putting the puzzle pieces back together. FASEB J 2007; 21:1633-1634; author reply 1635.
41. Sigalov AB. Transmembrane interactions as immunotherapeutic targets: Lessons from viral pathogenesis. Adv Exp Med Biol 2007; 601:335-344.
42. Vlasak R, Luytjes W, Spaan W et al. Human and bovine coronaviruses recognize sialic acid-containing receptors similar to those of influenza C viruses. Proc Natl Acad Sci USA 1988; 85:4526-4529.
43. Higa HH, Rogers GN, Paulson JC. Influenza virus hemagglutinins differentiate between receptor determinants bearing N-acetyl-, N-glycollyl- and N,O-diacetylneuraminic acids. Virology 1985; 144:279-282.
44. Weis W, Brown JH, Cusack S et al. Structure of the influenza virus haemagglutinin complexed with its receptor, sialic acid. Nature 1988; 333:426-431.
45. Gentsch JR, Pacitti AF. Differential interaction of reovirus type 3 with sialylated receptor components on animal cells. Virology 1987; 161:245-248.
46. Paul RW, Choi AH, Lee PW. The alpha-anomeric form of sialic acid is the minimal receptor determinant recognized by reovirus. Virology 1989; 172:382-385.
47. Paul RW, Lee PW. Glycophorin is the reovirus receptor on human erythrocytes. Virology 1987; 159:94-101.
48. Greve JM, Davis G, Meyer AM et al. The major human rhinovirus receptor is ICAM-1. Cell 1989; 56:839-847.
49. Mendelsohn CL, Wimmer E, Racaniello VR. Cellular receptor for poliovirus: Molecular cloning, nucleotide sequence and expression of a new member of the immunoglobulin superfamily. Cell 1989; 56:855-865.
50. Staunton DE, Merluzzi VJ, Rothlein R et al. A cell adhesion molecule, ICAM-1, is the major surface receptor for rhinoviruses. Cell 1989; 56:849-853.
51. Tomassini JE, Graham D, DeWitt CM et al. cDNA cloning reveals that the major group rhinovirus receptor on HeLa cells is intercellular adhesion molecule 1. Proc Natl Acad Sci USA 1989; 86:4907-4911.
52. Dalgleish AG, Beverley PC, Clapham PR et al. The CD4 (T4) antigen is an essential component of the receptor for the AIDS retrovirus. Nature 1984; 312:763-767.
53. Klatzmann D, Champagne E, Chamaret S et al. T-lymphocyte T4 molecule behaves as the receptor for human retrovirus LAV. Nature 1984; 312:767-768.
54. Maddon PJ, Dalgleish AG, McDougal JS et al. The T4 gene encodes the AIDS virus receptor and is expressed in the immune system and the brain. Cell 1986; 47:333-348.
55. McDougal JS, Kennedy MS, Sligh JM et al. Binding of HTLV-III/LAV to T4+ T-cells by a complex of the 110K viral protein and the T4 molecule. Science 1986; 231:382-385.
56. Sattentau QJ, Weiss RA. The CD4 antigen: Physiological ligand and HIV receptor. Cell 1988; 52:631-633.
57. White JM. Membrane fusion. Science 1992; 258:917-924.
58. White JM. Viral and cellular membrane fusion proteins. Annu Rev Physiol 1990; 52:675-697.
59. Daniels PS, Jeffries S, Yates P et al. The receptor-binding and membrane-fusion properties of influenza virus variants selected using anti-haemagglutinin monoclonal antibodies. EMBO J 1987; 6:1459-1465.
60. Hoekstra D, Kok JW. Entry mechanisms of enveloped viruses. Implications for fusion of intracellular membranes. Biosci Rep 1989; 9:273-305.
61. Lamb RA. Paramyxovirus fusion: A hypothesis for changes. Virology 1993; 197:1-11.
62. Marsh M, Helenius A. Virus entry into animal cells. Adv Virus Res 1989; 36:107-151.
63. Underwood PA, Skehel JJ, Wiley DC. Receptor-binding characteristics of monoclonal antibody-selected antigenic variants of influenza virus. J Virol 1987; 61:206-208.
64. Wiley DC, Skehel JJ. The structure and function of the hemagglutinin membrane glycoprotein of influenza virus. Annu Rev Biochem 1987; 56:365-394.
65. Tyler KLaF, Bernard N. Fundamental Virology. 3rd ed. Philadelphia: Lippincott—Raven Publishers, 1996:161-206.

66. Center RJ, Leapman RD, Lebowitz J et al. Oligomeric structure of the human immunodeficiency virus type 1 envelope protein on the virion surface. J Virol 2002; 76:7863-7867.
67. Weiss CD, Levy JA, White JM. Oligomeric organization of gp120 on infectious human immunodeficiency virus type 1 particles. J Virol 1990; 64:5674-5677.
68. Zhang CW, Chishti Y, Hussey RE et al. Expression, purification and characterization of recombinant HIV gp140. The gp41 ectodomain of HIV or simian immunodeficiency virus is sufficient to maintain the retroviral envelope glycoprotein as a trimer. J Biol Chem 2001; 276:39577-39585.
69. Alkhatib G, Combadiere C, Broder CC et al. CC CKR5: A RANTES, MIP-1alpha, MIP-1beta receptor as a fusion cofactor for macrophage-tropic HIV-1. Science 1996; 272:1955-1958.
70. Choe H, Farzan M, Sun Y et al. The beta-chemokine receptors CCR3 and CCR5 facilitate infection by primary HIV-1 isolates. Cell 1996; 85:1135-1148.
71. Deng H, Liu R, Ellmeier W et al. Identification of a major coreceptor for primary isolates of HIV-1. Nature 1996; 381:661-666.
72. Doranz BJ, Rucker J, Yi Y et al. A dual-tropic primary HIV-1 isolate that uses fusin and the beta-chemokine receptors CKR-5, CKR-3 and CKR-2b as fusion cofactors. Cell 1996; 85:1149-1158.
73. Feng Y, Broder CC, Kennedy PE et al. HIV-1 entry cofactor: Functional cDNA cloning of a seven-transmembrane, G protein-coupled receptor. Science 1996; 272:872-877.
74. Trkola A, Dragic T, Arthos J et al. CD4-dependent, antibody-sensitive interactions between HIV-1 and its coreceptor CCR-5. Nature 1996; 384:184-187.
75. Wu L, Gerard NP, Wyatt R et al. CD4-induced interaction of primary HIV-1 gp120 glycoproteins with the chemokine receptor CCR-5. Nature 1996; 384:179-183.
76. Quintana FJ, Gerber D, Kent SC et al. HIV-1 fusion peptide targets the TCR and inhibits antigen-specific T-cell activation. J Clin Invest 2005; 115:2149-2158.
77. Call ME, Pyrdol J, Wiedmann M et al. The organizing principle in the formation of the T-cell receptor-CD3 complex. Cell 2002; 111:967-979.
78. Sigalov A. Multi-chain immune recognition receptors: Spatial organization and signal transduction. Semin Immunol 2005; 17:51-64.
79. Sigalov AB. Multichain immune recognition receptor signaling: Different players, same game? Trends Immunol 2004; 25:583-589.
80. Quintana FJ, Gerber D, Bloch I et al. A structurally altered D,L-amino acid TCRalpha transmembrane peptide interacts with the TCRalpha and inhibits T-cell activation in vitro and in an animal model. Biochemistry 2007; 46:2317-2325.
81. Preston BD, Poiesz BJ, Loeb LA. Fidelity of HIV-1 reverse transcriptase. Science 1988; 242:1168-1171.
82. DeMeritt IB, Milford LE, Yurochko AD. Activation of the NF-kappaB pathway in human cytomegalovirus-infected cells is necessary for efficient transactivation of the major immediate-early promoter. J Virol 2004; 78:4498-4507.
83. Mocarski E Jr, Hahn G, White KL et al. Myeloid cell recruitment and function in pathogenesis and latency. In: Reddehase M, ed. Cytomegaloviruses: Pathogenesis, Molecular Biology and Infection Control. Norfolk: Caister Scientific Press, 2006:465-482.
84. Mocarski E, Shenk T, Pass RF. Cytomegaloviruses. In: Knipe D, Howley PM, eds. Fields Virology. Philadelphia: Lippincott Williams and Wilkins, 2007; 2:2702-2772.
85. Browne EP, Shenk T. Human cytomegalovirus UL83-coded pp65 virion protein inhibits antiviral gene expression in infected cells. Proc Natl Acad Sci USA 2003; 100:11439-11444.
86. Paulus C, Krauss S, Nevels M. A human cytomegalovirus antagonist of type I IFN-dependent signal transducer and activator of transcription signaling. Proc Natl Acad Sci USA 2006; 103:3840-3845.
87. Wiertz E, Hill A, Tortorella D et al. Cytomegaloviruses use multiple mechanisms to elude the host immune response. Immunol Lett 1997; 57:213-216.
88. Arnon TI, Achdout H, Levi O et al. Inhibition of the NKp30 activating receptor by pp65 of human cytomegalovirus. Nat Immunol 2005; 6:515-523.
89. Jones KS, Fugo K, Petrow-Sadowski C et al. Human T-cell leukemia virus type 1 (HTLV-1) and HTLV-2 use different receptor complexes to enter T-cells. J Virol 2006; 80:8291-8302.
90. Manel N, Taylor N, Kinet S et al. HTLV envelopes and their receptor GLUT1, the ubiquitous glucose transporter: a new vision on HTLV infection? Front Biosci 2004; 9:3218-3241.
91. Manel N, Kim FJ, Kinet S et al. The ubiquitous glucose transporter GLUT-1 is a receptor for HTLV. Cell 2003; 115:449-459.
92. Kraft S, Kinet JP. New developments in FcepsilonRI regulation, function and inhibition. Nat Rev Immunol 2007; 7:365-378.
93. Pinon JD, Kelly SM, Price NC et al. An antiviral peptide targets a coiled-coil domain of the human T-cell leukemia virus envelope glycoprotein. J Virol 2003; 77:3281-3290.

94. Andersen PS, Geisler C, Buus S et al. Role of the T-cell receptor ligand affinity in T-cell activation by bacterial superantigens. J Biol Chem 2001; 276:33452-33457.
95. Igakura T, Stinchcombe JC, Goon PK et al. Spread of HTLV-I between lymphocytes by virus-induced polarization of the cytoskeleton. Science 2003; 299:1713-1716.
96. Daenke S, Booth S. HTLV-1-induced cell fusion is limited at two distinct steps in the fusion pathway after receptor binding. J Cell Sci 2000; 113(Pt 1):37-44.
97. Jones PL, Korte T, Blumenthal R. Conformational changes in cell surface HIV-1 envelope glycoproteins are triggered by cooperation between cell surface CD4 and coreceptors. J Biol Chem 1998; 273:404-409.
98. Wilson KA, Bar S, Maerz AL et al. The conserved glycine-rich segment linking the N-terminal fusion peptide to the coiled coil of human T-cell leukemia virus type 1 transmembrane glycoprotein gp21 is a determinant of membrane fusion function. J Virol 2005; 79:4533-4539.
99. Wilson KA, Maerz AL, Poumbourios P. Evidence that the transmembrane domain proximal region of the human T-cell leukemia virus type 1 fusion glycoprotein gp21 has distinct roles in the prefusion and fusion-activated states. J Biol Chem 2001; 276:49466-49475.
100. Cianciolo GJ, Copeland TD, Oroszlan S et al. Inhibition of lymphocyte proliferation by a synthetic peptide homologous to retroviral envelope proteins. Science 1985; 230:453-455.
101. Ruegg CL, Monell CR, Strand M. Inhibition of lymphoproliferation by a synthetic peptide with sequence identity to gp41 of human immunodeficiency virus type 1. J Virol 1989; 63:3257-3260.
102. Yaddanapudi K, Palacios G, Towner JS et al. Implication of a retrovirus-like glycoprotein peptide in the immunopathogenesis of Ebola and Marburg viruses. FASEB J 2006; 20:2519-2530.
103. Bukrinsky MI, Sharova N, McDonald TL et al. Association of integrase, matrix and reverse transcriptase antigens of human immunodeficiency virus type 1 with viral nucleic acids following acute infection. Proc Natl Acad Sci USA 1993; 90:6125-6129.
104. Farnet CM, Haseltine WA. Determination of viral proteins present in the human immunodeficiency virus type 1 preintegration complex. J Virol 1991; 65:1910-1915.
105. Miller MD, Farnet CM, Bushman FD. Human immunodeficiency virus type 1 preintegration complexes: studies of organization and composition. J Virol 1997; 71:5382-5390.
106. Turlure F, Devroe E, Silver PA et al. Human cell proteins and human immunodeficiency virus DNA integration. Front Biosci 2004; 9:3187-3208.
107. Mansky LM, Temin HM. Lower in vivo mutation rate of human immunodeficiency virus type 1 than that predicted from the fidelity of purified reverse transcriptase. J Virol 1995; 69:5087-5094.
108. Freed EaM, MA. HIVs and their replication. In: Knipe DaH, PM, eds. Fields Virology. 5 ed. Philadelphia: Lippincott Williams and Wilkins, 2007; 1:2107-2185.
109. Gibellini D, Vitone F, Schiavone P et al. HIV-1 tat protein and cell proliferation and survival: A brief review. New Microbiol 2005; 28:95-109.
110. Amarapal P, Tantivanich S, Balachandra K et al. The role of the Tat gene in the pathogenesis of HIV infection. Southeast Asian J Trop Med Public Health 2005; 36:352-361.
111. Seelamgari A, Maddukuri A, Berro R et al. Role of viral regulatory and accessory proteins in HIV-1 replication. Front Biosci 2004; 9:2388-2413.
112. Strebel K. Virus-host interactions: Role of HIV proteins Vif, Tat and Rev. AIDS 2003; 17(Suppl 4):S25-34.
113. Keppler OT, Tibroni N, Venzke S et al. Modulation of specific surface receptors and activation sensitization in primary resting CD4+ T-lymphocytes by the Nef protein of HIV-1. J Leukoc Biol 2006; 79:616-627.
114. Schrager JA, Marsh JW. HIV-1 Nef increases T-cell activation in a stimulus-dependent manner. Proc Natl Acad Sci USA 1999; 96:8167-8172.
115. Simmons A, Aluvihare V, McMichael A. Nef triggers a transcriptional program in T-cells imitating single-signal T-cell activation and inducing HIV virulence mediators. Immunity 2001; 14:763-777.
116. Baur AS, Sawai ET, Dazin P et al. HIV-1 Nef leads to inhibition or activation of T-cells depending on its intracellular localization. Immunity 1994; 1:373-384.
117. Djordjevic JT, Schibeci SD, Stewart GJ et al. HIV type 1 Nef increases the association of T-cell receptor (TCR)-signaling molecules with T-cell rafts and promotes activation-induced raft fusion. AIDS Res Hum Retroviruses 2004; 20:547-555.
118. Ahmad N, Venkatesan S. Nef protein of HIV-1 is a transcriptional repressor of HIV-1 LTR. Science 1988; 241:1481-1485.
119. Garcia JV, Miller AD. Downregulation of cell surface CD4 by nef. Res Virol 1992; 143:52-55.
120. Garcia JV, Miller AD. Serine phosphorylation-independent downregulation of cell-surface CD4 by nef. Nature 1991; 350:508-511.
121. Schwartz O, Marechal V, Le Gall S et al. Endocytosis of major histocompatibility complex class I molecules is induced by the HIV-1 Nef protein. Nat Med 1996; 2:338-342.

122. Cohen GB, Gandhi RT, Davis DM et al. The selective downregulation of class I major histocompatibility complex proteins by HIV-1 protects HIV-infected cells from NK cells. Immunity 1999; 10:661-671.
123. Le Gall S, Erdtmann L, Benichou S et al. Nef interacts with the mu subunit of clathrin adaptor complexes and reveals a cryptic sorting signal in MHC I molecules. Immunity 1998; 8:483-495.
124. Swigut T, Shohdy N, Skowronski J. Mechanism for down-regulation of CD28 by Nef. EMBO J 2001; 20:1593-1604.
125. Munch J, Janardhan A, Stolte N et al. T-cell receptor: CD3 down-regulation is a selected in vivo function of simian immunodeficiency virus Nef but is not sufficient for effective viral replication in rhesus macaques. J Virol 2002; 76:12360-12364.
126. Munch J, Rajan D, Schindler M et al. Nef-mediated enhancement of virion infectivity and stimulation of viral replication are fundamental properties of primate lentiviruses. J Virol 2007; 81:13852-13864.
127. Schaefer TM, Bell I, Fallert BA et al. The T-cell receptor zeta chain contains two homologous domains with which simian immunodeficiency virus Nef interacts and mediates down-modulation. J Virol 2000; 74:3273-3283.
128. Swigut T, Greenberg M, Skowronski J. Cooperative interactions of simian immunodeficiency virus Nef, AP-2 and CD3-zeta mediate the selective induction of T-cell receptor-CD3 endocytosis. J Virol 2003; 77:8116-8126.
129. Bell I, Ashman C, Maughan J et al. Association of simian immunodeficiency virus Nef with the T-cell receptor (TCR) zeta chain leads to TCR down-modulation. J Gen Virol 1998; 79 (Pt 11):2717-2727.
130. Schindler M, Munch J, Kutsch O et al. Nef-mediated suppression of T-cell activation was lost in a lentiviral lineage that gave rise to HIV-1. Cell 2006; 125:1055-1067.
131. Williams M, Roeth JF, Kasper MR et al. Human immunodeficiency virus type 1 Nef domains required for disruption of major histocompatibility complex class I trafficking are also necessary for coprecipitation of Nef with HLA-A2. J Virol 2005; 79:632-636.
132. Ye H, Choi HJ, Poe J et al. Oligomerization is required for HIV-1 Nef-induced activation of the Src family protein-tyrosine kinase, Hck. Biochemistry 2004; 43:15775-15784.
133. Arold S, Hoh F, Domergue S et al. Characterization and molecular basis of the oligomeric structure of HIV-1 nef protein. Protein Sci 2000; 9:1137-1148.
134. Liu LX, Heveker N, Fackler OT et al. Mutation of a conserved residue (D123) required for oligomerization of human immunodeficiency virus type 1 Nef protein abolishes interaction with human thioesterase and results in impairment of Nef biological functions. J Virol 2000; 74:5310-5319.
135. Kienzle N, Freund J, Kalbitzer HR et al. Oligomerization of the Nef protein from human immunodeficiency virus (HIV) type 1. Eur J Biochem 1993; 214:451-457.
136. Fenard D, Yonemoto W, de Noronha C et al. Nef is physically recruited into the immunological synapse and potentiates T-cell activation early after TCR engagement. J Immunol 2005; 175:6050-6057.
137. Fultz PN. Replication of an acutely lethal simian immunodeficiency virus activates and induces proliferation of lymphocytes. J Virol 1991; 65:4902-4909.
138. Dehghani H, Brown CR, Plishka R et al. The ITAM in Nef influences acute pathogenesis of AIDS-inducing simian immunodeficiency viruses SIVsm and SIVagm without altering kinetics or extent of viremia. J Virol 2002; 76:4379-4389.
139. Petit C, Buseyne F, Boccaccio C et al. Nef is required for efficient HIV-1 replication in cocultures of dendritic cells and lymphocytes. Virology 2001; 286:225-236.
140. Piguet V, Trono D. The Nef protein of primate lentiviruses. Rev Med Virol 1999; 9:111-120.
141. Albrecht B, Collins ND, Burniston MT et al. Human T-lymphotropic virus type 1 open reading frame I p12(I) is required for efficient viral infectivity in primary lymphocytes. J Virol 2000; 74:9828-9835.
142. Albrecht B, D'Souza CD, Ding W et al. Activation of nuclear factor of activated T-cells by human T-lymphotropic virus type 1 accessory protein p12(I). J Virol 2002; 76:3493-3501.
143. Bindhu M, Nair A, Lairmore MD. Role of accessory proteins of HTLV-1 in viral replication, T-cell activation and cellular gene expression. Front Biosci 2004; 9:2556-2576.
144. Hollsberg P, Ausubel LJ, Hafler DA. Human T-cell lymphotropic virus type I-induced T-cell activation. Resistance to TGF-beta 1-induced suppression. J Immunol 1994; 153:566-573.
145. Mann DL, Martin P, Hamlin-Green G et al. Virus production and spontaneous cell proliferation in HTLV-I-infected lymphocytes. Clin Immunol Immunopathol 1994; 72:312-320.
146. Albrecht B, Lairmore MD. Critical role of human T-lymphotropic virus type 1 accessory proteins in viral replication and pathogenesis. Microbiol Mol Biol Rev 2002; 66:396-406.
147. Ding W, Albrecht B, Kelley RE et al. Human T-cell lymphotropic virus type 1 p12(I) expression increases cytoplasmic calcium to enhance the activation of nuclear factor of activated T-cells. J Virol 2002; 76:10374-10382.
148. Ding W, Kim SJ, Nair AM et al. Human T-cell lymphotropic virus type 1 p12I enhances interleukin-2 production during T-cell activation. J Virol 2003; 77:11027-11039.

149. Nicot C, Mulloy JC, Ferrari MG et al. HTLV-1 p12(I) protein enhances STAT5 activation and decreases the interleukin-2 requirement for proliferation of primary human peripheral blood mononuclear cells. Blood 2001; 98:823-829.

150. Johnson JM, Mulloy JC, Ciminale V et al. The MHC class I heavy chain is a common target of the small proteins encoded by the 3′ end of HTLV type 1 and HTLV type 2. AIDS Res Hum Retroviruses 2000; 16:1777-1781.

151. Collins ND, Newbound GC, Albrecht B et al. Selective ablation of human T-cell lymphotropic virus type 1 p12I reduces viral infectivity in vivo. Blood 1998; 91:4701-4707.

152. Guyot DJ, Newbound GC, Lairmore MD. Signaling via the CD2 receptor enhances HTLV-1 replication in T-lymphocytes. Virology 1997; 234:123-129.

153. Von Bonin A, Ehrlich S, Fleischer B. The transmembrane region of CD2-associated signal-transducing proteins is crucial for the outcome of CD2-mediated T-cell activation. Immunology 1998; 93:376-382.

154. Wild MK, Verhagen AM, Meuer SC et al. The receptor function of CD2 in human CD2 transgenic mice is based on highly conserved associations with signal transduction molecules. Cell Immunol 1997; 180:168-175.

155. Kimata JT, Palker TJ, Ratner L. The mitogenic activity of human T-cell leukemia virus type I is T-cell associated and requires the CD2/LFA-3 activation pathway. J Virol 1993; 67:3134-3141.

156. Tsukahara T, Ratner L. Substitution of HIV Type 1 Nef with HTLV-1 p12. AIDS Res Hum Retroviruses 2004; 20:938-943.

157. Fukumoto R, Dundr M, Nicot C et al. Inhibition of T-cell receptor signal transduction and viral expression by the linker for activation of T-cells-interacting p12(I) protein of human T-cell leukemia/lymphoma virus type 1. J Virol 2007; 81:9088-9099.

158. Huynh NT, Ffrench RA, Boadle RA et al. Transmembrane T-cell receptor peptides inhibit B- and natural killer-cell function. Immunology 2003; 108:458-464.

159. Gerber D, Quintana FJ, Bloch I et al. D-enantiomer peptide of the TCRalpha transmembrane domain inhibits T-cell activation in vitro and in vivo. FASEB J 2005; 19:1190-1192.

INDEX